PoliTO Springer Series

Springer, in cooperation with Politecnico di Torino, publishes the PoliTO Springer Series. This co-branded series of publications includes works by authors and volume editors mainly affiliated with Politecnico di Torino and covers academic and professional topics in the following areas: Mathematics and Statistics, Chemistry and Physical Sciences, Computer Science, All fields of Engineering. Interdisciplinary contributions combining the above areas are also welcome. The series will consist of lecture notes, research monographs, and briefs. Lectures notes are meant to provide quick information on research advances and may be based e.g. on summer schools or intensive courses on topics of current research, while SpringerBriefs are intended as concise summaries of cutting-edge research and its practical applications. The PoliTO Springer Series will promote international authorship, and addresses a global readership of scholars, students, researchers, professionals and policymakers.

THE SERIES IS INDEXED IN SCOPUS

More information about this series at http://www.springer.com/series/13890

Giacomo Chiesa
Editor

Bioclimatic Approaches in Urban and Building Design

With a Keynote Chapter by Prof. Mario Grosso

POLITECNICO DI TORINO

Springer

Editor
Giacomo Chiesa
Department of Architecture
and Design (DAD)
Politecnico di Torino
Turin, Italy

ISSN 2509-6796 ISSN 2509-7024 (electronic)
PoliTO Springer Series
ISBN 978-3-030-59327-8 ISBN 978-3-030-59328-5 (eBook)
https://doi.org/10.1007/978-3-030-59328-5

This Springer imprint is published by the registered company Springer Nature Switzerland AG
The registered company address is: Gewerbestrasse 11, 6330 Cham, Switzerland

Foreword

This book is not the first to be written on bioclimatic architecture. For that distinction, we would have to look back at least two millennia: in his treatise known as *De Architectura libri decem* (Ten Books on Architecture), Vitruvius enumerated not only the prime architectural virtues of *firmitas* (firmness), *utilitas* (commodity) and *venustas* (delight), but also the architectural principles that in his view governed each climatic region of the Roman world. Having studied the writings of Socrates and Aristotle before him, Vitruvius bequeathed to civilisation a "multi-voice contribution" detailing the proper solar orientation of rooms and streets, the cooling effects of wind and water, and the leveraging of "passive" technologies based on the precept that "all machinery is derived from nature".

Since *De Architectura* was published, the geographical scope of our world has become global in scale and our society has witnessed technological revolutions that Vitruvius could scarcely have imagined. But still it seems fitting to recall one of his most basic observations: "It is obvious," he wrote, "that designs for homes ought to conform to the diversities of climate". This dictum went without saying in pre-industrial times, as vernacular builders instinctively followed the limitations imposed by nature on the materials that could be accessed and the strategies which could be drawn upon to preserve thermal stability. But by the middle of the twentieth century, it was no longer self-evident. Fossil-fuelled feats of engineering had allowed a universalist ethic to take hold, telling us that anything could be built anywhere.

There were, however, those who questioned this anti-climatic paradigm. Victor Olgyay, a graduate of the *Scuola Superiore Di Architetture* in Rome who had spent the 1950s scrutinising the relationship between a building's form and its thermal environment, compiled the considerable evidence that he and his brother Aladar had collected over the years in a volume called *Design with Climate: Bioclimatic Approach to Architectural Regionalism*. According to Olgyay's approach, different climatic circumstances did in fact require different design responses—but at the same time, certain principles could be applied in any location with a given *type* of climate, whether it be cool, temperate, tropical or arid. Vitruvius, it turned out, was right after all.

Still, a solid physical basis was needed to support the bioclimatic approach. In the preface to his 1963 book, Olgyay asserted that architecture had so far been in a subjective trial-and-error stage, and in order to mature properly it must "adopt the techniques of analytical reasoning". He called for a new principle of architecture, to "blend past solutions to the problems of shelter with new technologies"—and this required, above all, "insights into the effects of climate on the human environment".

Olgyay was probably not aware that in the very year his book was published, a doctoral thesis was submitted to the Hebrew University in Jerusalem entitled *Estimation of the Effect of Climate on Man: Development of a New Thermal Index*. Its author was Baruch Givoni, an architect-turned-scientist who over the following years would conduct exhaustive experimental work on the biophysical relationships between energy, buildings and their inhabitants—and in 1969, he would summarise this work in the first edition of *Man, Climate and Architecture*. While even today this book is considered by some to be the most authoritative volume ever produced in the field of "building climatology", it did not initially generate much attention.

This changed in 1973, when the global energy crisis altered the public's sensibilities. Following the newfound demand for scientifically rigorous information that could help society use energy more efficiently, Givoni released an expanded second edition in 1976. Echoing the voices of earlier theorists and distilling the empirical results that his research group had attained, Givoni examined the fundamentals of heat transfer in the context of buildings and raised new possibilities for their application. Thermal mass and thermal insulation, night ventilation, evaporative cooling and even passive solar heating were concepts that may have been unfamiliar at the time, but which formed the basis for half a century of innovation in bioclimatic architecture—resulting in countless volumes filled with experimental findings, novel inventions and noteworthy demonstration projects.

So why another book now? When asked in 1974 about the impact of his research, Givoni lamented that architecture is perceived mainly as art, and that its scientific sides are almost entirely ignored—and a decade later, he wrote that existing scientific knowledge on the subject of building climatology was still not reflected in the common practice of planning and construction. Indeed, the gap between accumulated knowledge and its practical application was all too familiar to early proponents of bioclimatic architecture, who often felt like lone wolves howling at the moon. But today? "Green building" is everywhere, building energy codes and standards are widespread, and indeed the efficiency of the building stock has in many cases improved dramatically.

At the same time, the "bioclimatic approach" has not become the norm. How often is a building's conceptual design really driven by a deep understanding of the local climate? And to what extent are "passive systems" for heating and cooling truly integrated in the building envelope, not as an exception, but as the rule? As global climate disruption forces a change in attitudes, we are encouraged to see more roofs and even walls covered with solar PV panels—but are the buildings underneath these facades wasting the electricity they generate, and more? Indeed, a plurality of voices is needed more than ever to remind planners, policy makers and the public at large that skin-deep prescriptions for repairing our environment will not suffice.

In this volume, we will hear some of these voices. We will get a fuller picture of how bioclimatic architecture originated, how it has evolved, and where it may be headed. We will see that in an urbanised culture, the same type of rigorous analysis and sensitive design that are required for creating comfortable indoor spaces are also need for the streets and squares outside. And in an age of exponential change, we will get a sense of the twenty-first century innovations in digital technology that are revolutionizing the design and production of post-carbon buildings and cities. These voices tell us that the future remains to be designed. But they also tell us that what was obvious to Vitruvius has not changed—our designs ought to conform to the diversities climate.

Prof. David Pearlmutter
Ben-Gurion University
Beer-Sheva, Israel
davidp@bgu.ac.il

Preface

The idea to write this book first emerged between 2008 and 2010, when as an early stage researcher I was trying to understand the multifaceted approach to bioclimatic design developed over recent decades. However, it was only on the recent retirement of Prof. Mario Grosso, who has been my tutor, colleague and friend over the last 10 years as well as the leading expert at the Politecnico di Torino on bioclimatic architectural design thanks to his 48 years of experience in the field, that the original idea was transformed into reality. This publication aims at defining a critical vision of bioclimatic and green design by combining a large set of contributions which focus on international and Italian approaches, including a key chapter by Mario about his research experience.

This book is targeted at architects, engineers, scientists and building designers who want to learn more about bioclimatic design. It focusses on the correlations between climate and architecture in the past, present and future based on the experiences of researchers in the field. The specific proposed analysis is grounded in the architectural technology approach and adopts a performance-driven methodology which is perfectly compatible with advanced digital design visions. This approach to bioregionalism, which has included from the very beginning the adoption of passive, low and renewable energy systems for the built environment, was and is the object of continuous innovation since it is an area where different disciplines and research fields pool their knowledge in the direction of a sustainable and green design vision. In particular, architectural technology is the research field that since the 70s has introduced bioclimatic design approaches to Italian architectural culture, thus uniting various fields of research. Consequently, several authors in this book come from this field. Although the earliest environmentalist discussions concentrated on the relationship between energy and the environment, debates now include the need to curb negative impacts of climate change and local overheating phenomena, such as urban heat islands, together with the need to develop and think of building and urban renovations in a comprehensive and critical design vision. Furthermore, even if bioclimatic architecture is an approach that has been with us at least since the 60s of the last century, climate-related approaches to building/urban design and full integration between passive systems and envelope elements are still more one-offs rather than the norm. The inclusion of a full urban dimension and the expansion of digital tools and computational power are other essential technological and innovational aspects of current green design visions. It is clear that these new challenges require attentive readings and critical discussion. Hence, the book is more informative than highly focussed in detailed calculation methodologies. It aims at correlating current multi-visions of the bioclimatic topic by describing open issues and by referring to authors' specific publications that delve more deeply into certain topics. The large number of contributions is an essential aspect of the book because bioclimatic design, after more than 60 years of existence, is characterised by a complex multi-voiced vision. For this reason, this work sets out to overcome certain shortcomings in arriving at shared cultural-definitions of some aspects of bioclimatic design and also to define potential connections between specific outcomes of recent researches.

The book is divided into six parts:

I. Environmental and Technological Design Framework;
II. Roots and Evolution;
III. Environmental Building Design Concepts and Methods;
IV. Urban Environmental Design;
V. Passive Climate Control Strategies and Technologies;
VI. Technological Representation and Digital Innovation.

In particular, Part I focusses on analysing the bioclimatic design framework and approaches. The first chapter introduces some current open issues on sustainable design, e.g. the anthropogenic background which includes climate change, urban dimensions and growth in cooling needs. The correlation between certain aspects of technological and environmental research is also introduced at this early stage. The keynote chapter written by Mario Grosso is an essential contribution that defines the main topics in bioclimatic architectural design and practice over 50 years. Thanks to the autobiographic experience here reported, it is possible to debate on the evolution of bioclimatic and technological-connected design issues. The contribution of Rossana Raiteri and Fausto Novi helps to detail the technological vision in bioclimatic and sustainable design by giving a critical conceptualisation of the issue. The role of technological research, which includes bioclimatic design, is connected with the theme of the environment. Fabrizio Tucci details the current strategies and requirements which underpin current and future bioclimatic and environmental design issues, including judgment criteria. The chapter discusses bioclimatic design opportunities and possibilities in the context of currently open environmental issues. Mat Santamouris discusses the main current and future issues in the built environment as he focusses on climate change, the increase in world population, energy poverty and future energy needs by analysing building consumption trends. The chapter also deals with the urban dimension and connected challenges in energy and environmental technological terms.

Part II mainly analyses the origins and evolution of bioclimatic architecture. The chapter by Mario Grosso includes a study of international references and key figures in bioclimatic architecture, from the origins. Maria Luisa Germanà analyses past, present and future aspects of bioclimatic design. She deals with rural and urban built environments as a continuum and proposes a technological reading, which overcomes the often opposing approaches to urban and rural environments in the past. Furthermore, she critically reflects on current bioclimatic approaches and vernacular characters from an epistemological point of view. The contribution of Michela Toni details bioclimatic approaches at different design scales thanks to her personal research experience from the 70s, when bioclimatic design was an emerging issue, until current technological research pulsations. The chapter by Adriano Magliocco analyses the essential relationship between bioclimatic design and environmental impacts by critically revisiting famous architectural urban design processes developed over the last four decades. The past present and future impacts of bioclimatic design in solar cities are examined. Special attention is given to the contradictions and the oversimplifications of bioclimatic approaches that aim towards a solely energy-driven vision, thus running the risk of losing the human dimension in sustainable building design.

Part III deals with environmental and bioclimatic design concepts and methodologies. Maria Cristina Forlani analyses recent energy challenges in building design, including correlated ambiguous implications. If the field of architecture is to face up to its contemporary responsibilities and accept the fundamental importance of climate issues in design choices, design practice must systematically and comprehensively take into consideration embodied energy, waste and emissions. Andrea Giachetta and Maria Canepa propose a critical vision of the ecological cultural approach to building design. They discuss the spread of sustainable issues in architecture that have generated positive results, but have also produced a simplification in cultural approaches with the consequent risk to pursue false results. Rossella Franchino analyses the cultural landscape domain in Mediterranean areas, re-balancing antropogenic issues with natural ones and analysing the connected bioclimatic implications. Consuelo Nava addresses the new quality of building design and considers advances in sustainable design issues including new paradigms which are able to use to great effect advanced information, resource management and KET in the attempt to develop an agile design approach. The critical discussion includes a reflection on resilient scenarios including sample design applications and experimental prototyping with additive manufacturing. Ernesto Antonini talks about bioclimatic connections with various environmental impacts, dealing with the embodied energy dimension. In particular, his chapter analyses EPBD implications together with other sustainable rating systems including energy and life-cycle assessment impacts.

Part IV includes contributions which focus on the environmental design aspects at urban scale. The urban dimension is one of the main contemporary challenges if we consider the need to develop sustainable and resilient cities and the continuous innovations in modelling tools that support bioclimatic and sustainable design actions at urban scale. In particular, Alessandra Battisti deals with requalification processes in urban open spaces to align with EU guidelines on climate change. The chapter focusses on materials, urban vegetation, water, shading and other environmental and technological design aspects that connect energy needs and bioclimatic comfort issues with urban morphologies. Renata Valente studies environmental design implications for open urban spaces in consideration of the complexity of environmental issues which correlate both scientific and human studies. The contribution includes an analysis on flooding events and other climate change challenges and suggests resilient regeneration approaches for urban districts in an adaptive design vision. The contribution of Elena Garcia-Nevado and Benoit Beckers reports an ongoing study on the connection between architecture and urban physics, a new research area that is expected to grow in the coming years. In relation to advanced monitoring approaches and thanks to the increase in the computational power of simulation digital tools, the authors study radiative exchanges in dense urban areas and support the development of a simulated advanced thermography design approach. Focussing on different aspects of urban microclimatic design, Valentina Dessì describes on urban materials and water usage and management at urban and district scale in an adaptive design approach. The evaporative cooling potential of water in outdoor open spaces is investigated with international examples given and the community role that urban structures may play in citizen involvement is described.

Part V deals in particular with climate and passive control strategies and technologies. A chapter by Claude-Alain Roulet introduces passive means that can control and improve indoor thermal comfort in low-energy buildings. Taking into consideration passive architectural systems and active technologies, the chapter examines the issues related to passive space cooling and heating to reach indoor environmental quality without consuming energy, thanks to proper architectural design measures. The chapter by Giacomo Chiesa reports a focused study on the development of geo-climatic applicability indicators that support local choices among different low-energy cooling technologies. Climate-correlated maps that define the early design potential of different passive dissipative technologies are reported for the Italian peninsula including climate change implications. Katia Perini analyses the impact of greening building envelope technologies and considers the potential of greenery to improve microclimatic conditions by reducing space cooling energy demands and improving outdoor comfort. Vertical greening systems are studied in detail from an economic and environmental point of view, including biomaterial development. Giovanni Scudo deals with the potential contribution of bioclimatic design approaches to developing liveable spaces considering multi-comfort issues and guaranteeing natural rhythms. Green technologies are also analysed considering their ability in improving urban microclimates, by proposing a toolbox for preliminary green design.

Part VI introduces aspects of digital innovation and technological representation in bioclimatic design. Chiara Tonelli investigates the implications that digital and energy transitions have on our evolving society and building architecture. This chapter focusses on the potential implications that smart technologies may have on our houses and cities. Furthermore, the chapter written by Maria Antonietta Esposito and Filippo Bosi studies the impact of BIM and digital innovation aspects on the green design approach and describes new industrial processes. The analyses include an historical study of bioclimatic implications from the 60s onwards and connects them with the related development of calculation methodologies and tools that define energy and physical environmental simulation. The chapter considers the evolution from CAD to BIM platforms in relation to green design issues. Alessandro Rogora discusses simplified tools and methodologies to represent energy issues connected to architectural design projects. Given that, architects are responsible for construction and design processes, the possibility to have proper design tools and simulation software is essential to support the development of counterintuitive approaches allowing to achieve comfort conditions, since architectural design is the combination of both technical and artistic challenges. The contribution of Rosa Romano focusses on reviewing kinetic adaptive facades. The development of smart and automatic control systems allows for the development of a new approach to building envelope in which the façade itself becomes a movable and adaptable system which is able to react to environmental conditions and optimise indoor comfort.

Turin, Italy

Giacomo Chiesa
giacomo.chiesa@polito.it

Contents

Part I
Environmental and Technological Design Framework

Chapter 1
Design with Climate: An Unconventional Introduction

Giacomo Chiesa

Environmental and technological design choices are affected and hybridised by other fields of knowledge and visions. Everything fertilising architecture (Piano 2002), including visual and artistic approaches capable of describing and interpreting our present and/or our future. After all, our cities are a representation of our cultures, historical matrixes of the way people live in a place in time-layered moments, while our future cities deal with the logical perpetuation of our present (Matteoli and Pagani 2010). This is why it is important to correlate bioclimatic design with current challenges, marking out potential future paths. This chapter roughly describes an unconventional introduction to bioclimatic backgrounds aimed at tackling some of the major challenges of urban and architectural design with a multi-hybridised approach. As remembered by prof. Pearlmutter in the book preface and demonstrated by prof. Grosso in his key chapter, bioclimatic design is still an urgency. It needs to become part of our common way of thinking of architectural and urban design, as the matter of sustainability and climate issues become increasingly important in relation to current and future human space-living needs. The focus here is on the anthropogenic background, touching on climate change and massive urbanisation, trying to suggest potential visual and cultural connections for bioclimatic design. Furthermore, a reflection on how to localise the aspects of environmental and bioclimatic design research in the approach to technological design is introduced early on (Fig. 1.1).

G. Chiesa (✉)
Department of Architecture and Design, Politecnico di Torino, Turin, Italy
e-mail: giacomo.chiesa@polito.it

G. Chiesa (ed.), *Bioclimatic Approaches in Urban and Building Design*, PoliTO Springer Series, https://doi.org/10.1007/978-3-030-59328-5_1

Fig. 1.1 Sample picture representing the effect of urban areas on the surrounding territory (from project "(es)senza notte", G. Chiesa, 2009)

1.1 Anthropocenic Background: "Human Beings Are like a Virus, a Cancer for This Planet"

The title of this Section, a quotation from the well-known film The Matrix (Wachowski and Wachowski 1999), summarises the urgency of turning our approach to life towards a sustainable development rationale, wiping out the planet's disease generated by humans, by paraphrasing Lovelock, the father of the Gaia hypothesis. Staying with iconographic visions suggested by films, the representation of our disintegrating crazy life is the subject of the famous *quatsi* trilogy by Godfrey Reggio, set to music by Philipp Glass. This work includes *Koyaanisquatsi* (Reggio 1983), which, in the Hopi language, means life in turmoil, out of balance, disintegrating, a film about the disruptive impact of our industrial civilisation on the environment, *Powaqqatsi* (Reggio 1988), which focuses on the severe imbalance between industrialised and under-developed worlds, and *Naqoyqatsi* (Reggio 2002), which looks at the information age and reflects on the loss of the power of language in describing an explosive civilised violence and money-driven society. Furthermore, the impact that industrialised human society—including the built environment and human landscape morphing—has on the environment was also the subject of a short film by Rhomer (1964) describing the anthropic metamorphosis of landscape in the industrial era. Rohmer's out-of-time black and white frames and the massive and minimalistic colour repetitions of Reggio's films are a visual representation of the conceptual idea that considers technology as an instrument to dominate nature, following the philosophical vision defined centuries ago by Francis Bacon (1620), by orienting the scientific method of Galileo. Furthermore, the impact of human activities on the environment has become the topic of specific territorial photography research. Environmental photography in particular, which is directly connected to previous documentary and topographic works (e.g. Eggleston 2015; Guidi 2003; Becher and Becher 2002;—Luigi Ghirri—Mussini 2001; Basilico 1981), focuses, especially in recent decades, on two main issues. On one hand, it studies contradictions, challenges, expectations, and risks in megacities—see for example (Bialobrezeski 2011, 2004; Jodice 2004; Barbieri 2002), while, on the other, it represents human-morphed manufactured and extracted landscapes, including social aspects, e.g. studying people and territory connections taking a landscape perceptive approach, but also analysing vanishing natures—see for example (Burtynsky et al. 2018; Soth 2017; Botto 2017; Barth 2008; Fabiani 2007; Pauli 2003; Hütte 2001). A filmic representation of some of these aspects can be seenin the Burtinsky work Manufactured Landscape (Baichwal 2007). Here, Rhomer's vision is replaced by a crude reality dominated by mega-construction projects, among others, almost infinite production chains—not too different the iconographic Little Tramp representation in Modern Times (Chaplin 1936)—, abandoned skeletons of mega-boats, and mountains of junk, sifted through by the inhabitants of slums for things they can use. Not only are we affecting human society and its inequalities, we are also having an increasing impact on the planet as a whole, consuming resources quicker than the Earth can reproduce them and producing more waste than can be assimilated. This condition can be represented by that well-known synthetic indicator, the Ecological Footprint (Wakernagel and Rees, 1996), which can be roughly defined as the amount of land needed to feed a society and assimilate the waste produced in the long term. In 2016, thet average global Ecological Footprint was calculated to be about 1.7 times our planet (Lin et al. 2019), highlighting our unsustainable way of living. This index is of course not evenly distributed across. There is a considerable difference between countries in term of economic and ecological impacts. We are living at a time in which human beings are becoming the major geological force, engendering direct impacts on the biosphere, defining our new geological era: the Anthropocene (Crutzen and Stoermer 2000). This great challenge requires us, as human beings, to take responsibility for the environment. And this is a topic that is becoming more and more important to define policies and approaches that include the issues of sustainability and resilience in technological society. This great need for personal and collective responsibility towards society and the environment can be seen in several philosophical attitudes of the last century. Examples include Arendt's loss in centrality of the traditional meaning of work and policy in the human condition (Arendt 1958), the flexible man and the loss in technical expertise envisaged by Sennet (1999), the end of nude humanity envisaged in Anders's works (Anders 2007a, b [1956]), human virtualisation and modularisation (Benasayag 2010), and Das Prinzip Verantwortung, in which Jonas introduces an ethical vision of technological society including the environment (Jonas 1979). This last work defines an introduction of human (and homo faber) responsibility for nature and its potential rights, considering the new dimension of humanity and artificial science in impacting the environment. Since Sophocles' Antigone (Sophocles, about 442 BC), we have known that "Wonders abound in this world yet no wonder is greater than man. None!", as men pass through raging storms, trapping birds, ruling over mountains with their machines, mastering wind, and governing cities. It has also been clear since ancient times that the arrogant man straying from the righteous path is lost. In current times, the traditional way of thinking of technology as an approach to dominating nature, in the way that pioneers do, has been overtaken by biomimetic approaches in architecture, for example (Gruber 2011; Chiesa 2010). Consequently, it is essential to include the need to preserve our living space and environment, avoiding arrogance, as technology is "a highly sophisticated instrument, but like every instrument it can be used badly" (Cardwell 1972). The pioneer myth was mentioned by Norbert Wiener (1950), who defined the pioneer's unsustainable way of thinking of nature, considering it as an element to be depleted. The father of cybernetics introduced a perfect metaphor of unsustainability, referring to the literary image of Alice in Wonderland, when she arrives at the Mad Hatter's tea party, where everyone drinks tea, eats cakes and consumes the food set out at each place at the table and then moves on to the next one. An unsustainable vision in which the table, like the Earth-s natural resources, is seen as an infinite banquet. The human approach to nature's ability to cope with disturbance is not unique and can be described according to the four sociological myths defined by Thompson (1983) and Douglas and Wildavsky (1982). The approaches mentioned include a fatalistic model, based on individualistic views, in which nature's behaviour is considered to be very unpredictable, depriving man of having any control over natural dynamics. In other words, "I don't know". The second approach can be described as an expansionist model in which nature is an extra-strong system, unaffected by human actions, in which environmental problems are rejected and ignored—"I don't care". Then there is a collectivist approach or community model, in which nature is extra-fragile—as conceived by several early environmental movements—and unable to withstand any kind of disturbance, losing its balance. Lastly, the fourth approach, the controlled model, is possibly that which best describes current scientific views, based on the concept of threshold, considering nature as strong within certain limits, although the exact quantification of thresholds is an open debate. Paleoclimatological studies clearly report examples of ancient society collapses under perturbing phenomena. One such case was the biggest ancient city of all, Ur (sixth-third centuries BC), with other Mesopotamian sites, such as Tell Leilan. These cities dominated huge areas, thanks to their technical ability to control rivers and agriculture, and to cope with more common short-term perturbations, but they were unable to withstand major environmental changes, i.e. a volcanic eruption and consequent centennial aridity and increased wind circulation (Fagan 2004; Weiss et al. 1993). This makes it clear that correct design and technological actions are needed to preserve the environment and our living spaces. Alternative ways of life are possible and a scientific and technical approach to building and urban design has been established in line with bioregionalism and local climatic conditions. Nevertheless, when we consider the human environment, it cannot be divided by the cultural dimension that permeates the outlook of our living space (Magnaghi 1994; Vitta 2008), as the human territorial palimpsest is characterised by both changing and permanent characteristics. If we consider the long term situation, the main determinants that drive urban plans and the organisation of living space correlate to the cultural, political, economic, energy, climate, and information issues that characterise society (Chiesa 2015, 2020). This complexity deals with the integration of different interdependent dynamics ranging from cultural to ecological, like a bio-cultural approach to design (Girardet 2008), which is essential to define proper ecological design actions, including critical considerations, a topic that is at the base of this book. Especially in the last two decades, green design approaches have become a key issue for architecture and urban development, although a fully bioclimatic design approach is still far from being reached. The potential of passive and low-energy HVAC solutions is generally overtaken by the adoption of alternative visions focused solely on optimising the use of equipment instead of the other two main building aspects, including site and envelope, which are linked to a free-running comfort approach. Bioclimatic design focuses on maximising comfort levels by optimising the suitability and exploitation of the full potential of passive solutions connected to local climate, thanks to the correct design of building envelopes, considering climate and site, and the control of the natural heat gain balance, including dissipation, mitigation, and prevention of heat gains. The correct control of free-running strategies could considerably reduce the use of mechanical systems in terms of activation hours and intensity of energy needs, without being exclusive, but being part of a space-operational-comfort control based on climate issues. This approach to building design, driven by climatic aspects, has to be further developed and applied until becomes common practice, considering the considerable environmental challenges of our time, which include anthropological impact on the planet, climate change and urbanisation.

Bioclimatic design is based on local climatic conditions assuming them to be the fundamental parameter influencing design choices. Climate is the time-averaged weather conditions of a place, which characterise it in the long term. Climate change is having an ever-greater impact on climatic conditions on different scales, making the characteristics of a location a time-dependent variable, rather than a fixed unchanging condition—see for example (Chiesa and von Hardenberg 2020). If climate change is the central focus of a growing number of researches, several of which connected with building issues, this chapter remembers the need to consider this challenge essential in order to tackle present and future bioclimatic architectural choices. To give an idea of the potential magnitude of climate change, two examples are reported. The first is based on Eurostat statistics (2020) and analyses multi-year variations in rising and falling degree-days, a climate-related cumulative index linked directly to building energy requirements. The second shows the potential impact of climate change on bioclimatic charts by introducing monthly average 24 h temperature and humidity values, assuming TMY (typical meteorological years) referring to different time periods. In both cases, the analyses are based on monitored databases and are not driven by climate simulation models. Of course, the adoption of future prediction models is expected to increase the afore-mentioned effects. Focusing on the first analysis, two non-consecutive 15-year periods—period 1 (1975–89) and period 2 (2004–18)—were assumed as references to analyse the changes in climate monitored. The annual data available are defined at country level (NUTS 0—Nomenclature of Territorial Units for Statistics) and are averaged to define typical 15-year values. The map shown in Fig. 1.2 shows the European distribution of average (a) heating and (b) cooling DD indexes in the two periods considered. These maps show evident changes in DD values, highlighting a general reduction in HDD values between the first and

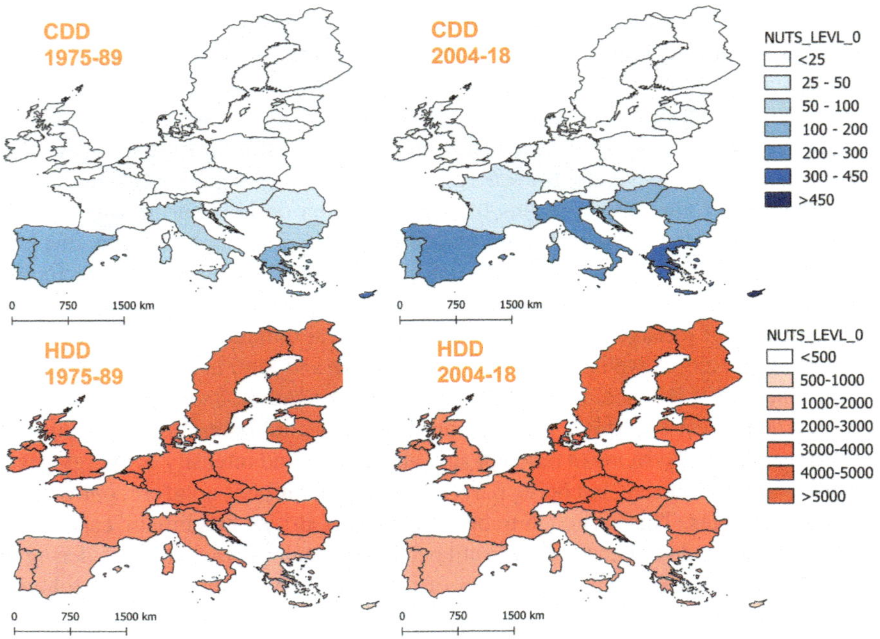

Fig. 1.2 Average CDD (above) and HDD (below) for reference periods one (1975–89) and two (2004–18) based on Eurostat data defined at Country level (NUTS 0)

the second period, while an opposite trend is evident for the CDD index. Figure 1.3 also shows the changes between the two periods in absolute values. It is evident that higher increases in CDD are found in those Countries in which the original values are higher, while a greater reduction in HDD is highlighted for those countries with higher starting values. Nevertheless, some countries, like Italy and Croatia have big differences in both HDD and CDD indexes, revealing a high-national-average impact

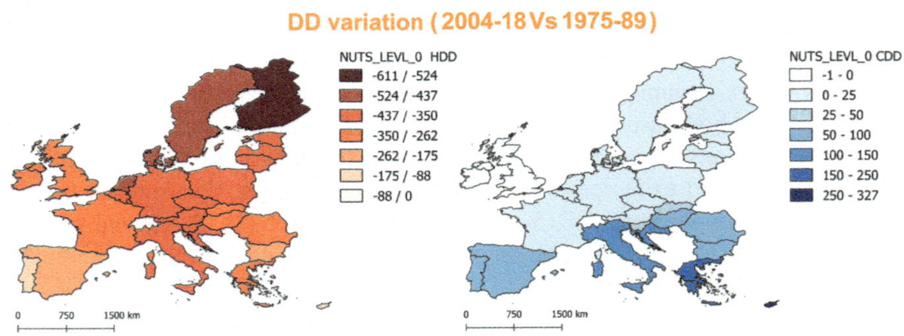

Fig. 1.3 Absolute variations in HDD (left) and CDD (right) between the two 15-year reference periods (2004–18 Vs 1975–89)

of climate change. These effects need to be taken into account when retrofit or new building design are considered in order to define bioclimatic strategies that can absorb changes in climatic conditions during their expected lifespan.

The second analysis illustrates the impact of climate change on bioclimatic design tools, such as bioclimatic charts. These tools allow the assessment of local climatic conditions, indicating the characteristics of an area. In this introduction, climate assessments are based on the average monthly 24 h distribution plotted on a Givoni-Milne bioclimatic chart. Results are shown for two locations, assuming two different TMYs in all sites, calculated for the same meteorological stations, but in different reference periods. The chosen index is a well-known approach to assess local bioclimatic needs and review local climatic conditions in bioclimatic design methodologies, in line with the Olgyay approach (Olgyay 1963). The chosen locations are Oslo and Cordoba, while climate data refer to two monitoring periods—the first (1981–90 irradiation; 1961–90 temperatures); the second (1991–2010 irradiation; 2000–09 temperatures). Both locations are characterised by evident changes as shown in Fig. 1.4. Oslo shows an increase especially in winter temperatures, while thermally neutral periods indicate an increase in daily temperature variations. This evident change in winter temperatures for high latitude locations was also highlighted in other studies and is correlated with the large change in albedo values due to a reduction

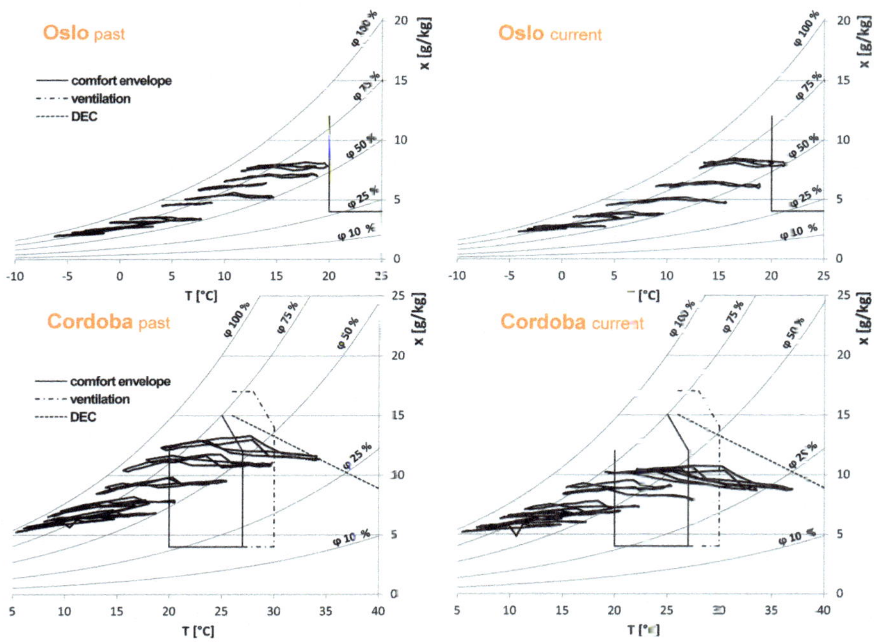

Fig. 1.4 Representation on bioclimatic chart of past (ref. temperature 1961–90) and present (ref. temperature 2000–09) TMY conditions using the monthly 24 h average for Oslo (FIN)—upper charts—and Cordoba (SPA)—lower charts

in the polar ice caps—known as the Artic Amplification (Serreze and Barry 2011). Cordoba, on the other hand, endures a a high rise in temperatures and drop in absolute humidity values particularly during the summer months, indicating a change in local climate towards a hotter and drier condition, correlated with a potential increase in summer cooling requirements. Even if climate change is not evenly distributed, it has the potential to influence typical local weather conditions and consequently the bioclimatic needs of a location, affecting technological design choices and the consequent potential of bioclimatic technology to guarantee comfort and reduce energy requirements. This clearly shows the importance of including climate change issues in building design and operational management, particularly to guarantee a time-resistant bioclimatic approach to building comfort.

In addition to globally and regionally-driven changes, urban environments too are capable of morphing local climatic conditions, due to phenomena such as the urban heat island effect, which is becoming one of the major challenges in sustainable urban design. Several studies are demonstrating how this phenomenon is effecting local temperatures, i.e. showing an urban increase of above 7 °C compared to rural air temperatures,—see, for example, the works of Peng et al. (2020), Santamouris (2019b), and Akbari and Konoacki (2005). The importance of describing and including this phenomenon in sustainable and bioclimatic design is evident. Building design and operational issues on urban sites require specific methodologies to be able to consider the urban climate dimension and correlate it with bioclimatic choices and tools—see, for example, the books by Erell et al. (2011) and Ng (2010). Fortunately, tools devoted to supporting green design choices in dense urban environments are becoming increasingly popular. These include visual parametric programming interfaces such as Grasshopper (Rutten 2007) for Rhinoceros—see, for example, sample applications reported by Chokhachian et al. (2020) and Chiesa and Casetta (2020). Several tools, such as Envi-Met,—see, for example, the research outputs reported by Perini et al. (2017)—and UWG-Urban Weather Generator (Bueno et al. 2013), a tool capable of morphing typical weather data considering urban morphological aspects—see, for example, studies by Salvati et al. (2020) and Morganti et al. (2017), have also been developed. Using this tool in the grasshopper environment, Fig. 1.5 roughly indicates the impact of urban morphologies on local TMY data by simulating the effect of a dense city (surface coverage 75%; building height 10 floors; building morphology of courtyard plus internal buildings) and morphing the airport (rural) TMY of the City of Turin. Effects on bioclimatic requirements are shown on the aforementioned bioclimatic chart of Givoni. Results highlight a general contraction in average monthly days with higher temperatures, with a considerable increase in temperatures at night and limited changes in the middle of the day. The two graphs on the right offer a better indication of this effect for a representative winter month (January) and one in the summer (July). This analysis does not include single obstacle effects, leaving architects to apply the full bioclimatic design approach based on the morphed urban climate TMY. Results indicate the positive effect of the urban heat island in reducing heating energy requirements and the negative counteraction on the demands of energy required to cool buildings.

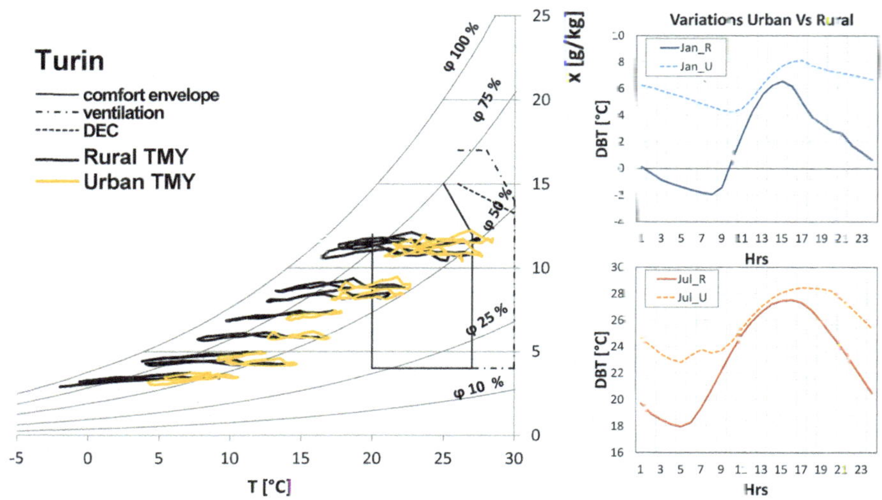

Fig. 1.5 Rough estimation of the Urban heat island effect on an airport TMY considering a virtual dense urban area with a surface coverage ratio of 75%, an average building height of 10 floors and an urban morphology based on courtyard buildings with additional crossed geometries inside the court. Urban morphed and rural TMYs are analysed to extract average monthly days, plotted on the left of the bioclimatic chart. The daily variations in local dry bulb temperature are shown on the right, for representative days in winter (January) and summer (July)

Thanks to the rise in computational power of our computers and in the parallel dissemination of standardised modelling approaches to building design, we are able to start envisaging a new horizon in urban analyses considering climatic, architectural, environmental and energy issues—think, for example, of the new approach to radiation simulations in densely populated areas mentioned by Beckers (2013). The challenge of developing bioclimatic urban design actions is very urgent considering the growing impact of urbanisation and the quick spread of megacities. At the moment, more than half the global population lives in cities and urban aggregations, a percentage which is expected to reach 68% according to the UN World Urbanisation Prospects report (United Nations 2019). The growing trend is not evenly distributed, as some specific cities are undergoing a population decline (low-fertility countries, economic problems, natural disasters). Nevertheless, at general level, by 2030, the number of megacities (more than 10 million inhabitants) in the world is expected to reach 43, although the largest growing trend is for cities with around one million inhabitants, especially in Asia and Africa. Every week, about 1.5 million people move from the countryside to urban spaces (West 2017)—That is about half the population of the Municipality of Rome, more than that of Milan, about a 6th of New York's population and 0.5 times the population of Paris (the city alone), being about 0.136 times the population of the extended urban area of Paris (urban agglomeration). It is, therefore, clear that this great phenomenon requires tremendous effort to support a sustainable approach to urbanisation. Furthermore, cities are often resistant to small perturbation phenomena, but still exposed to natural and anthropogenic risks, including unexpected black swans (Taleb 2007), such as pandemics. Considering the major natural disasters, i.e. cyclones, droughts, earthquakes, floods, landslides and volcanic eruptions, in 2018, the UN remembered that about 59% of cities with more than 500,000 inhabitants are highly exposed to at least one of these calamities (United Nations 2018), highlighting the need for resistance and stronger approaches to urban design in order to reduce our weaknesses. It is clear that sustainable and green design approaches are essential to tackle these challenges, supporting the adoption of a more responsible design vision with respect to local environment and the definition of a correct correlation with local resources, energy vectors and climate-related issues.

The previously-mentioned issues are linked to the average growth in local temperatures, a trend that is expected to rise in the years to come. It is, therefore, clear that this will have an impact on cooling energy needs to guarantee the comfort of spaces (IEA 2018; Santamouris 2016; Koizumi 2007). The growing intensity in cooling energy loads and in cooling discomfort hours in buildings, a trend that has been evident for at least two decades, is linked to several additional causes. These causes include the rise in indoor heat due to the use of electrical appliances, the rise in expectations in terms of comfort and the way we consider it, and, last but not least, the style of buildings, the design of which is exactly the opposite of bioclimatic (Santamouris 2007). A visual representation of the international style of the buildings that characterised the fossil-city era is indirectly reported in the impressive B&W photographic work of Basilico, especially in his Scattered City book (Basilico 2005). Unless you read the captions below the pictures in this book, it is not possible to know whether a picture portrays Munich, Milan, Barcelona, Buenos Aires or even Beirut. Fossil-cities are populated by buildings that look the same, supporting Droege's drastic vision of current fossil-architecture as a production of a standardised approach to design, in which user needs are bypassed by the need to create low cost buildings that depreciate within a short space of time, reducing "residences to ordinary consumer products fuelled at low cost" (Droege 2006). These apartments "have become monothematic scripts" representing a cynical approach to building design that are lacking on all fronts (economic, social, and environmental) in terms of sustainability. This vision is not all that far from the aforementioned loss in knowledge of craftsmanship (Sennet 2008) due to contemporary flexibility which, in the architectural world, supports the creation of monothematic landscapes dominated by CAD-driven homogenous buildings (Pagani and Chiesa 2016). A need for a different, local and user-need related approach to building design is evident, and innovations in bioclimatic design may be the right way to drive these changes toward a newer vision for sustainable architectures and cities.

Furthermore, the spread of standards and regulations to reduce the energy required to heat a space has led to the adoption of technologies which, in several cases, have generated an increase in the risk of overheating as a by-product, even in cold climates, and a consequently of the amount of energy used for cooling in buildings. See the international debate introduced in IEA EBC Annex 62 (Plesner 2018; Heiselberg 2018; Kolokotroni and Heiselberg 2015). Heating consumption is currently the dominant element in building energy needs. However, as we are reminded by several researchers—e.g. (Isaac and van Vuuren 2009)—cooling requirements are expected to overtake those for heating at European level as we approach 2050, with global energy consumption for cooling having already reached 1.25 PWh in 2010 (Harvey et al. 2014). In Europe, the cost of electricity consumed for cooling to offset climate change alone was estimated to be between 22 and 89 billion euros a year (Mima et al. 2011), while the consumption of cooling systems is expected to exceed 44,430 GWh, corresponding to 18.1 million tonnes of CO_2 (Adnot 1999) in Europe alone by 2020. This change is related not only to energy flows, but also to other risks directly connected to human comfort and the body's thermal adaptability. It has been proven that extreme weather conditions are directly related to a rise in mortality rates, but also that humans are less able to adapt to hot conditions than they are to the cold (Santamouris 2019a; Baccini et al. 2008). This is a risk for our society that is also connected with our expectations and our cultural adaptations in terms of comfort. An inclusive vision of cooling standards and regulations and of architectural practice is essential to reduce mechanical consumption and integrate passive and low-energy technologies, considering their local climatic feasibility and maximising free-running potential. This has to be taken into account in future nZEB applications, making bioclimatic design an essential aspect to reduce mechanical system energy requirements, and the number of hours of use of equipment. Envelope-site-climate systems must be correctly designed, both with regard to their actual design and also their operation, considering both summer and winter conditions. This is a great challenge for architects and architectural schools.

1.1.1 Environmental and Technological Design: A Multi-voice Background

This book focuses on the various aspects of bioclimatic design, ranging from design implications, which are more closely linked to architectural project definitions, through to system design and integration in buildings, connecting different scales and technical backgrounds, including advancements in design thanks to the use of digital tools. Taking the advantages of pre-fossil architectures and cities, based on non-pedigree buildings (Rudofsky 1977), innovative bioclimatic approaches to design will lead to the development of visions that promote the definition of renewable and resilient cities (Hopkins 2008; Niewman and Jennings 2008; Register 2006; Jack Todd and Todd 1993). This will overcome the lack of regionalism and connection between architecture and climate that has characterised the fossil-city era, and progress towards ecological and climate-designed cities. The authors of the book refer to several international backgrounds, including the Italian Architectural Technology research field. This field has been working since the 70 s to develop an environmental approach to building design based on a new and sympathetic relationship between climate and the culture of buildings. To support the international reader in framing

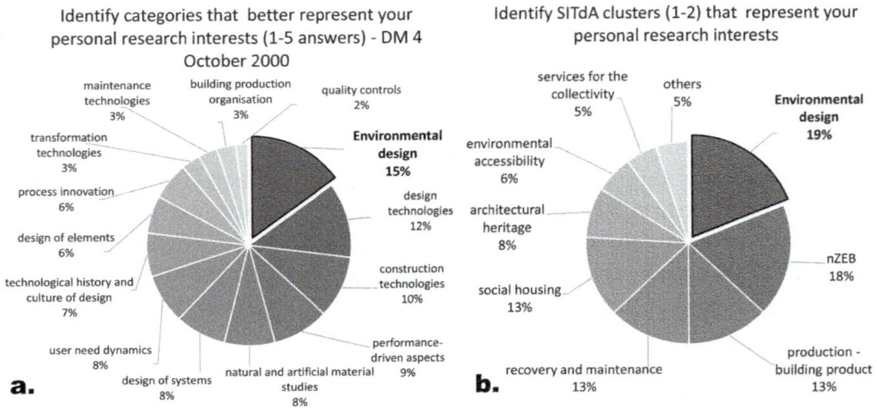

Fig. 1.6 Architectural technology—distribution of current research interests on the basis of categories defined **a** in DM 4 October 2000, and **b** by the scientific society SITdA (research clusters)—data acquired late 2014

the links between bioclimatic design and the Architectural Technology sector, reference can be made to an extract of results from a survey conducted in 2014 by the author, together with prof. Pagani and with prof. Bosia (Chiesa et al. 2015). The definition of this analysis is based on a previous series of interviews conducted by the author at the Politecnico di Torino in 2010, the results of which help define the survey and its goals. The aforesaid work does not aim to be a structural defining work. It does, however, draw up potential conceptual maps based on experience, interest and disciplinary osmoses and hybridisations. Thanks to a representative response by the entire research sector, the analysis allowed the mapping of some of these disciplinary interactions and the connection of the main subjects of research. The results, however, are indicative and cannot be considered as an accurate description of the sector. A representation of personal research topic categories is shown in Fig. 1.6a. It follows the definition of the scientific contents of the Architectural Technology research field, as covered byresented in Annex B to D.M. 4 October 2000 (D.M. 4 Ottobre, 2020). The figure shows that the main topics of research are: environmental design (15%)—which is the category including design with climate aspects—, design technologies (12%), construction technologies (10%), performance-driven aspects (9%), user-need dynamics (8%), natural and artificial material studies (8%), and system design (8%). It is clear from this graph, that one of the most important topics for the Architectural Technology sector is environmental design and, consequently, bioclimatic aspects. Figure 1.6b shows a similar distribution but based on categories (clusters) defined by the national Scientific Society representing the Architectural Technology research field (SITdA). This distribution further highlights the importance of environmental and sustainable design, the two most representative topics being "environmental design" (19%) and "nZEB building" (18%).

Figure 1.7 analyses the correlations between the two categories and the connections between architectural technologists and other scientific research sectors to

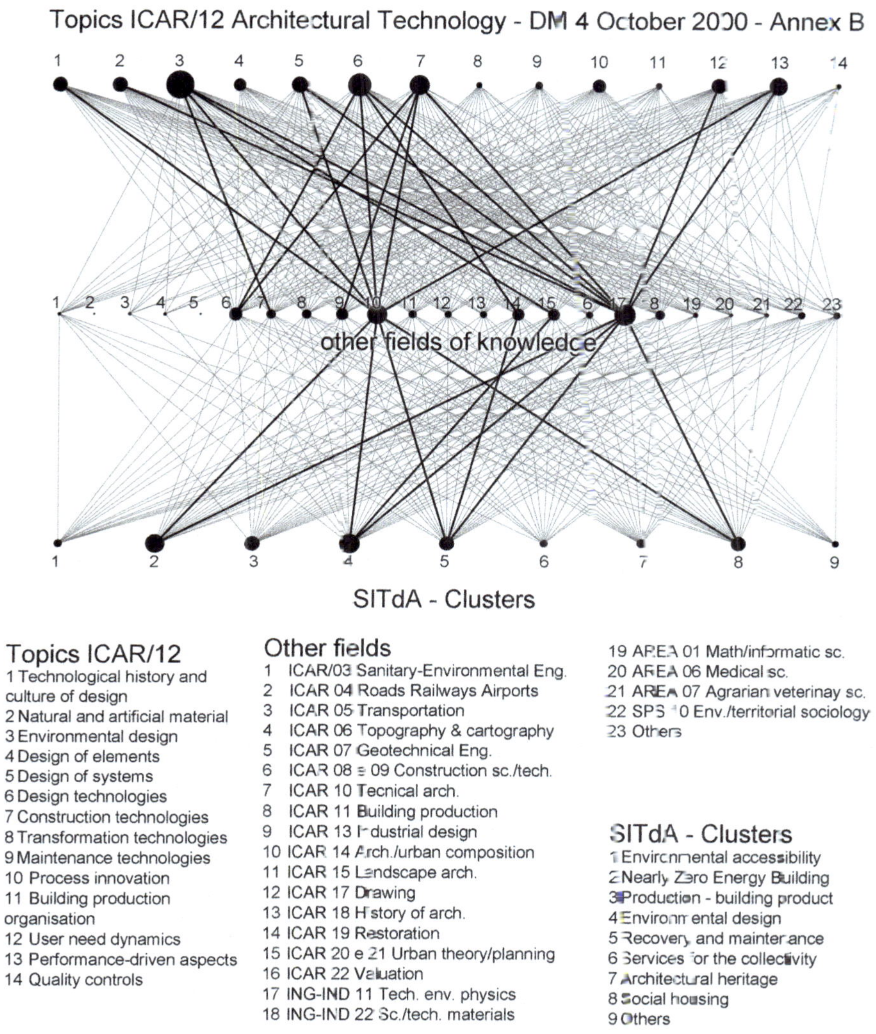

Fig. 1.7 Correlation between Architectural Technology topics (upper line DM 4 October 2020, Annex B; lower line SITdA clusters) and other research sectors

define a map that localises environmental (and bioclimatic) design issues within the whole scientific research community. In this figure, point dimensions depend on the number of replies, while the thickness of links depend on the number of researchers, emphasising the specific inter-sectoral connection. From this graph, developed by combining spreadsheets with Grasshopper visual scripting, it is clear that the Environmental Design category (considering both upper and lower lines) is mainly linked to architectural and urban design (composition), urbanism, urban theory & urban planning, and environmental technical physics. The same intersectional links are shared with the "recovery and maintenance" research cluster while, focusing on the building scale alone, similar research frontiers are highlighted for "nZEB building" and "social housing" clusters, suggesting potential connections.

Details of this analysis can be found in a study on free keywords that describe personal research interests with a disciplinary cloud visualisation. A simple network analysis has been developed in Gephi, as shown in Fig. 1.8, by running a gravitational study capable of geolocalising each network node in relation to the others within the given space. The picture shows the existence of specific research sub-areas with higher internal interconnections and lower external links, and broad research topics characterised by a more articulate network full of different, smaller sub-aspects. For the purpose of this chapter, the keywords related to bioclimatic design (see also the specific word, central lower part) and comfort (central left position), which is a topic

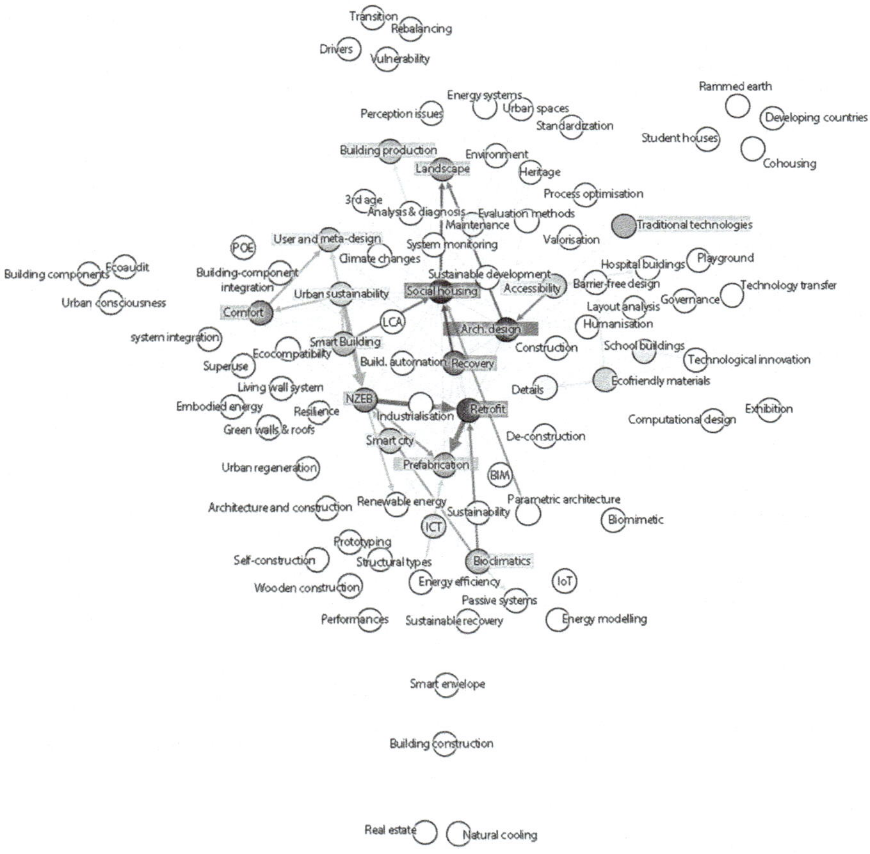

Fig. 1.8 Network analysis of freely mentioned keywords. Gravitational attraction forces are used to produce this graph

strongly correlated to bioclimatic charts, are characterised by a high impact (grey colour). These two words are characterised by a large number of links, connecting nZEB, passive systems, energy efficiency and modelling, urban sustainability and other words from the green design paradigm. The proximity between nodes is mainly due to first or second connection orders.

This survey analysis can help the reader localise the chapters of this book in the international research domain, allowing for personal comparison between given contents and the reader's local context. Furthermore, in this book the key chapter by Mario Grosso (Grosso 2020) will help the reader to follow the path taken by bioclimatic design over the past 50 years, following the personal research experience of one of the masters of Italian bioclimatic design.

1.2 Conclusions

This chapter takes an un-conventional approach to connect certain some key aspects of the bioclimatic design approach, including climate change and massive urbanisation climate phenomena. These aspects are connected with the need to move our fossil-architectural and urban vision towards a renewable future, an issue that bioclimatic design approaches may face. A mapping analysis is presented to help the reader localise current bioclimatic researches in branches of research connected to the building sector, even though results are only a partial representation of a complex and dynamic context. The networks indicated help analyse the correlations between sectors and suggest future research and design lines. Climate and energy issues, both of which are strongly related to sustainable and bioclimatic design, are, in fact some, of the characters capable of influencing, in the medium and long-term, human civilisations, urban forms, and architectural issues.

References

Adnot J (Ed) (1999) Energy efficiency of room air-conditioners (EERAC), study for the Directorate general for energy (DG XVII) of the Commission of the European Communities, Luxemburg
Akbari H, Konopacki S (2005) Calculating energy-saving potentials of heat-island reduction strategies. Energ Policy 33:721–756
Arendt A (1958) The human condition. University of Chicago Press, Chicago
Baccini M, Bigger A, Accetta G, Kosatsky T, Katsouyanni K, Analitis A, Anderson R, Bisanti L, D'Ippoliti D, Danova J, Forsberg B, Medina S, Paldy A, Rabczenko D, Schindler C, Michelozzi P (2008) Heat effect son mortality in 15 European cities. Epidemiology 19(5):711–719
Bacon F (1620) Novum Organum Scientiarum [ed. 1996, Nuovo Organo. RCS, Milano]
Baichwal J (Director) (2007) Edward Burtynsky. Manufactured Landscapes. Documentary film, Ontario Inc., Canada, Duration 83 minutes
Barbieri O (2002) Notsofareast. Donzelli, Milano
Barth N (ed) (2008) Vanishing landscapes. Frances Lincoln Limited, London
Basilico G (2005) Scattered city. Baldini Castoldi Dalai, Milano

Basilico G (1981) Ritratti di fabbriche. SugarCo, Milano
Becher H, Becher B (2002) Industrial landscapes. Mit Press, Boston
Beckers B (ed) (2013) Solar energy at urban scale. Wiley, Hoboken
Benasayag M (2010) Organisme et artefacts: vers la virtualisation du vivant?. La Décuverte/JanPaul Bayol, Paris
Bialobrezeski P (2011) The raw and the cooked. Hatje Cantz Verlag, Berlin
Bialobrezeski P (2004) Neontigers. Hatje Cantz Verlag, Berlin
Botto A (2017) KA-BOOM the explosion of landscape. Bessard, Paris
Bueno B, Norford L, Hidalgo J, Pigeon G (2013) The urban weather generator. J Build Perform Simul 6(4):269–281
Burtynsky E, Baichwal J, De Pencier N (eds) (2018) Anthropocene—Burtynsky. Steidl, Göttingen
Cardwell DSL (1972) Technology, science and history. Heinemann, London
Chaplin C (1936) Modern times. Silent Film, U.S., Duration 87 minutes
Chiesa G (2020) Technological paradigms and digital eras: data-driven visions for building design. Springer, Cham
Chiesa G (2015) Paradigmi tecnologici ed ere digitali: il dato come parametro di innovazione in architettura e urbanistica. Accademia University press, Torino
Chiesa G (2010) Biomimetica, tecnologia e innovazione per l'architettura, Celid, Torino (Chiesa G (2017) Biomimetics. Technology and innovation for architecture, Celid, Torino)
Chiesa G, Casetta E (2020) Scripting and parametric CAD modelling for performance-driven design, In: Chiesa G, op. cit. Springer, Cham, pp 65–100
Chiesa G, Pagani R, Bosia D (2015) ICAR/12: evoluzioni e innovazioni nella disciplina, unpublished document, p 11
Chiesa G, von Hardenberg J (2020) Including climate change time-dimensions in bioclimatic design. Techne 20:204–212
Chokhachian A, Perini K, Guilini S, Auer T (2020) Urban performance and density: generative study on interdependencies of urban form and environmental measures. Sustainable cities and society 53:101952
Crutzen PJ, Stoermer EF (2000) The "Antropocene". IGBP Newsletter 41:17–18
D.M. 4 Ottobre 2020, Allegato B (1999) Declaratorie descrizione dei contenuti scientifico-disciplinari dei settori di cui all'art. 1 del D.M. 23 Dicembre 1999, accessible at: https://www.istruzione.it/archivio/alfresco/d/d/workspace/SpacesStore/f4794b36-8e19-4f2f-ac74-60baee0885ff/Declaratorie_Descrizione_Contenuti_Scientifico_Disciplinari.pdf. Last view April 2020
Douglas M, Wildavsky A (1982) Risk and culture: an essay on the selection of technological environmental danger. University of California Press, Berkeley
Droege P (2006) The renewable city—a comprehensive guide to an urban revolution. Wiley-Academy, Chichester
Eggleston W (2015) The democratic forest. Steidl, Göttingen
Erell E, Pearlmutter D, Williamson T (2011) Urban microclimate: designing the space between buildings. Earthscan, London
Eurostat, Energy statistics—cooling and heating degree days, available on line at: ex.europa.eu/Eurostat/cache/metadata/en/ngr_chdd_esms.html (viewed Feb 2020)
Fabiani F (ed) (2007) Atlante italiano^{007}—rischio paesaggio: ritratto dell'Italia che cambia. Electa, Milano
Fagan B (2004) The long summer: how climate changed civilization. Basic Book, New York
Girardet H (2008) Cities people planet: urban development and climate change. Wiley, Hoboken
Grosso M (2020) Keynote chapter—bioclimatic design in architecture: a research and didactic experience. In Chiesa G (ed) Bioclimatic approaches in urban and building design. Springer, Cham
Gruber P (2011) Biomimetics in architecture: architecture of life and buildings. Springer, Wien/New York
Guidi G (2003) In between cities. Electa, Milano

Günther A (2007a) L'uomo è antiquato 1. Bollati Boringhieri, Milano [or. Ed. 1956, Die Antiquiertheit des Menschen. Band I: Über die Seele im Zeitalter der zweiten industriellen Revolution. Beck, Munich]
Günther A (2007b) L'uomo è antiquato 2. Bollati Boringhieri, Milano [or. Ed. 1980, Die Antiquiertheit des Menschen. Band II: Über die Zerstörung des Lebens im Zeitalter der dritten industriellen Revolution. Beck, Munich]
Harvey LDD, Korytarova K, Lucon O, Roshchnka V (2014) Construction of a global disaggregated dataset of building energy use and floor area in 2010. Energ Build 75:488–496
Heiselberg P (ed) (2018) Ventilative cooling design guide, IEA EBC Annex 62. Aalborg University, Aalborg
Hopkins R (2008) The transition handbook: from oil dependency to local resilience. GreenBook, Foxhole
Hütte A (2001) As dark as light. Schirmer/Mosel, Berlin
IEA (2018) The Future of Cooling. IEA, Paris. available at: https://www.iea.org/reports/the-future-of-cooling (last view April 2020)
Isaac M, van Vuuren DP (2009) Modeling global residential sector energy demand for heating and air conditioning in the context of climate change. Energ Policy 37:507–521
Jack Todd N, Todd J (1993) From eco-cities to living machines: principles of ecological design. North Atlantic Book, Berkeley
Jodice F (2004) What we want. Skira, Milano
Jonas H (1979) Das Prinzip Verantwortung. Insel Verlag, Frankfurt
Koizumi S (ed) (2007) Energy efficiency of air conditioners in developing countries and the role of CDM, IEA information paper. OECD/IEA, Paris Cedex 15
Kolokotroni M, Heiselberg P (eds) (2015) Ventilative cooling: state-of-the-art review, IEA EBC Annex 62, Aalborg University, Aalborg
Lin D, Hanscom L, Martindill J, Borucke M, Cohen L, Galli A, Lazarus E, Zokai G, Iha K, Eaton D, Wackernagel M (2019) Working guidebook to the national footprint and biocapacity accounts. Global Footprint Network, Oakland
Magnaghi A (1994) Il territorio dell'abitare: lo sviluppo locale come alternativa strategica. Franco Angeli, Milano
Matteoli L, Pagani R (eds) (2010) Cityfutures: architecture design technology for the future of the cities. Hoepli, Milano
Mima S, Criqui P, Watkiss P (2011) The impacts and economic costs of climate change on energy in the European Union: summary of sector results from the climate cost project. Technical Policy Briefing Note Series, Stockholm Environment Institute, Oxford
Morganti M, Salvati A, Coch E, Cecere C (2017) Urban morphology indicators for solar energy analysis. Procedia Eng 134:807–814
Mussini M (ed) (2001) Luigi Ghirri. Federico Motta, Milano
Newman P, Jennings I (2008) Cities as sustainable ecosystems: principles and practices. Island Press, Washington
Ng E (ed) (2010) Designing high-density cities: for social and environmental sustainability. Easthscan, London
Olgyay V (1963) Design with climate. Bioclimatic approach to architectural regionalism. Princeton University Press, Princeton and Oxford
Pagani R, Chiesa G (eds) (2016) Urban data: tools and methods towards the algorithmic city. FrancoAngeli, Milano
Pauli L (ed) (2003) Edward Burtynsky: manufactured landscapes. Yale University Press, New Haven
Peng J, Hu Y, Dong J, Liu Q, Liu Y (2020) Quantifying spatial morphology and connectivity of urban heat islands in a megacity: a radius approach. Sci Total Environ 714:136792
Perini K, Chokhachian A, Dong S, Auer T (2017) Modeling and simulating urban outdoor comfort: Coupling ENVI-Met and TRNSYS by grasshopper. Energ Build 152:373–384
Piano R (2002) La responsabilità dell'architetto, 3rd edn. Passigli, Firenze

Plesner C (ed) (2018) Status and recommendations for better implementation of ventilative cooling in standards, legislation and compliance tools, IEA EBC Annex 62. Aalborg University, Aalborg
Reggio G (2002) Naqoyqatsi: life as war. Documentary film, U.S., Duration 89 minutes
Reggio G (1988) Powaqqatsi: life in transformation. Documentary film, U.S., Duration 86 minutes
Reggio G (1983) Koyaanisquatsi: life out of balance. Documentary film, U.S., Duration 83 minutes
Register R (2006) Ecocities: rebuilding cities in balance with nature. New Society Publisher, Gabriola Island
Rhomer E (Director) (1964) Vers l'unité du monde. Les métamorphoses du paysage: l'ère industrielle. Documentary film of the series En profil dans le texte, France, Duration 22:23
Rudofsky B (1977) The prodigious builders. Harcourt Brace Janovich, New York/London
Rutten D (2007) Grasshopper, visual programming. Robert McNeel and associates, grasshopper3d.com (last view April 2020)
Salvati A, Palme M, Chiesa G, Kolokotroni M (2020) Built form, urban climate and building energy modelling: case-studies in Rome and Antofagasta. J Build Perform Simul 13(2):209–225
Santamouris M (ed) (2019a) Cooling energy solutions for buildings and cities. World Scientific, Singapore
Santamouris M (2019b) Urban heat island and local climate change, In: Santamouris M (ed) Minimizing energy consumption, energy poverty and global and local climate change in the built environment: innovating to zero. Elsevier, Amsterdam, p 65
Santamouris M (2016) Cooling the buildings—past, present and future. Energ Build 128:617–638
Santamouris M (ed) (2007) Advances in passive cooling. Earthscan, London
Sennet R (2008) The craftman. Yale University Press, New Haven
Sennet R (1999) The corrosion of character: the personal consequences of work in the new capitalism. Norton & Company, New York-London
Serreze M, Barry R (2011) Processes and impacts of arctic amplification. Global Planet Change 77(1–2):85–96
Sophocles (about 442 BC) Antigone, Chorus
Soth A (2017) Sleeping by the Mississippi. Mack, London
Taleb NN (2007) The black swan: THE IMPACT OF THE HIGHLY IMPROBABLE. Random House and Penguin Books, New York
Thompson M (1983) A cultural basic for comparison. In: Kunreuther H, Linnerooth J (eds) Risk analysis and decision process: the siting of liquefied energy gas facilities in four countries. Springer, Berlin
United Nations, Department of Economic and Social Affairs, Population Division (2019) World urbanisation prospects. The 2018 revision. United Nations, New York
United Nations, Department of Economic and Social Affairs, Population Division (2018) The world's cities in 2018—data booklet (ST/ESA/ SER.A/417)
Vitta M (2008) Dell'abitare: corpi, spazi, oggetti, immagini. Einaudi, Torino
Wachowski A, Wachowski L (1999) The matrix. Science fiction Film, U.S., Australia, 136 minutes
Wackernagel M, Rees WE (1996) Our ecological footprint: reducing human impact on the Earth. New Society, Gabriola Island
Weiss H, Courty MA, Wetterstrom W, Guichard F, Senior L, Meadow R, Curnow A (1993) The genesis and collapse of third Millennium North Mesopotamian civilization. Science 261(5124):995–1004
West G (2017) Scale: the universal laws of growth, innovation, sustainability, and the pace of life in organisms, cities, economies, and companies. Penguin Press, London
Wiener N (1950) The human use of human being. Houghton Mifflin, Boston

Chapter 2
Keynote Chapter—Bioclimatic Design in Architecture: A Research and Didactic Experience

Mario Grosso

Abstract This essay intends to outline a synthetic excursus on my research, didactical and consultancy experience in the field of bioclimatic design for architecture. The aim is to highlight the milestones of my academic and professional career with regard to interrelationships between the bioclimatic approach and the design process in building design as well as landscape and urban design.

2.1 Academic Experience

2.1.1 Theoretical Background

Before entering the research team formed by Lorenzo Matteoli, a professor who was a pioneer in Italy for the integration of environmental and bioclimatic issues in architecture, at the Institute of Built Environment of the School of Architecture, Polytechnic University of Turin in 1977—an action which initiated my experience on the bioclimatic issues—I had been contract researcher since 1974, and teaching assistant since 1972, for Leonardo Mosso. He was Professor of Architectural Composition and Social Architecture as well as the main Aalto's scholar in Italy and founder of the still active Aalto's Institute. This activity was permeated by a 'structuralistic' approach, based on the philosopher Pareyson's 'Theory of formativity' (Pareyson 1954) as well as an ecological and social attitude towards the field of architecture in its widest meaning including historical compound renewal and regional planning. The combination of Pareyson's aesthetic theory, for which a work of art needs to be analysed and perceived in its autonomous and intrinsic values out of the influence of its author's historical context, with the ecological approach to the analysis and planning of the territorial transformations, had facilitated my conversion to the bioclimatic field in architecture. Mosso's ecological thinking was pioneering for that

M. Grosso (✉)
Department of Architecture and Design, Politecnico di Torino, Turin, Italy
e-mail: mario.grosso@formerfaculty.polito.it

G. Chiesa (ed.), *Bioclimatic Approaches in Urban and Building Design*, PoliTO Springer Series, https://doi.org/10.1007/978-3-030-59328-5_2

time within the University faculty and rather consistent with the '68 student movement, which I experienced fully since my second year of study in 1967, when the school of architecture in Turin was occupied. During those years of protest—I have to highlight that it was very peaceful in our School at that time, and only later on it would have changed to violent actions, sometimes leading to rioting against the police and the Government—there was a lively exciting debate on the urban development model driven by unregulated industrialisation and land use, and causing environmental and social distress. A few years later, the awareness of an incumbent ecological crisis became apparent in the first of a series of UN assemblies dealing with the environment(1).

In addition, that students' movement set the foundation for a radical transformation of the method of teaching and learning in the architecture field, from a traditional dogmatic strictly disciplinary approach to a multi- and interdisciplinary one, by which theoretical knowledge, social aspects, and professional practice would have been interconnected.

A witness of that unrepeatable period was an exhibit-roundtable held in Bologna in October 1972, which saw the participation of people from any part of the world and all sort of fields—moviemakers, theatre actors, artists, sociologists, politicians, musicians, scientists, teachers, architects, farmers, citizens. My contribution was a written text on the role of the architectural designer as an intellectual figure influencing the ongoing social and territorial transformation (Grosso 1972). My main point-of-view was to emphasize the need for overcoming both the prevailing ideological and utopian approaches through a science-based methodological and structuralistic one, implying the involvement of all actors in the incumbent revolutionary (so we thought to be at that time) social and territorial transformation of our environment.

This approach was my reference line in both didactical and research activities I carried out in collaboration to Leonardo Mosso from 1972—just after been graduated—to 1974, when I won a 4-year contract research position at the School of Architecture. During that period, my activity was more autonomous and was characterise as well by a collaboration with Prof. Ciribini, one of the founder of Architectural Technology as a transformation of the previous "Construction Components discipline. In that period, my research activity was strictly interconnected to a professional practice as a free-lance architect. This experience had been focus-sing on ecological and social territorial analysis and planning as well as on a historical urban recovery project(2). However, even in the professional activity I always tried to apply a 'scientific' methodological approach and, reversely, I have taken the impact on the real social and territorial context into account in my research and didactic experience.

[1]The 1972 UN summit in Stockholm set for the first time rules and recommendation for the recovery of the Earth environment, endangered by unregulated urban sprawl and industrial development.

[2]In 1973, just two years after graduation, I won, together with my colleagues Guido Laganà and Carla Barovetti, the first prize in a competition for the restoration planning of the historical centre of the city of Vimercate, in the Milan Province. That prize represented the beginning of a professional appointment by the Municipality of Vimercate, which lasted up to 1980, having as a objet the completion of an executive planning framework for the refurbishment of all historical centre as well as a detailed project related to a specific urban block of that area.

2.1.2 *Origin of the Bioclimatic Approach*

After the above-described experience, I entered Matteoli's research team, where there was an exciting and growing development of activities linked to both the National Plan for Energy, just implemented at that time after the '73 oil crisis, and the lab for testing air infiltration windows performance founded by Matteoli himself. Various projects were developed since, leading to national seminars and conferences[(3)] as well as publications. Amid the latter, it is worth to mention the following: two books on the interrelationships between energy and buildings, authored collectively (Matteoli et al. 1981a, b); an article on the development of windows' frame industry (Grosso 1980); a contribution in a special issue of a Journal on the integration of solar energy in school buildings (Matteoli et al. 1981c); a paper for a Conference on the application of renewable energy sources in the Pantelleria Island (Matteoli et al. 1981d).

Within that environment, two personal lines of research were present in a nutshell: the first regarding solar radiation, the second related to air infiltration and natural ventilation. Both topics would later been developed in a wide spectrum of aspects, all connected to their interrelationships with building and urban design as well as regional planning and landscape architecture. Key objectives of my research work have been, since the beginning, modelling and simulate the effects of those two main climate factors—solar radiation and air movement—on the built environment as well as developing simplified methods and tools to support the building, urban and landscape design processes towards a bioclimatic approach.

In addition, even after the conclusion of my practice as an architect and urban planner in 1984, all my research work have been always integrated to a professional activity as a consultant of architecture and engineering offices. This have been carried out in relation to my competence sectors, in particular, energy efficiency in buildings, natural ventilation control, passive ventilative cooling, solar energy applications, and use of sustainable materials. The most significant of these experiences are described in the last part of this contribution, after the description of the main line of research just following.

2.1.3 *Research on Solar Aspects and Didactic-Related Tools*

Dynamics of shadowing

My first approach to the solar-related research was a study on the dynamics of shadowing cast by any obstacle on a horizontal surface. An article published on Solar Energy (Appelbaum and Bany 1979) was the base of this study, and its aim

---

[3]One of the most interesting event was the Sogesta Summer School, where experts from various Universities and ENEA (an Agency for Alternative Energy sources, re-founded from the previous Nuclear Energy Agency)—including prof. Federico Butera, as the coordinator of the task group "energy for limited resources' areas"—gathered to discuss about the new perspective of the Energy National Plan as well as to practice an experimental study workshop on a bioclimatic project.

was to develop a method to draw shadows, cast from a peg on a horizontal surface at any hour of the day and day of the year, while changing location, i.e., latitude and longitude of the given place. It was a method analogous to the one applied for constructing a sundial, but referred to a horizontal instead of a vertical surface, and with a different aim: not time measuring, but modelling the effect of obstacles in a built environment on the effective solar radiation of a surface at a given time and location.

The outcome of this study was my first published book (Grosso 1982). This included an original approach, never developed before: a calculation routine, made using a pocket calculator by Bruno Caudana—a brilliant just graduated student at that time—able to simulate the horizontal projection of the shadow cast by a peg high one unit of measurement at any given time and location (Fig. 2.1). Through this calculator was possible to draw the hyperbolic-shaped shadows' daily profiles for each of the twelve monthly-reference days of the year at the local time of any lo-cation in the Earth, by imputing latitude, longitude and hourly angle. Its output considers the solar local time, including the effect of the equation of time ('analemma') (Fig. 2.2). Using the graphical shadows monthly profiles was then possible to draw the hourly shadow field cast by a solid parallelepiped obstacle on the horizontal surface by considering its envelope surfaces as made by the sum of infinite vertical parallel segments with the same dimension and position as the peg of the profiles. An evolution of that method, developed manly for didactical purposes, allowed for the graphical elaboration of the shadow cast by an obstacle on any vertical surface intercepting that projection. This was made based on the horizontal shadows profiles, by a technique overturning the triangle peg-beam-shadow from the vertical to the horizontal plane and again on the vertical plane containing the receiving surface.

An algorithm was further developed to calculate the area of the shadow projected from a solid parallelepiped obstacle to another parallel surface, such as the one of a building façade intercepting that projection. This simplified calculation method was presented at an international energy-related congress (Grosso and Pacini 1984).

Solar irradiation at territorial scale

A calculation programme at territorial scale was developed in collaboration to the CSI—a Piedmont Regional Centre for Information Services—with the aim of mapping the daily monthly-averaged solar irradiation of mountain land areas, taking their surfaces slope orientation and tilt into account. As a reference case study, the Val Chisone valley was considered. Monthly maps with zoning classification of solar irradiation were produced (Fig. 2.3) and shown at an exhibition-congress (Grosso et al. 1984). Methodology and results were presented in papers for both a Journal (Grosso 1984) and an International Conference (Grosso et al. 1983).

The main use of such a tool would have supposed to be for regional framework planning of building development location at Municipality scale, based on solar access criteria. However, this approach was too advanced for that time, and it had no

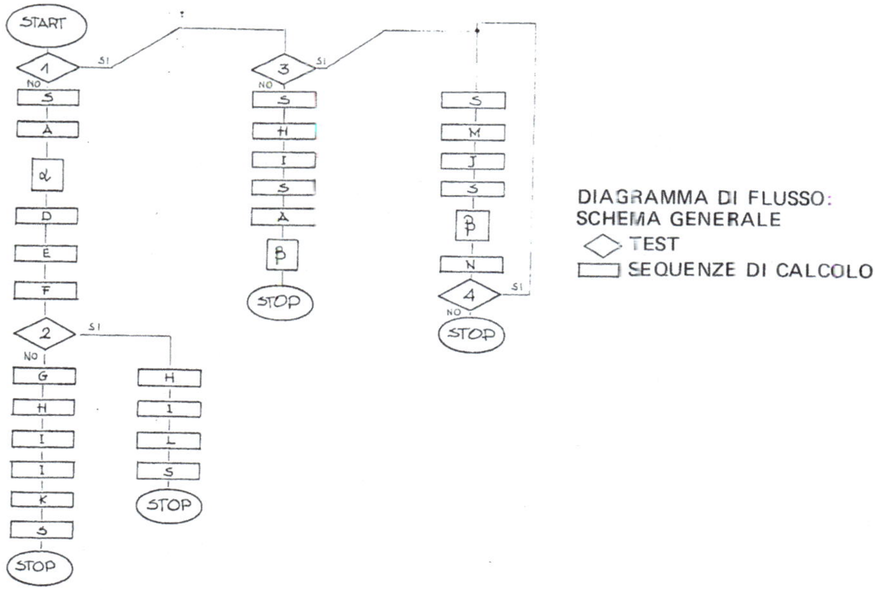

Fig. 2.1 Flow-chart of the procedure developed on a Texas Instrument Calculator for calculating the shadow relative length at a given location and time (Grosso 1982, p. 88)

impact on the urbanisation plans elaborated since by the Municipality administrations of Piedmont.

Solar residential buildings

The most concrete and effective application of the above mentioned technique on solar dynamic shadowing was on the EC co-funded Project UPSE comprising the design and constructing 457 solar dwellings in 17 locations on Piedmont. The aim of this project was to demonstrate that it would have been possible to reduce consistently the space heating energy consumption in multi-storey residential buildings, which used current construction technology, by applying solar active and passive systems as well as increased insulation with respect to the standards of that time. In particular, a mechanical ventilation system with supplied air heated by solar collectors was installed, in addition to sunspaces and, in a few cases, solar storage walls (Trombe-Michel type). Main result of this project was an energy saving for space heating ranging from 35 to 55% of the consumption of similar residential buildings, which did not have the systems implemented by the project.

The shadow dynamic method was used for placing the passive systems on the South-oriented façades of the buildings, in such a way that they would not be on the shadow of sur-rounding solid obstacles during the winter season, hence exploiting solar radiation as much as possible. An example of this application is the façade design of Grugliasco's building (Fig. 2.4), which was based on the analysis of the

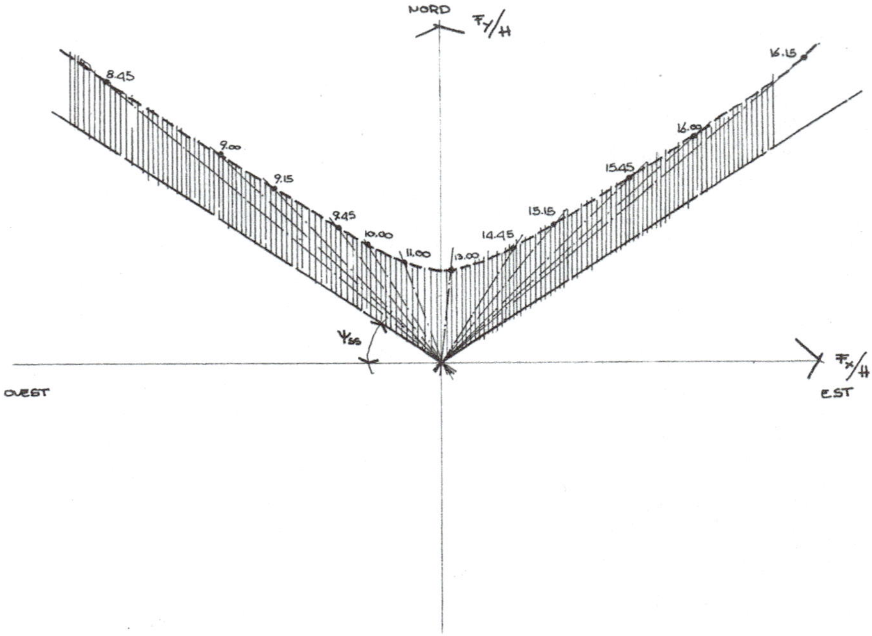

Fig. 2.2 Daily profile of the shadow cast by a peg on winter solstice at a location of Latitude 45°N (Grosso 1982, p. 40)

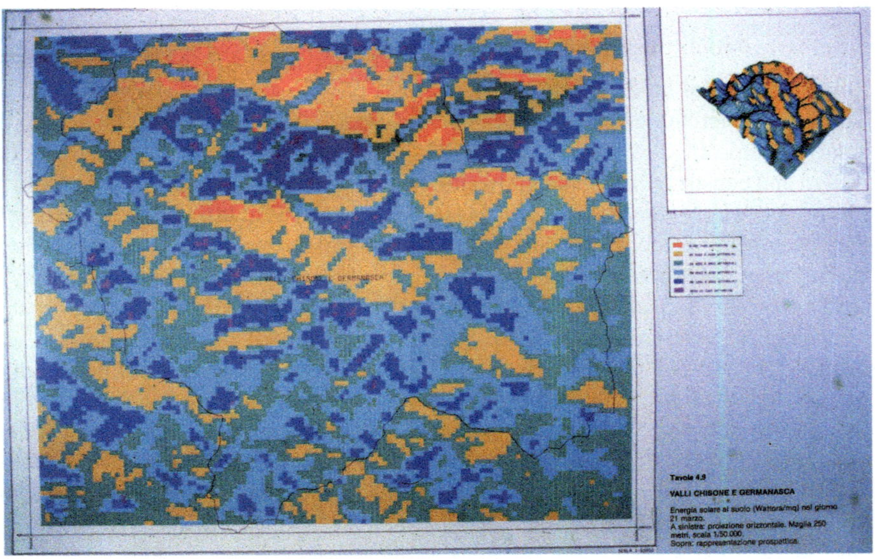

Fig. 2.3 Map and 3-D view, representing classes of solar radiation for March 21 on Val Chisone (Piedmont, Italy) (Grosso 1984)

Fig. 2.4 The UPSE building in Grugliasco (TO) (picture by the author)

shadows cast in the winter solstice by surrounding buildings, as represented in the graphical representation of Fig. 2.5.

A didactical tool: the Heliodon

While graphical shadowing analysis methods, as the one above described, are very useful for bioclimatic design purposes, so as the computer-aided simulator developed commercially after the early '90 s of the XX Century—see, as an example, the ECOTECH model, later incorporated by the AUTOCAD software—, a physical model as the Heliodon is far more effective for didactical aims. As Lechner wrote (2009), "Experience has shown that although computer models are powerful, physical models are still better. It is well known that a picture is worth a thousand words, but it is less well known that a model is worth a thousand pictures".

There are three types of Heliodons (Lechner 2009, Appendix I): Sun Simulator, Sun Emulator, and Tabletop. The first simulates the sunrays by lamps installed at regular intervals on 180°-span parallel metal circular arches, tilted as to simulate a specific latitude, although it can be adjusted for a range of latitudes plus or minus 5° from the constructed latitude. At the centre of these arches, a plane is located to place a scale model. The second is similar, but the lamps are installed on a 360°-circular span parallel arches, which can rotate along the diameter to simulate latitudes from equator to the pole. The third has a fixed lamp, reflecting light on a parabolic foil in order to simulate the parallel solar beams, and a movable separate table composed of a plane, which can rotate to simulate hourly angle and tilt to simulate latitude and solar declination, on which a scale model is placed.

Fig. 2.5 Drawing of the shadows cast by surrounding obstacles on the Southern façade of the UPSE building in Grugliasco (drawing by the author)

After having supervised a thesis on solar design tools, within which the student developed an ingenious Leonardo's-like tabletop Heliodon (Fig. 2.6), built in wood, I decided to design and have constructed a prototype of tabletop Heliodon, in 1987. It was based on a technical draw-ing table, a tool that had been in use by all designers using graphical output until the CAD software replaced it in any offices. This was used for a thesis, but remain abandoned during my study period abroad.

Following that application, this tabletop Heliodon was replaced by a more complex tool of the sun simulator type, design and made built by Valentino Manni, a technician at that time of the DINSE Department.

The tabletop Heliodon was composed of two following main elements.

(a) A concave circular Aluminium plate with a parabolic section, in the focus of which a lamp projected light towards the inner surface of the plate in order to reflect light beams as much parallel as possible, within a radius able to include the normal projection of the table on which the models were paced.
(b) A round wooden table (Fig. 2.7) supported by a steel structure, characterised by a motorised mechanism allowing three types of movements: (1) vertical tilt to simulate Latitude from 0° (pole) to 90° (equator); (2) adjustment of this tilt to

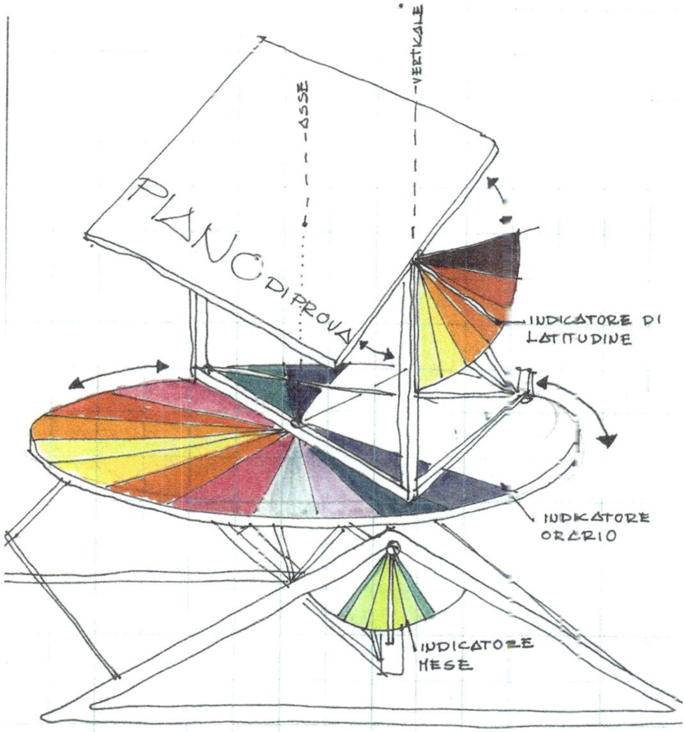

Fig. 2.6 Drawing of a table top Heliodon by a student in Architecture, 1986

consider the sun declination angle, i.e., the day of the year; (3) a 180° rotation around a vertical axis to simulate the hourly angle (solar hour of the day).

The sun simulator Heliodon (Fig. 2.8) is still working in the LaSTin (Laboratory of Technological System Innovation) as a didactic support for the courses of both the Master Degree Programme on Architecture for a Sustainable Design and the Bachelor Degree in Architecture, at the Polytechnic University of Turin.

It is composed of the following elements.

(a) A 3-m long steel arm rotating 180° for hourly movement around a motorised hub, which, in turn, can tilt vertically to simulate latitude and sun declination angles.
(b) A concave circular Aluminium plate with a parabolic section with a lamp on its focus, similar to the one described above of the tabletop Heliodon, but with a special finishing of the inner surface for an optimal light beams reflection; it is attached to the top extreme of the arm with the concave plate facing the table on which the model is placed.

Fig. 2.7 A picture by the author of the table top Heliodon design by him, 1987

Fig. 2.8 The sun simulator Heliodon working at LaSTin, DAD (picture by the author)

(c) A round table with a diameter of 180 cm, supported by a steel structure connected in solidarity with the motorised hub by a horizontal rod.

Amid others, this sun simulator Heliodon has been used in the course Environmental Technological Design of the Bachelor Degree Programme in Architecture, English curriculum, which I was initiated in 2015 and taught until my retirement (November 2018). My former teaching assistant, Giacomo Chiesa, have held this course since then. In particular, the Heliodon has been used as a test exercise for checking the location of virtual building units and outdoor activities in the design of a students' residential building, in relation to optimal winter solar access and summer solar control. Pictures of the hourly shadowing conditions of the proposed initial project solution within the built context, in the two reference days—winter and summer solstices—for the latitude of Turin, were compared to the shadows simulation made using a CAD software. This comparison helped students not only to validate the simulation analysis, but also to better understand, through the 3-D close-to-reality Heliodon visualisation, the dynamics of the solar irradiation on the envelopes of the designed buildings as well as on the outdoor spaces of the project lot. This check allowed for changing the initial design solution depending on the critical points highlighted by the Heliodon results (Fig. 2.9).

A calculation tools for solar active systems

Consistently to my proneness towards modelling the effects of the main climate factors on building and urban design, back home after the international COMIS experience (see below), I developed, with the collaboration of a brilliant student whose final Master Degree thesis I supervised—Luca Raimondo—an EXCEL sheet for calculating the energy contribution of active thermal and Photovoltaic solar systems to the total load of a building. This tool (SolDim), which had been upgraded several times during the years, is based on monthly average temperature and solar radiation data—direct and diffuse—to be inputted for a given location. The energy calculator uses simplified algorithms for space heating in stationary regime as well as parametric values for the thermal losses through the envelope (for the main 4 orientations and the horizontal roof), and DHW and electricity needs dimensioning.

SolDim was applied to both my didactic activities and the consultancy practice, when appropriate.

Fig. 2.9 Test of the dynamic shadowing for a preliminary design layout on the sun simulator Heliodon at LaSTin: winter solstice (left, 10 a.m.; right, 2 p.m.) (picture by the author)

2.1.4 Research on Airflow Aspects and Didactic-Related Tools

Window draught elimination

My first airflow-related research experience was the scientific and technical direction of a project funded by an oil-based utility Company in Rome, under the general responsibility of Lorenzo Matteoli. That project aimed at evaluating the energy saving effect of the reduction of draught through windows in an existing school building, taken as a reference case study representing a large majority of the existing buildings at that time (1983–85). Those buildings had a rather high air infiltration rate in winter due to old wooden frame windows without weather-stripping. The retrofit operation consisted on the application of silicon mastic along the joint between any operating window sash and its frame, using a special technique in order to avoid the bonding between the two.

This technique consisted on the placement of silicon mastic on the stop bead of the frame and a light gluing scotch tape on the corresponding rabbet bead of the sashes (Fig. 2.10). Hence, the windows are kept closed for 24 h, allowing the mastic to fit any impervious gap between sash perimeter and frame. Once the silicon mastic hardened, windows could be opened thanks to the tape, which avoided the two wooden elements bonding. The result was a perfect fitting built-in weather-stripping, much more effective than a commercial strip with constant section.

The school building had a traditional hot water heating system with a central boiler and wall cast-iron radiators in any classroom. To demonstrate the effect of draught reduction, a heat-flow meter was installed on the boiler and measurements were taken before and after the above-described operation. The outcome was an energy saving of approximately 20% of the total consumption for space heating. The methodology and results of this project were published in a scientific international Journal (Grosso et al. 1985) as well as in a research summary book (Grosso 1985).

The C.O.M.I.S. experience

As a consequence of the published results on the draught elimination project, I was invited in 1987 by Helmut Feustel to join an year-long international workshop—Conjunction Of Multi-zone Infiltration Specialists (C.O.M.I.S.)—to be held at the Lawrence Berkeley Laboratory of the University of California, Berkeley, CA, USA, starting from September 1988. I was able to accept this invitation and participate to the workshop due to the possibility, foreseen by the Italian law for senior researchers, of spending a sabbatical period abroad of a maximum of two years. I exploited all this possibility and, in addition, I got a research grant by the CNR (National Research Council) and a financial contribution by an Italian windows manufacturing Company, allowing for an extension of 6 months of my stay in California.

The aim of the COMIS workshop was to develop a Multizone infiltration and ventilation model, the first of the kind, through the contribution of top International experts in the field, coming from various Countries: Helmut E Feustel (USA), Francis Allard (France), Viktor B Dorer (Switzerland), Eduardo Rodriguez Garcia (Spain), Magnus K Herrlin (Sweden), Liu Mingshen (Peoples Republic of China), Hans C Phaff (Netherlands), Yasuo Utsumi (Japan), Hiroshi Yoshino (Japan). The

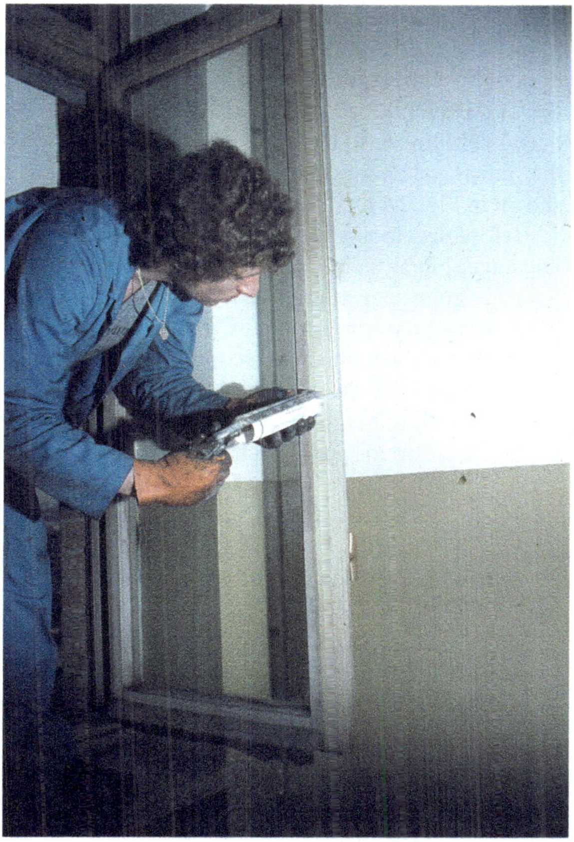

Fig. 2.10 Sealant application for the window draught reduction in a school building (picture by the author)

results of the COMIS workshop were published by the Air Infiltration and Ventilation Centre (AIVC), with the support of the International Energy Agency (IEA), Energy Conservation in Buildings and Community Systems Programme (Feustel et al. 1990, 1993).

My specific contribution was the development of a module of the COMIS model related to wind pressure distribution around buildings. Wind pressure coefficients are an essential input of any multizone ventilation programmes as well as related energy simulation software. Up to the date of the COMIS workshop, such an input were usually taken from tables of façade-average values, dependent on wind angle and building aspect ratios, or needed to be measured on purpose for any specific project by wind tunnel tests or calculated using CFD simulations. Both latter solutions are very time consuming and costly and need expert competences not usually present in a building design office even today.

Instead, my aim was to develop a procedure to calculate automatically wind pressure coefficients (CpCalc), which did not exist at that time and could be applicable to any project in a simple and straightforward mode. The intent was to supply a general tool to support both scientists and building/urban designers in getting a necessary input for assessing the effect of air movement on air infiltration as well as the potential of natural ventilation for IAQ and passive cooling. CpCalc was deemed to be part of the COMIS model under development, but also to be run stand-alone. It would have been lately linked to the main international energy simulation programme: DOE—the 'father' of EnergyPlus—TRNSYS, and ESPr. A flow-chart of the programme is shown in Fig. 2.11.

Before developing the routines CpCalc is composed of, using the FORTRAN 77 language, I analysed thousands of pressure coefficients data from wind tunnel tests on rectangular shaped models, with the intent to perform on them fitting regressions. This was necessary to correlate Cp values to several wind environmental and building parameters, which characterise any design context and process: terrain roughness, plan area density, average height of surrounding buildings, wind incident angles, relative building height, frontal and side aspect ratios, and element positioning on the facades. For this purpose I had to select wind tunnel test Cp databases with results

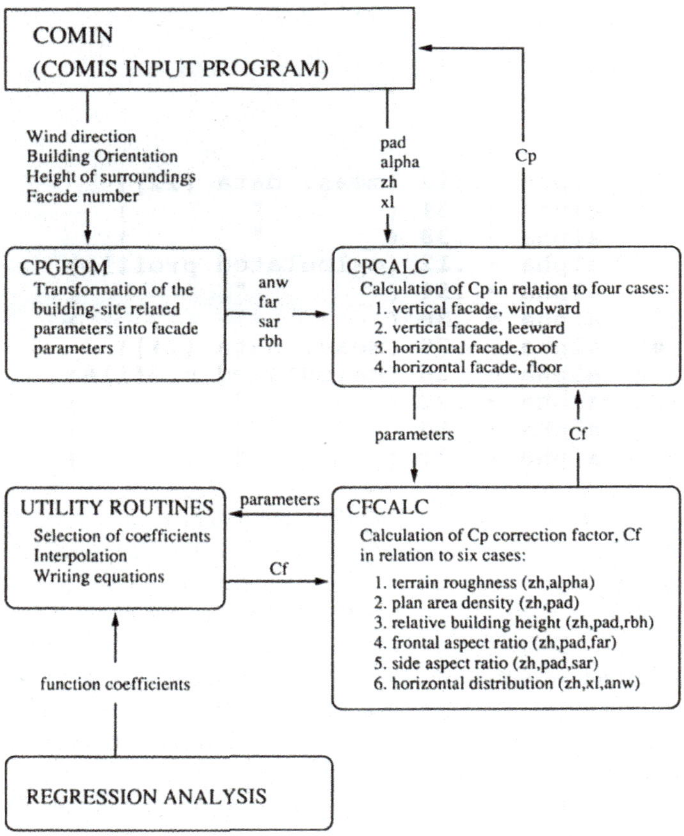

Fig. 2.11 Flow-chart of the CpCalc programme

for modular models with a range of values as wide as possible. The Cp databases from the following authors were then chosen: Hussein and Lee (1980), Akins and Cermak (1976) (Fig. 2.12). In addition, the equation for calculating surface-averaged pressure coefficients for low-rise buildings by Swami and Chandra (1987) was used as a reference evaluation tool in same cases. The methodology used, the flow-chart and structure of the algorithms developed, and all results from the regression analyses were presented in an AIVC workshop, held at LBL, and lately published, after an

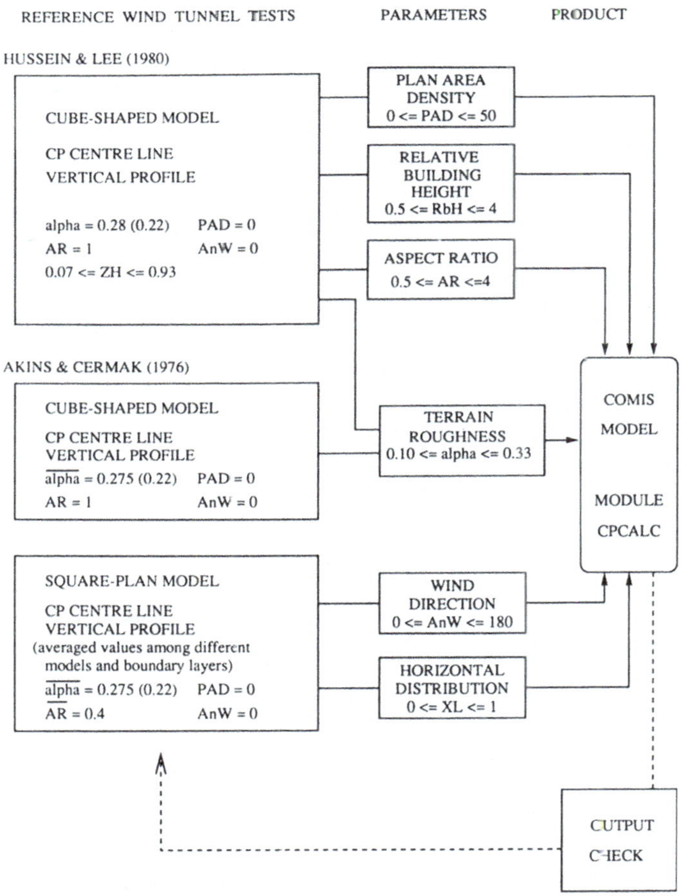

Fig. 2.12 Allocation of the reference data sets in the CpCalc programme development

extensive and thorough peer review by three reviewers, on the International Journal Energy and Buildings (Grosso 1992).

European Projects using CpCalc

Thanks to the development of CpCalc, I had the opportunity to participate as a unit research responsible to the following EC-funded projects: PASCOOL, AIOLOS, and PRECIS.

PASCOOL (PASsive COOLing of buildings) was co-funded by the European Commission, DG XII, Joule II Non Nuclear Energy Programme, Sub-programme: Energy Conservation and Utilisation. Its main objective was to develop a model for simulating the energy performance of passive cooling systems in buildings. The project was coordinated by M. San-tamouris and A. Argiriou of the University of Athens, Physics Department and was carried out from 1992 to 1995.

My contribution was related to the development of an upgraded version of CpCalc (CpCalc+), which was translated in VISUAL Basic Language and was completed with the correlation of wind pressure data on roof (Figs. 2.13 and 2.14). This was made possible by wind tunnel tests per-formed at the INETI Laboratory in Lisbon (Marques da Silva and Saraiva 1994).

My contribution in PASCOOL was included in the sub-group Climate—coordinated by Servando Alvarez—whose main topics were: (a) procedure and evaluation of methods for generating TMYs (Typical Meteorological Years) and their application to selected European localities; (b) characterisation of environmental heat sinks (sky temperature depression, psychrometric, sol-air temperature, ground temperature); (c) wind effect—pressure distribution around buildings; (d) performance of natural cooling techniques (underground pipes, direct and indirect evaporative cooling, radiative cooling).

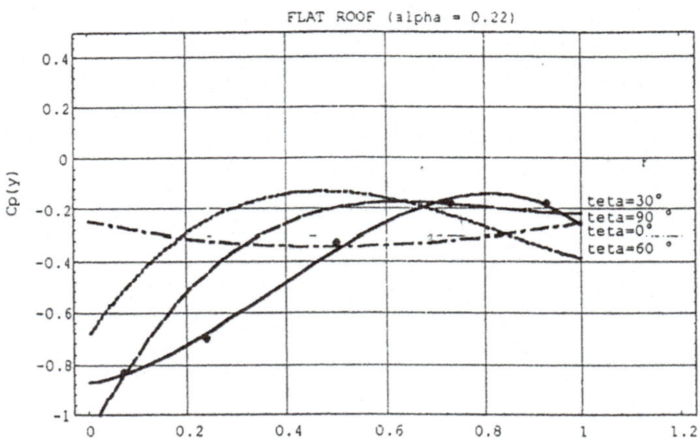

Fig. 2.13 Cp variation with wind direction on the centreline of a flat roof along the wind flow

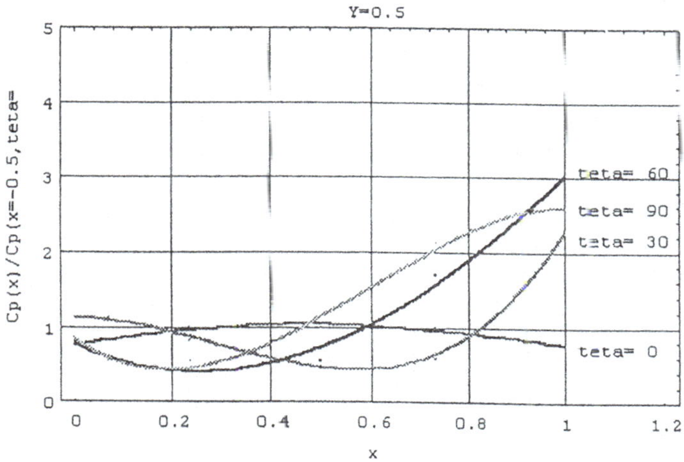

Fig. 2.14 Normalised Cp lateral deviation from the centreline along the wind flow, on a flat roof, when varying wind incident angle

The results y of my work were presented in various international Conferences (Grosso 1993; Grosso et al. 1994, 1995; Grosso and Parisi 1995), and laid the foundation for the publication of my second single author book, a handbook on passive cooling in buildings for Italian students, researchers, and professionals (Grosso 1997, and further editions).

AIOLOS was a project carried out within the framework of the ALTENER Programme of the European Commission, DG XVII for Energy. Its main goal was to create a specific material on the efficient use of passive ventilation for buildings than could be transferred to education activities and be used by all the professionals involved in the field of buildings.

My contribution was to write design guidelines and technical solutions for natural ventilation as well as to include CpCalc+ in the software AIOLOS, which calculates multizone natural ventilation airflow rates. A publication, edited by Francis Allard, includes the outcome of the project, including a chapter on my contribution and a CD on AIOLOS software (Allard 1998).

PRECis (assessing the Potential for Renewable Energies in Cities) was co-funded by the European Commission, DG XII, Joule III Non Nuclear Energy Programme, Sub-programme: Energy Conservation and Utilisation. The Project was coordinated by The Martin Centre for Architecture and Urban Studies, University of Cambridge, UK, and involved: the research Institute for CFD in Trondheim, Norway; the Fribourg School of Architecture, Switzerland; and my research unit at Polytechnic University of Turin. It was carried out from 1997 to 2000.

The main project goal was to develop methods and tools for assessing the potential use of renewable energy sources in urban planning and regulations. It included solar energy and daylight as well as wind effect as a potential natural ventilation driver. I was involved in the latter, through the utilisation of CpCalc+ as a tool for calculating the wind pressure coefficients on the envelope of a reference building in relation to its location in a theoretical urban block with varying form and layout. Hence, the correlation analysis of the CpCalc+ output allowed for evaluating the wind-driven airflow rates due to cross natural ventilation through a double room unit of the reference building, by applying Ansley's equation (Ansley 1982) with the calculated Cp differences as an input.

The results of my research for the PRECis Project were published in a paper on an International Journal (Grosso 1998) and in various International Conferences (Grosso et al. 2000a, b, c; Grosso and Banchio 2000).

Within the PRECIs Project, I commissioned to TNO—an Institute and Laboratory of the University of Delft, Netherlands—an analysis of the effect of a single obstacle in various positions on the wind pressure distribution on a building's envelope. This analysis was carried out using their Cp-generator based on wind tunnel test results. My intent was to complete the CpCalc+ programme with a correlation of data on the effects on the wind pressure distribution on a building's envelope of a single obstacle, which I could not include neither in the original COMIS CpCalc version nor in the upgraded PASCOOL's one, both considering only the effect of terrain roughness and plan area density. However, the amount of data and their characteristic stochastic variation would have not made possible a fitting linear analysis as had done before, but a different probabilistic approach should have been applied. Unfortunately, I have not be able since to find resources and personnel to carry out this task and that database is still there waiting to be elaborated and linked to CpCalc+. A hope comes from the perspective of the development of a future new version of the programme using the Python language (Chiesa and Grosso 2019a).

A didactic method: the wind-wake core projection and microclimate matrix

Based on the wind tunnel test diagrams representing the wind wake cores (calm zones) on the leeward side of parallelepiped modular models, elaborated by Boutet (1987), I developed a simplified method to draw such calm zones in any urban context. I did it by correlating the varying values of the relative width and length of the models with respect to their height, to the depth of the wind wake core of each model. The resulting fitting curves allow for determining the depth of the calm zone of any parallelepiped-shaped building within the range of aspect rations variation of the Boutet's models. Correction factors were then applied for different wind incident angles, with steps at 0°, 30°, 45°, and 90° and interpolating for values in between.

The combination of this method to the dynamic shadowing projection technique yielded the microclimate matrix tool, whose development was inspired by Brown and Dekay.

The microclimate matrix (Fig. 2.15), allows for assessing the optimal position of buildings and outdoor activities for a given project included in an urban context in relation to wind and solar access/protection, depending on seasonal thermal comfort and IAQ needs. This method was described in various publications (Grosso 2008b; Grosso et al. 2015; Nigra et al. 2016; Chiesa and Grosso 2017, 2019b).

Both the wind wake core and microclimate matrix graphical techniques have been applied in the course of Environmental Technological Design, which I taught in the Bachelor Degree Programme in Architecture, English curriculum, until my

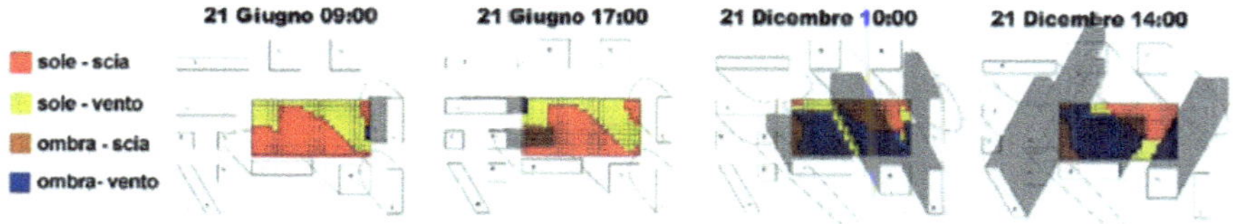

Fig. 2.15 An example of microclimate matrix: red (sun + wake), yellow (sun + wind),brown (shadow + wake),blue (shadow + wind); from left to right: June 21, 9 a.m.; June 21, 5 p.m.; December 21, 10 a.m.; December 21, 2 p.m. *Source* Grosso (2017, p. 180)

retirement, and afterword by Giacomo Chiesa. I used as well this method in the same course within the Atelier on Sustainable Design of the Master Degree Programme in Architecture for a Sustainable Design, which I taught last year and I will be teaching next academic year.

A tool to calculate the energy saving due to ventilative cooling

From 2006 to 2009, I collaborated to the Municipality of Turin for the development and implementation of an Annex to the Building Code dealing with mandatory and voluntary requirements, aimed at improving energy efficiency, thermal comfort, and indoor air quality in buildings. The voluntary part of the scheme was to assign scores related to the application of the voluntary requirements in construction or refurbishment projects, presented for authorisation by the Municipality, and apply relevant reduction of the urbanisation charges.

In particular, I proposed and wrote the requirement related to passive cooling through con-trolled natural ventilation, i.e., controlling wind-driven and stack-driven airflow by letting it in a building through automatic opening devices, applied to windows and vents, at given indoor and outdoor air temperature conditions.

In order to check the energy saving performance related to the application of that requirement, I proposed and developed a software (EcoWind), calculating the potential energy reduction of the cooling annual need of a double-room building unit due to the application of controlled natural ventilation for the climate of Turin. The software was developed thanks to the collaboration of a team composed of an energy expert (Luca Raimondo) and two informatics professionals (Luca Colombaro and Andrea Deganutti) (Grosso et al. 2009). It was based on the semi-dynamics hourly calculation of the sensible heat transfer through a building's envelope, de-scribed by the standard EN ISO 13790:2008, Energy performance of buildings—Calculation of energy use for space heating and cooling, and empiric formulas for the airflow rate prediction (Grosso 2008b).

This software was further upgraded for various locations representative of the climates of Italy in both a web version (SPERAVent) (Grosso et al. 2010–11) and a compacted version on CD (WindChill) (Grosso et al. 2011), attached to the third edition of my book on passive cooling.

2.1.5 Research Projects Integrating All Aspects of Bioclimatic Design

Environmental evaluation for the 2006 Winter Olympic Games in Turin
The Turin Organising Committee for the XXth Olympic winter games (TOROC), held in 2006, appointed the Polytechnic University of Turin, Italy, Dept. Settlements Science and Environmental Technology (DINSE)—which converged into the Dept. Architecture and Design with the reform of 2012—to evaluate the environmental quality and compatibility of the projects included in the Olympic Programme (OP). This was done according to the recommendations issued on the basis of a Strategic Environmental Assessment (SEA), whose application represented the first case in Italy, and one of the first in Europe, for this type of events.

The projects assessed concerned land transformation and constructed assets, construction of new buildings, and refurbishment of old unused industrial and commercial structures.

Gabriella Peretti, a colleague of mine, and me, were the co-ordinators of the work, which was carried out through:

(a) guidelines for project sustainability as regards the construction and management of Olympic and Multimedia Villages;
(b) a system aimed at evaluating the Eco-compatibility Projects within the Quality Settlement Indicator as provided in the Monitoring Plan foreseen by the SEA.

Objectives, requirements and indicators used for the eco-compatibility assessment regarding use of climate resources, outdoor spaces, and control of resources consumption—all aspects related to the application of bioclimatic principles, are described in Tables 2.1.

The methodology applied was based, with adaptation to the client requirements, on the one elaborated within the Working Group 'Sustainability in buildings' of the Constructions Committee of UNI, instituted in 2000, and coordinated by Peretti from the inception to 2006, and by me from 2006 to 2019. It consisted of a multi-criteria method inspired to the one developed by iiSBE (international initiative for a Sustainable Built Environment): GBCtool, lateron SBTool. This is a spreadsheet for calculating several indicators, subdivided in issues, categories, and criteria, to which a weighted system is applied in order to obtain a final aggregated performance score. The UNI's WG method—never converted into a standard due to the overlapping of an analogous CEN standard development process—was, instead, based on a subdivision of the performance indicators in 'needs' (objectives), 'classes of requirements', and 'requirements' as well as on the main phases of a building life cycle (Fig. 2.16). An example of the weighting system in Fig. 2.17.

Table 2.1 (a) Requirements and relevant indicators related to the use of climate resources and outdoor spaces, (b) requirements and relevant indicators related to the control of resources consumption

(a)

Objectives	Requirements	Indicators
Use of climatic resources (UCR)	R1. Winter solar radiation use	(a) shape efficiency for solar access (b) ratio of shadows length to distance between buildings
	R2. Natural ventilation use	(a) ratio (%) of the number of dwellings with potential wind-driven and/or stack effect-driven cross ventilation to the overall number of dwellings (b) natural ventilation air flow rates in sample dwelling units
Environmental quality of external spaces	R3. Winter wind dynamics control	winter wind protection factor
	R4. Summer wind dynamics control	summer wind exposure factor

(b)

Objectives	Requirements	Indicators
Resources consumption control (rcc)	R8. Use of thermal insulation	Weighted average U-value of the buildings' envelope
	R9. Substitution of fossil fuel with renewable sources in HVAC and electricity production	Efficiency Factor for primary energy loads related to HVAC and electricity production systems, defined as the ratio of the difference between a reference load and the project load to the reference load (range 0–1)
	R10. Summer radiation control	(a) percentage of the number of windows with fixed shaded glazed area grater than 60% of total glazed area between 15 h and 17 h, summer solstice, on the total number of windows oriented to SE, S, SW, W (sample buildings) (b) area-weighted average shading coefficient of the windows oriented to SE, S, SW, W (all buildings)

(continued)

Table 2.1 (continued)

(b)

Objectives	Requirements	Indicators
	R11. Use of thermal mass	(a) area-weighted thermal decrement factor for external massive walls, roof, and internal slabs of all buildings (summer use) (b) area-weighted thermal decrement factor for external massive walls and roof of all buildings (winter use)

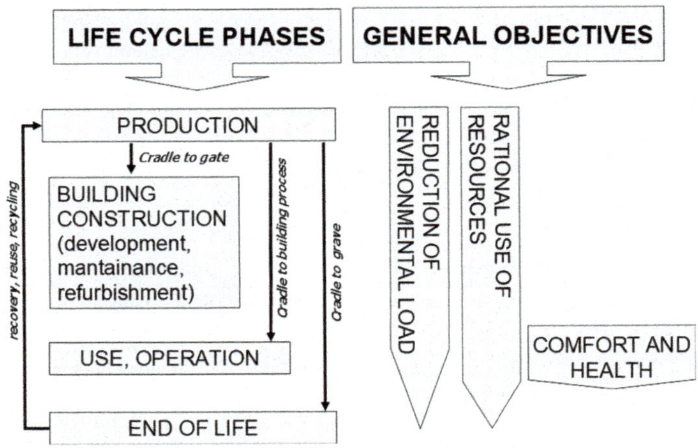

Fig. 2.16 Scheme of the requirements allocation system used for the TOROC project

The methodology and results of the work carried out were described in a paper presented to the International Conference on Sustainable Buildings, held in Warsaw, 2004 (Peretti and Grosso 2004).

PR.I.M.E.[3]

In 2010, I was contacted by Maria Irene Cardillo, an architect with office in Rome, who was looking for an expert on natural ventilation for answering to a "Call for proposals on projects of energy efficiency and use of renewable energy sources in urban areas", opened by the Italian Ministry of Environment. My name was given to her by a German expert she contacted earlier.

I accepted her request for collaboration and we elaborated together a project proposal called PR.I.M.E.[3] (Innovative procedures for energy-efficient and eco-compatible building modules).

The project was funded. It involved the Consortium So.Ri.Ser. (Society for Research and Services), which included Cardillo's SME AE.C.I. (Architecture and

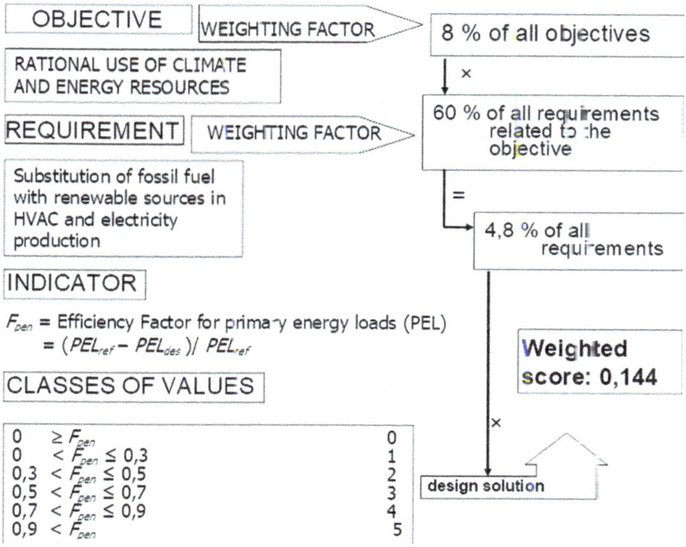

Fig. 2.17 Example of the weighting system used for the TOROC project

Construction), COR-MATEX—a company making machinery to treat recycled material, located in Prato (Florence)—and the Engineering firm INGE.CO, Rome. The Polytechnic University of Turin was involved through a research unit I coordinated, while holding also the role of scientific responsible of the entire project.

The project was developed from 2011 to 2013. Its main objective was to design and test a prototype of a movable prefabricated building module, with an Aluminium bearing structure, a highly insulated envelope and sustainable low-energy environmental control systems (ECS). The insulation layers were made of materials coming from recycling process of waste tires and industrial textile scraps, enclosed in Aluminium panels. The ECS included a hybrid passive air conditioning wall element, called H-NAC (Hybrid-Natural Air Conditioning) and a sanitary wastewater heat recovery system. The H-NAC was tested in the Laboratory of Technology System Innovation (LaSTIn) of the Dept. Architecture and Design, Polytecnic University of Turin, of which I was the scientific Director and which was just instituted at that time as a transformation of the a previous lab with had a different name, inherited by its founder (Laboratory "R. Mattone").

The design and test of the H-NAC system represented the most innovative environmental-technological feature of the PR.I.M.E^3 project. Its main purpose was to demonstrate that it could be conditioned indoor air both in winter and in summer, using a natural source as the outdoor air with a very low energy consumption. The main challenge was to apply the concept and functioning of a heat recovery system (HRS), whose commercial applications worked—and still work—only with controlled mechanical ventilation, to a system applying very low pressures, close to

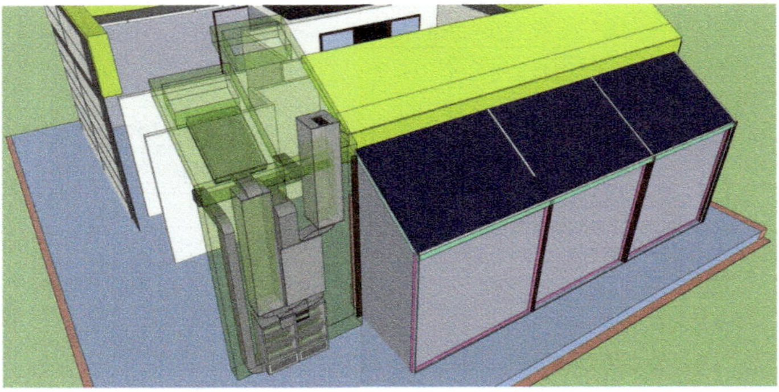

Fig. 2.18 3D representation of a hypothetical integration of the HNAC system to the PR.I.M.E^3. Building module

the ones of natural airflow. For that purpose a special metal 'box' with crossing channels, having sections much higher than the ones of mechanical HRS, was built and tested. After the first tests, spiral 'turbolators' were inserted in the channels in order to increase the heat exchange efficiency, which reached eventually a good value (50%) even not comparable to mechanical HRSs. The entire H-NAC system was conceived as a wall element (Fig. 2.18), composed of various elements, in addition to the HRS: a latent-heat adsorption transfer (LHAT) component for air de-humidification, where two types of adsorbing material were tested (zeolite and silica gel); a Direct Evaporative Cooling (DEC) system, called 'rain-shaft'; and solar collectors, both air- and water-fuelled to heat air as well regenerate the LHAT system (Fig. 2.19).

2.1.6 *Teaching on the Overall Bioclimatic Spectrum*

Institutional Teaching at the Polytechnic University of Turin

Back from the experience in California, in 1992 I had the chance to start teaching courses at the School of Architecture of the Polytechnic University of Turin, as a course holder and no longer as an assistant like in the pre-American period. Since then and up to the curricula re-form of 1995, I held two courses on Architectural Technology, one in the School's headquarter, at the Valentino Castle, the other in the decentralised site of Mondovì (CN). In both experiences, I applied all what I had learned during the period abroad, not only within the COMIS workshop, but also in the two years afterword, when I had carried out consultancy activities (see below in the relevant paragraph).

In that initial teaching experience as a course holder, I applied concepts and methods inspired by researchers and scholars, whom I mentioned in my other chapter of this book on the origin and evolution of bioclimatic approach in architecture. It was of a great help my association to the Society of Building Science Educators (SBSE), which includes teachers of ECS (Environmental Control Systems). They

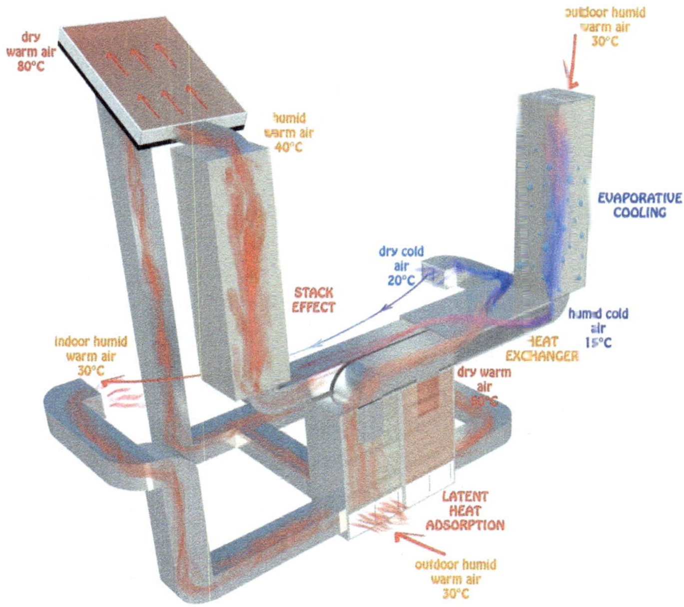

Fig. 2.19 Scheme of the complete cycle of airflows and relevant heat exchanges in the HNAC system

come from architecture schools in the USA, rather than from engineering schools as it occurs, generally, in Europe. Since ECS at the Polytechnic University of Turin was, and has always been, a field held by Building Physics teachers, my approach at that time caused a conflict with the main representative of that field. He criticised, e.g., the use I made of the balance point temperature as a parameter for evaluating the contribution of passive systems to the energy need for space heating and cooling in a building, which I thought to be a good synthetic criterion more useful for architecture students than the more analytical one applied by the Building Physics teachers. The quarrel was then solved thanks to the mediation of the School's Dean and based on the principle of autonomy and responsibility of teachers granted by the Italian Constitution. However, my belief on that interrelation between Architectural Technology and Building Physics as a plus for the effectiveness of a knowledge and methodological transfer of the bioclimatic approach to students would have had negative consequences on my career progress. This was, and still is, due to the characteristics of the Italian academic recruitment and career advancement, based on closed scientific disciplinary sectors, penalising candidates belonging to inter-boundary fields of research.

To be honest, my experience abroad had an important value for my career advancement when I obtained the role of associate professor on Architectural Technology and Environmental Design in 1998 by winning a national competition for 10 openings to be allocated in all Italian Schools of architecture. That was the last call at national level. Afterword, all calls for openings were managed at local level.

Since the curricula reform of 1995 several other reforms were instituted for the Italian University, leading eventually, in 2003, to the 3 + 2 format (Bachelor + Master degree), consistent to the European setting. Since 1995 up to my retirement, and now as a Contract Professor, I have been teaching mainly in architecture studios. From 1995 to 2002, I taught 'Construction in Architecture', in the 2nd year of the 5-year degree programme, a studio integrated to architectural design and structural engineering, with the last two academic years at the Mondovi site. Within the same Degree Programme, I taught in a 'Final Synthesis Studio' 'Environmental Design (and Eco-Design)' from 1998 to 2002, and an analogous studio at Mondovì in the academic year 2001–02. From 2002 to 2007, I taught 'Technology Innovation' and 'Environ-mental Technology', studios at the third year of the Bachelor Degree Programme Architecture for Design, with an integration of Material Science, respectively, in Turin and Mondovì. From 2004 to 2009, I taught 'Eco-compatible Technology' in a Studio Architecture and Urban Landscape Design for the Master Degree Programme Architecture for the Environment and Landscape at the Mondovi site, and 'Environmental Technology' for a studio on 'Local Planning' for the Maser degree Programme 'Territorial, Urban, and Landscape Planning' in the academic year 2004–05. From 2007 to 2011, I taught 'Technology of Architecture II' in a studio 'Technological Design in Architecture' for the Master Degree Programme in Architecture, with an integration of Structural Engineering. For the new Master Degree Programme 'Architecture for a sustainable design', I taught 'Technology Innovation' in a studio 'Sustainability in the design of building-systems', integrated with 'architectural design', from 2011 to 2014, and from 2014 to 2016, 'Advanced Environmental Technological Design' in a Final Design Studio, in English, integrating architectural design and building physics. All these studios were held in the 2nd (final) year of the Programme. In the first year of the same Programme, I have been teaching 'Environmental Technological Design' in the Atelier 'The architectural sustainable Design E' since 2018. In all these studios, I have based my teaching on a strict integration between knowledge, methodology, and tools related to environmental and technological aspects, with particular focus on the bioclimatic approach, and the architectural design process, through all its phases, from building programming to preliminary, schematic, and design development (Fig. 2.20).

I taught single courses on: 'Environmental Technological Design', in the third year, from 2012 to 2018, and 'Fundamental of Architectural Technology', in the first year, in the academic year 2016–17, both in English, for the Bachelor Degree Programme in Architecture; 'Elements of Environmental Urban Design', in the third year of the Bachelor Degree Programme 'Territorial, Urban, and Landscape Planning'; 'Technology for Building and Environmental Hygiene', from 1998 to 2000, in the third year of a University Diploma I Building Constructions, an Architecture-Engineering interfaculty experience, ended with the 3 + 2 reform.

Fig. 2.20 An example of maquette of a building designed in the Final Design Studio, 2015

At the third Degree level, I taught a Course 'Design of indoor climate-control passive systems in buildings' for the Ph.D. Programme 'Management, Production and Design', SCUDO, Polytechnic University of Turin.

Thesis development supervision
Thesis supervision had been always the most favoured didactic activity I carried out in my entire academic career. It allowed a special and deep link with a single student, who had chosen to be supervised by me because of my specific competence in relation to his/her scientific interest. Although several of the theses were of research nature, I generally preferred topics involving an integrated design development, including both technological (environmental) and architectural aspects. To cover the latter, I usually asked for a co-supervision by a professor on architectural design. This integration led most of the times to optimal results in demonstrating the potential and effective application of bioclimatic principles and methods to real project case studies. In this regard, amid the about 500 theses I supervised in my entire academic career, I would like to mention the ones supervised, in the academic years 2014/and 2015/2016, together with Marianna Nigra, a contract professor of architectural design at the time of the theses supervision, and later on a Ph.D student of mine. The theses' subjects were about various design projects related to new building construction and refurbishment as well as infrastructure and urban scale, with location spanning all over the world. A synergic and deep supervising support action allowed those theses to be based on a common technological-architectural methodological approach, but with a specific original imprinting characterising each student. Due to the extraordinary results of this experience, 15 of those theses were selected for publication (Grosso and Nigra 2019).

Bioclimatic approach to landscape architecture

I was involved, from 2004 to 2009, in the organisation—together with my colleague Gabriella Peretti—and teaching of a Master Degree Course on Design of Parks, Gardens, and Landscape, established by the 2nd Faculty of Architecture, Polytechnic University of Turin, and the Faculty of Agriculture, University of Turin. I taught a Studio on Environmental Urban Design, focussed on the interaction between climate factors and urban design, concerning their impact on outdoor comfort, air quality, and energy conservation.

In 2009, that experience lead to the inception of a new agreement between the above-mentioned Faculties, extended to the Faculty of Agriculture of the University of Milan and the School of Architecture of the University of Genoa, which already had a Master Degree on Landscape Architecture. This inter-university Master Degree Course, which was titled Design of Green Areas and Landscape, went through a cumbersome Institutional process to reach an agreement amid the Faculties involved, with a strong opposition of the Territorial and Regional Planning Department of the Polytechnic University of Turin. It claimed to have a paternity on that field having a landscape related teaching module in its Master Degree Course on Regional Planning.

However, the battle I personally lead in my Faculty to win—successfully, at the end—the approval of that agreement, was based on the belief that landscape architecture had the legitimation as an autonomous discipline, being grounded on a design scale typical of urban and micro-territorial contexts, quite different from a regional planning scale. In addition, the new Programme was also focussed on gardens design, including restoration of historical gardens, a topic stranger to the regional planning course.

Within that inter-university Course, I taught a module on Environmental Technology within a 2nd year Studio on Landscape Design, held at the University of Genoa. This teaching was focussed on the bioclimatic effect of vegetation on outdoor comfort as well as on special materials, which could be used in landscape design to control and mitigate microclimate outdoor conditions. I used for that teaching also the outcome of a work carried out by Gianni Scudo and his team at the Polytechnic University of Milan—for which I had done a consultancy activity—within the European project RUROS (Rediscovering he Urban Realm and Open Spaces), funded under FP5-EESD and coordinated by the Centre For Renewable Energy Sources, Greece (2001–2004). I carried out that teaching up to 2013, when the Programme's coordination was transferred to the Territorial Planning Department, within the new depart-mental asset of the Polytechnic University of Turin.

That experience lead as well to writing contributions for publications (Grosso 2012, 2014).

Teaching in other Universities

In my entire academic and professional career, I have always been inclined to apply my scientific and didactic expertise beyond the borders of my Institutional role with the aim of disseminating a 'bioclimatic' way of thinking within the professional world of architects and engineers. This have implied both, an active consultancy activity as described in the next para-graph, and hundreds of lectures and courses, I held in other Universities as well for Public and private non-academic organisations. The following are some of the most significant and long lasting of them related to bioclimatic design for post-graduate Master Programme (MP).

For the MP "Bio-ecologic Architecture and Technological Integration in the Environment", managed by the Inter-University Consortium A.B.I.T.A. (*Architettura Bio-ecologica e Integrazione Tecnologica Ambientale*) at the Dept. Architectural Technology and Design, University of Florence, I taught 'Natural Ventilation and Passive Ventilative Cooling of Buildings' from 2002 to 2013. For the MP "RIDEF (Renewable Energies, Decentralization, Energy Efficiency)—Energy for Kyoto", at the Dept. BEST—Building Environment Science and Technology, Polytechnic University of Milan, I taught 'Passive Cooling Systems' from 2003 to 2014. From 2002 to 2006, I taught 'Natural Ventilation and Passive Ventilative Cooling of Buildings' for two MPs coordinated by late prof. Cristina Benedetti, an Italian Pioneer in Bioclimatic Design and friend of mine: 'House-Climate', at the School of Economy, Free University of Bozen; 'Enviornmental Design', at the School of Architecture 'Valle Giulia', University of Rome 'La Sapienza'.

2.2 Extra-Academic Experience

As I already mentioned above, in my entire life-long work activity I always tried to combine theoretical investigation and practical application, possibly connected to the real professional design world. This came from a firm belief that the theoretical concepts I developed in the activities above described needed to be verified and tested through experiences concerning direct building and urban design applications. This occurred in both teaching and consultancy activity in support of architectural design offices. A synthesis of them are outlined as follows.

2.2.1 Teaching

I held hundreds of lectures and seminars on bioclimatic, energy and environmental related topics in professional updating and long-life learning courses, managed by public and private organisations. Amid them, it worth to mention the following: ANAB (National Association of Biological Architecture); INBAR (National Institute of Bio-architecture); ISES (International Solar Energy Society); WWF (World Wildlife Fund); Environment Park SpA, Turin; iiSBE (international initiative for a Sustainable Built Environment); CELIM Onlus, Milan, within a collaboration agreement Italy-Albany; the Italian Ministry of the Environment Rome; the Chinese Ministry of Science and Technology, Beijing; and the Order of Architects, Planners, Landscapers, and Restorers, of Turin, Genoa, Aosta Valley, Lecco, Reggio Calabria.

2.2.2 *Main Consultancy Projects*

I will describe synthetically, in the following paragraphs, some of the main activities and projects I was involved in, the most significant from the point of view of thoroughness of the knowledge and competence applied. It will be clear from this description that my involvement was not only the mere application of a specific knowledge such as the one on airflow and so-lar control, but was a holistic experience characterising the complex interrelationships be-tween science and technology expertise and the design process in the professional practice.

Architect Liebermann's design

Just after the COMIS workshop in Berkeley, CA, USA, I had the chance to meet with the last F.L. Wright's assistant, Daniel Liebermann. He was a fond of Borromini's architecture and a theorist of the application of Fibonacci's number series to architectural design. This theory was a way to give a sort of scientific validation to the form of his architecture within the tradition of Wright's organic approach. I recall Dan, who died at 84 year old in 2015, as both a brilliant and unconventional man, even eccentric. He would comport himself in layers of wool clothing during summer heat, and kept a trove of artefacts—including old cars—at his home in Berkeley, and later, in Inverness. He was very intelligent, so multifaceted in his life experiences; sometimes you'd go crazy because he would kind of go off on tangents. But he was very passionate about his work and about doing things right. He was a pioneer of recycling by using scraps such as tires and metals in his architecture. I could visited some of his work, mostly residential houses, and I was particularly hit by the Radius House, designed in 1962 for his parents on two contiguous parcels in Mill Valley, Marin County, CA. The interior is completely open without separation between the two storeys, whose only connection is a spiral stair. All structure is made of timber beams getting off on a radius form (Fibonacci's curve) by a central stone core. In addition, he introduced me to a completely unknown (for me) world of Californian architects followers of Bruce Goff, one of the major Wright's disciple. Amid those, I was astonished by a tree house in the woods above the Big Sur cliffs, in the way from San Francisco to Los Angeles.

However, my collaboration with him ended up, in Berkeley, to a design of a rammed earth house, never built, and to the writing of a book (Liebermann was also a self-made publisher) on fire resistant architecture (Grosso 1991), where I could apply some of the results of the tests I carried out at the wind tunnel of the University of California, Berkeley. When I came back to my work as a researcher and teacher at the Polytechnic University of Turin, I was able to manage to invite him as a visiting lecturer for a period of two months, in 1995. His lectures were very successful among students of both technology and architectural design courses. All were excited by viewing the examples and learning the organic approach to architecture from an architect who had experienced directly Wright's way of working. He was also surprise by Turin's townscape, saying that "it was the best kept secret of Italy". In particular, he was struck by Turin's Baroque architecture and by the Nervi 'Labour Palace', which Daniel Liebermann considered an extraordinary example of application of Wright's organic theory with its structural reinforced concrete grid of modular tree-like pillars sustaining radial beams on their tops within a square framework.

The Consalud Headquarter Building in Santiago, Chile

In 1998, I participated through the Department of Science and Techniques for Settlement Processes, in which I worked as a senior scientist at that time, to an International Competition for the bioclimatic consultancy to the design of the headquarter building of Consalud, a Health Insurance Company in Santiago, Chile, which was commissioned to the architecture office May y Soler. In spite of the participation of much stronger competitors in terms of profession-al experience and staff dimension, such as Buro Happold Engineering from UK, I won, probably also for a higher proneness of the appointed Architects to a consultant coming from a "Mediterranean" area, culturally closer to the Chilean context.

This consultancy activity was carried out on site as a full time charrette lasting a week, during which I discussed continuously with the designer, with the help of a student of mine, Mario Voerzio, who would have lately developed his thesis on the same topic together with another student, Gianfranco Dell'aquila. The comprised two parts: a central gathering space and cafeteria, and an office wing. My contribution was focussed on the latter.

The first action I overtook was a sensitivity analysis on the thermal effect of the building orientation, considering an overhanging shading made of horizontal tilted slats at roof level on the top of p placed all around the building, according to a concept inspired to the Chilean traditional of country houses (Fig. 2.21). This simulation, as all the others made in that experience, were carried out using the thermal model QUICKII (Mathews and Joubert 1989). Results from the simulation led to choose 20° East orientation as the optimal one in terms of balancing heating and cooling. In a second series of simulation, the width of the overhanging pergola was increased of one meter at each step, from the value considered in the above-mentioned sensitivity analysis, keeping the orientation at 30° from North eastward (the worst among the previously analysed as far as cooling loads are concerned). The advantage of increasing the overhang width resulted quite low in terms of both comfort and cooling loads. Since those decreases do not offset the in-crease in costs due to the relevant higher widths, a 6-m width for both northward and southward pergola was suggested. All other analyses and simulation regarded technologies for both the envelope and the environmental control systems (ECS).

As far as the latter are concerned, the most original and innovative solution I suggested was to combine the vertical bearing function and airflow transportation in one building element, a steel pillar-duct. This solution, whose only precedent dated back in a building of the late XIX Century in England, appeared to be rational and efficient, as long as all relevant joint thermal and structural stress problems were solved.

Based on the maximum airflow rates requirement for ventilation cooling, the pillar-duct constant inner cross section was 0.25 m^2. For thermal and acoustical insulation, a double insulation layer was applied: a rock wool matte of 5 cm thickness and an outer 2 cm thick gypsum board panel. A resilient material was inserted in

Fig. 2.21 A view of the office wing of the Consalud Headquarter building (picture by the author)

joints between pillar-ducts and floor slabs in order to avoid risks of sound structural transmission. The thickness of the pillar's steel profile

had to be increased as compared to a similar conventional structural vertical framework of the same resistance, due to the holes foreseen for the attachment of the horizontal ducts at each floor. The overall ventilation and air conditioning system was conceived as a hybrid natural-artificial one, with a chiller and a heating boiler dimensioned only for occasional pick energy demand. It was realised through a 4-ways air flux mechanical ventilation system, whose Air Handling Units (AHUs) were connected to an Earth Heat Exchange (EHX) battery, made of horizontal buried pipes, for space cooling and pre-heating as well as air solar collectors on roof for space heating (Fig. 2.22). The horizontal airflow network was made of supply air ducts within suspended ceiling and extract air directly from under-floor plenum to vents in the pillar-ducts.

This configuration did not allow for flushing directly at night by external air the only massive elements of the building, the concrete floor slabs, being the envelope made of curtain all-glazed walls. Hence, a free-cooling was conceived by using the mechanical ventilation system, when outdoor air temperature is below a value set as a function of indoor comfort, net of the daily solar and internal heat gains.

The bioclimatic strategies and technologies applied in the office building of the Consalud facility can be summarised as follows:

1. linear building form;
2. building orientation of 20° from North towards East;
3. 6 m deep overhanging pergola;
4. double glazing with external reflecting pane;
5. thermal mass for floor slabs;
6. mechanical night ventilation of floor slabs by a flow rate of 3 air change per hour (ACH);
7. mechanical free-cooling when appropriate;

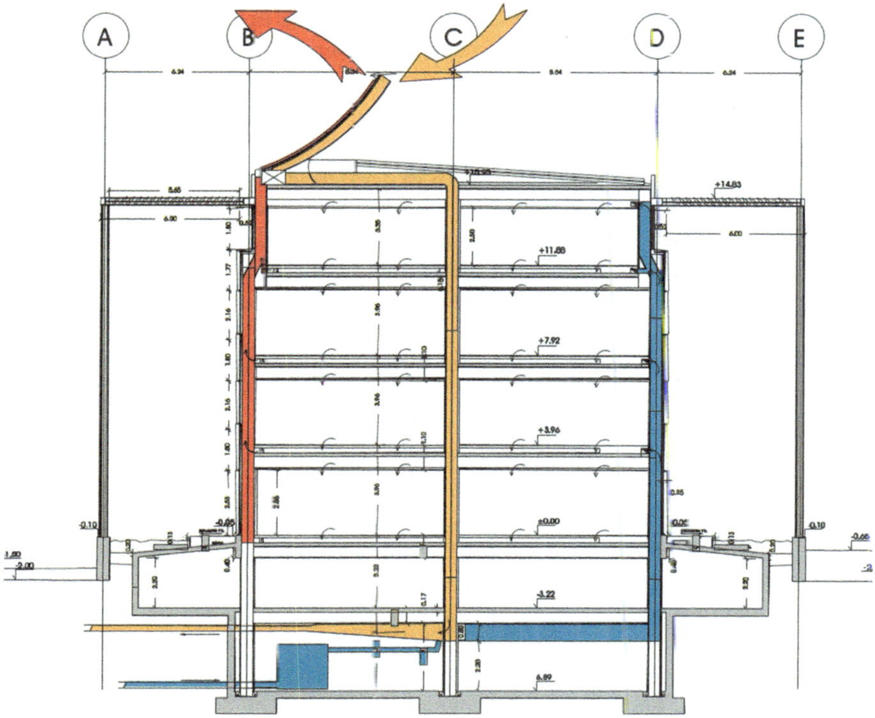

Fig. 2.22 Scheme of the hybrid indoor climate control system of the office wing of Consalud building (cooling mode): red = exhausted hot air; orange = outdoor warm air; blue = cold air (on bottom-left: pipes coming from the EAHX system)

8. cooling by ventilation through EHX buried pipes of 20 cm radius;
9. heating by ventilation through solar collectors located on roof;
10. stack ventilation of entrance hall.

The overall annual energy saving for ventilation and air conditioning (cooling) due to the above listed measures related to cooling, was of 510.4 MWh, i.e., the 58% of the energy needs of a reference configuration of the same building, but with a conventional HVAC system, no shading, and clear single glass on the curtain walls, equal to 880 MWh.

The high school building "L. Orsini", Imola (BO), Italy
In 2005, I submitted—together with colleagues operating in the professional fields of architectural design, mechanical engineering, structural engineering, geology, and construction works safety—an offer to a tender opened by the Municipality of Imola (BO), Office for Public Works, for the design development of the High School "L. Orsini", located in the Pedagna district. This call was opened on the basis of a previously approved project at a preliminary design phase, and required the contribution of an environmental consultancy with a particular focus on airflow modelling and

natural ventilation. My competence and experience in this field helped the on-purpose temporary associated professionals to win the tender.

This experience was characterised by an accurate and innovative design process integrating different architectural and engineering figures and technical knowledge, in order to design a building that was fully functional as well as a model of sustainability for students, teachers, and citizens as a whole. In order to achieve this goal, strongly required by the Municipality, the associated professionals were engaged since the design program in a collaborative process aimed at conceiving an innovative building-equipment system. As a consultant with a know-how bridging architectural design and mechanical engineering sectors, I plaid a fundamental role for the successful, not granted, outcome of that integration process.

Various design options were assessed from the performance and environmental point of views during the entire process. In particular, innovation involved construction technologies, materials, functions' distribution, use of renewable sources, integration between mechanical and passive systems aimed at increasing energy efficiency, attention to maintenance and to waste disposal and recycling, design and installation of the monitoring sensors' system.

The "L. Orsini" high school building, located in a lot used as a city park and has three stories plus a basement, was designed to accommodate 425 students. The shape of the building was conceived to optimise uptake of solar radiation as a source of direct heat gain in winter for classrooms as well as of energy collection for thermal water solar systems on roof and Solarwall® elements on the south façade (Fig. 2.23). The school building is composed by two bodies linked by a glazed atrium placed along the E-W axis, even if the body facing north opens like a fan towards the east. The atrium exerts various architectural-environmental functions, i.e., horizontal and vertical distribution, buffer space, stack-driven ventilation. The Southern body houses classrooms, while in the North side are located laboratories, offices and other complementary services. The building envelope is mostly opaque in the façade facing north, while, to the south, large floor-to-ceiling windows are fragmented by opaque vertical wall supporting the Solarwall® system and embedding vertical air ducts which connect batteries of Air Handling Units (AHU) from the technical basement to the roof. A shading system made of large metal blades movable mechanically in two position according to seasons (winter and summer) is located between the opaque elements in front of the windows and integrated by rolling blinds. Coat insulation layers are made of natural material, i.e., cork and wood fibres, and dimensioned to exceed standard transmittance requirements: 0.14 W/m^2K on the North wall, a considerable performance considering the year of construction/design of this building. The roof is ventilated in order to disperse heat accumulated during summer (Fig. 2.24). Designers paid particular attention to compliance with regulations on sound insulation, using eight different types of wall to maximize acoustical performance in different combinations and functional modes. In particular, dry-assembled types were preferred for internal partitions. The lighting appliances installed in classrooms were subjected to a careful design, optimised by CFD analyses in the atrium-classroom system to assess the compatibility between lighting equipment and night cooling airflows (Fig. 2.25). The suspended lamps perform, in fact, several functions:

Fig. 2.23 View of the South façade of the 'Orsini' School building, with the Solarwall vertical elements

artificial lighting to compensate daylighting through photo-sensors, distribution of mechanically treated air inside classrooms, and improvement of sound absorption. The lighting system is also equipped with sensors of occupants' presence to avoid waste of electricity. A great attention was paid to the design of the air conditioning system, in order to integrate it to the passive indoor climate-control techniques. In particular, it was adopted an approach able to maximise integration between the different systems of space heating, cooling and ventilation.

Equipment for the technical building systems are localised on the roof and in the basement. In particular, a 20 kWp photovoltaic array, vacuum solar collectors', exhaust-air towers from the atrium, an outside air intake grid for AHUs, and two AHUs dedicated to the Solarwall® system, are located on roof. Other four AHUs connected to the Earth-to-Air Heat Exchangers (EAHX) are placed in the basement. Furthermore, the school building is equipped with a space heating distribution system through radiant floor panels, connected to Imola's district heating, coupled with the mechanical ventilation system that allows for treating the latent heat component. In the heating period, the AHUs use, depending on the boundary conditions, air pre-treated by the EAHX system, mixed with outside air, or air heated by the Solarwall® elements (about 268 m^2 of collection surface). In the summer period, cooling is provided by the EAHX system injecting air into the mechanical ventilation network and the controlled natural ventilation by floor slab flushing and stack-driven flow through the atrium. In addition, an absorption chiller fed by the solar water collectors which over produced heat in summer, generate cold water to be brought to the radiant floor pipes in pick demand periods (Fig. 2.26).

Fig. 2.24 Inside view of the atrium of the 'Orsini' School building, with the top opening on the glazing roof

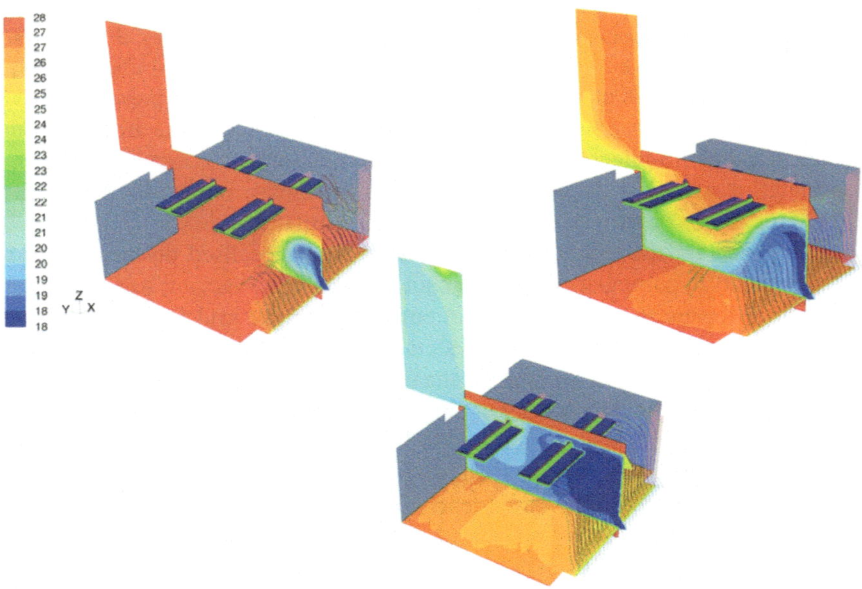

Fig. 2.25 3-D Representation of the CFD modelling of the controlled natural stack-driven airflow between classrooms and atrium in the 'Orsini' School building: after 5 s (top-left); after 35 s (top-right); after 1 h (bottom). *Source* Grosso (2017, p. 467)

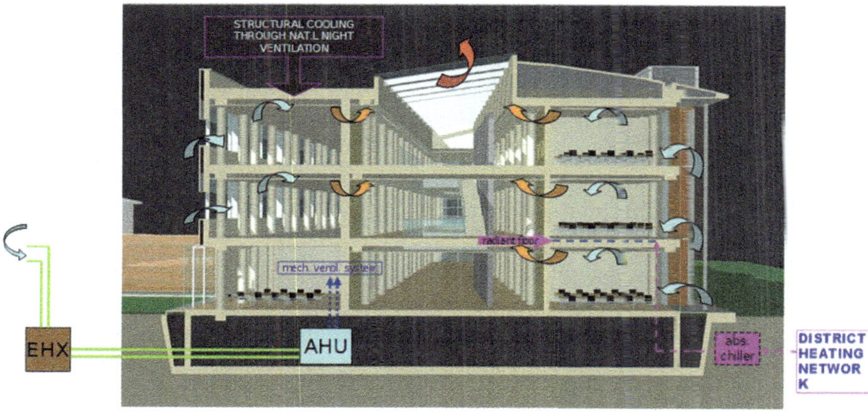

Fig. 2.26 Scheme of the hybrid cooling systems applied in the 'Orsini' School building: (elaborated by the author)

The floor slabs, in pre-tensioned reinforced concrete, are left in view at ceiling in order to act as an exposed thermal mass to maximise the performance of the natural night cooling system, provided in the summer period. Cooler outside air enters the Southern classrooms through motorised openings embedded in the windows and flushes both concrete ceiling and floor slabs hence lowering their surface temperature and the relevant heat exchange toward indoor spaces the day after. Air is naturally moved by the temperature gradient between inside and outside and by the effect induced by the temperature stratification inside the glazed central atrium. Acoustically insulated grids designed on-purpose and placed above doors allow for the passage of air between classrooms and the atrium, while glazed clerestories on atrium's roof facilitate air exhaustion through motorised openings. This system was optimised through CFD simulations of airflows in typical boundary conditions. All CFD simulations were elaborated by Marco Simonetti, an Engineer designer of HVAC systems at that time, and currently tenure-track researcher at the Polytechnic University of Turin.

An energy evaluation was carried out to estimate the energy saving potential of the various technologies, compared to a benchmark using conventional systems current at that time. This work was carried out using the dynamic simulation software TRNSYS, by Luca Raimondo, a brilliant student of mine just graduated in architecture at that time and now a successful professional in the field of building energy evaluation and certification.

After its construction and operation, the school building was interested by an energy and environmental monitoring campaign aimed at assessing operation performance with relevant control optimisation of the installed systems. In particular, a monitoring campaign was carried out on the EAHX system. This system is composed of three distinct fields that count respectively 12-12-8 parallel buried pipes of about 70 m in length. The three fields are located, respectively, 6, 30 and 6 meters from the building. Each field is organised in: air catchment head, air inlet chamber, a distribution duct, parallel tubes of rigid polyethylene with a diameter 0.25 m at a 1.1 m distance between each other, a collector duct connected to a condensation drain chamber, a connecting channel (Figs. 2.27 and 2.28). Furthermore, before the AHUs on basement, there is a mixing chamber collecting the pre-treated air coming from the three different fields. The system was dimensioned based on the airflow rate required for classrooms and its thermal performance simulated using the software GAEA (Grosso and Raimondo 2008).

The monitoring campaign was conducted by Giacomo Chiesa, a Ph.D. student under my supervision at that time, during the period May 2011–April 2012 (Chiesa et al. 2014). Results from that field measurements demonstrated the effectiveness of the system (Fig. 2.29). Measurements were also planned to assess the effectiveness of

Fig. 2.27 Luca Raimondo and myself during the sitework for the installation of the EAHX system (picture by Marco Simonetti)

Fig. 2.28 The connections between main collector and ducts in the EAHX system of the 'Orsini' School Building

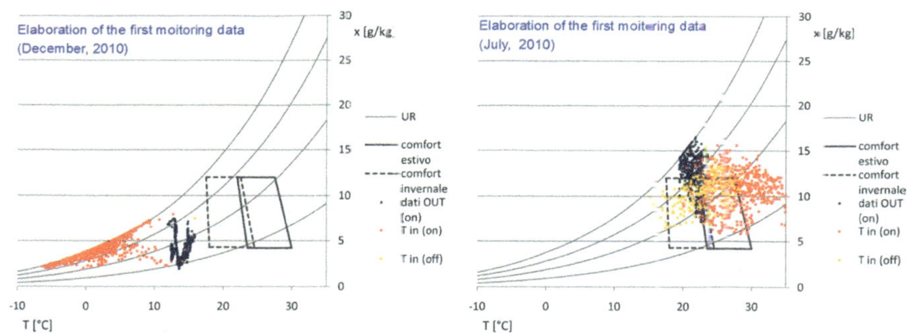

Fig. 2.29 Graphs showing the results of the monitoring campaign on the EAHX system of the 'Orsini' School building (elaborated by Giacomo Chiesa)

the night cooling through ventilation of the ceiling slabs (Fig. 2.30), but this campaign could not be implemented due to failure in the management of the School building. A critical review of operational process of the building is reported in a journal paper Chiesa and Grosso (2015).

MOOM Hotel Complex, Olgiate Olona (VA), Italy

In 2006, I was contacted by Massimo Riboldi, an entrepreneur in the hospitality sector, who asked me if I could help him to make a project for a construction of a new Hotel/Motel complex in Olgiate Olona (VA), Italy, as much eco-sustainable as possible. The project was commissioned, for the architectural part, to Beppe Riboli,

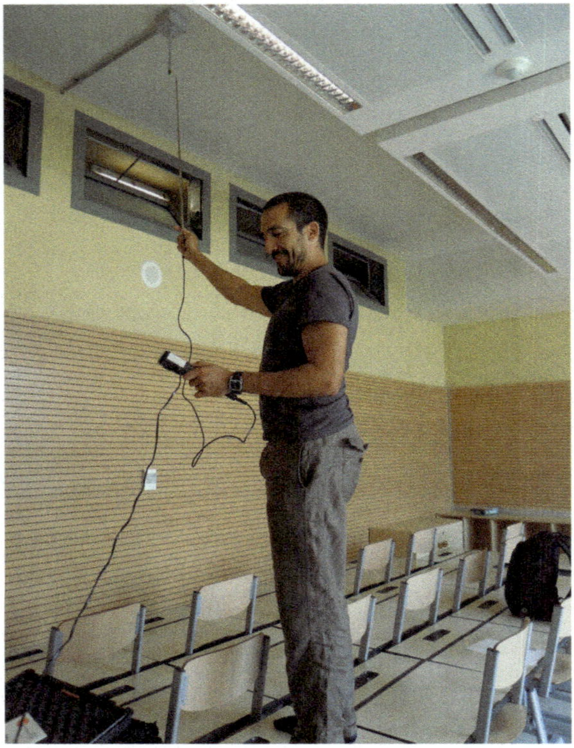

Fig. 2.30 Marco Simonetti checking the temperature sensor placed on the ceiling slab for the monitoring of the night cooling by natural ventilation (picture by the author)

a designer with office in Crema (CR), who had worked mainly in the field of dance clubs and other leisure facilities. He had already drawn concept schemes and was quite strong in defending his choices with regard to form, layout, structure, and materials, all based on a prevailing criterium of high communication impact targeted to potential users. These were expected to be mainly travellers on the way from the International Malpensa Airport to Milan, being the building located just along the highway connecting these to locations. As I saw those schemes and listened Riboli's ideas, I soon realised that my collaboration with him would have been not so easy in trying to convince him to accommodate architectural and technological choices to bioclimatic principles. In particular, the main critical aspects were the following.

- The complex has a U-shape layout, with the 3-storey Hotel building as the shorter side, laid down parallel to the highway, oriented to North-East, and the two longer perpendicular sides as two storey terrace-type wings with Motel function (Fig. 2.31). Given that functionally rational form, the orientation was a must, even if not necessarily consistent to a bioclimatic approach.
- The designer proposed as a strong communication feature the installation of an ostensible screen tailored-made of composite metallic material, shaped with sinusoidal curved edges to create a hollowed impressive surface, on both main façades

Fig. 2.31 View from South-West of the MOOM complex: the main Hotel building with the two Model wings laid down perpendicularly (picture by the author)

of the Hotel building and detached from them of about 30 cm (Fig. 2.32), oriented North-East (facing the highway) and South-West (facing the Motels courtyard). In spite of a doubtless commercial meaning, their solar shading function was either

Fig. 2.32 View of the main façade of the MOOM Hotel building with the special screen (picture by the author)

Fig. 2.33 View of the atrium inside the MOOM Hotel building with the concrete 3-D orthogonal structure (picture by the author)

unnecessary (the N-E facing screen) or ineffective (the S-W one), being fixed and hollowed.

- The bearing structure of the Hotel building was conceived as an orthogonal 3-D grid of rein-forced concrete elements of the same size, having them the function of pillars or beams. This allows for an intriguing view of the atrium interior, which is full-height open for three storeys (Fig. 2.33). However, it did not allow for applying stack-driven natural ventilation through the atrium for night cooling the hotel rooms by inletting external air, since all walls and doors separating the rooms from the atrium needed to be fireproof compartmented to comply with fire safety requirements. For the same reason, a skylight with an automatically operable opening had to be provided to be shut-off in case of fire inside the atrium.

At the beginning of the consultancy work, various bioclimatic strategies were proposed, including, in addition to the above-mentioned stack-driven ventilation, the installation of solar thermal and PV collectors as well as the installation of an EAHX system (Fig. 2.34). None of them were applied. The only eco-compatible technological choice was the use of a hollowed brick, made of clay with an embedded insulation layer of cellulose material, for the external bearing walls of the Motel wings.

Instead, being energy saving one of the main objective of the property, I proposed the installation of a three-generation system based on gas-fuelled micro-turbines, taking advantage of a fiscal reduction for gas consumption given to Hotels by the National Energy Authority. The micro-turbines generate electricity and heat recovered from exhaust fumes, used for heating water for both sanitary purposes all year, and space heating in winter. In summer, an integrated absorption chiller cools water

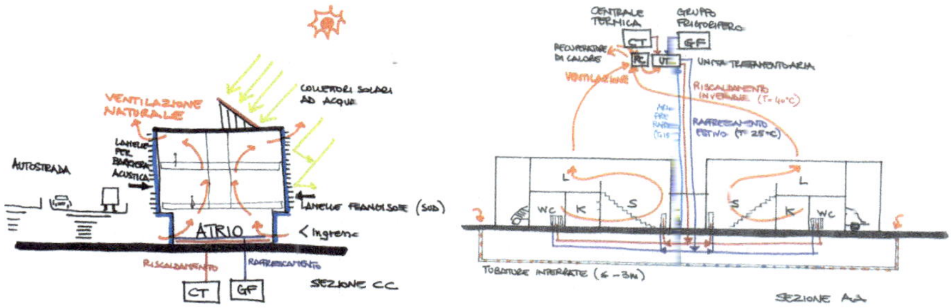

Fig. 2.34 Sketches of the bioclimatic strategies proposed for the MOOM hotel complex: Hotel (left), Motel (right) (elaborated by the author with the collaboration of Francesca Thiebat)

by using a fraction of the dissipated heat from the micro-turbines, hence fuelling a fancoil-based space cooling system (Fig. 2.35)

This three-generation system proved to be energy efficient particularly for cooling: by an energy certification I elaborated after completion of the project, the energy performance resulted of 83.77 kWh/m^2-year (Class C) for space heating, and 6,95 kWh/m^2-year (Class A) for space cooling; the relevant $CO_{2\text{-eq}}$ emissions is of 16,71 kg/m^2 year. This value corresponded to a reduction of approximately the 50% of carbon emissions from a comparable Hotel complex of the same dimensions and location of that period, using a conventional compression HVAC system and electricity from the grid.

As a conclusion, that experience was interesting and useful, regardless of the successful outcome in terms of application of bioclimatic principles, because it

Fig. 2.35 Detailed view of the micro-turbine for the MOOM hotel complex (picture by the author)

allowed me to experience a process of compromising integration between architectural/representation design and technological-environmental applications in a real project context.

Feasibility study for refurbishment and conversion of 'Emmanuel Anquetil' building into an eco-building
In May 2012, the Ministry of Public Infrastructure of the Republic of Mauritius initiated a public invitation for expression of interest for consultancy services for the feasibility study of the "Refurbishment and Conversion of 'Emmanuel Anquetil' Building into an Eco-building".

Eight Firms were short-listed and invited to submit their offer. The team lead by Design Forum Ltd, with office in St. Louis, Mauritius, was awarded in October 2013 the contract, signed on 22nd January 2014. In addition to the leader and other local consultants, the team comprised of: STEGET srl, directed by arch. Patrizia Giacomelli, ARTECH studio by arch. Giancarlo Pavoni, and me as the eco-building consultant. I had been collaborated previously with both STEGET and ARTECH in several projects, including the design development of the new Mauritius' Oceanography Institute.

The Emmanuel Anquetil Building (EAB), built in 1979, was outdated and not operating conveniently, hence, needed a thorough refurbishment and had to be put up to standards. The intention of the Client was to go beyond a simple refurbishment and make the project a "mod-el in environment friendly (green) sustainable retrofitting", by converting the EAB into an eco-building and setting the pace to a responsible set of actions. The decision to refurbish the building rather than to pull down and re-construct it was already a strong statement in favour of the will of the Client to start an ecological development.

Moreover, the EAB had historical values for the newly constituted Mauritian nation, having been built just after Mauritius obtained its independence. It was therefore worthwhile to keep as a historical footprint of achievements of that period (Fig. 2.36).

The main problems were the presence of asbestos panels, the effect of Pigeons' drops all over, the insufficient places of the vehicle park, high energy consumption due to services conceived when energy conservation was not a priority.

The work carried out included the following main phases:

(a) a thorough survey—conducted using both a direct visual inquiry and questionnaires to users—of the material state and functional performance of the envelope and the bearing structure, the health and comfort conditions, the state of maintenance, the energy consumption, the use of indoor spaces;
(b) proposal of alternative refurbishment solutions, including dynamic energy simulations and a payback time assessment for the energy-related ones;
(c) a proposed time planning for the execution of the refurbishment alternatives

As far as energy and bioclimatic strategies were concerned, the following solutions were proposed.

Fig. 2.36 View of 'Emmanuel Anquetil' building as it was in 2013 (picture by the author)

- Raise current ground floor functions to level 3 including courtyard with a glass covering, redesigning hall and entrance to the building to allow parking on ground level:

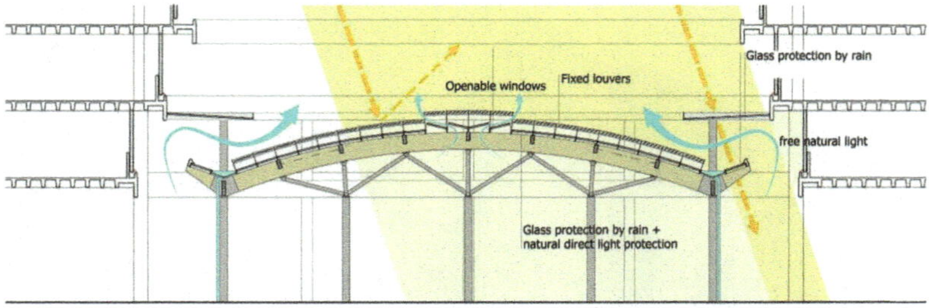

- Open up layout of office spaces at each floor eliminating partitions parallel to the longitudinal facades or use ones with height lower than ceiling level, if necessary, to facilitate wind-driven airflow through office rooms:

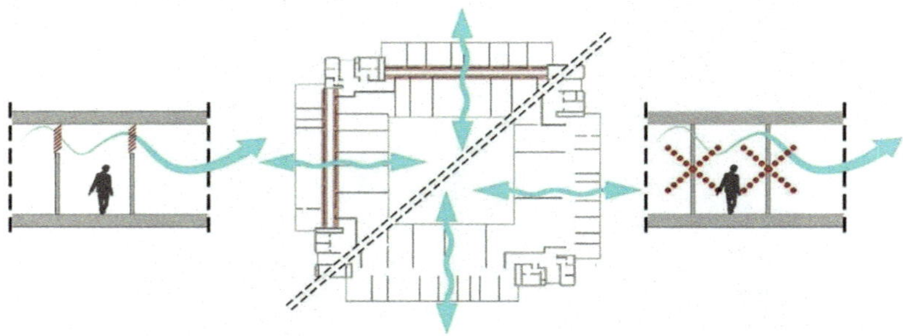

- Replacement of existing single-glass-pane + no-brake-frame windows with high thermal performance windows including: double-glazing with filled-in argon and low solar factor; low-e films; external operable Venetian blinds; top openings hopper automatically operated at set temperature conditions by sensor-driven motorised devices for natural night cooling.

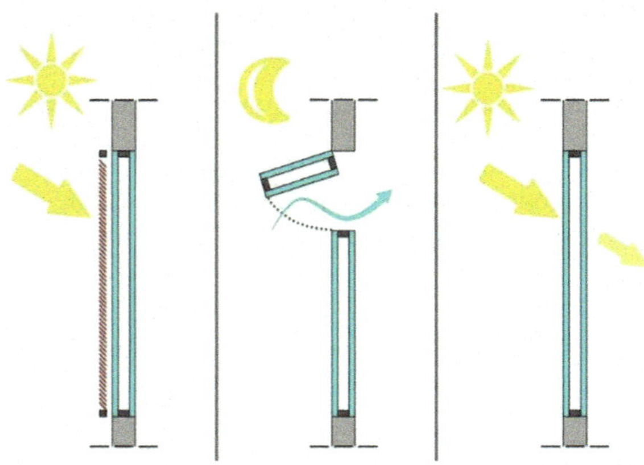

- Application of lighting dimming control devices connected to daylighting availability.

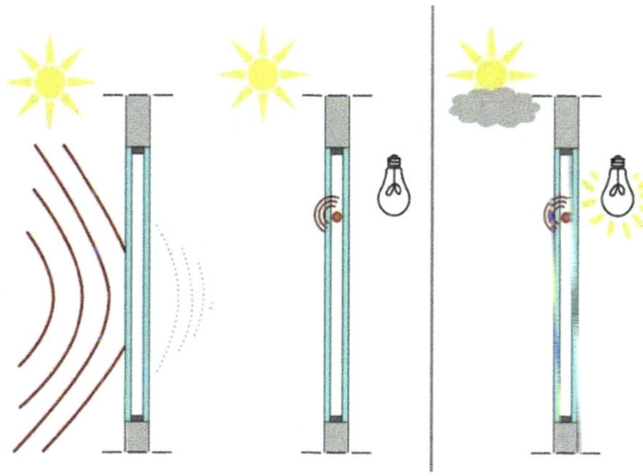

- Coat insulation layer added to the existing external walls, either externally and protected by a panel cladding or internally and protected by a plasterboard panelling.

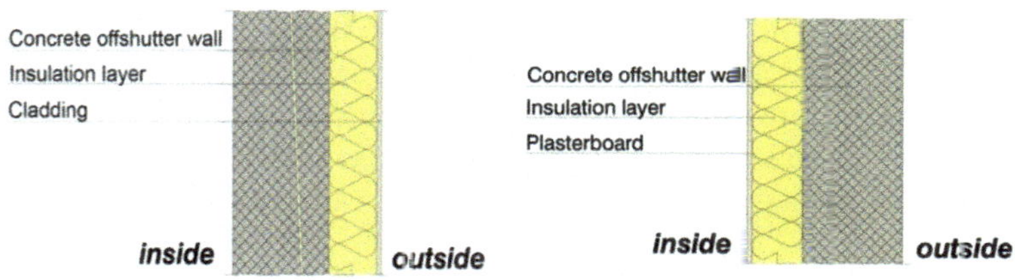

- Night cooling by natural ventilation of the high thermal mass of the concrete structure of EAB.

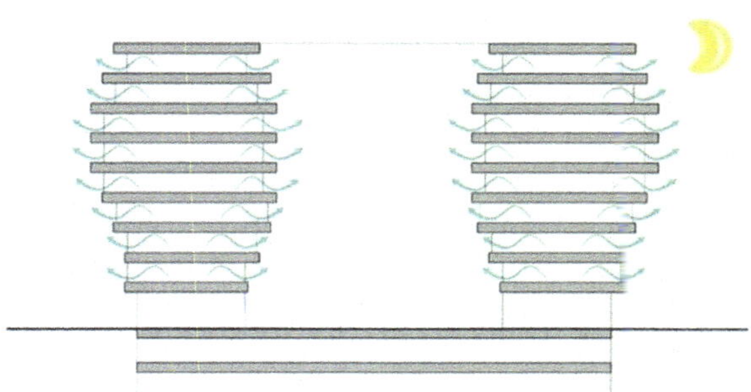

- Integration of photovoltaic panels on shed structures on the terrace roof for renewable power generation:

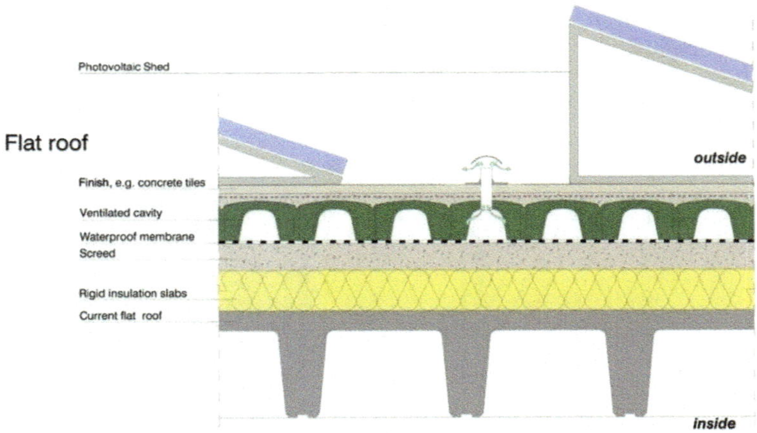

Dynamic energy simulations on different EAB envelope and HVAC system configurations, corresponding to the proposed refurbishment alternatives, were carried out using the software Design Builder with Energy + solver. The total simulated HVAC + lighting yearly electricity use for the whole EAB existing reference configuration was slightly less than the one measured from August 2012 to July 2013 (1307.15 MWh).

As a benchmark model for the simulations, a configuration was defined where the external glazed walls of office spaces were shifted to the outward edge of the overhanging floor slabs (with an offset of 20 cm for housing the outside shading devices), as a way to increase the net usable floor area, and conventional compression HVAC system applied (Fig. 2.37). Various simulations were then carried out to assess the possible saving potential of the various configurations versus the benchmark model (Fig. 2.38).

The assessment of the payback period of the investments for the energy-related refurbishment options was carried out using both simple and discounted methods.

As a conclusion, that experience was of special importance since it was a unique application of eco-sustainable and bioclimatic principles on a refurbishment project. This had been since then rare in general.

Minimum building energy performance standards and nearly zero energy buildings approach for Turkey

The United Nation Development Programme (UNDP), after a selection of companies invited to submit a proposal for the provision of services for the development of "Minimum Building Energy Performance Standards and nearly Zero Energy Buildings approach for Turkey", signed the relevant contract with a joint venture between STEGET S.r.l. and IRD Engineering S.r.l., on February 19, 2016. I was appointed as the Team Leader and energy expert of the project.

The work was particularly complex, given the ambitiousness of the objectives, the characteristics of the context, and the scale and geographical extension of its application.

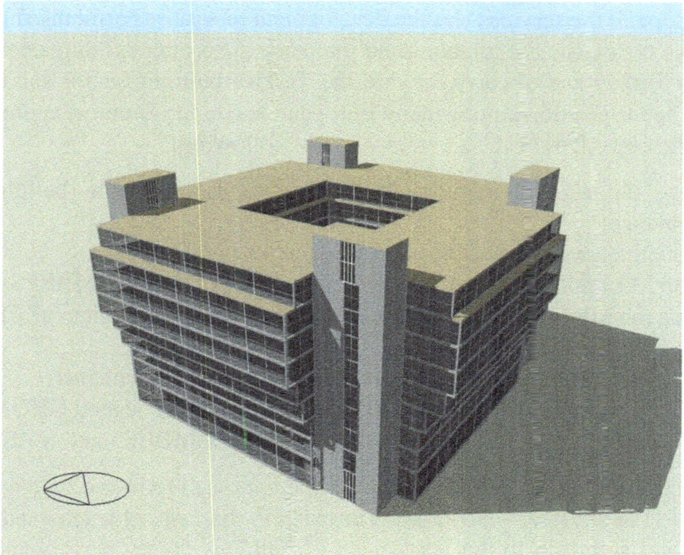

Fig. 2.37 EAB benchmark model in the design builder simulation: view from West

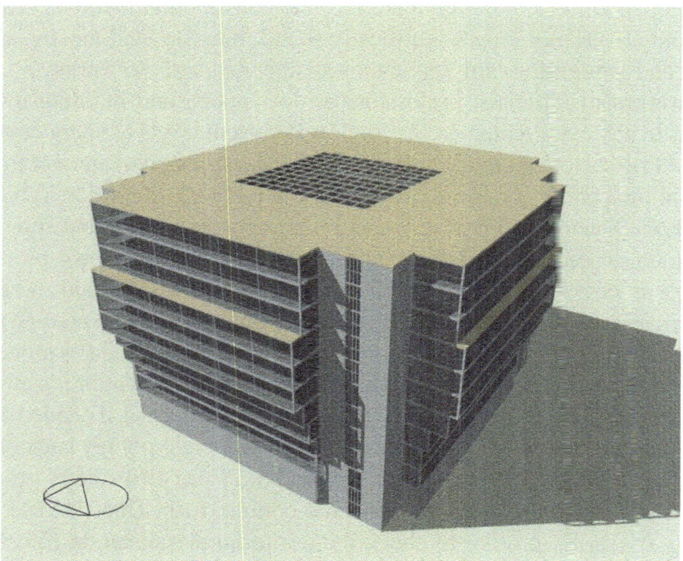

Fig. 2.38 Example of one of the simulated EAB alternative configurations: view from West

The scope of the services was the development of a loose harmonised framework to improve the existing regulations and standards governing building energy performance in Turkey and to carry forward the Turkish building sector and the energy sector to better position among many European and other foremost countries.

Specific objectives of the services were the following:

- improve the existing regulations and standards governing building energy performance;
- promote the use of renewable energy resources in buildings;
- develop Minimum Building Energy Performance Standards (MBEPS);
- upgrade the MBEPS by including nearly-Zero Energy Buildings (nZEB) requirements;
- implement energy efficiency in buildings in a sustainable manner;
- reduce the end-use primary energy consumption and associated GHG emissions;
- outline a programme for future implementations of MBEPS and n-ZEB.

The activities was carried out based on the approach followed by the Directives 2010/31/EU and 2012/27/EU of the European Parliament, characterised by both a cost-optimal approach to the assessment of building energy performance and the nearly-zero energy building target. The definition of nearly-zero energy building had to take into account Turkey's building cultural heritage, technology, climatic regions and conditions, building materials and construction methods. The aim was to carry out a work not only scientifically sound, but also, and mostly, aimed at an effective and straightforward application of the standards to Turkey's central and local Government Officials, professionals, developers, and building owners. The target was to achieve a reduction of 35–45% of current levels of energy consumption of the building sector and relevant CO_2 emissions, estimated to amount to the 35.4% of the total final energy consumption in 2012 (equal to 31.8 million TOE).

The applied methodology was characterised by a double-handed approach based on two main procedures: top-down and bottom-up. The top-down procedure was related to a process starting from a statistical analysis of building and energy data and ending to average values on energy consumption for heating and cooling divided by typology and climate zone. The bottom-up procedure was related to a process starting from the calculation of requirements for heating and cooling of building models based on data from the provided examples of projects, or assumptions if needed, and ending with new limits of energy requirements for each building typology and climate zone. In parallel, an analysis of meteorological data (TMYs) for all Turkey's provinces was carried out by Giacomo Chiesa, through a sub-contract to the Dept. DAD, Polytechnic University of Turin, in order to check the territorial classification fore-seen in the software BEP-TR used at that time in Turkey as a reference for building space heating energy performance assessment, and propose a new classification, taking into ac-count both heating and cooling needs.

In the typological analysis, the following main architectural and material construction types were considered: brick masonry; concrete block wall; wooden wall; stone wall. Some examples of traditional building types of Turkey are shown in Fig. 2.39.

The flow-charts of the process followed in the two main phases of the work are shown in Figs. 2.40 and 2.41.

As far as the bioclimatic approach was concerned, an interesting outcome of Chiesa's climate analysis was an assessment of the potential for various passive/hybrid indoor climate control strategies, evaluated in term of annual comfort-inducing hours for all Turkey's Province capital cities.

Fig. 2.39 Examples of typical buildings of Turkey, in wood (left, Istanbul) and stone (right)

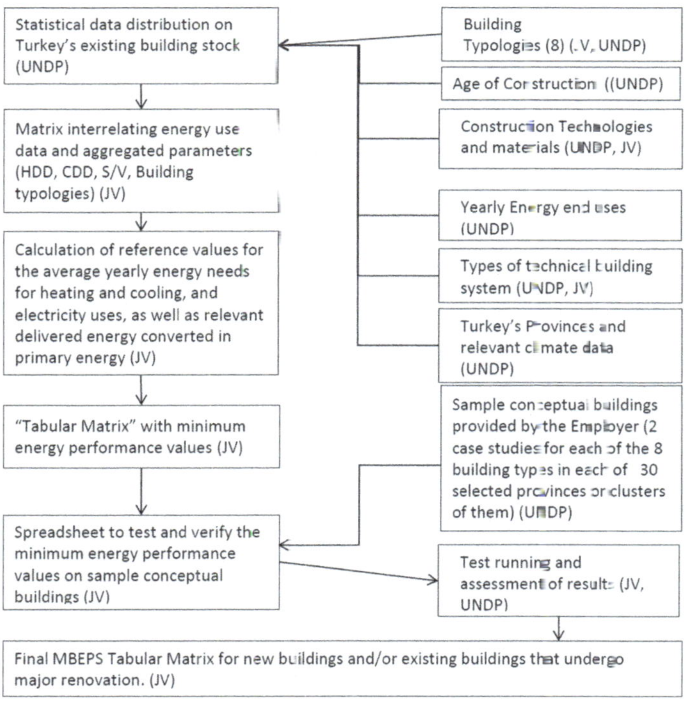

Fig. 2.40 Flow-chart of the process followed for the development of MBEPS

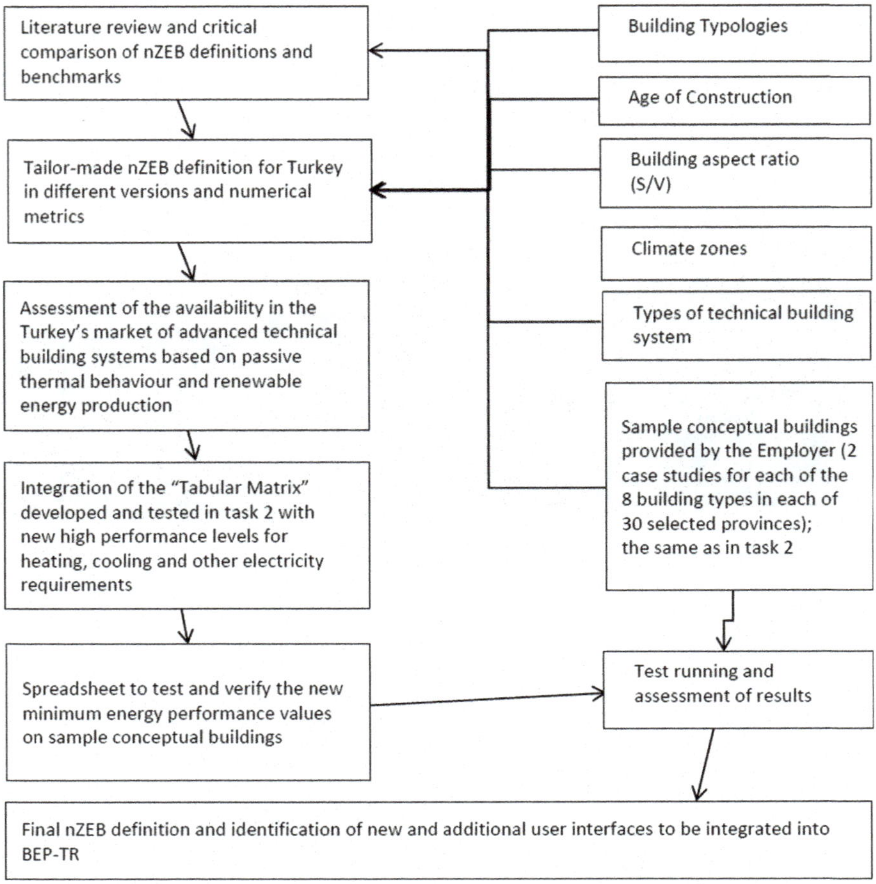

Fig. 2.41 Flow-chart of the process followed for the definition of nZEB for Turkey

The strategies considered were the following: High Thermal Mass (HTM); High Thermal Mass coupled with Night Ventilation (HTM and NV); Direct Evaporative Cooling (DEC); Controlled Natural Ventilation (CNV); Internal Heat Gains (IHG); Passive Solar Low Mass—Direct Gains (SG LM); Passive Solar High Mass—Direct Gains (SG HM); Dehumidification (DEH). An ex-ample of a bar-chart showing the potential comfort-inducing hours in a given location by using the above-listed strategies is shown in Fig. 2.42.

Collaboration to the ARTECH studio

My professional collaboration with the ARTECH studio of architect Giancarlo Pavoni, dates back to 2004, although I had known him since the early '80s of the XX Century, when he graduated within Matteoli's group. Giancarlo is an architect particularly skilful in combining technological and architectural design with a

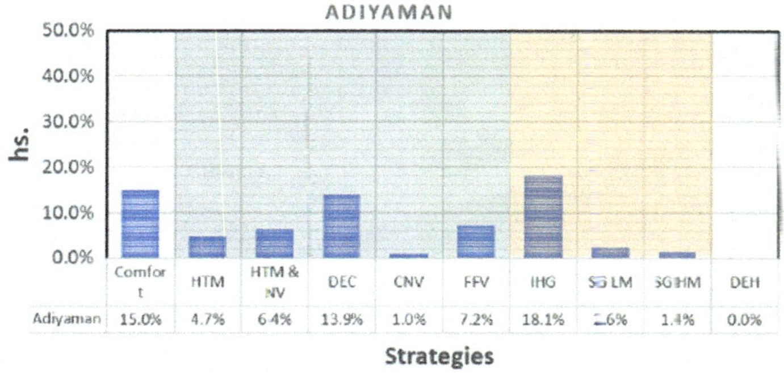

Fig. 2.42 Example of a bar-chart showing the potential comfort-inducing hours (% of the total annual hours) in a given location by using the above-listed passive/hybrid indoor climate control strategies

constant attention to the bioclimatic approach. My collaboration have been mostly applied to design competitions, unfortunately rarely won, for which temporary associations of professionals are formed. In them, I have always offered my competence as an environmental, technological, and energy expert and certificate assessor. Amid them, it is worth to mention the participation to the call opened by the Organising Committee of World EXPO Milan 2015 for the design of its services facilities. In addition, I collaborated to the design development of various projects, mostly located in the territory of the Municipality of Grugliasco, at the Western boundaries of Turin, with which Giancarlo has a permanent professional collaboration as an architect: a new City Hall building (never realised), a new Library, and the reconstruction of a High School complex.

But, amid the contribution to the design development of one of the few really constructed works, I would like to cite the Nursery Building 'Anna Frank", located in Nichelino, another city in the outskirts of Turin, at the Southern boundary. It is an example of straightforward and simple application of energy-conservation systems as well as of a harmonious synthesis between them and architectural form (Fig. 2.43). The applied energy-saving systems include mainly active and hybrid systems, the only passive being the sun-space on the Southern façade of the building:

- a controlled mechanical ventilation with heat recovery, assisted by solar air collectors placed on the roof at 55° tilt, to pre-heat air in winter;
- a set of solar water collectors for heating sanitary water, placed on the same plane of the air collectors;
- a radiant underfloor system for peak needs of heating and cooling
- a mono-bloc water chiller, composed of alternating air-condensers multi-compressors, working continuously at a nominal power of about 9 kW, allowing for a partial de-humidification of air and coolth storage in the floor slab as well as in a 2000 l supplementary inertial tank.
- daylighting-controlled dimming devices applied to the artificial lighting appliances.

Fig 2.43 3-D representation of the 'A. Frank' Nursery building, located in Nichelino (TO). (Courtesy of Giancarlo Pavoni). Collaboration to AE.C.I

The external masonry walls are made of a double layer of hollowed bricks (brick veneer out-ward and blocks inward), with interposed insulation rigid panels of mineral fibres. The terrace roof has a double layers of insulation on both sides of the air cavity, with rigid panels of mineral glass on the outer surface (below the tar and gravel layer) and boards of expanded poly-styrene o the inner surface. The ground floor is made of multiple layers alternating concrete slabs, air cavities, and rigid panels of mineral fibre, below the outer floor surface and above a thick substrate of gravel, pebbles, and sand-cement. All glazed elements have shading devices: the sun-space is shaded by a portal-shaped concrete structure projected from the building Southern side, with horizontal overhanging slats added to its windows; the windows facing West and East are shaded by external vertical movable louvres.

I also carried out an energo-economic analysis to evaluate the energy reduction, and the related cost savings and payback periods, with respect to the application of conventional systems and standard insulation requirements in force at that time. Results yielded an average payback period of 11.45 years, ranging from: 3.8 year (62.44% of saving) for the increased thermal insulation; 4.17 (19.73%) for heat recovery on the mechanical ventilation system; 7.75 (30.65%) for the solar water collectors (both DHW and space heating); 8.96 (44%) for the cooling system; 19.56 (30%) for the lighting control system; 19.71 (20%) for the sun-space; 19.99 (40%) for the shading devices; 20.51 (19%) for the solar air collectors (heating).

Collaboration to AE.C.I.

The professional collaboration to AE.C.I., a design and construction firm owned by architect Maria Irene Cardillo, started after the end of the PR.I.M.E3 research project (see above). It has ranged from the participation of several design competitions to the elaboration of pro-posals to be submitted to calls from the European Commission, all dealing with the application of the GREEN-BLOCK system as an outcome of the above-mentioned research project.

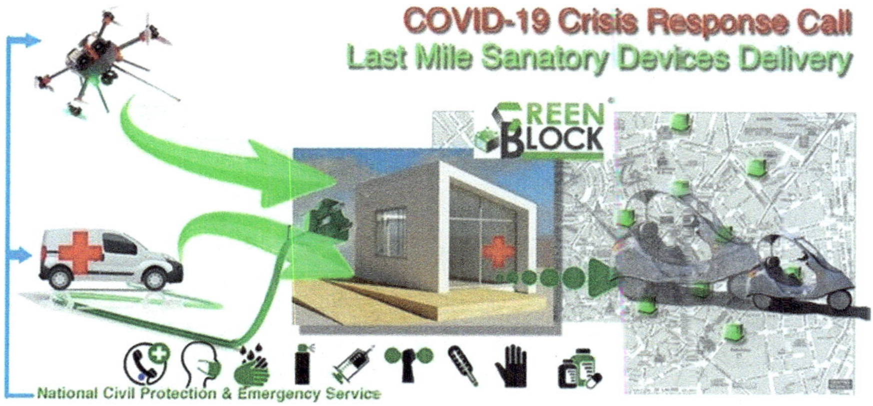

Fig. 2.44 A panel for the presentation of a proposal including GREEN-BLOCK submitted to a COVID 19 EU call

The prefabricated GREEN-BLOCK system, still at a TRL 7 phase of prototyping, is com-posed of a bearing structure made of Aluminium profiles designed on purpose and to be pro-duced according to a process patented by METRA and an envelope also composed of Aluminium panels with an embedded insulation layer made of recycled materials from scraps of textile and tire industries. This modular system, whose basic unit can be carried on a light truck and can be built up to four storeys, is very flexible in its composition options as well as versatile in its potential functions. It has been proposed for various applications, including social housing, luxury residential compound, tourist resorts, emergency facility, and, most recently, as a storage hub for last-mile e-commerce delivery, including transportation of sanitary devices related to the COVID-19 contagion emergency situation (Fig. 2.44).

So far, the proposals presented for calls under the SME Instrument European Programme have been very positively assessed but never funded due to budget restrictions. However, the GREEN BLOCK® system was selected as a EU Ambassador amid the "Best 15 EU Top Technological Innovators" for being presented to the following International fairs: 'TechInnovation 2017' at Singapore; 'CES 2018' at Las Vegas; 'CIIF 2018' at Shanghai. In addition, a Seal of Excellence was assigned to GREEN BLOCK® by the European Commission in 2019 for having "Scored as a high-quality project proposal in a highly competitive evaluation process". The GREEN BLOCK® system obtained the following certifications: (a) Sustainability certificate—iiSBE 2019; (b) Assessment of the energy performance; (c) LCA—en 15978; (d) CE—control GB certificate for Aluminium profiles; (e) LAPI Certificate and Test for insulation panels.

The most significant project elaborated for design competitions, in collaboration with arch. Maria Irrene Cardillo, are described in my book on passive cooling (Grosso 2017, pp. 401–425).

In addition, I contributed to the elaboration of various other projects and proposals, from which I mention here an innovative and original one presented to a Congress-Exhibit Arc13 in Do-ha, Qatar, 2013 (Figs. 2.45 and 2.46). The basic idea was

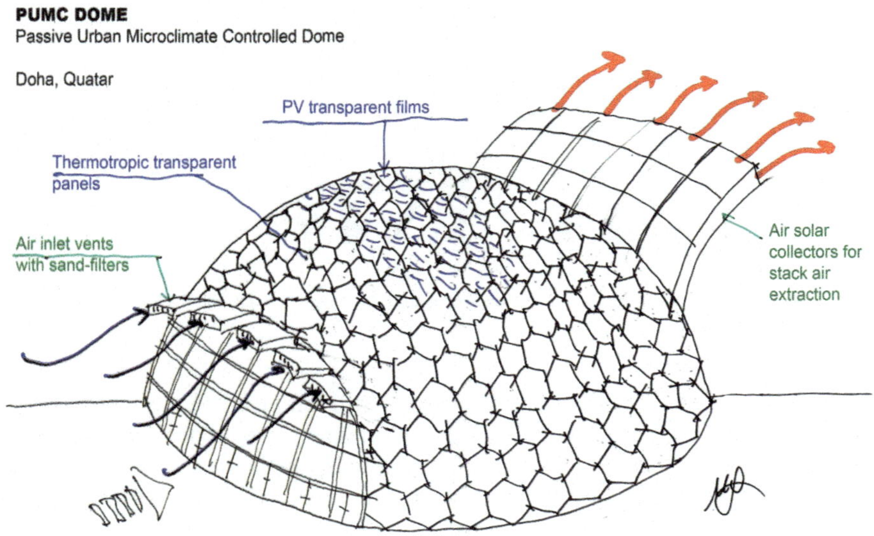

Fig. 2.45 Sketch of a 3-D representation of the PUMC Dome for Doha with the main airflows (elaborated by the author)

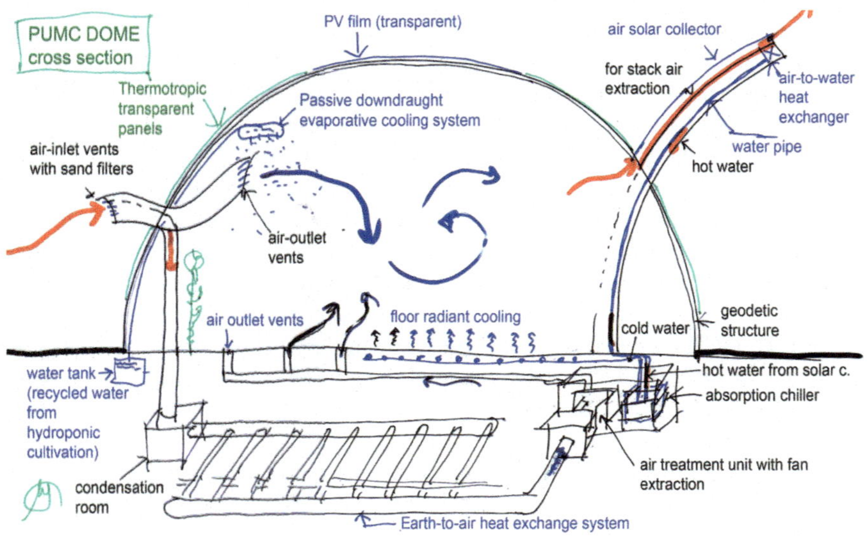

Fig. 2.46 Sketch of a section of the PUMC Dome for Doha with the indoor climate and air quality control strategies (elaborated by the author)

to create a large urban space—a passive urban microclimate e-controlled (PUMC) dome—within with various leisure, commercial, and agricultural activities could be exerted. The indoor microclimate is controlled by passive techniques, hence, minimising non-renewable energy consumption and related carbon emissions as well as avoiding exposure to sand-borne winds. The structure is composed of the following elements.

- A geodetic dome-shaped structure is made of Aluminium hexagonal cells enclosing trans-parent panels made of thermotropic material.
- Transparent hexagonal panels made of thermoplastics materials such as poly-olefin, polycarbonate or polymethacrylate, with added thermotropic polymers, autonomously switching from a clear to a scattering state when the temperature rises (see Fraunhofer IAP technology), to reduce high altitude solar radiation transmission causing rising ground surface temperature and, hence, radiative thermal discomfort in urban spaces.
- A multiple controlled natural ventilation system including: (a) rectangular or circular-section air ducts, made of metal foils, crossing the geodetic structure from outside and from under-ground to inside the dome; (b) duct air-inlet southwest-oriented vents placed at various height above ground following the geodetic-shaped structure (from a minimum of 6 m) with embedded dust and sand filtering; (c) duct air-outlet vents placed inside the dome underneath the geodetic-shaped covering at various height (from a minimum of 10 m); (d) dumpers placed at the inlet and outlet vents to automatically control airflow rate depending on microclimate as well as Indoor Air Quality (IAQ) conditions.
- Air solar collectors placed on the top of the leeward Northeast-oriented side of the geodetic structure, with the solar radiation absorbing selective surface and relevant glazing facing Southwest: air-inlet grid are placed at the bottom of collectors towards the inside of the dome and are provided with automatically-controlled dumpers; air-outlet grid are placed on the top of collectors and provided with dumpers as well. Air solar collectors function as air extractors through the stack effect generated by the rising temperature inside the collector air cavities as well as for preheating water to feed absorption chillers.
- A water circulation piping system connecting air solar collectors to absorption chillers—place on technical rooms—which have a function of cooling water flowing in a radiant floor network placed on spots where overheating is more likely, i.e., where the foreseen activity is characterised by people crowding.
- Passive Downdraught Evaporative Cooling (PDEC): water in closed loop is sprinkled or misted through nozzles placed at the under ceiling outlet mouths of the air ducts connecting the outside to the inside space to cool the entering air; the system is automatically switched on when air temperature is above a set-point threshold (generally, between 26 °C and 28 °C) and air humidity below e defined value (usually, 60% of Relative Humidity).
- Earth-to-Air Heat eXchange (EAHX) system, made of horizontal buried ducts made of reinforced polyethylene material, supplying passively-conditioned air in the dome. Air inlet occurs through vertical duct connected to the external vents

as part of the above described ventilation system; air outlet from the buried pipes goes to Air Treatment Units—placed on technical rooms—which provide for fan-extracting the air as well as controlling air humidity and temperature before the air enter the dome-enclosed space. This automatic control sys-tem acts synergistically with the PDEC system in order to guarantee the proper microclimate comfort conditions.

- Vegetation growing by a hydroponic technique on vertical walls confining people assembly places within the dome, thus cooling air by vapour-transpiration and reducing radiant-exchange discomfort by low-emissivity irradiation as well as absorbing air pollution generated by people and equipments.
- Cultivation of plants for both ornamental purposes (flowers) and food production (vegetable and fruits) by a water-saving hydroponic technique in open-air dedicated spaces within the dome. A AEROSEKUR-patented hydroponic water treatment system based on a photo-ozonolysis process was proposed; it destroys metabolites like oxalic acid converting them into CO_2 as well as all other noxious micro-organisms which may grow in the hydroponic water preventing the undesired biofilm build-up.
- In addition to air-filtering at the inlet-vents of the ventilation system and the air outlet of the EAHX ATUs, air treatment for IAQ is provided in spaces where crowd presence is fore-seen; for this purpose a AEROSEKUR-patented technology derived from aerospace applications is proposed. It involves the use of a laminar flow through a device composed of plates on which the active sorbent is deposited. The airflow containing the BVOC is completely purified after crossing the laminar flow device with a minimum of energy consumption in comparison to a traditional fixed bed adsorption system. Furthermore, the laminar flow adsorption device can be fully regenerated and the BVOC released with proper flow conditions and vented in the external atmosphere.

As a general comment, my collaboration with Maria Irene Cardillo has been, and still is, of a nature able to emphasise and apply in a synergistic way all best achievements of my academic and professional career, with particular focus on environmental and technological design, including the bioclimatic approach.

References

Akins RE, Cermak JE (1976) Wind pressures on buildings. CER76-77REA-JEC15, Fluid Dynamic and Diffusion Laboratory, Colorado State University, CO, USA

Allard F (ed) (1998) Natural ventilation in buildings—a design handbook. James & James Ltd, London

Ansley RM (1982) Natural ventilation model studies. In: Reinhold TA (ed) Wind tunnel modelling for civil engineering application. Proceedings of the international workshop on wind tunnel modelling criteria

Appelbaum J, Bany J (1979) Shadow effect of adjacent solar collectors in large scale systems. Solar Energy 23(6):497–507

Boutet TS (1987) Controlling air movement—a manual for architects and builders. McGraw-Hill Book Company, New York, USA
Chiesa G, Grosso M (2015) Sustainable school buildings: design-management-monitoring, results and weaknesses. The case study of the high school "L. Orsini", Imola. J Technol Architec Environ 09:247–255 (Techne)
Chiesa G, Grosso M (2017) Environmental and technological design: a didacti-cal experience towards a sustainable design approach. In: Proceedings of the XV International Forum le Vie dei Mercanti: World Heritage and Disaster. Knowledge, culture and re-presentation. La scuola di Pitagora, Capri 15-17 Luglio, pp 10
Chiesa G, Grosso M (2019a) Python-based calculation tool of wind-pressure coefficients on building envelopes. J Phys Conf Ser 1343:012132
Chiesa G, Grosso M (2019b) A parametric tool for assessing optimal location of buildings according to environmental criteria. In Sayigh A (ed) Sustainable building for a cleaner environment. Springer International Publishing, Berlin, p 15
Chiesa G, Simonetti M, Grosso M (2014) A 3-field earth-heat-exchange system for a school building in Imola, Italy: monitoring results. Renew Energy 62:563–570
Feustel H, Grosso M et al (1990) Fundamentals of the multizone air flow model—COMIS. Technical Note # 29. IEA-AIVC, Coventry, pp 1–114
Feustel H, Grosso M et al (1993). User guide of the multizone air flow model-COMIS. Technical Note # 30. IEA-AIVC, Coventry, pp 1–137
Grosso M (1972) "A contribution to a debate/meeting", Proceeding of the Exhibit/Round table Tra rivolta e rivoluzione: immagine e progetto (Between uprising and revolution: imagine and design). Ente Bolognese Manifestazioni Artistiche, Graphis Edizioni d'Arte, Novembre 1972 January 1973, Bologna, Italy, pp 738–743
Grosso M (1980) Incremento dei costi energetici e mercato del serramento metallico. In: Nuova Finestra, n. 9. Aedilmedia, Milano, pp 113–116
Grosso M (1982) Dinamica delle ombre (Dynamics of shadows), 2nd edn. CELID, Torino
Grosso M et al (1983) Sun/land: software for the evaluation of solar radiation in a territorial scale. In: Proceedings of the 6th miami international conference: alternative energy sources, C.E.R.I., School of Engineering. and Architecture, University of Miami, Coral Gables, Florida
Grosso M (1984) Il sole e la montagna: analisi dei siti e pianificazione della residen-za. Neve International, Anno XXVI(2):23–43 (Torino)
Grosso M et al (1984) L'interazione sole/territorio: analisi comprensoriali e regionali. In: Catalogo della mostra-convegno: Analisi dei caratteri e dell'uso del suolo con tecniche automatiche, Castello del Valentino. Politecnico di Torino—CSI-Piemonte, Torino, pp 18–27
Grosso M (1985a) Technological and energetic evaluation of existing buildings: a retrofit technique on external windows to decrease heat load due to air infiltration. Research summaries, #REF IT2. Air Infiltration Centre, Berkshire, pp 41–42
Grosso M et al (1985) Monitoring the effects of draught elimination. Appl Energy 20(1):59–83 (Elsevier Applied Science Publishers, Oxford, UK)
Grosso M (1991) Fire resistant architecture: criteria for a forest fire and wind storm resistant dwelling design. D. B-H Liebermann & Associates Publisher, Berkeley, p 57
Grosso M (1992) Wind pressure distribution around buildings—a parametrical model. Energy Build 18(2):201–231 (Elsevier, Amsterdam)
Grosso M (1993) Modelling wind pressure distribution on buildings for passive cooling. In: Proceedings of the international conference: solar energy in architecture and urban planning, Florence, 17–21 May. CEE, H.S. Stephens & Associates, Bedford, pp 342–344
Grosso M (1997, 2008a, 2011, 2017) Il raffrescamento passivo degli edifici–in zone a clima temperato (Passive cooling of buildings—for temperate climate zones), I, II, III, and IV editions. The III and IV editions have an attached CD with WINDCHILL, a two-zone dynamic simulation software calculating the energy saving potential due to the application of wind and stack-driven ventilative cooling

Grosso M (1998) Urban form and renewable energy potential. Renew Energy 15:331–336 (Elsevier Pergamon, Oxford)
Grosso M (2008b) Progettazione Bioclimatica (Bioclimatic design). In: Castelli L (ed) Architettura sostenibile (Sustainable architecture). UTET, Turin, Wolters Kluver Italy, Milan, pp 5–40
Grosso M (2012) "L'aria: protezione ed elemento dinamico di mitigazione termica" (Air: protection and dynamic element of thermal mitigation). In: Piano R, Bucci F, Bute-ra F, Calabresi G, Casiroli F, Jurina L, Majowiecki M, Ottolini G, Sala M, Scudo G (eds) Al-manacco dell'Architetto, vol 1. Proctor Edizioni S.p.a, Bologna, pp 965–971
Grosso M (2014) Il miglioramento della qualità ambientale in ambiente urbano e su-burbano (Improvement of environmental quality in urban and suburban contexts). In: Bal-letti F, Ghersi A (eds) Paesaggio Urbano. FrancoAngeli, Milano, pp 43–48
Grosso M, Banchio G (2000, July) Cities of wind: natural ventilation access in urban design. In Architecture, city, environment. Proceedings of the PLEA 2000 conference, Cambridge. James & James, London, pp 506–511
Grosso M, Nigra M (2019) Progettare la complessità in architettura (Designing the complexity in architecture), 15 theses in the Master Degree Programme "Architecture for a Sustainable Design", Polytechnic University of Turin. Maggioli Publisher, editorial series politecnico, Santarcangelo di Romagna (RN)
Grosso M, Pacini F (1984) Relative shading of buildings in an urban context: a simplified calculation method. International Congress: energy 1984—energy for small & medium-sized countries, Section A3: solar energy in buildings, Minister of Energy and Infrastructure, The Israel Trade Fairs Center Ltd., Tel Aviv, Israel
Grosso M, Parisi E (1995, September) The effects of urban context and openings' position on wind-driven cross ventilation. In: Proceedings of the European conference REBUILD: integration of renewable energies in existing urban structures, Corfù, Greece, pp 11
Grosso M, Raimondo L (2008) Horizontal air-to-earth heat exchangers in Northern Ita-ly: testing, design, and monitoring. Int J Vent 7(1):1–10 (Veetech Ltd.)
Grosso M, Marino D, Parisi E (1994, Novembre 24–26) Wind pressure distribution on flat and tilted roofs: a parametrical model. In: Proceedings of the European conference on energy performance and indoor climate in buildings, Vaulx en Velin, France. Ecole Nationale des Travaux Public de l'Etat, Lyon, pp 167–172
Grosso M, Marino D, Parisi E (1995, August 14–16) A wind pressure distribution calculation program for multizone airflow models. In: Proceedings of "building simulation '95", fourth international conference, international building performance simulation association, Madison. Madison, Wisconsin, USA, pp 105–118
Grosso M, Dotta S, Giordano R (2000a, July 9–12) Evaluation of natural ventilation potential in urban design: an interactive learning tool. In: Sustainable buildings for the 21st century: teaching issues, tools and methodologies for sustainability. Proceedings of the TIA third international conference, Oxford, UK. EC DG XVII, Oxford, UK, pp 1.10–1.20
Grosso M, Parisi E, D'Elia I (2000b, July) The effect of urban form on wind pres-sure drag. In: Proceedings of the World Renewable Energy Congress-VI, Brighton, UK. Elsevier, London, pp 444–449
Grosso M, Giordano R, Dessì V (2000c, July 1–7) A procedure for evaluating natural ventilation potential of urban sites. In Proceedings of the World Renewable Energy Congress-VI, Brighton, UK. Elsevier, London, pp 450–455
Grosso M, Raimondo L, Colombaro L, Deganutti A (2009) ECO-wind—a hourly-based semi-dynamic software calculating the cooling energy potential reduction due to a passive ventilative technique of a two-zone building unit. A tool associated to the Energy Attachment of Turin Building Code aimed at supporting the eligibility of a fiscal deduction related to building projects applying natural ventilation as an energy saving indicator. Department of Science and Environmental Technology, Polytechnic University of Turin for Turin's Energy Agency
Grosso M, Raimondo L, Colombaro L, Deganutti A. (2010–11) SPERAVent—a hourly-based semi-dynamic online software calculating the cooling energy potential reduction due to a passive

ventilative technique of a two-zone building unit. Upgrade and extension of ECO-wind on an Internet platform by ATCode

Grosso M, Raimondo L, Colombaro L, Deganutti A (2011) WINDChill—a hourly-based semi-dynamic online software calculating the cooling energy potential reduction due to a passive ventilative technique of a two-zone building unit. Conversion and reduction of SPERAVent for CD version attached to the book "Il raffrescamento passivo degli edifici", 3rd edn. Maggioli. ATCode

Grosso M, Chiesa G, Nigra M (2015) Architectural an environmental compositional aspect for technological innovation in the built environment, In Gambardella C (ed) Heritage and technology, mind knowledge experience, Le Vie dei Mercanti XIII International Forum, La Scuola di Pitagora, Napoli, pp 1572–1581

Hussein M, Lee BE (1980) An investigation of wind forces on three dimensional roughness elements in a simulated atmospheric boundary layer. BS55, Department of Building Science, University of Sheffield, UK

Lechner N (2009) Heating, cooling, lighting: sustainable design methods for architects. Wiley, New York, p 142

Marques da Silva F, Saraiva J (1994) Determination of pressure coefficients over simple shaped building models under different boundary layers. PASCOOL-CLI Wind Tunnel Test Report, INETI, Lisbon, Portugal

Mathews EH, Joubert PH (1989) Quick TEMP-A thermal analysis program for de-signers of naturally ventilated buildings. Build Environ 24(2):155–162

Matteoli L, Grosso M et al (1981a) Energia/Involucro (Energy/envelope). CELID, Torino, p 98

Matteoli L, Grosso M et al (1981b) Energia/Progetto (Energy/Design). CELID Torino, p 248

Matteoli L, Grosso M et al (1981c) L'integrazione dell'energia solare negli edifici scolastici, Studi e Documenti degli Annali della Pubblica Istruzione, n. 12, Casa Editrice Le Monnier, Firenze, pp 211

Matteoli L, Grosso M et al (1981d) Studio per l'integrazione energetica e ambientale dell'isola di Pantelleria. In Atti del Quarto seminario informativo CNR/PFE: Risparmio di energia nel riscaldamento degli edifici, PEG editrice, Milano, pp 197–279

Nigra M, Grosso M, Chiesa G (2016) Integrated approach to sustainable building de-sign programming. NEWDIST, DIST, Polytechnic and University of Turin, Vol. Special Issue, p 7

Pareyson L (1954) Estetica. Teoria della formatività, Torino, Edizioni di «Filosofia», 1954, 2nd edn. Bompiani, Milan

Peretti G, Grosso M (2004) The Italian national standard approach to assessing eco-compatibility of building and urban design. Regional Eastern and Central European Con-ference "Sustainsable Buildings 2004, Varsawa, October, pp 81–82, Instytut Techniki Budowlanej

Swami MV and Chandra S (1987) Procedures for calculating natural ventilation airflow rates in buildings. Final report FSEC-CR-163-86, Florida Solar Energy Center, Cape Ca-naveral, FL, USA.

Chapter 3
Technological Concept in the Culture of Sustainable Architectural Design

Rossana Raiteri and Fausto Novi

Abstract Research on bioclimatic architecture has been one of the first significant contributions of architectural technology disciplines in the more general framework of the design/environment relationship. Technology disciplines have often been characterized by their ability to anticipate the main changes that were looming on the horizon; changes whose impact on the training and planning processes of the anthropized environment are still proving to be of crucial interest, with particular reference to the training of new skills capable of creating a link between the knowledge of scientific/technical research and the field of choices belonging to the design domain. This contribution intends to draw a picture that outlines the character and role of technological research (of which bioclimatic architecture is a part) in relation to the culture of design. It is not a question of facing a complete chronological excursus, but of focusing on the environmental theme "for architects" through some application examples.

Keywords Architectural technology · Environment · Project design

Key Concepts Mediation of technology in the conversion of scientific · Technical research products towards the design domain

> It is necessary to understand that life is a consequence of the history of life on earth. The relationship of the human being with nature and the planet cannot be conceived either reductively or separately, as the spring of the concept of globalization, since the Earth is not the sum of separate elements, the physical planet, plus the biosphere plus humanity, but rather the relationship between the Earth and humanity, which must conceive itself as a planetary and biosphere entity.
>
> Edgar Morin in "Educare per l'era planetaria" Armando editore, 2005; (translation by the authors)

Research in bioclimatic architecture has been one of the first significant contributions of architectural technology disciplines in the more general framework of the design/environment relationship.

R. Raiteri · F. Novi (✉)
Università di Genova, DAD, Genoa, Italy
e-mail: fausto.novi@gmail.com

G. Chiesa (ed.), *Bioclimatic Approaches in Urban and Building Design*, PoliTO Springer Series, https://doi.org/10.1007/978-3-030-59328-5_3

The disciplines of technology, since their establishment, (the discipline was introduced into the university system in 1969, by Eduardo Vittoria, who founded the first Institute of Architecture Technology in Naples, in 1970) have often been characterized by the their ability to capture the main changes (social, economic, productive, etc.) that were looming on the horizon and whose impact on the training and planning processes of the anthropized environment are still proving to be of crucial interest, particularly on the development of new skills capable of creating a link between the scientific-technical research knowledge and the field of the choices belonging to the design domain.

These notes want to propose a briefly outline of the ways in which this tendency occurred and with which outcomes in the world of education and research.

A paradigmatic example of the recent past is certainly the approach to the theme of industrialization processes which, in architecture, have developed late and with distinctive modalities compared to all the other production sectors.

The relative technological poverty, the scarcity of investments, the great organizational and productive complexity, the high labor intensity and low profitability, if we exclude the real estate profit, have always slowed down the industrialization process. The process had to made its way to adapt to the cultural and productive environment by inventing its own route, different from that of classic serial production which dictated its own strict rules.

The most elderly readers of these notes among the "technologists" will remember the wind of optimism that animated many of us at the time of the experimentation of the programs for industrialized building methods. At that time it seemed that many problems would have been solved by programming the entire life cycle of the building for its entire duration: the standardization of the component systems became a way to get out of a rigid freezing of spaces in a definitive configuration: it could be possible to reshape the spatial layout when the changing of the housing needs occurred. The obsolescence of everything could be delayed by separating what had to last for a given time from what had to last much longer (programmed obsolescence), guaranteeing the possibility of bringing the building back to its maximum performance for several phases during its life, etc. The systems were able to vary their framework adapting to each other from time to time and over time.

Research and experiences were interesting but, while the conceptual results have marked the design culture within the technological disciplines—just think of the concept of performance—the concrete outcome was after all quite unsuccessful if compared to expectations.

The great organizational effort that got its sense (also economic) if applicable to a large number of constructions belonging of the same program, did not have the imagined result.

The gap between what is feasible from what is really possible is always difficult to assess.

The approach to the building industrialization and to the concept of project-process which it involved, determined, in a certain sense, an early ecological awareness and, in any case, a vocation to extend the attention to the complex system of relationships that the architectural design must establish with the global environment (cultural, geographical, climatic, economic-productive, etc.) where it takes place. This approach already contained the germ of "sustainability" declined taking into account the dictates of industrial production, while avoiding, at the same time, its obtuse and atopic repetition.

It was a period full of excitement, during which research programs that had their roots at the end of the war were intertwined, continued in various programs of studies, researches and architecture competitions, which also started up educational experiments and comparisons with what was happening in Europe and the United States, where industrialization processes had already started, according to very different strategies.

In Europe, the various paths to building industrialization undertaken were characterized by a strong political administrative centralism, with the production of vast housing agglomerations built to cope with the lack of housing after the war destructions and the immigration of workers to cities. Linear housing blocks built with heavy concrete technology and closed systems, simple in distribution, rapid in construction on site and with certified standard quality (French Grand Ensemble, German Plattenbau, suburbs of cities in Eastern Europe and the Nordic countries).

In the United States, programs with large funding were developed, established on light technologies, "catalog" produced from a multiplicity of industries, all aimed at the construction of building programs for basic or university education where the demands of organizational flexibility and restraint of production, construction and energy costs, constituted the most important demand (SCSD and URBS for schools and universities in California, RAS and SEF for Canadian universities, ABS for academic buildings in the north-east states).

The only interesting exception is the Dutch SAR, which offered a method of designing public housing starting from a set of rules that covered the whole design process, from the general planning to the design of single subsystems, components and joints, within an open administrative, social and productive environment.

In Italy the building programs, which started late, were already characterized by environmental issues (at that time limited almost solely to the energy saving) and had their maximum development on the occasion of the start of the "Piano Decennale" for housing construction (L. 457–1978) and with the "Progetto Finalizzato EDilizia" of the CNR (1989–1995), which dealt with the entire building sector in general.

Two lessons of awareness remain, among everything, of this season. One disquieting and one positive: the first concerns the observation that the effort to foresee any event that may occur throughout the life of the building, up to its disposal, is fated to fail, in whole or in part, not for incidents along the way, but by definition. The variables involved are too aleatory, moreover in relation to the long duration of time which, in most cases and in our culture, characterizes the life span of what is built.

The second lesson, which is basically a positive reading of the previous one, a sort of reverse of the same coin, is that in any case it is not possible to limit one's design forecasting ability to the simple generation of the "object", but that it is necessary to design the "process" (or rather, the entire set of processes) that leads to the object existence and its persistence, until its destruction.

This statement may seem contradictory to what has just been said before and therefore lacking in logic: we cannot depend on the reliability of our foresights but we cannot avoid to do them. If the foresights are converted into determinist PROGRAMS where one tries to predict and control a hypothetical sequence of events, the chances of failure are very high.

If we replace the concept of STRATEGY capable of withstanding wide margins of uncertainty (the replacement of the concept of program with that of a method for implementing a strategy), then we can try to face the lack of certainties, being aware of the risks that inexorably arise along a bumpy path full of traps.

On the other hand, if certainties are pursued we fall into schematic, loosing the objective and disregarding the very nature of the problem. The desire to increase the possibilities of control leads to the temptation of simplification and to impotence.

The complexity of decision-making problems in design is a beautiful paradigm of the general complexity of contemporary culture. But we'll come back to this later.

Similarly, returning to the present, the heightened awareness of the finitude of energy resources, in all fields of human activities, has stimulated, in the architectural field, the proliferation of many research and experimentation on to the so-called bioclimatic architecture, in a sort of "return to the future" also based on the review of the functioning of vernacular architecture.

The culture of architectural technology (not only in Italy) has made a decisive contribution to the growth of a new awareness, in an initial climate full of enthusiasm and optimism about the possibility of promoting a change of route and the formulation of new paradigms.

At this moment, everything that can be referred to ecology and sustainability re-emerges with a particular strength. This new trend can give some hope, but we need to be aware that whenever a complex topic is translated into a slogan, the goal to be achieved moves away. Moreover, when the slogans meet with the populist tendency to show that we are going in the "right" direction with the result of obtaining a safe byproduct from the commercial and consensus point of view, which is the result of extensive "green" MARKETING operations and consumerist substitutions.

But let's get back to our theme: bioclimatic architecture uses the natural elements of the site (sun, wind, water, vegetation and soil) to create thermally efficient buildings, capable of satisfying the thermal comfort requirements, independently from the use of air conditioning systems.

The bioclimatic approach is linked to the principle of self-sufficiency and to the awareness that the main phenomena that negatively affect the environment are caused by the consumption of large quantities of non-renewable energy. Construction is a highly polluting sector since most of the greenhouse gases and CO_2 emissions come from the air conditioning and heating systems. The purpose of bioclimatic architecture is the control of the internal microclimate, with "passive" design strategies that, minimizing the use of mechanical systems, maximize the efficiency of thermal exchanges between building and environment. (Wikipedia).

While bio-architecture consists of all the architectural disciplines that assume an ecologically correct attitude towards the ecosystem. In a vision characterized by the broader interdisciplinary and economical use of resources, bio-architecture tends to integrate human activities with environmental pre-existences and natural phenomena, in order to improve current and future quality of life. Bio-architecture, an architectural practice that respects the principles of sustainability, aims to establish a balanced relationship between the natural environment and the built environment, satisfying the needs of current generations without compromising, with the indiscriminate consumption of resources, the needs of future generations. (Wikipedia).

Therefore, bioclimatic architecture is a subspecies of bio-architecture as it privileges aspects that seek to optimize microclimatic control without calling on energy consumption.

The advent of the term "bioclimatic architecture" dates back to the 70s, but has its roots in the 50 and 60s of the last century.

The adjective "bioclimatic", especially considered in the present, is a pleonasm since it indicates a quality that should be intrinsically included in the word "architecture": it should not be necessary to specify that what is planned should not be irrationally energy-intensive and polluting.

The same can be said of all other similar adjectives as sustainable, ecological, bio, recyclable etc. that have multiplied, showing the need to focus attention on the environmental effects of design and construction practices that have gradually lost their ability to take into account the environmental conditions, as a whole, in which architecture fits. Skills that require the use of some of these conditions to their advantage, like, first of all, the climatic characteristics of the place where a certain architecture will fit.

However, these adjectives reveal a need for a paradigm shift in the design culture which, in a rather recent past, has seen the rapid and progressive abandonment of "traditional" architecture based on the links with resources and with the rooting in a specific place of territory. The origins of this mutation are well known and we will not remember them here. The most evident effect of this mutation was the emptying of what had been designed and built of many of its "passive" resistance capabilities, replacing these same capacities with the inconsiderate use of energy consumption, when energy was inexpensive and seemed to be unlimited.

But this beginning must be contextualized in a panorama with much wider horizons.

Today, those who are preparing to carry out a design activity that has to do, in some way, with the built environment, is faced with a situation whose complexity is likely to overwhelm since its birth any possibility of controlling the quality of the results of its decision-making process, despite the growing availability of scientific analysis and verification tools made available by the digital world (or perhaps, paradoxically, also because of this). The environmental problem, no longer avoidable, has also contributed to adding elements of complexity, as well as the great changes that have occurred in the world of the production of new materials, of new mechanical technologies and, above all, as has already been said, of the exponential growth of tools made available by the digital world. About this, one may consider the proliferation of databases, tools, analysis, simulation and verification techniques currently available. These are precious resources for the search for more reliable solutions that, at the same time, require an organizational apparatus that is able to manage them in the most reasonable manner and according to a hierarchy variable from time to time.

The field of so-called sustainability (in which all the "eco", "bio" and other meanings could be found) expands itself to include various references: economic, social, environmental, production sustainability, etc. All these types of sustainability should be analyzed simultaneously, without separating them from each other, but integrating them into the design generation process, to demonstrate that what is possible is also concretely feasible.

After the season of research and experimentation on the theme of building industrialization, the architectural technology disciplines dedicated themselves to the broad theme of the relationship between architecture and environment in a multiplicity of ways.

Let's try to briefly describe what were the ways to relate to this topic and to outline their effects on teaching and experimentation.

Since the beginning of the 80s, lessons on the role of the environment in architectural design (Rome Sapienza, Milan, Florence, etc.) have begun to be offered in a few architecture degree courses, and then gradually many teachings appeared at the beginning of the 2000s in all the degree courses of the architecture schools. Their heading led back to the theme of environmental sustainability, all or almost were activated on the initiative of the technology sector. Finally, recent degree courses, curricular courses and master courses have been dedicated explicitly to sustainability; just to name a few: Sustainable Architecture and Landscape Design at the Politecnico di Milano. Architecture for sustainable design at the Politecnico di Torino, Sustainability in the Master's degree in Architecture at IUAV, etc.

It was a signal of the interest shown by the technological field to a problem that still needs to be fully explored, especially in terms of the effects on the design domain, but also (perhaps less nobly) of a sort of desire to "mark" a new territory of skills.

Within our schools, however, there has not been a transversal dissemination including the other design disciplines. Every disciplinary sector has closed up inside its own statute and no significant integration has occurred.

"Sustainability" in architecture has remained a side theme, a specialist point of view not integrated with the whole of training; at best, it was considered a key to verify the correspondence of a design issue to new requests not considered, from the beginning, as generator of the design itself.

On the other hand, the more or less experimental applications that have distinguished the achievements of "sustainable architectures" have lent themselves to inevitable criticisms due to the excessive prevalence given to some aspects badly or poorly integrated in a balanced synthesis with all the other design variables. The same procedure of paradigmatic experiments and demonstrations has repeated, that were not very livable (and also less marketable) that distinguished the first applications of technologies in general. Someone will remember for example a few "solar" demonstration buildings from the mid-80s, that were the result of technical-design zeal, probably very effective but strongly marked by the presence of active solar appliances and with their shape deformed by passive adaptations.

The activation of teachings and courses related to environmental sustainability issues has indicated the need to control a complex technical knowledge, difficult to face up to in a synthetic way and, at the same time, it has produced the awareness of the need to increasingly apply to collaborations with disciplinary fields unrelated to the field of architecture.

Many of the contributions presented in this publication are the direct testimony of a new way of introducing transdisciplinarity into design teaching and research.

Some interesting experiences, which show the will to overcome the distance of the individual teachings from their use during the design process, have developed in recent years within the master's degree courses, where it was possible to propose to the students a design approach that would allow them to tackle the issue of sustainability (according to the variations required by the topics in question) right from the first stages of work. In general, students have shown to be very sensitive and motivated to the proposed approach.

In these cases, the presence of interdisciplinary contributions has been of central importance for the development of work and, beyond the specific theme, has allowed to emphasize a practice that should normally be pursued to face the growing complexity of the design problems.

Another interesting field is that of research doctorates and, more generally, of research products.

The doctorate in Architectural Technology, of which we have direct experience, was born in 1983–84 by consorting several universities (Milan Polytechnic, Turin Polytechnic, Naples Federico II and Genoa) headed by the Milan Polytechnic.

Currently the doctorates, which have become internal to the universities, combine different disciplinary sectors that often, as happens fatally in faculties of a very composite nature as in the case of architecture (coexistence of scientific, technical-scientific, humanistic and design disciplines), cannot find easy points of contact. This nature has created a certain confusion and indeterminacy in defining the singularity of the objectives to be pursued and many difficulties in the composition of the different interests represented by the members of the faculty (we leave out here to talk about the indecent lack of economic resources).

Clearly this situation could only give rise to a great loneliness of each doctoral student and to a chronic lack of possibility of carrying out research work that could draw resources from international contacts, study trips, international tutoring, etc. But all this, as we know, concerns the painful situation in which the entire sector of education and research in our country has been dealing for years.

Reflecting today to what happened in those early doctorates, one feels a certain nostalgia for the work climate that was created then. Just take a look at the titles of the doctoral theses of that period and at the following engagements in the research of many of the Ph.D. students of that time: in the first doctorates generation (those with several universities consorted) the environmental theme entered quite early.

Already in the theses of the first cycle the theme of the relationship between architecture and energy appears, then gradually the interests of individual doctoral students and the complicity of tutors have led to the exploration of issues generally related to sustainability, applicability in degraded urban contexts, in the building and urban redevelopment, in the role of vegetation, of natural light, etc.

It is true that these were messy doctorates, without too many borders and with some dispersion of resources and some difficulties in deepening the inhomogeneous themes of the theses. Complex topics have often been approached with a certain amount of superficiality and improvisation, (some weird or unrealistic themes were even addressed). But perhaps this very confusion, due also to the relative novelty of the approach to sustainability, has allowed an opening to complexity and transdisciplinarity and to the breaking of the banks of the disciplinary perimeters (stiffened by the history of their academic origins), even within of the broad domain of architecture (from the territorial to the building scale, from the "philosophical" and methodological definitions to the instrumental deepenings made accessible by the digital world, etc.).

Observing the various contributions collected in this same volume, which range from technical-specialist contributions, to the analysis of the realized cases, to the tools and methods that can be used at different scales, to the management aspects, up to the critical reinterpretations of the sustainability culture, some final remarks can be made.

First of all, the variety of the different research lines brings out a consolidated tendency to create people with new skills that increasingly tend to be placed in intermediate positions between scientific-technical disciplines outside the architectural field and architecture. The increasingly unavoidable transdisciplinary approach, especially in a complex and variable field such as design, makes the contribution of other knowledge essential and decisive. However, the complexity and multiplicity of disciplinary contributions requires a sort of intermediate transformation that makes them useful in the design process. In other words, we are observing the progressive presence of people able to act as an interface between technical-scientific sectors (such as botany, geology, building physics, the various materials sciences, etc.) and the design context, to facilitate the dialogue between these different worlds. This is a kind of "translation" and transformation work, an intermediation that facilitates the possibility of enlarging the design domain to different knowledge, right from the early stages of the design development. On the other hand, these "translators" are increasingly present in large and medium professional offices.

This metamorphosis in the field of architectural technology has been taking place for years and is fated to grow more and more.

The thought goes to the first remarkable cases of technological intermediation: to the mythical Ove Arup and to the colossal company that today bears his name, which has contributed to the realization of the most famous architectures and engineering works, enhancing, in symbiosis with the architects, the feasibility conditions, translating and transferring their technical knowledge, often searching for new ones, extending their engineering skills from the structures (in the first works during and after the war) to covering any aspect of the technology applicable to construction, and involving almost any discipline, crossing the world of scientific knowledge in all directions, but always in such a way as to enter the architectural design in an almost indistinguishable way (in a recent article in the Corriere della Sera, which celebrated the inauguration of an exhibition in Denmark, Arup has been called "the man in the shadow of architecture").

To close: what is happening (and has yet to happen) in our disciplinary microcosm reflects what is happening in the contemporary worlds of philosophy, science and technology. On the one hand, we are witnessing the need to proceed by splitting knowledge. This is necessary to study so specific phenomena as to require very sophisticated tools of knowledge. On the other hand, we are increasingly aware that the disjunction, the splitting up, the narrow point of view, produce the impossibility of integration into a whole. Edgar Morin says that "The ambition of the complex thinking is to account for the articulations between the fractured disciplinary domains of the disintegrating thinking", suggesting that the strictly disciplinary knowledges (which are simplifier because lacking complex correlations) should not be rejected but should be given a new place to them.

On the other hand, if we look at the reasons for the Nobel Prize awards in recent decades, we are surprised that there are still acknowledgements by discipline (for physics, chemistry, medicine, etc.) and for one or two individuals. The topics addressed by the winners highlight fields of research, no longer subject areas, and a network of shared knowledge and no longer isolated individuals.

Chapter 4
Bioclimatic Approaches and Environmental Design. Strategies, Criteria and Requirements for an Evolution of Experimentations

Fabrizio Tucci

4.1 Introduction: Framework Remarks and Questions Definition

It is incontrovertible today that the greatest contemporary challenge for the future of humanity is characterised by the irreversible process of impoverishment and diminution of resources available on the Earth (UNEP 2017), and by the growing occurrence of situations of extreme environmental and climate emergency all over our Planet (IPCC 2019). Moreover, we reached the climax of this epochal critical period: we know well that, in the world, the number of people living in cities exceeds the number of people living in the country; the production of food will be more and more difficult and insufficient on world scale; available water is constantly diminishing; population thirst is constantly increasing; most of world energy needs are still met through non-renewable sources; those resources are available for a few more years or decades at the most; the use of those sources implies real climate and environmental disasters such as first of all, the scientifically ascertained global warming process; the health of all living bodies is harmed to a greater and greater extent; the earth, water and air of this progressively impoverished little Planet are increasingly polluted (BSUG 2011; EC 2016; IEA 2018; UNEP 2018; ISPRA 2019a; CE 2019).

We designers, technologists and planners are asked to meet this challenge, together with the necessary, desirable, very late action of mankind's ruling class, for we are the main actors of the whole territory process of change; in particular, as entities regularly and widely operating in different sectors, and in the most complex fields, we are actually some of the potential protagonists of a possible, concrete change of trend (SGGE 2017a; GCN, FSS 2018; ISPRA 2019b).

It is exactly in this kind of questions—those that are the subject of the "Environmental Design" and the "Bioclimatic Approach to the project"—that it is necessary to

F. Tucci (✉)
PDTA Department, Sapienza University of Rome, Rome, Italy
e-mail: fabrizio.tucci@uniroma1.it

G. Chiesa (ed.), *Bioclimatic Approaches in Urban and Building Design*, PoliTO Springer Series, https://doi.org/10.1007/978-3-030-59328-5_4

Fig. 4.1 We designers, technologists and planners are asked to meet the 'environmental challenge', together with the necessary, desirable, very late action of mankind's ruling class, for we are the main actors of the whole territory process of change. It means that we need to protect and harmonize architecture's passive performance components, adjusting them to the environmental factors of the micro and macro-climatic context; and it means that we have to propend towards a careful layout of the project that optimizes its life cycle form an ecological point of view and that leans towards the enhancement of eco-compatibility and environmental sustainability qualities

try and ask ourselves a few questions about the methods of approaching the conception of operating processes, paying attention, first of all, to three (plus one) dominant issues:

Which logic-cognitive phases is it necessary to bear in mind to set a proper changing action, more and more often meant to answer emergency situations, and taking into consideration the limits of resources?
Which objectives and criteria is it convenient to pursue or adopt in order to direct process, project and product choices?
Which is the ideal framework of reference requirements which a changing action—though meant for basic environmental preservation and protection—should always respect and bear in mind?

Finally, a crucial element for 'small' actors: even though one is able to act consistently with the proper logic-cognitive phases, in agreement with the meta-criteria and respecting the requirements recalled by the first three questions:

Which opportunities of real and effective contribution to change would be created by a series of definite (and coordinated) actions in the territory? May the stochastic and synergetic principle help us to motivate and stimulate the proper action of the operator as well?

I will make the first method reflections starting exactly from the latest question.

4.2 Stochastic and Synergetic Principle in the Environmental Design and Bioclimatic Approach

The basic thesis to be faced, and to reflect on, under the particular point of view triggered by the matter, is awareness that the multiplying effect of the interactions activated on one side of the town, or territory, implies reorganizational potentialities beyond the original point in which such process begins. In this respect, H. Haken is illuminating. In his reflections on the approach to the problem of complex systems by synergy science (Haken 2013), he says: "In case of many natural systems, as well as a series of systems produced by man, the macroscopic state is obtained through a process of self-organisation of microscopic elements; therefore, the system obtains a specific space, a temporal and functional structure, without a specific global action from outside".[1]

In the light of the observations by the German physician, it is important to point out that in the change project, to avoid the object of change becoming easily unstable in the course of time, the project cannot involve the whole system right from the beginning: first of all, it has to stabilise itself in a limited region, so to say, 'isolated' in the system, in which it will be gradually acknowledged by the more or less rapid communication among parts. Then it is necessary to start to determine the relations and interactions with wider fields of the system, gradually, changing and adapting its state up to the stochastic dimension which von Bertalanffy defined the "constantly dynamic balance", or the "constant dynamism of perpetual balance seeking" (von Bertalanffy 2004).

As Alberti and his colleagues pointed out dealing with the question of 'stochasticity',[2] it is exactly from the classic theory of state changes that the interpretation of his phenomenon originates—as fundamental and necessary phase to 'decode' fluctuation properly, in order to avoid it being neutralised in the new environment built without being acknowledged (Alberti et al. 2009).

In other words, it is exactly such framework that fosters the theoretical approach by parts, and increases the value of real organic action over a small time-space segment of the ecosystem hosting it. The model, characterising and distinctive role of this approach needs time to adjust to the local maturation conditions allowing the macro-system itself 'to accept' the action, and to understand its scope and importance, above all, if it is experimental and innovative, until output assimilation in terms of environment; we might say, in a sense, the above is in harmony with Quaroni's theory about the extraordinary potentialities of 'widespread project development',

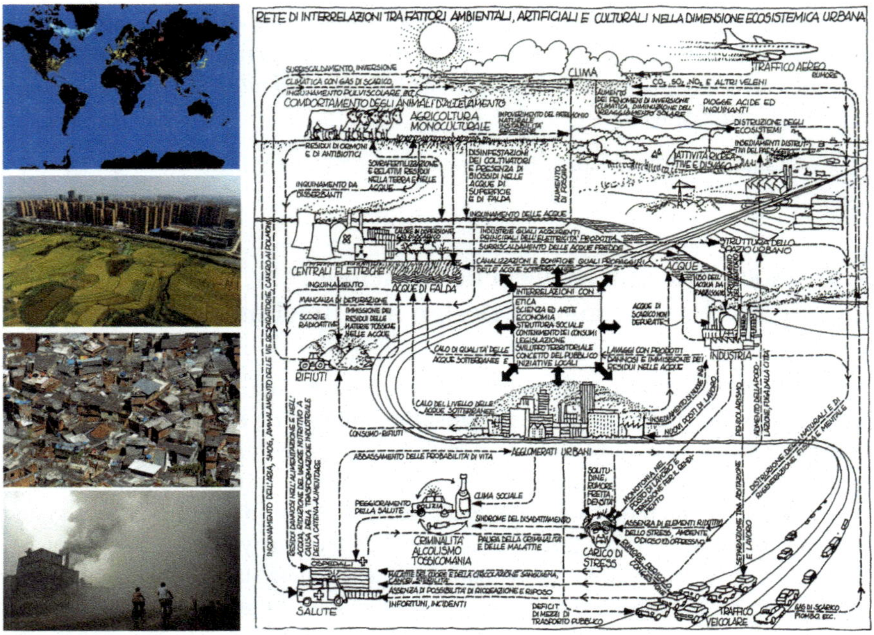

Fig. 4.2 A bioclimatic whole or bioclimatic system is a unit including all the organisms living together in a given area, interacting with the physical environment, in its natural and artificial components, in such a way that an energy flow leads to a well-defined biotic structure and cyclization of living or non-living materials within the system

according to which many people are able to envisage and operate 'small' changes and 'partial' modifications of one's world.

"An accurate, integrated, radical action focussing different entities, artifacts and ideas on a given point of the environment-space is more profitable, in terms of sustainable reconversion of living, than a policy by segments and different times",[3] stated Thomas Herzog recently, with explicit reference to the profitability of the theories by von Bertalanffy for the proper setting of Bioclimatic Design to which formulation he himself contributed (Herzog 2005).

4.3 Logic-Cognitive Phases

The latest reflections are an important package of reference to go back to the first of the three questions I asked myself, and to focus on the field of conceiving the possible logic-cognitive phases which would be necessary to define (and implement) the whole requirements emerging more and more clearly today. For those requirements (we'll deal with them through the third question) planners, and on the whole, the actors of

Fig. 4.3 In the last two decades, the vision or conception of the citizen or the architect with respect to Nature turned upside down. Now Nature is not ignored, or 'raped' any longer; on the other hand, it has to be observed and studied with respect. Nature is to give us teachings and directives as a real 'Baumeister'

changing and protection actions should be able to face different levels in different ways:

(a) Interactive detection of conduct characteristics (in terms of the ability to establish interactions effectively) of the basic elements of the 'portion' of the architectural, urban or territorial system which is the subject of the action, regardless of the scale, such as: 'Spaces in': the single spaces of the actions, adjusting transitions selectively and in a polyvalent way, in relation to the ways, roles, performances and configurations observed; 'Spaces between': the intermediate spaces 'between' the ones which are the subjects of the actions, governing interaction and adaptivity processes; 'Spaces out': the external spaces 'out' of intermediates and of the ones which are the subjects of the actions.

(b) The understanding of the overall behaviour of the 'portion' of city or territorial system analysed, in terms of: detection of the dynamic balance of the continuous flows of internal interactions characterising the micro processes of the city or territorial portion studied; detection of the dynamic balance of the interaction flows which the city or territorial portion studied establishes with its immediate ecosystem macro-area around; detection of the dynamic balance of the continuous flows of interactions which the urban or territorial portion studied establishes with the broader macro-sphere of climate and environmental factors on the whole.

In this first series of remarks, the importance of the meaning of the systemic vision for a correct approach to the formulation of the strategies and the instruments of the bioclimatic project in the context of experimentation is quite clear. But what is the meaning, in our opinion, of 'bioclimatic' and 'bioclimatic system'—words which we have been using since the title of this contribution?

We find an answer in the prestigious words of Olgyay, according to whom "a bioclimatic whole or bioclimatic system is a unit including all the organisms living together (biotic community) in a given area, interacting with the physical environment, in its natural and artificial components, in such a way that an energy flow leads to a well-defined biotic structure and cyclization of living or non-living materials within the system"[4] (Olgyay 1962).

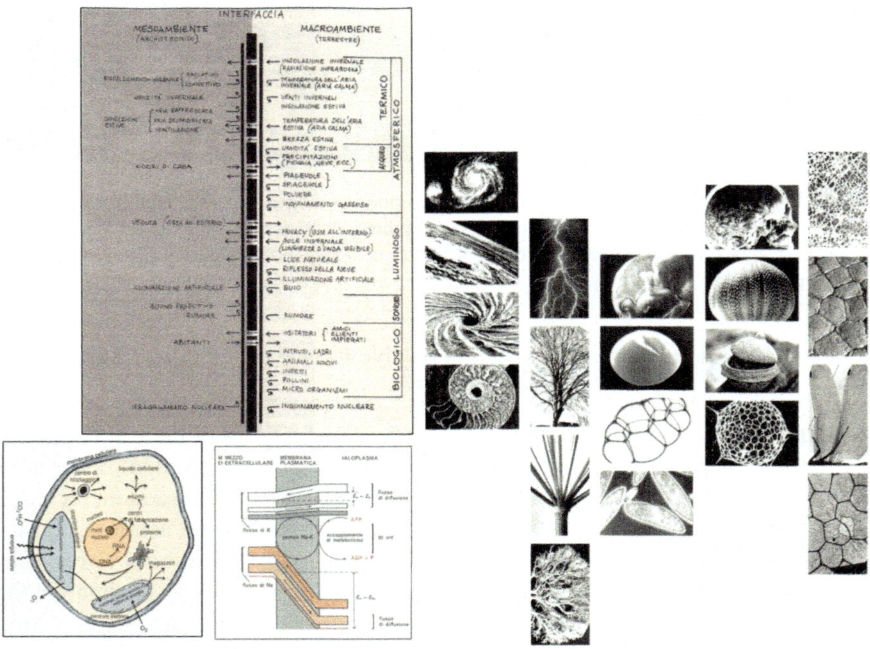

Fig. 4.4 The problem of the quality control of the environment-space changes encourages experimenters to explore the borders of innovation and modification, of the rule and project reasons, of conceiving and controlling; it is also an opportunity to investigate the roles eventually played by key elements of the built up area such as the intermediate space or the crucial, strategic "system of technological systems" – as someone defined it - the architectonic case, the building skin, the smart skin

4.3.1 Ascalarity and Fluctuation of Relations Between Parts and the Whole

The concepts of 'cyclization' and 'fluctuation' of relationships between living and non-living materials in the definition of Olgyay encourage us to focus on the key of the relation between partial elements and the whole; a basic principle of the systemic thought, perhaps the most general one, is exactly the one of the continuous shift of the point of view from the parts to the whole and vice versa; in short: the skill to shift attention among the several system levels.

In the whole living world, we find systems included in other systems; by applying the same concepts to the different levels of an architecture or one part of the same – for instance, the concept of stress to a building entity—important results were often achieved, on the theoretical level as well as on the project and application level.

On the other hand, it is also necessary to realise that, in general different system levels represent different degrees of complexity. At each level, the phenomena observed have properties which are not owned by lower levels. It's no accident that the systemic properties of a certain level are 'emerging' properties, since they emerge among the numerous ones potentially present at that given level (Ceruti 2009).

Therefore, the logical switchover to the dimension of the 'non-scalar transversal character' of bioclimatic planning is rapidly gaining ground. This affects the taking into consideration of the characteristics and properties of the several levels and, as we have just said, strengthens the principle of diversification against the well-known one of homologation (Herzog and Steckeweh 2010).

The analytical cognitive investigation on the characters and properties of the different levels of interpretation of the action area has to be understood in this sense: every part is analysed in its intrinsic qualities as well as, and above all, in relation to its skill to be in relation with the whole context. However, by now, it is clear—as quantum physics undoubtedly shown us—that there are actually no 'parts' to be regarded as independently or separately. What we call 'one part' is nothing but one scheme in an indivisible plot of interactions: we can thus regard the shift from the parts to the whole as a necessary shift of our attention from the objects to the interactions among themselves.

4.4 Objectives and Criteria of Judgement

The explicit reference to the crucial role of the 'interactions' paves the way to the reflections on the second question. In order to create the right conditions for the planner to operate properly and effectively, following the aforesaid phases, it is convenient to provide an 'adaptive' principle to set the ecologic parameters of the intrinsic qualities of existing or planning phases, in order to determine proper approaches to solve complex questions such as the one of evaluating and organising the 'interactions' going on in the territorial and settlement ecosystems. Those logics are based on a complex vision of the local, socio-economic, political, cultural, geographical, climatic, environmental variables and dynamics; but at the same time, they are meant to provide solutions to paradigmatic questions; logics are different in relation to the choice of specific actions—different spaces for configurations, functions and performances—but comparable in relation to the common criteria of judgement followed.

The question of determining those judgement criteria, to be followed in the choice of specific actions, is an important and necessary phase, above all, if we take into consideration it as part of a context characterised by resource scarceness and increasing emergency situations.

In order to set this second series of remarks, we recall what we could define some sort of shared fundamental vision (and from which all the other considerations can unravel) of architecture's ancestral task: in the confinement of an indoor space, in the articulation of an in between space, in the characterization of an outdoor space, or in the connotation of a more complex series of indoor, intermediate and outdoor spaces in the urban field, its role has always been the one to protect, interact and self regulate itself in respect to the microclimatic conditions of the context. Conditions such as sun radiation we need to protect ourselves from and the lack of natural light we need to overcome; the outdoor temperatures of hot or cold climates that need to be mitigated and dominated; different forms and nature of rainfall that ought to be controlled or the lack of the latter that needs to be overcome by applying different forms of hydrological supply; the air mass fluid dynamics in their seasonal manifestations that ought to be handled and applied to architecture, or the access of ventilation phenomena that we need to protect from (CoR 2011; Stern 2015; EEA 2016; Grosso 2017; Santamouris 2017; Lendager et al. 2018).

But architecture's role has also been to manage, in the most effective and efficient way, the energy aspects functional to the way we live in respect to the available resources; and to guarantee an environmental comfort that is the outcome, the closest one to the optimal thresholds hoped for, of the problematic factors and limitations of contextual conditions, which are rarely naturally close to such thresholds; becoming bearer and vouch for social and justice needs and deliver effective and efficient answers to the functional requirements and to related distribution and use problems, by constantly searching for a research that is coherent and harmonious with the environmental characteristics of the context (Un Habitat 2011; Battisti 2014; Santamouris and Kolokotsa 2016; Tucci 2018a).

After all, the current tendencies in the European and International framework in terms of eco-sustainable building policies and relevant enforcement actions (in particular several cases—either carried out or being carried out—relating to Germany, Austria, France, Belgium, Denmark, Holland, England and Scandinavian countries) seem to privilege a new regain of such tasks and of the objectives that such tasks encompass, through strategies and experimental criteria, mostly in the urban and suburban areas' requalification actions, following two prevailing trends: through the introduction of innovative elements from an architectural, technological and energy point of view, within a diffused environmental recovery and regeneration activity; and by a systematic demonstration of a careful concern for the reconstruction of urban micro-landscapes, with a strong accentuation and integration of the relationship between nature and artifice, following a bio-ecological conception of physical space (OECD 2016; EC 2017; SGGE 2017b; WBCSD 2017; EEA 2018).

In each of the lines of action potentially adoptable, the mark of the nine meta-criteria of judgement must always be present and correlated, at any time or level of application to the space-environment. Through its different, at times sector-based contributions, the very scientific literature refers to them in terms of control and reference to a proper action in the direction stigmatised through words such as 'ecologic' and 'environmental'—once again, with an important contribution from our scientific and disciplinary sector:

- the criterion of 'liminality' of project development conditions, always (in dialectic reasoning) in between dwelling, social and health quality seeking on the one hand, and the need for eco-friendliness on the other;
- the criterion of 'ambivalence' of the border/frontier conditions of the element or the elements characterising the environmental action;
- the 'repetition ability' criterion, that is, of 'self-preservation' (time repetition–renewal) and 'reproduction' (space repetition–multiplication) of the environmental action system;
- the 'stochastic' criterion, recalled also in the first notes of this contribution, that is, of the ability to combine a 'casual' component with a 'selective' process, in such a way that just some results of the casual can last in the environmental action system;
- the criterion of 'controllability' and 'correctability' of actions, consistently with the need for continuous updating and adjusting in progress of temporary project results, in compliance with the conception of error friendliness;
- the two criterions, in relationship, of 'flexibility' and 'adaptability' of development and management processes, in a diachronic and synchronic way, of the results of the environmental action;
- the criterion of biunique, non-destructive 'interactivity', with the resources of the environmental system;
- the criterion of real use of systems and technologies, either traditional or innovative, capable of promoting and supporting a high level of ecological effectiveness of the whole behaviour of the city and/or territorial organism that is the subject of protection, safeguard or change actions.

4.4.1 Nature as Baumeister

The logic-cognitive phases and the characters of the judgement meta-criteria emerged during the reflections made up to now, prefiguring the real opportunity to follow an evolution path tending to become more and more complex, are about the approach to the artificial action in terms of epistemology (the meaning of the changing action) and instrument-methods (how to structure the changing action). However, we must concentrate on the objective of such process: directing the contents of innovation in terms of technology, performance, morphology and physiognomy, in a framework of environmental awareness; the numerous and different consequences should represent and grant trend reversal, towards quality and ability to provide effective and immediate answers to problematic situations for emergency cases or resource scarceness, more and more significantly dominated by chaos.

In this respect, James Gleick—the author of the famous text Chaos, Making A New Science—reminds us that "simple shapes are inhuman. They are not in tune with the way in which Nature organises itself, or with the way in which the world is perceived by humans".[5]

Let's bear in mind that, in terms of formal values, new expressions of mathematics brought physical sciences in harmony with the contemporary feeling of Nature free of expressing itself and organising itself, such as fractal geometry or predictive adaptive systems, born recently but already explosive in their interdisciplinary implications (Gleick 2011). In the last two decades, the vision or conception of the citizen or the architect with respect to Nature turned upside down. Now Nature is not ignored, or 'raped' any longer; on the other hand, it has to be observed and studied with respect.[6] Nature is to give us teachings and directives as a real Baumeister (Hausladen et al. 2016).

In any case, as pointed out by Tato Dierna, it is nevertheless necessary "to make Environment a big house, in the same way as one would like one's house to be a small Environment with its richness and complexities" (Dierna 2011), and it is fundamental to give Environmental Protection a role, not just in terms of increasing environmental awareness and participation in nature cycles, but also in terms of indentifying and applying possible method requirements for the actions meant to establish proper processes of development of the interactions impressed in the changing dynamics of artificial ecosystem reality.

4.5 Requirements for Experimentation and Project Practice

After pondering over possible logical phases, objectives and criteria which should characterise any action changing the Environment, we are now approaching the third series of reflections, as attempt to answer the third question which I faced at the beginning of this contribution: which Requirements should be owned by the systems used to give experimentation bioclimatic characteristics and, in the broadest environmental technological sphere, Project Practice?

Therefore, let's try to synthesise—in a few points—what we might define the system of requirements which should be owned by any action process, and which should be followed by any operational process based on the objective to answer the need of protection and safeguard actions, as well as recovery, change and environmental value increase, more and more often to be decided and supplied under emergency conditions:

- the awareness that any attempts—though complex for itself—to reduce the parameters of the whole, thick and interrelated actions and relations based on exchange continuity, close interference of technological-physical-spatial dominions and transformation fields, run the risk of rendering such cognitive process fragile and vacuous, unless constantly accompanied with a strong critical control of data processing;
- rejection of the opportunity to relate the reading of architecture-technology-environment interactions to mechanisms of cause-and-effect univocally determined, in the awareness, that every operation about reality decomposition and re-composition processes inspired by the establishment of structures with stable and final hierarchies would make the attempt of understanding the environmental ecosystem reality actually fruitless and merely instrumental, as well as without any really scientific foundations;
- need for the continuous balance of empiricism and Noumenological questions of reference, in the conviction, that for identifying the ecologic parameters of behaviour, and the intrinsic qualities of interacting technological aspects and environmental spaces, it is necessary to introduce logics of approach to the theme of evaluating systems based on the 'differentiation' of the choice of specific actions, as well as on the 'comparability' of those data in relation to the common criteria of judgement adopted;
- the shifting of logic-cognitive attention to the 'processes of material and immaterial relation or interaction' making up the essence of metabolism of the environmental part which is the subject of the action,[7] and which characterise the behaviour of the technological systems and of the organism itself on the whole, in place of the usual attention paid to the analyses-estimates of the solely functional or formal-spatial aspects of the environment on the whole (Sieverts et al. 2005);
- the 'vitality'—recalling Lynch[8]—of the environmental action morphologic and technologic system, always ready to question itself, and always ready for development, sensitive to the specific socio-economic-cultural-geographic characteristics of the environmental object of the changing application (Lynch 1984);
- the 'flexibility' of the environmental action system,[9] capable of changing coherently at the same time with the evolution of space, the interactions and the ways of living of the users of those spaces, and of the actors of those interactions (Daniels and Hammann 2009);
- the 'tolerability' of a certain margin of error in the environmental action process, the one which German biologist Christine von Weizsäcker called Fehlerfreundlickeit, "the serene awareness and acceptance of the error or malfunctioning case", and which is also contemplated in the law of evolution and existence of nature itself (Von Weizsäcker 2010);
- the 'ascalarity' of the approach to the accomplishment of the environmental action, based on the rejection of the possibility of identifying an exact and preferential scale of system application, consistently with the conviction that the real task of seeking an operational definition—as stated by Tiezzi, Ceruti and Funtowicz with different points of views[10]—is studying the interactions of different, possible reading levels, "rather than trying to discover the only real and proper time and

space scale for sustainable development, a scale which does not exist" (Tiezzi 2007);

- for its nature, the 'adaptivity' of the environmental action morphologic and technological system does not require (or denies) an 'absolute' or 'univocal' position about the structure generating data, and concentrates the whole research specification effort on the accomplishment of a 'system architecture', whose 'predictive' effectiveness is based on the 'Theory of Independent Judges'.[11] In compliance with such Theory, the use of a complex series of some predictive tasks can be the real turning point to direct properly any kind of research, experimentation and process conception for the actions in our realities, missing operating certainties to a greater and greater extent, more and more depending on frequent emergencies, and grappling with growing resource scarceness, towards the construction of an adaptive, evolutionary and flexible system architecture[12] (Buscema and Grossi 2007; Funtowicz and Ravetz 2012).

4.5.1 The Search for Constantly Dynamic Balances

In the light of such framework, it is clear that from the operational point of view, for a good environmental, technological approach to the project, you have to be skilful, intelligent and smart to receive all innovation inputs and contributions coming from different interdisciplinary sectors, but just as far as such approach represents a new complex and multiform identity, today more than ever; it is the advanced expression of the search for complex, diversified, flexible and adaptive models, for optimising the relation between technological innovation and quality of the peopled area, between evolution of the typological-formal aspects and global eco-efficiency of the built-up environment (Antonini and Tucci 2017).

For this reason, quality control of the environment-space changes can be carried out just if one is able to conceive and propose procedures and instruments to achieve adequate forms of balance in the stratified local interrelation between natural and anthropical components, between living and inert materials. This problem sphere encourages experimenters to explore the borders of innovation and modification, of the rule and project reasons, of conceiving and controlling; it is also an opportunity to investigate the roles eventually played by key elements of the built up area such as the intermediate space or the crucial, strategic "system of technological systems"—as someone defined it—the architectonic case, the building skin, the smart skin, which a few years ago, in one of my books dedicated to it, I facetiously called 'well tempered'[13]—paraphrasing the concept of Well Tempered Architecture by Reyner Banham,[14] in turn inspired by the Wohltemperierte Klavier (Well Tempered Clavier) by Bach—as finally capable of actively and efficiently interacting with the biophysical and microclimatic material or immaterial factors such as exposure to the sun in its multiple thermal aspects, natural illumination, external temperature of the air, natural ventilation, and many other elements which are now the basis of a possible innovation framework of requirements for bioclimatic architecture (Tucci 2014).

And now a short digression, before making conclusions.

In 1969, Banham himself already recorded the obtuse enthusiasm through which many architects welcomed the outboard motor on any kind of boat: with an outboard motor one feels almighty, because it is able to change any floating object into a maneuverable boat (Banham 1969). A small and concentrated mechanical packet, totally independent from the planning of the rest of the boat, is able to change an undifferentiated mass into an object with a function equalling the purpose assigned to it: a particular project art to conceive the shape of the boat is no longer necessary; and the same can be said for the hull and the sail, because the mechanical element solves any problems, in exchange for very high consumption and high pollution.

On the other hand, a sailing boat can make it with no engine, because the boat itself is conceived as a machine: the resistance of the hull to the current is minimal, the sail exploits the wind in an optimal way and can adapt itself to the different conditions of prevailing or frequent winds; the very passengers become an integral part of the whole system, because they can shift the boat from an oblique to a balanced position with their weight.

Designing in a bioclimatic, "well-tempered" way—that is, in harmony with the environment and meant to use its natural thermodynamic and fluid dynamic flows for achieving a passive bioclimatic conduct drastically reducing energy consumption and increasing internal room comfort—implies adopting the principle of the sailing boat with respect to the motorboat; fostering profitable exploitation of the environmental microclimatic and biophysical factors which nature offers us, rather than using factitious or artificial systems damaging environmental equilibrium and quite often, in the long run, our health, too; adopting the vision according to which the architecture is a real climate instrument capable of fostering perennial equilibrium in constant change according to the aforesaid Fließgleichgewicht theory of von Bertalanffy. Day after day, season after season, the architecture is able to interact through its 'skin', adapting itself through its overall technological and 'dynamic' system to the different external climate conditions, not just protecting itself from them as happened until now, but opening itself to them, filtering them, selecting them, interacting with them, keeping its energetic environmental trend as the sailing boat does during its constant interaction with the wind, the water and the sun, that is, exploiting the material or immaterial physical forces originating from the environment and available in it, without lighting an artificial and absolutely non-contextualised energy-eating impulse in a stupid way under the point of view of eco-sustainability, often leading the user of the building to paradoxical conduct, such as wearing a T-shirt in winter for it's too hot, or a long-sleeved sweater in summer for it's too cold inside rooms so irrationally managed in terms of temperature conditioning.

All those questions concern directly both technologies as instruments of knowledge and implementation of environmental changes and the type/morphological aspects of architecture and landscape, as chances to test the changes induced by taking into consideration the interactions examined in the numerous reflections made in this contribution, from the selection of the single technological product/component to the sustainable management of the whole cycle of life of the changing process typical of architecture, in a green building and dwelling vision (Tucci 2018b).

4.6 Conclusions (In Progress)

The architecture eco efficiency concept is no longer an additional variable of the construction process, because it becomes the key element upon which we can measure the interpretation and the respect of realistic contemporary needs instead: ranging from environmental quality valorization to the optimization of bioclimatic aspects; from the limitations on the ecological footprint to the reduction—or total annulment—of noxious emissions; from the implementation of renewable energy sources to energy demand reduction.

In the attempt to converge all these very different innovation vectors—some of which generated from a need for preservation and, maybe, survival (noxious emissions' reduction, ecological footprint's limitation, etc.), some others deriving from an explosive drive towards change (improvement of bioclimatic conditions, renewable sources' implementation, etc.)—design enters a structural condition which doesn't concern only the way in which we produce our material culture, nor solely on the way we give shape to the built environment, but it concerns the relationship between concept and fruition, proposal to increment wealth and realistic wellbeing quality, between productive functions considered as quantity expansion and "culture of the limit" considered as a way of controlling from the bottom the productive freedom and its qualities, in a direct relationship with new technological potentialities and through a sustainable use of the available resources.

An intrinsic complexity lies beneath the need to plan actions on the environment and to guarantee the safeguard of nature, like the need to affirm the evolutionary use of great potentialities expressed by innovative technologies—or also the innovative use of traditional technologies in evolution—in order to define the places we inhabit; but this complexity, promoting innovation and evolution, is a positive one because it can instantly become a project, more than any other element at stake.

I hope the work that many of us researchers, teachers and planners have been carrying on for so many years in the field of architecture technology, design, town and landscape planning may represent a small contribution towards this direction, in the belief that the future of an architecture finally with no need for adjectives (without 'bioclimatic', 'eco-efficient', 'bioecological', 'eco-friendly', 'sustainable', etc.) and in which virtuosity under the bioclimatic, ecological and energetic point of view is implicit, is not far away from us at all, on the contrary, it's a reality that we are beginning to experience firsthand.

Notes

1. Cf. Haken, H. (2003) Advanced Synergetics: Instability Hierarchies of Self-Organizing Systems and Devices, Springer-Verlag, New York.

2. See to this extent: Alberti, M., Coe, S. e Hepinstall-Cymerman, H. (2009), Using Urban Landscape Trajectories to Develop a Multi-Temporal Land Cover Database to Support Ecological Modeling. Remote Sensing, Seattle.
3. Cf. Herzog, T. (2005), Architecture + Technologie, Prestel Verlag, Munich London.
4. A forerunner that always remains an important reference text: Olgyay V. (1962). Design with Climate, Princeton University Press, New Jersey.
5. Cf. Gleick, J. (2011), Chaos: Making a New Science, Open Road Media.
6. See, inter alia, the text: Hausladen, G., Liedl, P. e De Saldanha, M. (2014), Building to Suit the Climate, Birkhauser Verlag, Basel, Munich.
7. Thomas Sievert's position regarding this point is interesting as espressed in the text: Sieverts, T., Koch, M., Stein, U. e Steinbusch, M. (2005), Zwischenstadt—inzwischen Stadt? Müller und Busmann, Wuppertal.
8. Particular reference should be made, within Lynch's extensive literature, to: Lynch, K. (1984), A Theory of Good City Form, MIT Press, Boston.
9. Of particular significance inherent to the subject is: Daniels, K. e Hammann, R. (2009), Energy Design for Tomorrow, Axel Menges, Kornwestheim.
10. Cf. texts: Ceruti, M. (2009), Il vincolo e la possibilità, Raffaello Cortina Editore, Milano; Tiezzi, E. (2007), La soglia della sostenibilità, Donzelli Editore, Roma.
11. For more on the "Theory of Independent Judges" and its potential effectiveness in "forecasting", refer to: Buscema, M. e Grossi, E. (2007), Introduction to artificial neural networks, Lippincott Williams & Wilkins, London.
12. Regarding implications of the expression "environmental awareness" in its transversality between science, architecture and society, see also: Funtowicz, S. e Ravetz, J.R. (2012), Environmental problems, post-normal science and estended peer communities, Etud. Rech. Syst. Agraires Dév., Paris.
13. Reference is made to: Tucci, F. (2006), Involucro ben temperato. Efficienza energetica in architettura attraverso la pelle degli edifici, Alinea Editrice, Firenze.
14. The Well-Tempered Architecture concept was first developed by Reyner Banham in his "historical" text: Banham, R. (1969), The Architecture of Well-tempered Environment, The University of Chicago Press, Chicago.

References

Antonini E, Tucci F (eds) (2017) Architettura, Città e Territorio verso la Green Economy. La costruzione di un Manifesto della Green Economy per l'Architettura e la Città del Futuro (Architecture, City and Territory towards a Green Economy. Building a Manifesto of the Green Economy for the Architecture and the City of the Future). Edizioni Ambiente, Milano

Alberti M, Coe S, Hepinstall-Cymerman H (2009) Using urban landscape trajectories to develop a multi-temporal land cover database to support ecological modeling. Remote Sensing, Seattle

Banham R (1969) The architecture of well-tempered environment. The University of Chicago Press, Chicago

Battisti A (2014) Rinnovare le Periferie. Riqualificazione bioclimatica e ambientale delle aree urbane periferiche. Alinea Editrice, Firenze
BSUG (Bayerisches Staatsministerium für Umwelt und Gesundheit) (2011) Leitfaden Energienutzungsplan. Technische Universität München, Landkreis München vertreten durch das Landratsamt München, StMUG, StMWIVT, OBB Verlag
Buscema M, Grossi E (2007) Introduction to artificial neural networks. Lippincott Williams & Wilkins, London
Ceruti M (2009) Il vincolo e la possibilità. Raffaello Cortina Editore, Milano
CE (Circle Economy) (2019) The circularity gap report. Greenhouse Publishing
CoR (2011) Adaptation to climate change: policy instruments for adaptation to climate change in big European cities and metropolitan areas. European Union, Committee of the Regions, Brussels
Daniels K, Hammann R (2009) Energy design for tomorrow. Axel Menges, Kornwestheim
Dierna S (2011) Perchè andare verso un'ecoefficienza in Architettura. In: Tucci F (a cura di) Efficienza ecologica ed energetica in Architettura. Alinea Editrice, Firenze
EC (European Commission) (2016) Quality of life in European Cities. Flash Eurobarometer
EC (European Commission) (2017) Implementation of the circular economy action plan. EU Publishing, Brussels
EEA (2016) Urban adaptation to climate change in Europe 2016. Transforming cities in a changing climate. Report 12/2016. www.eea.europa.eu/publications/urban-adaptation-2016
EEA (2018) Adapting to climate change: European countries assess vulnerability and risks. Available at https://www.eea.europa.eu/highlights/adapting-to-climate-change-european
Funtowicz S, Ravetz JR (2012) Environmental problems, post-normal science and estended peer communities. Etud. Rech. Syst. Agraires Dév, Paris
GCN Green City Network, FSS Fondazione per lo Sviluppo Sostenibile (2018) Linee Guida per le Green City. SUSDEF Pubblicazioni, Roma
Gleick, J. (2011), Chaos: Making a New Science, Open Road Media
Grosso M (2017) Il raffrescamento passivo degli edifici. Maggioli Editore, Santarcangelo di Romagna
Haken H (2013) Advanced synergetics: instability hierarchies of self-organizing systems and devices. Springer, New York
Hausladen G, Liedl P, De Saldanha M (2016) Building to suit the climate. Birkhauser Verlag, Basel, Munich
Herzog T (2005) Architecture + technologie. Prestel Verlag, Munich London
Herzog T, Steckeweh C (2010) StadtWende. Komplexität im Wandel, Jovis, Berlino
IEA International Energy Agency (2018) World energy outlook 2018. OECD/IEA Publishing, Paris
IPCC Intergovernmental Panel on Climate Change (2019) Special report on climate change and land. Cambridge University Press, Cambridge, UK and New York, USA
ISPRA Istituto Superiore per la Protezione e la Ricerca Ambientale (2019a) Annuario dei dati ambientali. ISPRA Edizioni, Roma
ISPRA Istituto Superiore per la Protezione e la Ricerca Ambientale (2019b) Consumo di suolo, dinamiche territoriali e servizi ecosistemici. Report SNPA, ISPRA Edizioni, Roma
Lendager A, Lysgaard Vind D (2018) A changemaker's guide to the future. Lendager Group, Copenhagen
Lynch K (1984) A theory of Good City Form. MIT Press, Boston
OECD (2016) Green cities programme methodology. ICLEI Local Governments for Sustainability, European Bank for Reconstruction and Development. EBRD Publishing, London, Paris
Olgyay V (1962) Design with climate. Princeton University Press, New Jersey
Santamouris M, Kolokotsa D (eds) (2016) Urban climate mitigation techniques. Routledge, London
Santamouris M (2017) Advances in passive cooling. Earthscan, London
SGGE Stati Generali della Green Economy (2017) La Città Futura. Manifesto della Green Economy per l'architettura e l'urbanistica. SUSDEF Pubblicazioni, Roma

SGGE Stati Generali della Green Economy, Tucci F (ed) (2017b) Verso l'attuazione del Manifesto della Green Economy per l'architettura e l'urbanistica. Obiettivi, ambiti di indirizzo, strategie prioritarie. SUSDEF Pubblicazioni, Roma

Sieverts T, Koch M, Stein U, Steinbusch M (2005) Zwischenstadt—inzwischen Stadt?. Müller und Busmann, Wuppertal

Stern N (2015) Why are we maiting? the logic, urgency, and promise of takling climate change. The MIT Press, Boston

Tiezzi E (2007) La soglia della sostenibilità. Donzelli Editore, Rome

Tucci F (2014) Involucro, Clima, Energia. Qualità bioclimatica ed efficienza energetica in architettura nel progetto tecnologico ambientale della pelle degli edifici (Envelope, climate, energy. Bioclimatic quality and energy efficiency in architecture in the environmental technological design of building skins). Altralinea, Firenze

Tucci F (2018a) Resilienza ed economie green per il futuro dell'architettura e dell'ambiente costruito (Resilience and green economies for the future of architecture and the built environment). Techne J Technol Archit Environ 153–164

Tucci F (2018b) Costruire e Abitare Green: Approcci, Strategie Sperimentazioni per una Progettazione Tecnologica Ambientale (Green building and dwelling: approaches, strategies, experimentation for an environmental technological design). Altralinea Editrice, Florence

Un Habitat (2011) Saving cities: adaptation as part of development. United Nations Human Settlements Programme Publishing

UNEP United Nations Environment Program, IRP International Resource Panel (2017) Assessing global resource use. A systems approach to resource efficiency and pollution reduction. Un Environment, UNESCO Publishing

UNEP United Nations Environment Program (2018) Resource efficiency: potential and economical implications. UNESCO Publishing

Von Bertalanffy L (2004) Teoria generale dei sistemi. Mondadori, Milano [nuova ed.]

Von Weizsäcker C (2010) Fehlerfreundlichkeit. In: Kornwachs K (a cura di) Offenheit Zeitlichkeit Komplexität. zur Theorieoffener Systeme. Campus Verlag, Francoforte [nuova ed. dig.]

WBCSD World Business Council for Sustainable Development (2017) How we drive sustainable development. WBCSD Publishing, Ginevra

Chapter 5
Present and Future Challenges and Opportunities in the Built Environment

Mat Santamouris

Abstract The present chapter aims to present and discuss the main challenges and opportunities in the Built Environment. The document analyses issues related to the impact of global and local climate change on the built environment, the challenges associated to the expected increase of the world population, social and technological issues related to energy poverty and future energy consumption. It also presents and discusses the major environmental problems in our cities and the expected energy and social problems related to the rapid urbanisation. The actual state of the art of the energy and environmental technologies for buildings and cities is presented. Futures priorities and ideas on the way to translate the current challenges into future opportunities are presented.

5.1 Introduction

The built environment comprises buildings, infrastructures and open spaces and it is the final result pf many social, environmental and financial processes that serve and fit with the needs and the standards of the local society. Several economic, financial pressures and environmental influences related to labour demand and offer, properties, equity and related investments, family income, fabrication distribution and consumption of goods, human and local identity and cultural heritage, security of the population and land use, energy and resources flow determine and define the characteristics of the urban human communities. The building sector has very dynamic characteristics and a quite complex nature. Its main objective is to protect and when possible improve the life quality of population, generate wealth and employment opportunities but at the same time it produces wastes and pollution, consumes tremendous quantities of materials, resources and energy and affects the climate of cities and the globe and compromises the survivability, vulnerability and poverty of human communities.

M. Santamouris (✉)
Anita Lawrence High Performance Architecture, Faculty of Built Environment, University of New South Wales, Sydney, Australia
e-mail: m.santamouris@unsw.edu.au

G. Chiesa (ed.), *Bioclimatic Approaches in Urban and Building Design*, PoliTO Springer Series, https://doi.org/10.1007/978-3-030-59328-5_5

Essentially, yje built environment and its components act as the catalyst promoting future developments and trends at the technological. Social and environmental level.

The built environment is facing several challenges. Among them, overpopulation, poverty, excessive energy consumption, pollution and waste management, regional and global climate change. Seems to be the most important one. Despite the important challenges, continuous progress and development transforms gradually challenge into opportunities supplying the dynamic development of the human communities.

This chapter aims to present in a brief way three of the main problems of the built environment, energy consumption, regional climate change and energy poverty. Also, to articulate ideas on the potential benefits for an eradication of energy poverty, mitigation of regional climate change and minimisation of the energy consumption. Most of the material is drawn from references (Santamouris 2016, 2019, 2010), where readers can get more detailed information and documentation.

5.2 Energy Consumption and Buildings

The present chapter concentrates on the characteristics and the energy consumption of the building sector in Europe. Residential buildings consume almost 27% of the total consumption in the Continent while the commercial and in general the tertiary sector is responsible for an other 14% of the total European consumption. The high share of the residential sector is mainly due to the total area of residences, 18 billion m^2 and almost 75% of the total building stock, with an average floor area around 87 m^2 per residence.

The average energy consumption of Buildings in Europe is around 220 kWh/m^2/y, and varies between a maximum around 320 kWh/m^2/y for Finland and 150 kWh/m^2/y for Bulgaria. In parallel, the average energy consumption of the tertiary buildings around to 295 kWh/m^2/y, while the corresponding average consumption of the residential building is close to 200 kWh/m^2/y. Tertiary buildings are characterized by a constant and increasing energy consumption trend due to the annual evolution of the services by 1.3%. The increase consumption rate of the tertiary sector is close to 1.1% during the last 10 years. In parallel, the electricity energy consumption of the tertiary buildings has increased by 60% during the recent years. Among the most consuming tertiary buildings are office and trade buildings, responsible each for about 26% of the total commercial consumption. It is estimated that the continued rise of the services sector in commercial buildings will surge a total consumption increase of the sector up to 93% by 2030. Despite the continuous decrease of its absolute energy consumption caused by the technical improvements and the strict legislative measures, space heating of premises is the most consuming energy component.

Strict energy legislative measures applied and implemented the recent years have contributed to a continuous increase of the energy efficiency in the residential sector. Despite the increase of the energy efficiency, the residential sector has increased its energy consumption the last 20 years by 14%. The observed Increase of the energy of the residential sector is due to the considerable increase of the residential stock, the rise of the attributed living space per person and also because of several social, political, technological and financial measures and reasons.

Because of the high heating performance of the new residential buildings in Europe, that represents almost 20% of the total residential stock, the specific consumption for heating purposes in the residential sector has decreased significantly, close to 15% the recent years. Consumption data shows that new residential buildings may consume up to 60% less heating energy than buildings built before 1999. The reduction f the specific heating consumption is much higher in the Nordic European countries, between 50 and 60%, while in the Southern countries the corresponding decrease of the heating demand is between 20 and 30%.

The consumption of energy in buildings is highly impacted by social, environmental and mainly on serious economic factors and variations. As revealed from the recent perturbations of the energy consumption, it is evident that the building sector is very sensitive in economic changes and variations. Because of serious economic problems, population may partly satisfy energy needs and finally reduce significantly the final energy demand. It is underlined that because the financial crisis of 2007–2015, in Greece, households reduce their energy consumption by 4%. Similar results and characteristics are also observed in other European countries with economic problems like Ireland, Portugal and Slovakia. In these countries, the decrease of the energy consumptions was as high as 22%.

5.3 Urban Regional Climate Change

Regional climate change, or urban overheating, is a major issue for most of the large size cities in the world. There are more than 400 cities in the world where the problem of urban overheating is experimentally documented. The magnitude of urban overheating may exceed 8–9 C with an average values between 5 to 6 C. Urban overheating is the result of the more positive thermal balance of cities compared to their surrounding suburban or rural environment. In parallel to the urban overheating, heat waves have increased their magnitude and frequency all over the world. The synergy of heat waves with urban overheating increase further the ambient temperature and intensify urban overheating.

Higher ambient temperatures have an important impact on the energy consumption for cooling purposes, while increase the levels of heat related mortality and morbidity, rise the levels of urban pollutants and in particular of the harmful ground level ozone, put in risk and decrease the quality of live and the survivability of the low income households while have an important economic impact in the quality of life of urban citizens.

Urban overheating is extremely well documented in most of the European cities. In parallel, its impact on the energy demand and heat related mortality seems to be tremendous. Urban overheating is more intense in the low income neighborhoods of the European cities and place under threat the life of vulnerable urban population.

5.4 Energy Poverty

Energy poverty refers to the situation in which families lack a materially, economic and socially level of energy consumption and global energy services and necessitated level of energy services. Energy poverty is a serious threat for the low income population in Europe and the world. According to the official statistics there are about 100 to 150 million of energy poor in Europe that can not afford to satisfy their basic energy needs and pay their bills. As a result, energy poor live under very low and very high indoor temperatures during the winter and summer period.

Energy poverty levels are quite high in many European States with quite low GDP. It may exceed 30% of the total population. Energy poverty results from a combination of factors like the excessive energy prices, the insufficient family income and the inadequate quality of dwellings. Many recent statistical studies have demonstrated that vulnerable European citizens use to live in dwellings of poor energy performance. While about the 10% of the population in Europe lives in dwellings with serious energy and environmental problems, the corresponding percentage increases up to 25% for the low income and vulnerable population. According to the European statistical services, between 20 and 55% of vulnerable and low income households in Southern and Eastern Europe live in residences with leaking windows. In parallel, statistics show that in the UK, almost 17% of the households of lower income use to live in completely inappropriate dwellings. Presenting mould and condensation. The ratio between low income to the high income population is close to three. Additionally. More than 1,966,000 dwellings, or 8% of the stock in the UK suffer from excess cold problems.

5.5 Minimizing to Zero—A New Sustainability Objective for Europe

Minimizing to zero is a concept aiming to develop a strategy able to minimize or eradicate completely the impact of harmful defects, emissions, wastes, etc. The main objective of the strategy is to minimize the energy consumption of buildings, eradicate energy poverty and mitigate as much as possible climate regional and global climate change. This can be achieved through the development and implementation of advanced innovative technologies combined with appropriate economic. Social and environmental policies.

It is evident that such a strategy is very radical and far from the actual reality and practice. However, it is evident that the idea is very appealing, it incorporates high level moral and humanistic features, is an unequivocal choice, and can be the ultimate road map of future strategies and developments.

The main objectives and goals of such a strategy deals with:

(a) the energy consumed by the building sector,

(b) the impact of buildings on the regional climate change and in particular with urban overheating
(c) the potential eradication of energy poverty in the built environment.

Such ambitious goals should be achieved gradually. Even a partial satisfaction of these objectives could contribute highly to improve the quality of life in human societies. However, the whole strategy needs an excellent planning and the development of an agenda on the necessary technological, social and economic innovations that have to be designed and implemented. This requires a considerable investment made in the global building sector that should result in the development of new future opportunities and will generate serious medium and long term societal benefits, and will alleviate people from the specific problems.

It is widely accepted that the energy consumption of buildings can be seriously reduced through the use of advanced technology. The necessary financial investments to minimize the buildings energy consumption in Europe by 2050 are estimated between 14.5 and 23.6 trillion of Euros. It is also estimated that the CO_2 emissions of the residential and tertiary sectors will be limited by 90.3% and 37.7% compared to 2005, respectively. In parallel, the implementation of innovative energy measures will create about 189 jobs per million of Euros invested

Future measures aiming to minimize the buildings energy consumption is propose to focus on the following:

(a) Rise the global energy performance of the building energy systems to considerably limit the energy consumption and the final energy needs
(b) to use renewable energy to cover the remaining energy demand.
(c) to optimize the operation of the environmental and energy building systems through the implementation of smart and intelligent technologies.

5.6 Conclusions

Recent research efforts aiming to improve technologies for the building sector, have resulted in the development and commercialization of building related technologies that minimize the energy consumption of buildings while contribute to mitigate the amplitude of the regional urban overheating. The development of advanced envelop protection systems, the high efficiency HVAC systems, the high performance technologies based on the use of renewable energies, as well as well as technologies of cool and super cool materials aiming to increase the albedo of cities are among the more promising technologies to improve the environmental and energy performance in the built environment. Photonic and plasmonic materials as well as materials of high reflectivity, can be used in roofs and pavements to decrease the surface temperature in cities and reduce the release of the sensible heat that contributes to increase the urban ambient temperature. Hundreds of large scale real implementations of advanced mitigation materials in cities have demonstrated that it is feasible to decrease the peak urban ambient temperatures up to 2.5–3 C. It is expected that the development and implementation of advanced photonic materials in cities may enhance the mitigation potential up to 4.0–5.5°.

It is well accepted that the nest possible policy to eradicate energy poverty and protect the well being of vulnerable population, is associated to the energy retrofitting of low income dwellings. All large scale retrofitting projects addressing the low income population in Europe has concluded that there are very serious financial, social and environmental advantages for the vulnerable households. Full understanding of the necessary technical characteristics of the retrofitted low income houses and assessment of the cost and benefits, are necessary conditions to design and implement an efficient program to eradicate the energy poverty problem in Europe.

Energy poverty, excessive energy consumption of buildings, and the regional climate change and urban overheating are among the most significant problems associated to the built environment in Europe. It is important to be understood that the above three problems are strongly interrelated characterized by very serious trade offs and synergies. Thus policies addressing the problems of energy consumption, regional climate change and energy poverty should be designed in an integrated and holistic way to address in parallel the three problem as a common issue. If not its is inevitable to face a continuous degradation of the thermal conditions in our cities, a new increase of the energy consumption for buildings and a surge of the energy poverty in Europe, while economic and social discrepancies may rise as well.

The objective of Innovating to zero the built environment in Europe is an ambitious target that aims to eradicate energy poverty, minimize the energy consumption of buildings and mitigate as much as possible urban overheating and regional climate change.

It is well accepted that a minimization to zero is an unequivocal objective that will generate considerable new opportunities for growth and development.

References

Santamouris M (2019) Minimizing energy consumption, energy poverty and global and local climate change in the built environment minimizing to zero the building sector

Santamouris M (2010) Recent progress on urban overheating and heat Island research. Integrated assessment of the energy, environmental, vulnerability and health impact synergies with the global climate change. Energ Build 207:109482

Santamouris M (2016) Innovating to zero the building sector in Europe: minimising the energy consumption, eradication of the energy poverty and mitigating the local climate change. Solar Energ 128:61–94

Part II
Roots and Evolution

Chapter 6
Origin and Evolution of the Bioclimatic Approach to Architecture

Mario Grosso

6.1 Introduction

This essay intends to outline a synthetic excursus on the origin and development of the approach relating architecture to climate in the works of researchers and architects, as an extensive interpretation of what is called in this book bioclimatic architecture. Indeed, 'bioclimatic' is an adjective "concerning the relations between climate and living organisms" (Collins Dictionary 2020), and buildings can be considered 'living organisms', according to the 'organic' approach characterising the way of thinking of the authors of this publication as well as of the subjects of this analysis. This excursus includes both theoretical approaches and paradigmatic examples of architecture. This issue, which does not claim to be exhaustive of such an extensive field of thoughts and experiences, is organised in three parts: (a) treatise writers and examples of archetype regional architecture, divided by geographical areas and spanning from ancient times to the renaissance period; (b) the contributions of the architects masters of the Modern Movement; (c) the contributions of scholars and design handbook writers of the XX century.

6.2 Treatise Writers and Regional Architecture

6.2.1 Mediterranean Area

The first architecture "theorist" who conceived an interrelation between climate and architecture is Vitruvius, whose work was "deeply influenced by climatic awareness and advice. This starts from principles of site choice and town layout—to avoid

M. Grosso (✉)
Department of Architecture and Design, Politecnico di Torino, Turin, Italy
e-mail: mario.grosso@formerfaculty.polito.it

G. Chiesa (ed.), *Bioclimatic Approaches in Urban and Building Design*, PoliTO Springer Series, https://doi.org/10.1007/978-3-030-59328-5_6

funnelling of prevailing winds; the avoidance of south winds and heat, as well as of excessive humidity, in the choice of sites as well as a whole chapter devoted to the climate as a determinant of the style of the house" (Vitruvius, ca.15 B.C.) (Markus and Morris 1980, p. 5). "Houses should conform to diversity of climate, being of southerly exposure, and roofed, in the north and of northerly exposure, and more open, in the south" (Markus and Morris 1980, ibidem).

Markus and Morris highlight the influence that Greek writers might have had on Vitruvius. Particularly: Aristotle, for his suggestion to laying out a town so that it faces East and be sheltered from the North (Aristotle ca. 330–322 B.C); Xenophon, who advised to provide porticoes for shading high summer so-lar beams, but allowing low-angle winter solar radiation to penetrate deeply into inner spaces, and to provide the south side higher than the north's 'to keep out the cold winds' (Xenophon post 371 B.C.).

However, in an apparent contradiction to Vitruvius's theory, the design and construction of buildings by the ancient Roman architects—particularly, villas, amphitheatres and baths—were characterised by a uniform style and space arrangement, "basically similar whether in North Africa or in central Gaul, some 15° Latitude further north" (Markus and Morris 1980, p. 6). This is probably due to an overwhelming importance of functional and symbolic aspects over the 'bioclimatic' inspiration, within the context of the Roman imperialism territorial expansion, which needed a strong centralised cultural imprinting. Of course, the relative mild climate characteristics of the whole territory around the Mediterranean Sea helped in favouring such a development.

Amazingly, this architectural uniformity contradictory with a bioclimatic approach would appear centuries later during the Modern Movement in the proliferation of buildings, whose concept and aspect were inspired by the so-called 'International Style'. Their over-glazed envelopes and undifferentiated façades exposure are just opposite of what a correct bioclimatic approach would suggest. On the contrary, a follow up of Vitruvius's approach was the inclusion of a solar access rule in the Justinian's code, in the sixth century (Justinianum and Tribonianum 529 A.D., 534 A.D.).

Other scholars who studied deeply a bioclimatic aspect of ancient Greek and Roman buildings—the solar architecture—were Ken Butti and Perlin (1980). They highlighted the drastic depletion of wood, before in the Greek area and later in the Roman's, due to both its use as a construction material and heating energy source. This phenomenon induced the potential use of solar energy, although not, of course, as energy fuel but as its exploitation as a passive source of heat depending of optimal exposure of building spaces. This influenced both the orientation and layout of houses—such as the classical Greek home (Butti and Perlin 1980, p. 5)—see Fig. 6.1—and Roman villas (Butti and Perlin 1980, p. 17)—and the layout of entire cities such as Olynthus (Butti and Perlin 1980, p. 6). In addition, the use of glass by the ancient Romans allowed for the exploitation of the heat trap generated by the relevant green-house effect, in the element called '*heliocaminus*', particularly applied in public baths (Butti and Perlin 1980, p. 18).

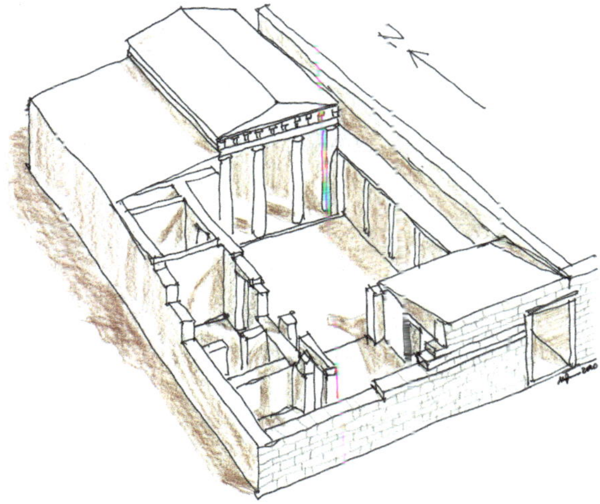

Fig. 6.1 Reconstruction of a Classical Greek home, from excavations of the city of Priene by Theodore Wiegand. Author's re-elaboration based on Butti and Perlin 1980, p. 5

In the renaissance period, in spite of inspiration to a 'classical order', based on symmetry and Euclidean geometry-based compositional rules, in two of the main treatise writers—Alberti's Ten Books on Architecture (Alberti ca. 1450) and Palladio's The Four Books on Architecture (Palladio 1570)—hints of a 'bioclimatic' approach can be found. In Alberti's depiction of beginning of architecture, he writes that men "(…) began to think of a covering to defend them from sun and rain; and in order thereto, they erected walls to place this covering upon. By this means, they knew they should be more completely sheltered from piercing colds, and stormy winds. Lastly, in the sides of the walls, from top to bottom, they opened passages and windows, for going in and out, and letting in light and air, and for the conveniency of discharging any wet, or gross vapours which might chance to get into the house." (Marcus and Morris 1980, p. 6).

"Palladio refers to Vitruvius's description of the original hut and elaborates on the change from flat to pitched roofs. He too advises against valley sites, on account of humidity, wind and the reflection of the sun's rays creating excessive heat." (Marcus and Morris 1980, p. 7).

However, from a practical viewpoint, the Renaissance architects—as their 17th and eighteenth Century follower—"did not realize to the full the implications of their climatic differentiation theories any more than the Roman architects had before them." (Marcus and Morris 1980, p. 8). In fact, a Palladian villa in London (e.g., Chiswick house 1725) "uses the same plan form, window pattern and volumetric relationships" as the Palladio's most famous 'Villa Rotonda' in Vicenza

The only feature somehow 'bioclimatic' of their buildings is the massive characteristic of the walls, which allows for a diurnal temperature swing, particularly useful to avoid summer over-heating.

On the contrary, in the same Palladio's time and region, an outstanding architectural application of 'bioclimatic' principles can be found in the Villas of Costoza

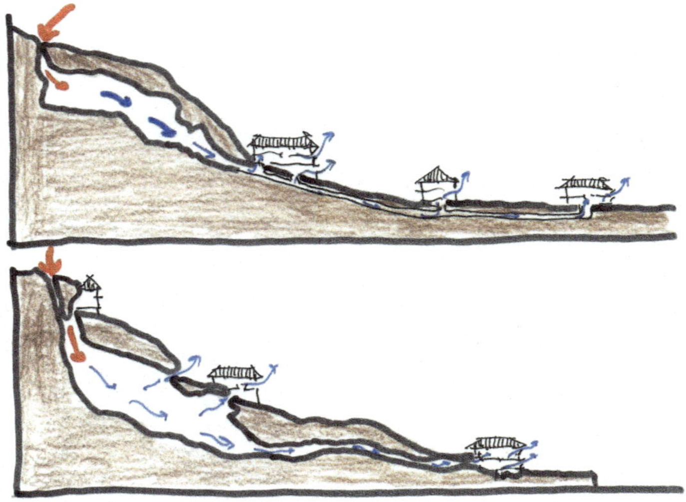

Fig. 6.2 Section of the 'Covoli', natural and artefact wind-ducts, above which villas were built in Costoza (Vicenza) during the XVI Century. Author's re-elaboration based on Grosso (2011, p. 173)

(Vicenza), where an ingenious natural cooling system exploited the 'covoli', underground wind ducts naturally present in the local limestone and partially used as quarries during by the ancient Romans (Grosso 2019) (Fig. 6.2).

In addition, the Italian territory is full of buildings of the kinds that various literature called 'vernacular' or 'spontaneous' or erected by 'prodigious builders' (Rudofsky 1977). These examples include the 'Scirocco rooms' in Sicily (Fig. 6.3), the Dammuso house in Pantelleria Island (Fig. 6.4), the Trulli in Puglia (Fig. 6.5, the 'at-rock-houses' in Piemonte (Fig. 6.6) (Grosso 2017).

6.2.2 Middle East Area

The Middle East territory is the richest part of the world for what the application of 'bioclimatic' elements and concepts in the building construction, from the ancient period to the middle age and further on, is concerned (Grosso 2017). Regarding Egypt, the main scholar and architect who studied and applied those elements and principles is Hassan Fathy (1986). He described: the *mashrabiya*, a wooden screen with five functions: daylight control, airflow control, reducing air temperature, increasing air humidity, and assure privacy (Grosso 2017, pp. 115–118) (Fig. 6.7); the *castrum*, with similar functions, but built of iron or stoneware (Grosso 2017, pp. 118–121) (Fig. 6.8). However, the most important architectural and technological bioclimatic features in the Middle East are related to the control of natural airflow through special vertical wind-catcher openings called *Malkaf*, in Egypt, Iraq, and Pakistan,

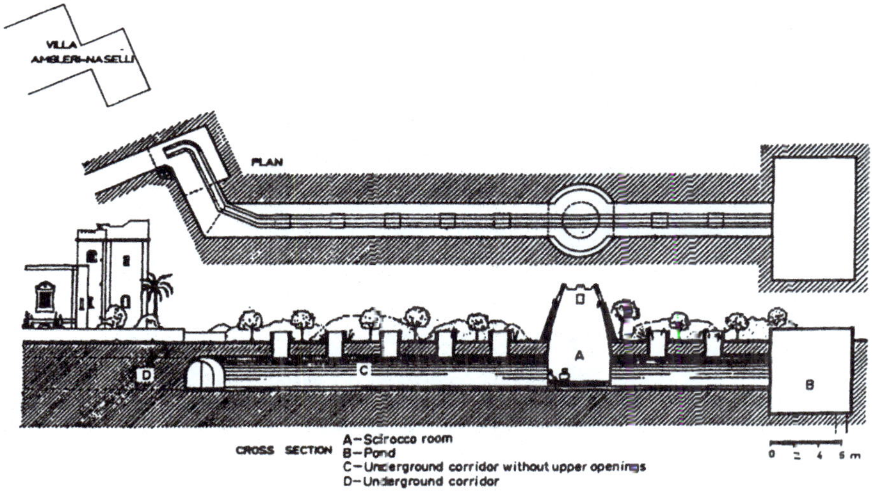

Fig. 6.3 Plan and longitudinal section of one ''Scirocco' room, and the connections to the attached Villa 'Naselli', near Palermo, Sicily, XVI Century. *Source* Grosso (1997, p 140)

and *Badgir*, in Iran (Grosso 2017, pp. 121–136) (Fig. 6.9). Those elements were studied, other than by Fathy (1986, p. 118), by Rudofky (1977) and other scholars (Fardeheb 1987; Bavar 1983). In addition, the most complex technical system for ventilating and cooling using the *badgir*, is the Iranian wind tower, whose origin dates some two thousands year B.C. and its use lasted up to the IX century (Fig. 6.10). Many researchers have studied the Iranian wind tower (Grosso and Ahmadi 2016), amid whom, one of the most prominent is Bahadori (1978).

6.2.3 *North-American Area*

The most interesting examples of a bioclimatic approach to 'vernacular' architecture in the North-American area are the 'pueblos', a heritage of the *Anazasi* people, in the South-western regions of USA, who disappeared in the XII Century. The first researcher who studied in a scientific way some of these examples is Ralph Knowles (1974). Particularly, he studied the following pueblos: the *Longhouse pueblo* at Mesa Verde, Colorado; the *Acoma Pueblo*, New Mexico; and the *Pueblo Bonito*, Chaco Canyon, New Mexico.

In the *Longhouse pueblo*, "The cave thus offers a high degree of protection from the summer sun, compared to a deep penetration of light during the winter (Fig. 6.11). Still more careful observations show that buildings were placed inside the cave in such a way that their vertical stone walls and horizontal terraces received great benefit from the low winter sun, while being protected during the summer by shadows cast

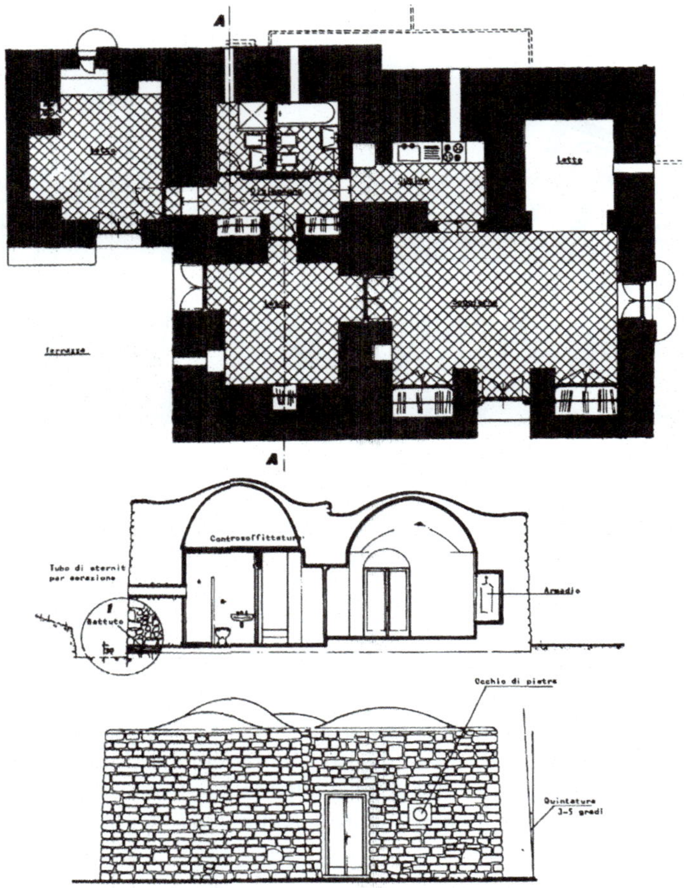

Fig. 6.4 Plan (top), section (middle), and elevation (bottom) of the 'Dammuso', a typical house of the Pantelleria island, Sicily. *Source* Grosso (1997, p. 145)

from the upper edge of the cave opening and by the high summer altitude of the sun" (Knowles 1974, p. 24).

These characteristics were also quantified by a measurement of the seasonal insolation through a test carried out in the Architecture Department at the University of Southern California: "An accurate model of Longhouse, based on Douglas Osborne's reconstruction of the pueblo (Fig. 6.12), was placed on a heliodon and photographed at hourly intervals." (Knowles 1974, *ibidem*). From those data, Knowles calculated the collection energy efficiency "by comparing the actual amount of energy received per hour, with the amount that might be received if all surfaces acting within the perimeter of the cave opening were totally effective, as if they all lay normal to the sun's rays". (Knowles 1974, p. 25). This calculation "(…) indicates that the building and cave arrangement is 56% more efficient as an energy collector in the winter than

Fig. 6.5 The 'Trulli', a typical house in Puglia. Picture by the author

Fig. 6.6 A house-at-the-rock, Angrogna Valley, Piemonte. Picture by the author

it is in the summer, which raises the question of purpose and how these builders went about realizing such an efficient energy system." (Knowles 1974, p. 26). To answer this question, Knowles wrote, "it is difficult to believe that they worked from a plan (…)" being "the arrangement too complex and the buildings too specialized in their siting. On the other hand, it is unreasonable to believe that they were not somewhat aware of the potential for shelter from rain and wind or that they did not hold some general notion of the advantages to be gained from a southern exposure." (Knowles 1974, *ibidem*).

Fig. 6.7 'Mashrabiya' of Zeinab Khatun's house, Cairo, Egypt—before 1486. *Source* Grosso (1997, p. 164)

Fig. 6.8 'Claustrum' of iron in the Mosque Al-Rifa'i, Cairo, Egypt. *Source* Grosso (1997, p. 171)

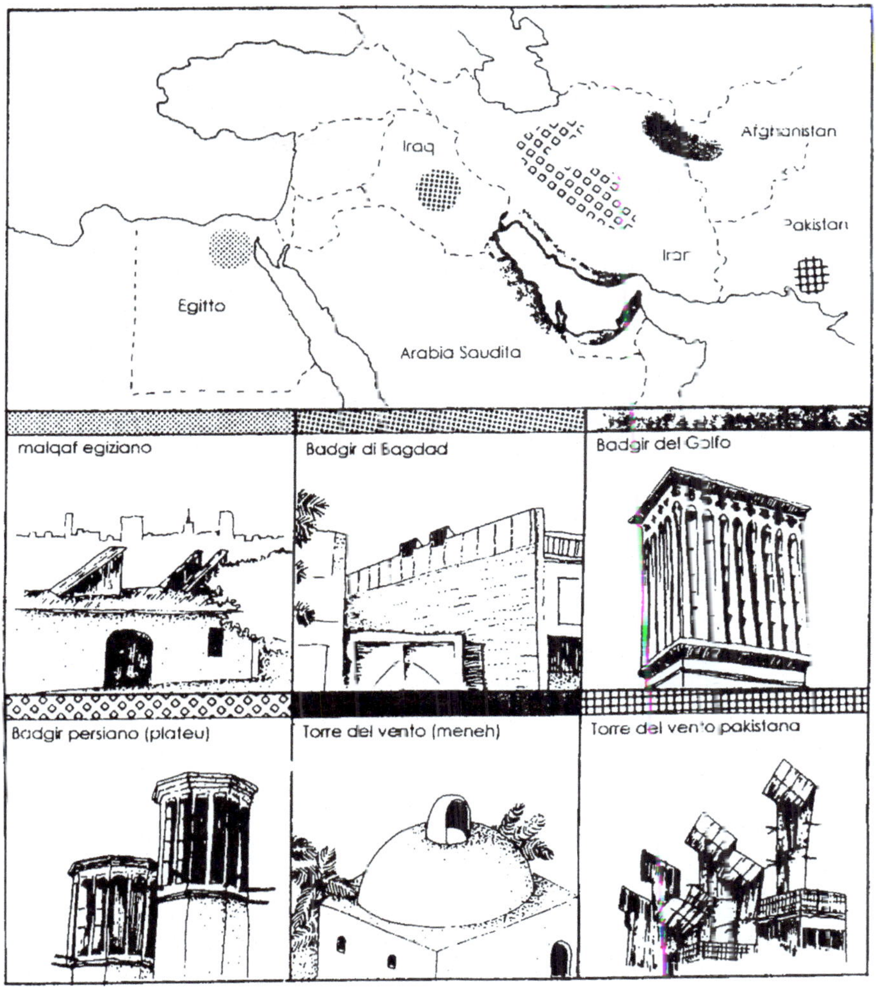

Fig. 6.9 Geographical distribution and schematic drawings of various types of wind-catcher in the Middle-East. *Source* Grosso (1997, p. 175)

Thus, Knowles's conclusion is, "If the builders of Longhouse did not make a plan and did not perceive ultimate form as a basis for organization, then they worked within the constraints of site and main recurrence in nature. These early builders were part of a self-organizing process from which purpose emerged slowly with the construction." (Knowles 1974, *ibidem*).

Acoma Pueblo "lies about fifty miles west of Albuquerque, New Mexico, at 35° north latitude. For defense, it was placed atop a flat and nearly inaccessible mesa that rises 400 feet from the plain below (Fig. 6.13). The climate is generally desertlike, but ambient air temperatures are modified by the mile-high altitude." (Knowles 1974,

Fig. 6.10 'Badgir'(wind-catchers) in Yazd, Iran. *Source* courtesy of Mehrnoosh Ahmadi

Fig. 6.11 Longhouse pueblo, Mesa Verde, Colorado, USA: picture of a detail (left) and scheme (right) showing the shading effect in winter and simmer of the cliff site. Author's re-elaboration from Grosso (1998, p. 136)

p. 27). "A plan of the pueblo shows three nearly parallel rows of buildings extending for about a thousand feet at just a few degrees off an east-west axis and terminating at the steep slopes of the mesa." (Knowles 1974, p. 28) (Fig. 6.14). "The typical Acoma house is generally tiered, as is the row itself. It steps down, exposing broad areas of roof and vertical wall to the south. In this arrangement, the ground floor was always used for storage and was the largest space in the house to provide the necessary reserves in case of crop failure or sustained attack. It extends its greatest length in a north-south direction of the plan, and is generally twelve to fifteen feet high. A second story averages eight to nine feet high, and in the most typical three-tiered sections, this second floor is stepped back. It originally contained spaces for sleeping and some storage in addition to a terrace upon the roof of the space below. The third story has

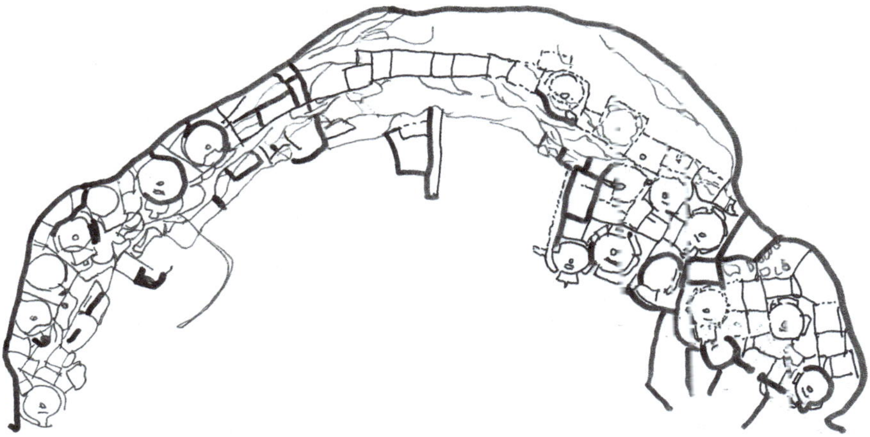

Fig. 6.12 Plan of Longhouse pueblo, Mesa Verde, Colorado, USA, based on Douglas Osborne's reconstruction. *Source* author's re-elaboration from Knowles (1974, p. 24)

Fig. 6.13 Pictorial view of Acoma Pueblo, New Mexico, USA. *Source* Grosso (2008, p. 134)

a slightly smaller vertical dimension of about eight feet. It contained the main living area, including cooking facilities." (Knowles 1974, p. 29). "Generally, the houses contain the larger spaces below and smaller ones above. In order to accommodate this arrangement structurally, the walls are generally sloped from a fifteen-inch thickness at the base to ten inches at top. They are composed of adobe-covered rubble for the first story and adobe-covered brick for the stories above. Floors and roofs are adobe packed on grass of sticks that are laid over small wood members that span timbers." (Knowles 1974, *ibidem*). Regarding bioclimatic features, "Both the regularity of the plan and the southerly orientation of the sections suggest a significant relation

Fig. 6.14 Plan of Acoma Pueblo, New Mexico, USA. Author's re-elaboration from Knowles (1974, p. 28)

between the building arrangement and the sun. Heliodon studies using a model of Acoma reconstructed by architecture students from aerial photographs prove to be true and demonstrate an interaction that is dynamic and complex (…)" (Knowles 1974, p. 30). Hence, as for the Longhouse Pueblo, a detailed quantitative analysis of the solar collected energy efficiency, "taken to be a ratio between the sum of energies transmitted through the verticals and horizontals compared to what might be transmitted if the sun were constantly normal to all surfaces" (Knowles 1974, p. 31), was carried out.

Pueblo Bonito, which now shows only its remains, "is located about one hundred miles south of Mesa Verde at 36° north latitude in the middle northwest section of New Mexico", and "(…) lies about 100 feet south of the sheer rock cliff of Chaco Canyon." (Knowles 1974, p. 35). "The history of the pueblo can be read in its changes of plan as interpreted and recorded by archaeologist N.M. Judd over a period from 1921–27." (Judd 1954; Fathy 1986). The Chaco Canyon area had been occupied long before the pueblo was begun. During this early time the occupants lived in small huts, isolated from one another. For some reason, perhaps for protection, they later decided to group their houses into a simple arrangement with common walls. Thus, the old pueblo was literally a curved version of Acoma. It had three tiers and opened to the southeast." (Knowles 1974, *ibidem*) (Fig. 6.15). "Judd reports that around A.D. 1050–1060, strangers to the old Bonitians, representing an unknown and

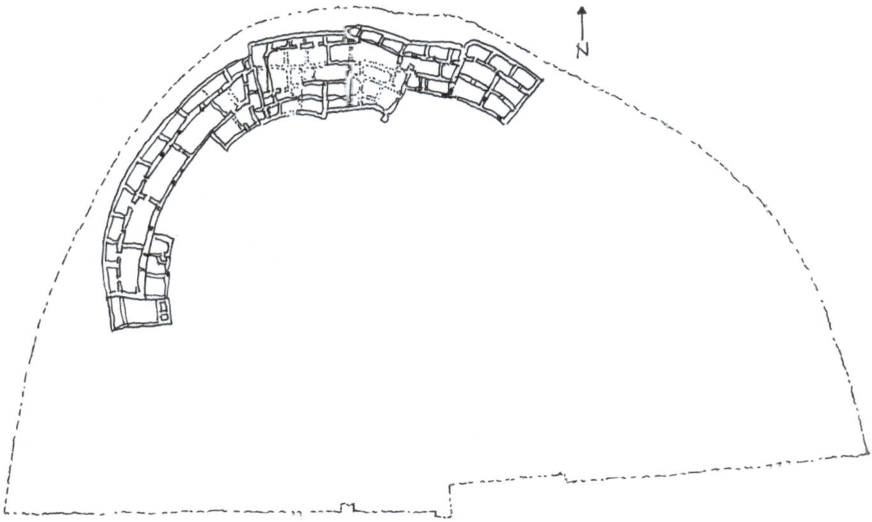

Fig. 6.15 Plan of Old Pueblo Bonito, New Mexico, USA. Author's re-elaboration from Knowles (1974, p. 35)

quite different culture, moved into the region and were apparently welcomed into the Chaco Valley". "(…) these self-confident and remarkably skilled builders moved into the community with no apparent resistance from the old Bonitians and immediately launched a vast new program of expansion." (Knowles 1974, *ibidem*). The result of this "new pueblo Bonito" is shown in the remains, which can be visited today (Fig. 6.16). "The techniques for construction were in advance of those employed by the old Bonitians and show a high degree of sophistication in the use of materials. The walls were no longer monolithic but were differentiated into rubble and adobe for the core of the wall with precisely shaped and fitted sandstone as the exterior facing. Roof timbers displayed fine craftsmanship, with pine and fir beams and poles supporting layers of dressed willows, pine boards, or cedar bark covered with adobe." (Knowles 1974, *ibidem*).

"A calculation of direct incident energy per hour on both the old and new pueblos indicates higher summer profiles for both Old and New Bonito; the old pueblo shows a summer increase of 188% over winter while the new pueblo increases 233% in summer." (Knowles 1974, p. 39).

6.3 Architecture Masters of the Modern Movement

The works of the Masters of Modern Architecture, i.e., Frank Lloyd Wright, Le Corbusier, Avar Aalto, Walter Gropius, and Mies Van Der Rohe, cannot be directly and completely associated to the development of a bioclimatic approach to architecture. However, in some of their architecture and writings, and not with the same

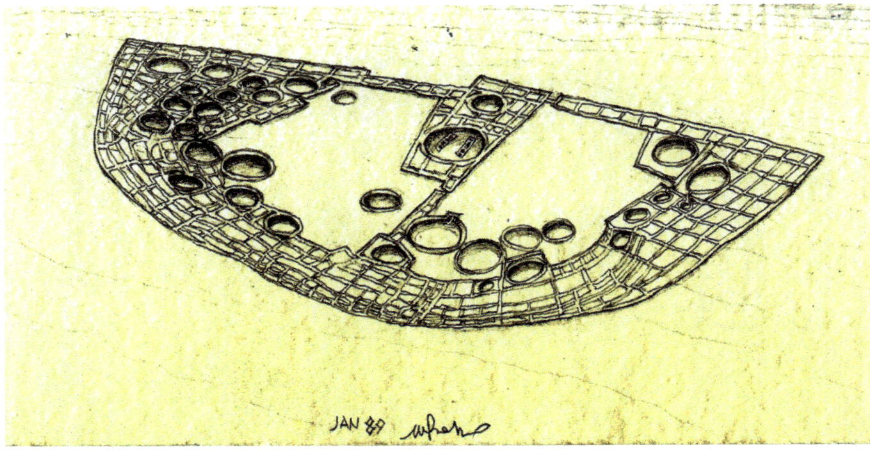

Fig. 6.16 Plan of Pueblo Bonito, New Mexico, USA, as it could be seen in 1989. Author's pencil live drawing, made during a trip from Berkeley, California

emphasis in all of them, hints of the importance of an interaction between natural environment and construction can be seen, as highlighted in the following paragraphs.

Frank Lloyd Wright

"Wright clearly enjoyed extrinsic advantages that were not available to his European contemporaries—the stimulus of great environmental needs, the adaptability of lightweight construction, the resourcefulness of his contractors, the intelligence of his patrons and the remarkable cultural climate in which they lived in Chicago at that time, enriched by the philosophers and critics of the Chicago school around the University of Chicago, but not burdened by too ceremonial and European a concept of culture". In addition, "His resourcefulness in the deployment of power technology and structure, together in the elaboration of domestic environments, seems inexhaustible in the first decade of the century" [XX century] (Banham 1969, p. 112).

The first of the Wright's projects in which some design features could be well associated to a' bioclimatic approach'—although he was unaware of that meaning—is The Frederick Robie House, in Woodlawn Avenue, Chicago, Ill. 1910.

"The first environmental feature of the house strikes the summer visitor even before he is through the door—the coolness of the shaded entrance court. It lies on the north side of the house and receives practically no sun; together with the basement floor, also profoundly shaded by the projection of the terrace above its windows on the sun-ward side, it provides a cool-air tank that works so efficiently, even on still thundery days of high humidity, that with all windows shut even the highest rooms under the roof are not uncomfortably warm." (Banham 1969, p. 116).

A much later project, the second commissioned to Wright by the journalist Herbert Jacobs—the Jacobs Herbert and Katherine second House, located at 3995 Shawn Trail in Madison, Wisconsin, built in 1946–48—was called by Wright himself the

"Solar Hemicycle" due to its semi-circular layout and use of natural materials to conserve solar energy. Contrary to Wright's previous works, it is explicitly 'bioclimatic' in its declared intention of optimising solar radiation access through the building shape and internal layout. The arch-like form, with convexity towards the sun, allows for all rooms having direct solar radiation all day through an entire glazed envelope, while opposing a thick masonry wall without windows northward will reduce winter thermal losses (Fig. 6.17).

Le Corbusier

Le Corbusier's "enquiring mind and his machine-age enthusiasms, his connection with industry, philosophers, medical men and *avant-garde* painters, gave him a range of intellectual contacts and sources of information far beyond that normal in an architect's professional life. The record of those contacts and sources is found throughout every one of the twenty seven issues of the magazine *L'Esprit Nouveau* which he and Amedée Ozenfant edited between 1920 and 1926" (Banham 1969, p. 145). But their records is rarely present in Le Corbusier's design. "For instance, a sequence of issues (nos. 6, 7 and 8) carried an extensive summary coverage of the researches of Charles Blanc into human physiological responses to light, colour and form (…) But whereas one would expect to find them nowadays [Banham's] in the context of studies of lighting levels and environmental colours (…) No parallels with responses

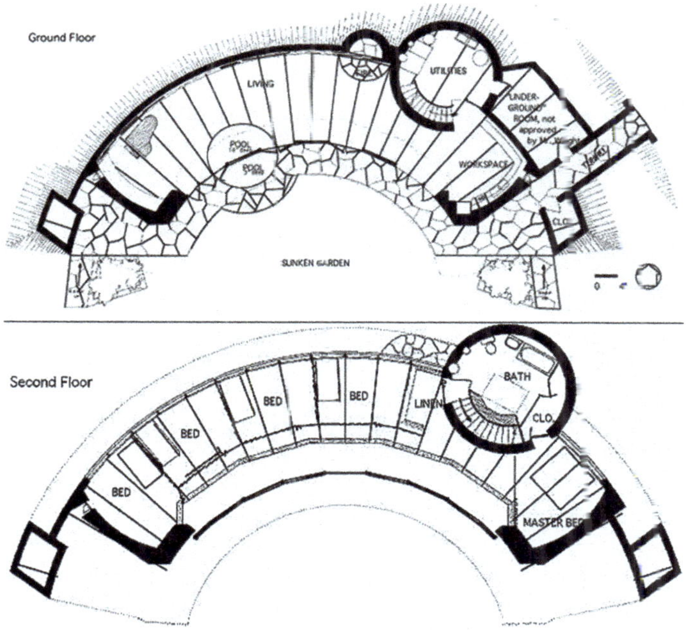

Fig. 6.17 Plan of Wright's 'Solar Hemicycle' House, Madison, Wisconsin, 1946–48: ground floor (top); second floor (bottom). Author's drawing

to coloured and illuminated spaces in buildings seem to be drawn or even implied (…)" (Banham 1969, p. 145).

I personally believe that Banham's statement is true, except for Le Corbusier's *Ronchamp* Chapel, of 1950, whose conception of diffuse natural light contributes to the overwhelming spiritual feeling one can experience when entering it. However, it is true that Le Corbusier's 'bioclimatic' approach to architecture was limited to the light issue, even if he "(…) was neither totally unaware of environmental problems, nor indifferent to the human need to solve those that he recognised. Much of the *Manuel de l'Habitation* refers directly to environmental problems and methods for resolving them" (Banham 1969, p. 147). But, Banham comments, "hardly any of these demands were satisfied in the houses that Le Corbusier designed in the years succeeding this pronouncement of 1921" (Banham 1969, ibid.). To be fair, a much later construction such as *l'Unité d'Habitation* in Marseille (1947), reveals some bioclimatic-related aspects as the building orientation and the presence of *brise-soleil*. However, even this choice proved to be wrong during operation, since the material used—concrete—reflected inward more radiation than the one reflected outward, hence counteracting the positive shading effect the *brise-soleil* were conceived for (Fig. 6.18).

Alvar Aalto

Alvar Aalto, after having embraced the principles of rationalism and functionalism for a short time, wrote in an article of 1940 (Aalto 1940), "the technical functionalism is right only if it is expanded to include the psychophysical field. This is the only manner to make architecture more human". Furthermore, in an essay of 1941, he wrote, "the biological system of nature contains in itself a clear example of what

Fig. 6.18 View of the 'L'unité d'habitation' by Le Corbusier, France, 1947. Picture by the author

should be the principle of a standardisation in architecture respectful of the proper modes of this art." (Mikkola 1976).

It is in his architectures that Aalto expressed his mastery in combining form, functions, indoor comfort, and natural light in an amazing synthesis of aesthetics and space efficiency (Fig. 6.19). In the sanatorium of Paimio (1929–33), a careful attention is made in the exposure of the hospital stay rooms, all facing South and SE as the solarium terrace, facing South, but with venetian blinds in windows to avoid overheating. In the student house (Seniors Dormitory) of the Massachusetts Institute of Technology (1947–48), the massive worm-shaped building allows for variegated changes in the daylight illuminance of all rooms exposed to the daily solar radiation. The Säynätsalo City Hall (1949–52) is a magisterial example of the application of zenital daylight in the corridor to the Council hall as for the convex roof of the Auditorium of the Helsinki Polytechnic at Otaniemi (1949–74). In addition, Aalto's peculiar folding fan-shaped plans of buildings—such as the Kulturi Tälo (House of Culture) in Helsinki (1952–56), the Tower House Neue Vahr in Bremen (1958–62), the Alvar Aalto Theatre in Essen (1959/1983–88), the Tower House Schönbühl in Lucerne (1965–68) (Fig. 6.20)—all represent a perfect synthesis of architectural composition geometry and environmentally sound inspiration, in their 'heliotropic' characteristics of sunward folding spaces.

If these features were never called 'bioclimatic', they were so in their substantial architectural conception and functionality.

Walter Gropius

Walter Gropius was the first architect who studied urban design based on the sun and sky exposure angles, as a solar access criteria for planning the built density

Fig. 6.19 The building of Aalto's Studio, currently Museum, Heksinki. Author's picture

Fig. 6.20 A typical folding fan-shaped Aalto's plan used in several buildings he designed. Author's sketch

and layout of social housing settlements (*Siedlung*) (Gropius 1929, 1956). These criteria were used for the Dessau-Törten *Siedlung*, realised in 1926. This is one of the strongholds of the rationalist town planning. It is composed of two stories buildings made of prefabricated elements and erected on site according to an assembly line technique. As quoted by Giedion (1954), Gropius conceived diagrams illustrating "the development of a rectangular building site with parallel rows of dwellings of different heights, from the one storey row houses to the ten-storey apartment blocks. On the same site, stablike, ten-storey buildings result in much broader green spaces between buildings than the narrow spaces between walk-ups or row houses" (Markus and Morris 1980, p. 8) (Fig. 6.21).

Ludwig Mies van der Rohe

Mies is the farthest from a bioclimatic approach to architecture amid the Masters of Modern Movement. This is due to his focus on the structural and functional aspects of architecture and his conception of the international style constructions in steel and glass, expressed since the competition projects for a skyscraper at *Friedrichstraße*, Berlin, in 1921, and the glass skyscraper (1922), never built, to the Seagram Building at New York, design with Philip Johnson. Nevertheless, his occasional meeting with F.L. Wright, in 1910, might have had an influence on some of his 'smaller'—but not less important—projects such as the Pavilion for the Universal Exhibition of Barcelona, designed in 1929, demolished after the end of the event, and reconstructed between 1983 and 1989, and Villa *Tugendhat* at Brno, CZ, in 1930. In the former, the lightness of the one-storey construction characterizes both the glazed walls and the thin flat roof, and is emphasised by the water surface mirroring the building, with an added—Mies probably unaware of—cooling effect. In the latter, the entirely glazed envelope of the two-storey building is overwhelmed and shaded by horizontal overhanging concrete flat roofs.

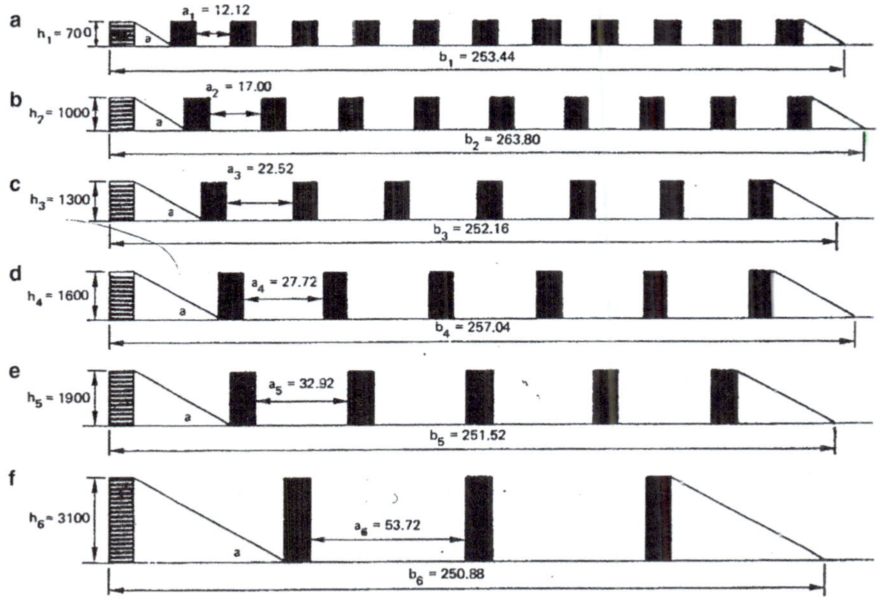

Fig. 6.21 Study by Gropius of buildings' layout according to solar access. Re-elaboration by the author from Marcus and Morris (1980, p 8)

6.4 Bioclimatic Research and Design Handbook Writers of the Twentieth Century

The selection of writers on bioclimatic research and design handbook and relevant publications presented below, was made according to the following criteria:

- English language and international diffusion;
- prominence in the field with original approach and precursor's features;
- personal knowledge and use in my didactical and research experience.

The Italian writers, some of them not less important than their foreign colleagues as contributors to the development of a bioclimatic approach to architectural design—such as Lorenzo Matteoli, Sergio Los, and Federico Butera—are mentioned in my other chapter of this book.

Vladimir Köppen

The first scholar who applied this word to climatology was Köppen. Born in St. Petersburg in 1846, he studied at St. Petersburg University, worked at the Main Physics Observatory from 1872 to 1875, and at the German Naval Observatory in Hamburg from 1875 to 1919. He lived in Graz from 1919 to the end of his life, in 1940. Köppen's principal works are devoted to climatology, general and synoptic meteorology, and marine meteorology. He studied the climatic conditions of many regions of the earth, particularly the oceans, the frequency and movement of cyclones and anticyclones, the effect of solar activity on weather and climate, and climates of the past. Köppen is especially well known for his classification of the world's climates, based on a bioclimatic approach since it takes vegetation and living beings into account (Köppen 1924, 1936). This classification was developed and updated many times until a final version elaborated with Rudolph Geiger (Köppen and Geiger 1961) (Grosso 2017, pp. 36–38). Along with the Russian climatologist A. I. Voeikov and the Austrian meteorologist J. Hann, in addition to the already mentioned R. Geiger, Köppen exerted a determining influence on the development of climatology from the end of the nineteenth century through the middle of the twentieth century.

Rudolph Geiger

Rudolph Geiger, born in Erlangen (Germany) in 1894 and dead in 1981 in Munich, is mainly renowned for his studies on the microclimate physical phenomena within the atmosphere near the ground (Geiger 1965). The results of these studies, based on both theoretical analyses and monitored actions, found applications mainly in the field of agriculture, but represent still nowadays a fundamental physical background for the climate-site analysis in architectural and urban design (Grosso 2017, Cap. II, pp. 53–69).

Viktor Olgyay

The first researcher who associated the word bioclimatic to architecture was Victor Olgyay, in an article of 1953 (Olgyay 1953), first of a series of publications, ended with his main contribution in 1963 (Olgyay 1963). However, his work found inspiration in an earlier publication by James Fitch, who laid the foundation of the environmental design (Fitch 1948).

In the preface of the book "Design with Climate", Olgyay set the concepts and aims of his approach. "With the widening spread of communications and populations, a new principle of architecture is called for, to blend past solutions of the problems of shelter with new technologies and insights into the effects of climate on human environment." (Olgyay 1963, p. V). Olgyay's systemic and multidisciplinary approach is amazingly up-to-date, being possible to be applied to the current context of architecture. He writes: "To meet the problem of climate control in an orderly and systematic way requires a pooling of efforts by several sciences. The first step is to define the measure and aim of requirements for comfort. For this, the answer lies in the field of biology. The next is to review the existing climate conditions, and this depends on science of meteorology. Finally, for the attainment of a rational solution, the engineering sciences must be drawn upon. With such help, the results may then be synthesised and adapted to architectural expression (…) But it is my hope that, in bridging these different fields, this study will combine both creative and analytical approaches to develop a unifying architectural concept." (Olgyay 1963, p. V). I find this wording exceptionally clear in defining bioclimatic architecture, and it is still valid today.

From a methodological point of view, the main Olgyay's contributions were: the bioclimatic chart (Fig. 6.22)—a diagram useful to define passive climate control

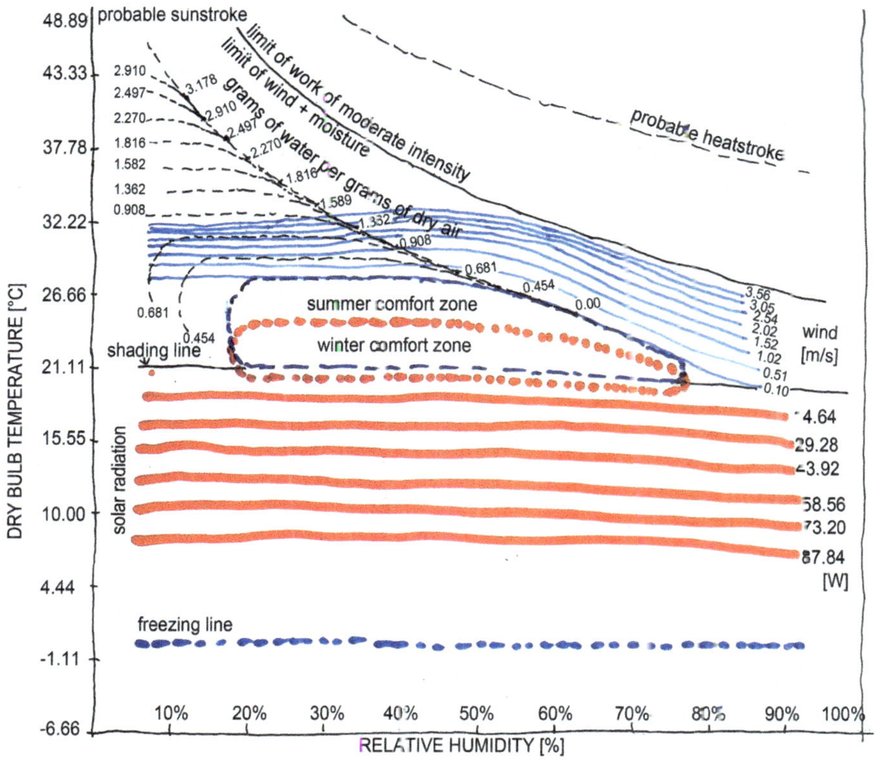

Fig. 6.22 Olgyay's bioclimatic chart. Author's re-elaboration, and conversion to SI metric, from Olgyay (1963)

strategies in relation to building location; the shading masks—to calculate the shading performance of various shading devices by applying a stereographic projection (Fig. 6.23); the interrelationship between building form and climate types (Fig. 6.24); simulation and visualisation of airflow around and inside a building.

Reyner Banham

Another founding father of the environmental approach to architecture, which somehow could include the bioclimatic architecture, is Reyner Banham, who wrote his concepts in a fundamental book published in 1969. His original contribution was mainly in the definition of environmental management of buildings according to three modalities: conservative, selective, and regenerative. The first mode characterises a house in a hot climate with "thick walls" holding "solar heat during the day, slowing down the rate at which the interior becomes hot, and then, after sunset, the radiation of that heat into the house will help to temper the sudden chill of evening. In more sophisticated forms, that use glass as a filter to discriminate between light-energy, which is allowed to pass, and heat energy, whose passage is barred, similar effects of

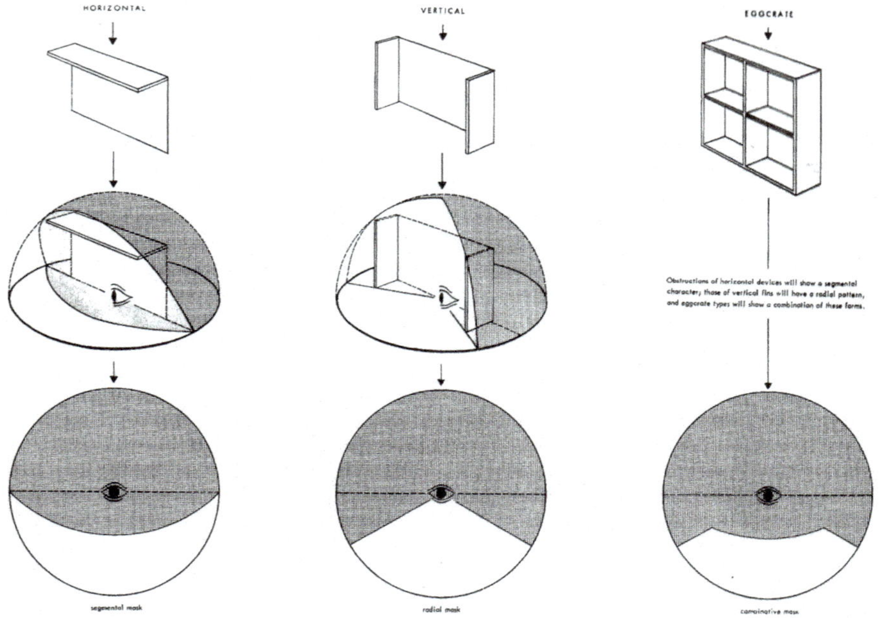

Fig. 6.23 Olgyay's shading diagrams. *Source* Grosso (2008, p. 317)

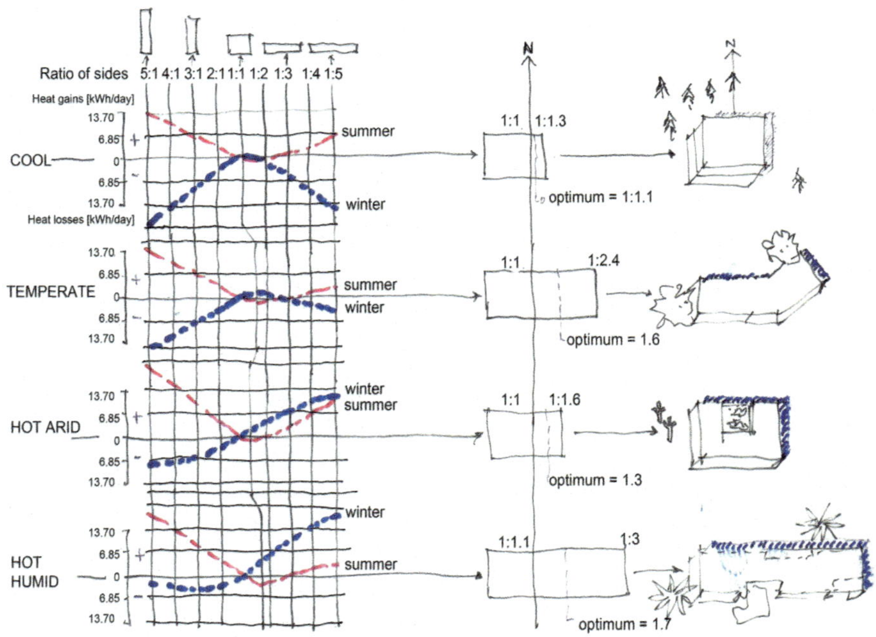

Fig. 6.24 Olgyay's diagrams relating energy needs and building shape for different climates. Author's re-elaboration, and conversion to SI metric, from Olgyay (1963)

thermal storage are used in the normal green-house, and the whole technique might well be termed the 'Conservative' mode of environmental management, in honour of the 'Conservative Wall' at Chatsworth, devised by the master-environmentalist Sir Joseph Paxton, in 1846." (Banham 1969, p. 22).

The 'Selective' mode "employs structure not just to retain desirable environmental conditions, but to admit desirable conditions from outside. Thus, a glazed window admits light but not rain, an overhanging roof admits reflected sunlight, but excludes the direct sun, a louvered grille admits ventilating air but excludes visual intrusions. The 'Regenerative' mode is the one using "applied power, without fully acknowledging its presence." (Banham 1969, p. 22).

According to Banham, these three modes "should not be too sharply distinguished in traditional practice", but "there is an important geographical or climate consideration that distinguishes solutions that are more conservative from those that are more selective, and an historical watershed...separates both of these from solutions that are primarily regenerative" (Banham 1969, pp. 22–23).

In the characterisation of these environmental control modes, humidity plays a critical role, being "the most pestiferous, subtle and elusive of control. While the deficient humidity of an overdried climate can be crudely made good by splashing water about and using shade to reduce evaporative loss, the removal of excess water from the atmosphere has so effectively defied all pre-technological efforts, that it has usually made better sense for those who could afford it to move elsewhere—the British in India retiring to hill-stations like Simla, New York business men with lung complaints to Colorado." Then, Banham concludes, "For excess moisture, only a regenerative solution, consuming power, has so far proven effective." (Banham 1969, p. 24).

Thus, it must be noted that this is true even nowadays, after 50 years from Banham's statement, although breakthroughs in the field of air conditioning have been made, mainly regarding energy efficiency. In addition, some recent experimental work has been carried out on the use of passive systems to dehumidifying air through adsorbing material in a cycle, where air is afterword cooled by low pressure heat recovery with direct and indirect evaporative cooling (Grosso et al. 2015).

Ralph L. Knowles

Knowles's first systematic contribution to an environmental approach to architecture and ecological planning (Knowles 1974) represents the result of a decade of studio work (1962–1972) undertaken with architectural students, first at Auburn University, Auburn, Alabama, and later at the University of Southern California. "The original intent of the study was not to involve natural forces but rather to improve the quality of the urban environment by designing buildings that offered clues to orientation. The impetus came from Kevin Lynch's book, The Image of the City, published two years before, in 1960". (Knowles 1974, preface).

Even before the full explosion of the oil crisis in 1973, Knowles showed to be a precursor of what would have been later conceived as a sustainable design approach to urban development. His book "advocates three new purposes for design: first, an economic purpose for urban growth that stresses the long-term costs of maintaining equilibrium in the built environment over the short-term costs of development; second, a large scale view of a community as a set of associations in which the diversity of community needs is met not by supplying ever increased mobility but by building closer contact diversity into the arrangement; and third, an aesthetic based on form as a natural adaptation for survival." (Knowles 1974, p. 1).

Knowles laid the foundation for a radical criticism of the urban sprawl characterising the USA way of living as an energy wasteful mode of growing. "The underline premise of this book is that urban growth, consciously directed and based on principles that correlate building form with natural variation for purposes of energy conservation, will exhibit a diversity of built forms that, when transferred into functional terms, will usually produce the close-contact diversity essential for a rich and humane community life without overreliance on mobility." (Knowles 1974, *ibidem*).

After the above-described analyses of the Pueblos and a further study on the morphological development of the settlements in the Owens Valley, Knowles elaborates, in the third part of the book, a method to define rules for a building growth aimed at optimising the collection of solar energy. Such rules are based on "a kind of 'clock-like' arrangement of planes placed on a heliodon. The planes can be moved to study different building configurations and so long as they stay within the volumetric limits, their shadows will remain within the site. Their interactions can be studied as the sun moves and cast shadows (…) during the critical periods of energy reception from the sun (which would vary with latitude.)" (Knowles 1974, p. 118).

This study will bring eventually Knowles to the development of the concept of the "solar envelope" (Fig. 6.25), which describes the volume within which a site may be developed without infringing upon the solar access of the adjacent parcels (Knowles 1981). From a designer's viewpoint, the development of the process represents an excellent example of visual thinking and graphic problem solving. More important is how the process recognizes and clearly articulates the spatial and temporal nature of designing for solar access.

Baruch Givoni

Baruch Givoni earned a B.Sc (Architect) at the Faculty of Architecture, Technion (Israel Institute of Technology), Haifa in 1953; a M.Sc (Hygiene) in the Graduate School of Public Health, University of Pittsburgh in 1959; and a Ph.D. (Public Health) in the School of Medicine, University of Jerusalem in 1963. He served as head of the Department of Building Climatology, Building Research Station at Technion, from 1970 to 1977, and taught Architecture and Urban Design at Technion (1973–1977), and UCLA (1977–1997). In addition, he held a chair at the Ben Gurion University, Beer Sheva, Israel (1977–1984). Dr. Givoni has been a visiting professor at a number of Universities in North and South America, Australia and Europe. His main contribution has been related to the influence of climate on architecture in terms of thermal comfort as realised through passive strategies (Givoni 1969). He drew a psychrometric chart, which can be somehow associated to Olgyay's bioclimatic chart, but with reference to the potential extension of the comfort zone by means of

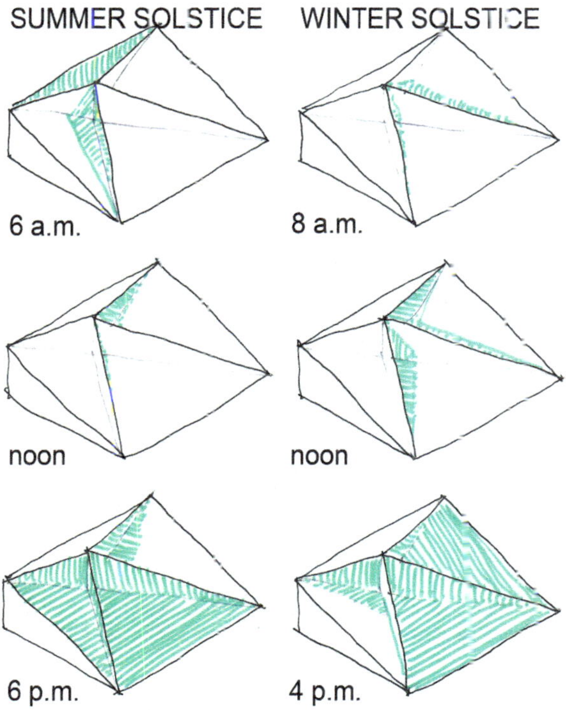

Fig. 6.25 Knowles's graphic representations of the 'solar envelope' concept. Author's re-elaboration from Knowles (1981)

application of various passive strategies (Milne and Givoni 1979) (Fig. 6.26). His further research focussed on passive cooling of buildings (Givoni 1994).

Edward Mazria

Mazria began a teaching and research career at the University of New Mexico in 1973. His architecture and renewable energy research at both UNM and the University of Oregon established his leadership in the field of resource conservation and passive heating, cooling and daylighting design. His design methodology, developed at that time and presented in the Passive Solar Energy Book, is currently in use worldwide (Mazria 1979).

Donald Watson

Based upon the preceding work of Olgyay and Givoni, Donald Watson wrote a climate design handbook particularly devoted to architects, being full of hand drawings representing schemes of solar radiation fluxes and airflow as they affect buildings of various types and architectural configuration (Watson and Labs 1983). In addition, this handbook includes diagrams and tables, representing data and graphs related to the application of Givoni's psychometric chart to various climate control strategies

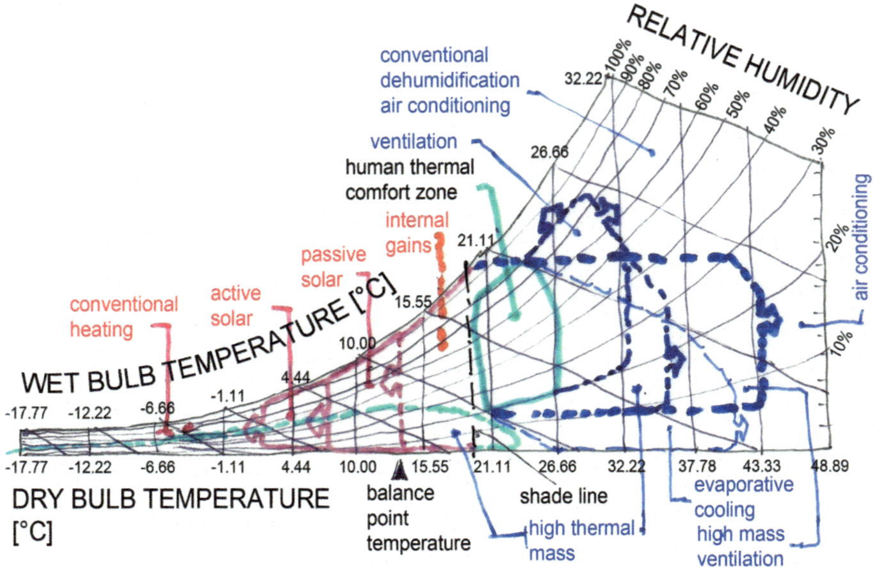

Fig. 6.26 Givoni's psychrometric diagram relating comfort and indoor climate control strategies. Author's re-elaboration with conversion to SI metric from Milne and Givoni (1979)

for the USA climate zones. This handbook was updated including more information and data on the requirements and performance of energy system, both passive and active, as well as a method for cost/benefit analysis (Watson 1993).

G.Z. Brown and Mark DeKay

Another successful handbook on bioclimatic design for architects was written by G.Z. Brown, of the Department of Architecture, University of Oregon, and Mark DeKay of the School of Architecture, Washington University in St. Louis (Brown and DeKay 1985). The purpose of this book was "to help architectural designers who are not energy experts understand the energy consequences of their most basic design decisions and to give them information so that they can use energy issues to generate form rather than simply as limits that must be accommodated." (Brown and DeKay 1985, preface). Many of the concepts developed in this book are based on a preceding publication by the same first author with others (Brown et al. 1982). Another important premise of this handbook is "that most decisions that affect a building's energy use occur during the schematic design stage of the project. Furthermore, the effort required to implement those decisions at the beginning of the design process is small compared to the effort that would be necessary later on." In my opinion, the most original contribution of this book is the method to analyse microclimates on a building site, which I have re-elaborated myself in what I would have called lately "microclimate matrix" method (Grosso 2008; 2017a, b, 2019; Nigra et al. 2016).

Terry S. Boutet

Terry S. Boutet, AIA, received both his bachelor of design and his master of architecture degrees from the University of Florida at Gainesville. While a graduate student there, he conducted research on natural ventilation in buildings, based on experimental work with a smoke airflow chamber. This research yielded results which ended up with the publication of a manual for architects and builders (Boutet 1987). Although dealing with a specific topic of the bioclimatic design—natural airflow control—I found this handbook particularly interesting and useful as a didactical tool, whereby I could developed a simplified method for determining the effect of obstacles to the wind airflow around buildings in an urban setting (Grosso 2014–19). The outcome of this method allowed for drawing wind wake cores on the leeward side of buildings surrounded by other buildings assimilated to rectangular shaped solid obstacles, hence providing one of the input for determining the above-mentioned 'microclimate matrix'.

Jeffrey Cook

Jeffrey Ross Cook was born Lunenburg, Nova Scotia 26 June 1934 and died in Paradise Valley, Arizona, 24 March 2003. Widely acknowledged as one of the pioneers of solar and bio-climatic design, he was an architect registered in two US states, a member of the American Institute of Architects, and an Honorary Fellow of the RIBA in London. From 1988, he was Regent's Professor of Architecture at Arizona State University, Tempe, where he had taught since 1962. He ran a Masters course in Solar Energy Design at ASU, which, largely due to his international reputation and dedication as a teacher, attracted students from many of the hot countries of the world. He was closely associated with the founding, and the running of Plea (Passive Low Energy Architecture) and its annual international conferences. Living just down the road from Frank Lloyd Wrigh's Taliesin West and virtually next door to Paolo Soleri's Cosanti Foundation, Cook was well acquainted with the demands of an architecture that responded to nature and to the wider environment. His own house, begun in 1968, was one of the first examples of solar passive energy in Arizona and its subsequent wide publicity catapulted him into prominence as the solar energy and design guru in the state.

In particular, Jeffrey Cook is mentioned here as the editor of the first comprehensive book on Passive Cooling (Cook 1989). In this book, he states that "'passive' was a term invented in the United States, where the subject evolved primarily as a product of a peculiar and uneven research and development programs of U.S./DOE and its predecessors. Passive cooling was the neglected stepsister of passive solar heating. The term passive cooling, however, embraces an enormous body of knowledge, derived both historically from the climate sensitive practices of indigenous culture worldwide and from the professional bioclimatic design disciplines developed internationally starting in the 1920s." (Cook 1989, p. 37). Cook's book is "organized by focusing chapters on each of the passive cooling resources and on their interactive integration as represented by whole building systems" (Cook 1989, p. 4). These resources are the "three heat sinks or nature", i.e., "the sky, the atmosphere, and the earth", as the sources of the four passive cooling techniques, dealt with in the book by different authors: ventilative cooling (Subrato Chandra); evaporative cooling (John I. Yellott); radiative cooling (Marlo Martin), and earth coupling (Kenneth Labs). Their integration in the building systems is described by Gene Clark. Cook himself is writing a chapter on the state of passive cooling research, in addition to the introduction. The basic concepts and the background of Cook's book are, actually, the same that inspired and characterised the publishing of my handbook on passive cooling, the first in Italy of this kind, although I wrote its first edition before reading Cook's Book (Grosso 1997).

As additional applications of passive cooling concepts, quoted by Cook, and not by the other authors analysed here, the following are worth to be mentioned.

The "'sky therm' roof, is an exemplary passive cooling and heating system, as a 'natural energy' system', developed by Harold Hay and John Yellott and monitored in Phoenix, 1966–1967" (Cook 1989, p. 3, Fig. 1.1).

The 'Noorwali House', Jeddah, Saudi Arabia, built in 'Hejazi' style, "represents a well developed architectural response to a hot and humid climate with winds off the sea and little climatic variation throughout the year" (Cook 1989, Fig. 1.4, p. 12). Its walls are built of native coral limestone, hence, allowing for a high thermal inertia. "The elaborate, cantilevered window screens or rowshin … provide a screen against harsh sunlight and a generous latticework to bring incoming air into the vertical stack of the ventilation core of the house." (Cook 1989, *ibidem*).

A "large-scale housing at Toa Payoh, built in the 1970s in the tropical climate of Singapore, demonstrates recent architecture based on bioclimatic strategies such as orientation, shading, and cross-ventilation" (Cook 1989, Fig. 1.5, p. 14).

Norbert Lechner

Lechner wrote the probably most comprehensive design handbook on environmental control systems for architects amid all previous Authors. With a foreword to the first edition (Lechner 1991) by the late James Marston Fitch and to the third edition (Lechner 2009) by Edward Mazria, his goal is "to provide the appropriate knowledge at the level of complexity needed at the schematic design stage."

The breakthrough of the third edition is mainly given by a move "from a shortage of information to a flood because of the Internet". The handbook "will aid the designer because it presents the information in a concise, logical, accessible, and useful arrangement" and "an additional goal of s book is to provide architects with the skills and knowledge needed to create low-energy and low-carbon-emission buildings (Lechner 2009, preface). This handbook is full of drawings, schemes, tables of data, pictures of architecture and case studies as well as appendixes with horizontal and vertical sun-path diagrams, solar altitude and azimuth angles, methods for estimating the height of trees and buildings, sundials, sun-path models, computer software useful for the schematic design stage, site evaluation tools, heliodons, and educational opportunities in energy-conscious design.

John R. Goulding, J. Owen Lewis, and Theo Steemers

Goulding, Lewis, and Steemers were the editors of the first comprehensive handbook on energy conscious design derived from a research funded by the European Commission, Directorate-General XII, within the SOLINFO Action of the Solar Energy Applications to Building Programme, coordinated by the Energy Research Group, University College Dublin. Original text and diagrams were prepared by the team *Architecture et Climat, Centre de Recherches en Architecture, Université Catholique de Louvain*, Belgium, under the direction of Professor André De Herde (Goulding et al. 1992). This handbook is divided in three parts: (1) the climate; (2) the building; (3) the occupants. The first part includes macro- meso- and micro-climate. The second part deals with heating and cooling strategies, and daylighting. The third part on thermal and visual comfort, and behaviour. It includes graphs, diagrams, and pictures without any formula, being mainly targeted to architects and to a very early conceptual design phase.

Fuller Moore

Moore's handbook (1993) had the goal "to introduce the concepts of controlling the thermal and luminous environment in buildings (…) from an architect's viewpoint with the comfort of the occupant as the central design determinant." The author's thought "that basic physical concepts must be thoroughly understood and internalized in order to intelligently apply environmental control issues" and, to this end, "he relied extensively on conceptual diagrams and analogies to help the reader develop an intuitive understanding of how heating, cooling, and lighting affect people and the design of buildings." (Moore 1993, preface).

However, in my opinion and use as a didactical tool, the most original part of this handbook is the Chap. 3 'Climate and shelter', where he described the climate-related characteristics of various 'traditional' constructions. These are classified by climatic regions and represented by the following paradigmatic types: the Eskimo *igloo* in cold regions; the North American Indian *tepee* (Grosso 2017, pp. 110–111) and the Mongolian *yurt* as examples of nomadic structures in temperate climates; the *Macrotermes* termite mound (Grosso 2017, p. 109) and the Anazasi's *pueblos* (Grosso 2017, pp. 112–114) in hot-arid regions; the Seminole house and the *Ma'dan* house (Iraq) in warm-humid regions.

Dean Hawkes

Dean Hawkes is an architect who has researched and taught at the Department of Architecture of the University of Cambridge from 1965 through 1995, where he was a founder member and, later, Director of the Martin Centre for Architectural and Urban Studies. He held the Chair of Architectural Design at the Welsh School of Architecture, Cardiff University from 1995 through 2002. He has held visiting professorships at the Chinese University of Hong Kong, the National University of Singapore, the Mackintosh School of Architecture, Glasgow School of Art, the Department of Architecture, Huddersfield University and the Leicester School of Architecture, De Montfort University. He is emeritus professor at Cardiff University and an emeritus fellow of Darwin College, University of Cambridge. As a researcher, his main contribution to the theory of environmental and bioclimatic design in architecture (Hawkes 1996) is, in my opinion, the definition of two modes of environmental control in a building—exclusive and selective—in some way recalling Banham's regenerative and selective modes. Mentioning Hardy and O'Sullivan (1967), for whom "in building designed for PSALI (Permanent Supplementary Artificial Lighting of Interiors) or PAL (Permanent Artificial Lighting), the influence of the external thermal environment can be reduced to such an extent that it becomes a minor factor" (Hawkes 1996, p. 38), Hawkes labels buildings designed in this manner as adopting the 'exclusive mode'. This means, "they use the building envelope to exclude the effects of the external environment upon the internal conditions" (Hawkes 1996, *ibidem*) (Fig. 6.27). Instead, he defines 'selective mode' "the approach to environmental control which has predominated historically in temperate climates" and "has depended upon the selective admission of substantial elements of the external environment into the building." (Hawkes 1996, *ibidem*) (Fig. 6.28). Hawkes then examines the extent to which the selective approach can provide a basis for energy-conserving design.

Matheos Santamouris

Currently 'Scientia Professor' at the University of South Wales, Sidney, Australia, I met him when he was Professor at the Department of Applied Physics, University of Athens, and coordinator of the Project PASCOOL, co-funded by the European Commission, Directorate General XII, Joule II Non-Nuclear Energy Programme, Sub-programme: Energy Conservation and Utilisation (1992–1995). He asked me to collaborate to this programme, when it was already under way, because He was interested in the *CpCalc* model I had developed at LBL during the COMIS workshop (see my other contribution to this book), as a tool to be updated and integrated in the passive cooling software under development within the PASCOOL project.

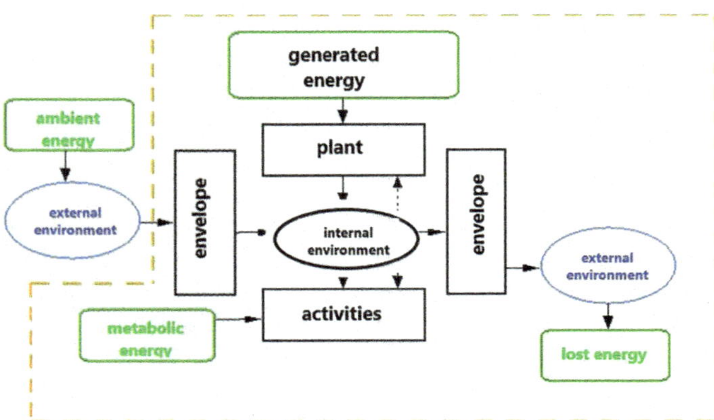

Fig. 6.27 An interrelations diagram representing Hawkes concept of 'exclusive mode' in the thermal control of a building. Author re-elaboration from Hawkes (1996)

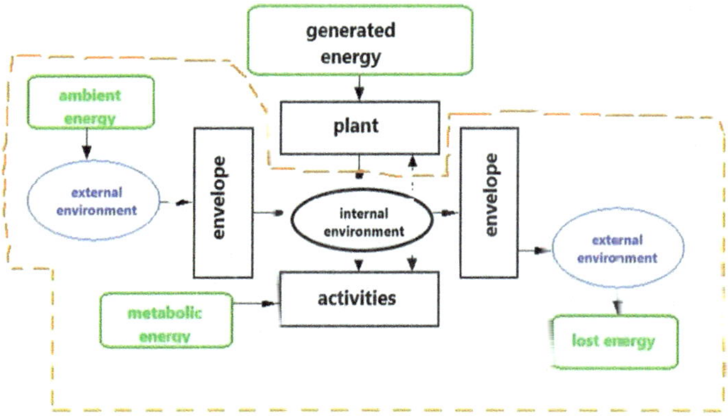

Fig. 6.28 An interrelations diagram representing Hawkes concept of 'selective mode' in the thermal control of a building. Author re-elaboration from Hawkes (1996)

Afterwards, He edited with D. Asimakopoulos the first comprehensive publication on passive cooling of buildings (Santamouris and Asimakopoulos 1996). This book was prepared by the Central Institution for Energy Efficiency Education (CIENE), within the framework of the SAVE Programme 'Short Educational Structure on Energy Efficiency in Buildings' of the European Commission, Directorate General XVII for Energy.

Based on a robust theoretical physical background, this book deals with all aspects affecting cooling load in buildings—from climate and microclimate to urban design and building characteristics—and described analytically all strategies and methods of passive cooling: heat attenuation, natural ventilation, solar control, ground cooling, evaporative cooling, and radiative cooling. In the natural ventilation section, the parametrical model *CpCalc* calculating the pressure coefficients on a building envelope, I had developed within the COMIS workshop, is also reported (Grosso 1992). In addition, the book includes various Appendixes with simplified methods for passive cooling applications.

Claude-Alain Roulet

Claude-Alain Roulet is retired from the Solar Energy and Building Physics Laboratory, École Polytechnique Fédérale de Lausanne. He did research in Experimental Physics. His last project (ended 2017) was 'BEEP Building Energy Efficiency—Indo-Swiss program'.

I met Claude-Alain during my experience in the EU Project PASCOOL, coordinated by Mat Santamouris. He worked there as an expert on natural ventilation and wrote as well about the COMIS model (Roulet et al. 1996). He was, other than a competent expert, a lovely friend, making the PASCOOL experience a nice one. His contribution ended up also in a conference paper (Florentzou et al. 1996). Although he has not published any handbook, he is the author of more than two hundreds articles and conference papers mainly on natural ventilation, thermal mass, and energy efficiency in buildings.

Francis Allard

Researcher and professor at the University of La Rochelle, France, I met him during the COMIS workshop at LBL, CA, USA. Afterwards, he edited the first comprehensive design handbook on natural ventilation, based on the results of the AIOLOS Project, co-funded by the European Commission, Directorate General XVII for Energy, Altener Programme, Coordinated by M. Santamouris (Allard 1998). This handbook includes fundamentals of natural ventilation, prediction methods, diagnostic techniques, critical barriers, design guidelines and technical solutions for natural ventilation (a chapter written by myself), and reference examples of natural ventilated buildings. In addition, the handbook has attached a CD of the network-based calculation programme AIOLOS, which includes the software *CpCalc*⁺ (Grosso et al. 1994), developed by myself within the above-mentioned PASCOOL Project, based on the previously elaborated parametrical Cp calculation model (Grosso 1990; 1993).

References

Aalto A (1940) Arkkitehtuurin lähentäminen ihmiseen (Architecture gets closer to man). In: The technological review, pp 50–51

Alberti LB (ca. 1450) De Re Aedificatoria. Ten books of architecture, translated into Italian by C. Bartoli and from Italian to English by J. Leoni, facsimile of 1755 edition. In: Rykwert J (ed). Tiranti, London, 1955, Book I, Chap. 2

Allard F (1998) Natural ventilation in buildings—a design handbook. James & James, London, UK

Aristotle (ca. 330–322 B.C.) Λα Πολιτιχα. Politics, translated by T.A. Sinclair. Penguin Books, London, 1962, Book VII, Chap. 11

Bahadori MN (1978) Passive cooling systems in Iranian architecture. Sci Am 238(2):144–154

Banham R (1969) The architecture of the well-tempered environment. In: The architectural press. The University of Chicago Press, London

Bavar C (1983) Urban form as physical expression of the social structure in the Arid Zones of Iran. In Golany G (ed) Design for arid regions. Van Nostrand Reinhold Company, New York, NY, USA, pp 55–96

Boutet TS (1987) Controlling air movement—a manual for architects and builders. McGraw-Hill Book Company, New York, USA

Brown GZ, DeKay M (1985) Sun, wind & light—architectural design strategies, 1st edn. Wiley & Sons, Inc., New York (2nd edn, 2001)

Brown GZ, Reynolds J, Ubbelohde MS (1982) INSIDE-OUT: design procedures for passive environmental technologies. Wiley & Sons Inc, New York

Butti K, Perlin J (1980) A golden thread. Van Nostrand Reinhold Company, New York, NY, USA

Chiesa G, Grosso M (2017a) Accessibilità e qualità ambientale del paesaggio urbano. La matrice microclimatica di sito come strumento di progetto, Ri-Vista XII I(1):78–91

Chiesa G, Grosso M (2017b) Environmental and technological design: a didactical experience towards a sustainable design approach. In: Proceedings of the XV international forum le Vie dei Mercanti: world heritage and disaster. Knowledge, culture and representation. La scuola di Pitagora, Capri 15–17 Luglio, p 10

Chiesa G, Grosso M (2019) A parametric tool for assessing optimal location of buildings according to environmental criteria. In: Sayigh A (ed) Sustainable building for a cleaner environment. Springer International Publishing, p 15

Collins Dictionary (2020) HarperCollins Publishers, UK. Available at: https://www.collinsdictionary.com/, last view May 2020
Cook J (ed) (1989) Passive cooling. MIT Press, Cambridge
Fardeheb F (1987) Examination and classification of passive solar cooling strategies. In: Middle Eastern vernacular architecture, Los Angeles, California, USA
Fathy H (1986) Natural energy and vernacular architecture. University of Chicago Press, Chicago, Illinois, USA
Fitch JM (1948) American building: the forces that shape it. Houghton-Miffling Company, Boston, Massachussets, USA
Flourentzou F, Msas J, Roulet C-A (1996) Experiments in natural ventilation for passive cooling. In: Proceedings of 17th AIVC conference, Gothenburg, Sweden, vol 1, pp 121–134
Geiger R (1965) The climate near the ground. Harvard University Press, Cambridge, Massachusetts, USA
Giedion S (1954) Walter gropius: work and teamwork. The Architectural Press, London, UK
Givoni B (1969) Man, climate, and architecture, 1st edn, Elsevier, Amsterdam New York. 2nd edn (1976), Applied Science Publishers, Ltd. London. 3rd edn (1981), Van Nostrand Reinhold, New York
Givoni B (1994) Passive and low energy cooling of buildings. Van Nostrand Reinhold, New York
Goulding JR, Lewis JO, Steemers TC (1992) Energy conscious design—a primer for architects. Batsford for the Commission of the European Communities, London UK
Gropius W (1929) Die Wohnformen, Flach-Mittel-order Hochnau. In Wagner M, Behne A (eds) Neues Berlin
Gropius W (1956) Scope of total architecture. Allen and Unwin, London, UK
Grosso M (1990) Wind pressure distribution around buildings. In Feustel H, Grosso M et al (eds) Fundamentals of the multi-zone air flow model—COMIS, IEA—air Infiltration and ventilation centre, Technical Note AIVC 29 p. 114, Coventry, UK
Grosso M (1992) Wind pressure around buildings: a parametrical model. Energy Build 18:101–131
Grosso M (1993) Wind pressure distribution around buildings. In Feuste H, Grosso M et al (eds) User guide of the multi-zone air flow model—COMIS, IEA—Air Infiltration and Ventilation Centre, Technical Note, p 187, Coventry, UK
Grosso (1997) Il raffrescamento passivo degli edifici—in zone a clima temperato (The passive cooling of buildings—in temperate climate zones). First edition. II edition 2008, III edition 2011, IV edition 2017. Maggioli, Santarcangelo di Romagna (RN), Italy
Grosso M (2008) "Progettazione Bioclimatica" (Bioclimatic design). In: Castelli L (ed) Architettura sostenibile (Sustainable architecture). Turin, Wolters Kluver Italy, Milan, UTET, pp 5–40
Grosso (2014–2019) A method to draw wind wake cores (calm zones). Didactical material for exercises carried out in the courses of Environmental Technological Design for the Bachelor Degree in Architecture and Master Degree in Architecture for a Sustainable Design at the School of Architecture, Polytechnic University of Turin
Grosso (2017) Il raffrescamento passivo degli edifici—in zone a clima temperato (The passive cooling of buildings—in temperate climate zones). Cap. IV, Archetipi Bioclimatici (Bioclimatic archetypes), pp 107–153. Maggioli, Santarcangelo di Romagna (RN), Italy
Grosso M (2019) Appendix to chapter 17: the Seven Root Sheets. In: Sayigh A (ed) Sustainable vernacular architecture: how the past can enrich the future. Springer, Cham, pp 398–405
Grosso M, Ahmadi M (2016) Potential cooling energy reduction by a one-channel wind tower: case study modelling in south-Mediterranean Climate. Int J Vent 15(3–4):267–287
Grosso M, Fracastoro G, Simonetti M, Chiesa G (2015) A hybrid passive cooling wall system: concept and laboratory testing results. Energy Procedia 78:79–84. ISSN: 1876-6102. https://doi.org/10.1016/j.egypro.2015.11.118
Grosso M, Marino D, Parisi E (1994) Wind pressure distribution on flat and tilted roofs: a parametrical model. In: Proceedings of the European conference on energy performance and indoor climate in buildings, Vaulx en Velin, France, 24–26 Novembre, pp 167–172, Ecole Nationale des Travaux Public de l'Etat, Lyon

Hardy A, O'Sullivan P (1967) Insolation and fenestration. Oriel Press, Newcastle, UK
Hawkes D (1996) The environmental tradition—studies in the architecture of environment. E & FN SPON, London, UK
Judd NM (1954) The material culture of Pueblo Bonito. Smithsonian Institution, Washington, D.C., USA
Justinianum, Tribonianum (529 A.D.) Codex Iustinianeus primus (lost)
Justinianum, Tribonianum (534 A.D.) Codex Iustinianeus repetitae praelectionis. First printed edition: Peter Schöffer Editio Princeps, Magonza, 1475
Knowles RL (1974) Energy and form—an ecological approach to urban growth. The MIT Press, Cambridge, Massachusetts and London, England
Knowles RL (1981) Sun rithm form. The MIT Press, Cambridge, Massachusetts, USA
Köppen W (1924) Die Klimate der geologischen Vorzeit. Berlin, Germany
Köppen W (1936) Das geographische System der Klimate. Berlin, Germany
Köppen W, Geiger R (1961) Handbuk der Klimatologie. Bd 1, Teil C, Berlin, Germany
Lechner N (1991) Heating, cooling, lighting: design methods for architects. Wiley and Sons Inc, New York, USA
Lechner N (2009) Heating, cooling, lighting: sustainable design methods for architects. Wiley and Sons Inc, New York, USA
Markus TA, Morris EN (1980) Buildings, climate and energy. Pitman Publishing Ltd, London, UK
Mazria E (1979) The passive solar energy book. Rodale Press, Emmaus, Pennsylvania, USA
Mikkola KFI (1976) Special Issue devoted to Alvar Aalto's memory of the Journal Arkkitehti, n. 7/8
Milne M, Givoni B (1979) Architectural design based on climate. In: Watson D (ed) Energy conservation through building design, architectural record books. McGraw-Hill Book Co., New York
Moore F (1993) Environmental control systems—heating cooling lighting. McGraw-Hill Inc., New McGraw-Hill Inc., New York, USA
Nigra M, Grosso M, Chiesa G (2016) Integrated approach to sustainable building design programming, vol Special Issue. NEWDIST, DIST, Polytechnic and University of Turin, p 7
Olgyay V (1953) Bioclimatic approach to architecture. In: BRAB conference report n. 5. National Research Council, Washington, D.C., USA, pp 13–23
Olgyay V (1963) Design with climate, bioclimatic approach to architectural regionalism. Princeton University Press, Princeton, New Jersey, USA
Palladio A (1570) I quattro libri dell'architettura. The four books of Architecture, translated by I. Ware, facsimile of 1738 edition. Dover Publications, New York, 1965, Book II, Chap. 12
Roulet C-A, Fürbringer J-M, Borchiellini R (1996). Evaluation of the multizone air flow simulation code COMIS
Rudofsky (1977) The prodigious builders. Secker & Warburg, London, UK
Santamouris M, Asimakopoulos D (eds) (1996) Passive cooling of buildings. James & James (Science Publishers) Ltd, London, UK
Vitruvius Pollione M (ca.15 B.C.) De Architectura. Of architecture, The Ten Books, translated by F. Granger. Heinemann, London and Cambridge, Mass., 1970. Book I, Chaps. 4 and 6, Book IV, Chap. 1
Watson D (ed) (1993) The energy design handbook. The American Institute of Archi-tects Press, Washington, D.C., USA
Watson D, Labs K (1983) Climatic design. McGraw-Hill Book Co., New York, USA
Xenophon (post 371 B.C.) Memorabilia, translated by E.C. Marchant, Loeb Classics, Heinemann, London, 1923, Book III, Chap. 8, Section 9

Chapter 7
The Urban-Rural Continuum. The Bioclimatic Approach to Design, Between Past and Future

Maria Luisa Germanà

Abstract The rural built environment, as a multiscale whole of human transformations aimed at accommodating agriculture-related activities, prompts various reflections on the bioclimatic approach in architectural design. This contribution, firstly, highlights the theme of the urban-rural continuum (an intermediate and contradictory place, which epitomizes on-going contemporary transformations) in order to demonstrate that the contrast between urban and rural—although it has been used, since ancient times, on both analytical and design level—has not only been superseded, but is also, theoretically, inadequate for identifying the rural built environment. The link between agricultural and construction activities confirms the overcoming of this contrast; a reading in technological terms is proposed, with reference to the spheres of the product and the process, whilst recollecting its remote and deeply rooted cultural origins. The possible dual and contradictory character (vernacular or scholarly) of the rural built environment is illustrated with reference to the specificity of the bioclimatic aspects recognized in traditional buildings ("ante litteram sustainability"), highlighting how the relationship with the site-factor (microclimatic characteristics and local construction culture) was almost annulled in the twentieth century, also creating the basis for a widespread emergency for the quality of the buildings employed in agricultural activities. Touching upon the question of whether the rural building can be a typology distinct from other production buildings, this paper highlights the importance of the relationships with the socio-economic aspects of the agricultural sector, underlining the fact that the main identity for rural buildings in every age, in addition to the role that they play in the landscape, should be a stronger bond with the environmental elements, showing the centrality of the bioclimatic approach in updating the primordial interactions between agriculture and construction. Furthermore, the distinction between vernacular or scholarly character of the rural buildings prompted a reflection on the contemporary epistemological bases of the bioclimatic approach to general architectural design, which risks being

---

M. L. Germanà (✉)
Dipartimento di Architettura, Università di Palermo, Palermo, Italy
e-mail: marialuisa.germana@unipa.it

G. Chiesa (ed.), *Bioclimatic Approaches in Urban and Building Design*, PoliTO Springer Series, https://doi.org/10.1007/978-3-030-59328-5_7

ineffective whether it remains on an empirical level or if it is confined to a hyper-specialized niche. Finally, the bioclimatic approach is indicated as a common analytical and design methodological basis, useful for addressing the contradictory nature of the urban-rural continuum, in the framework of contemporary challenges.

Keywords Rural built environment · Urban-rural continuum · Bio-climatic approach · Tacit and explicit knowledge · Agriculture/Architecture

7.1 Beyond the Opposition: The Urban-Rural Continuum

Demographics in recent decades have confirmed the widespread belief that the twenty-first century would be "the century of the city". What had already happened in the areas of the world touched by industrial production models in the previous two centuries, occurred globally on 23rd May 2007: the earth's population has become more urban than rural (NCSU 2007). The "urban tipping point" has driven general interest toward the city effect, resulting in a convergence of research and studies regarding high density settlements, especially in view of the very critical conditions caused by rapid and wide-scale urbanization: evidence of disparities and lack of social inclusion; chains of violence; poverty; political exclusion; pressure on natural resources; public health (an aspect dramatically highlighted by the emergency of COVID-19). The 2030 Agenda for sustainable development of the United Nations, with goal 11: "Make cities inclusive, safe, resilient and sustainable", in 2015 underlined that 60% of the world population would be concentrated within a few years in urban areas, especially in developing countries, accentuating the imbalance between the 3% of land area occupied by cities and 75% of the global CO_2 emission produced by this type of settlement (UN 2015). It is too early to understand whether the long-term effects of the COVID 19 pandemic, ratified by the WHO in March 2020, will really affect the trend towards the wide-scale growth of high density settlements; on the one hand, the demands for social distancing imposed by health precautions will inspire new models of use for spaces, especially public ones; on the other, the widespread critical financial conditions and consequent social tensions will affect urban issues that are already very problematic (also regarding access to food).

In the context of prevailing uncertainty, however, the consolidation of a trend that is already underway, geared towards overcoming the contrast between the rural and urban spheres, is sure to emerge quickly. In fact, it has been shown that the concepts of urban and rural, although they are still pervasive on the theoretical level, are controversial because they do not correspond to the profound social and environmental transformations of the last centuries and their contrast is less and less useful in contemporary reality, characterized by interconnections rather than clear separations (Dymitrow 2018, 2019). However, it could be argued that these concepts are still current and advantageous in their mutual interaction rather than contrasting, although they can be traced back to obsolete settlement models, in which the margins between urbs and rur were certain and clear. In fact, contemporary trends aim at an integrated,

Fig. 7.1 Despite their physical impact, the causal factors of the urban-rural continuum have, above all, an intangible dimension. Aerial view of a part of Po Valley, northern Italy. Ph. by F. Malucelli, av. in https://ambiente.regione.emilia-romagna.it/it/geologia/suoli/suoli-pianificazione/img/copy_of_foto_03.jpg/@@images/image.jpeg

polycentric and balanced development of the territory, leveraging on an urban-rural continuum, trying to smooth over the social, economic and territorial gap by reinforcing sustainable transport and mobility and technological and communication networks (Fig. 7.1).

These trends are grafted on to the fundamental question of the safety and sustainability of food resources; in fact, the production and responsible consumption of food are also linked to closer links between urban and rural areas (UN 2017, pars. 49:95). In addition, the orientation to encourage agricultural crops in urban and peri-urban spaces has now stabilized in much of Europe, becoming an opportunity for urban re-vitalization, while "enhancing a network of social, recreational and commercial activities, private and public space (especially neglected ones) revitalization, shared green spaces creation, integration between built environment and nature, awareness toward sustainable lifestyles, education, social cohesion and new possibilities for a green (productive) infrastructure boosting" (Gallo et al. 2016, p. 104). In order to have lasting and effective consequences, the concept of the urban-rural continuum "demands to be defined not only on a physical and functional level, but also on a cultural one, placing itself in a connection—both tangible and intangible—between tradition and innovation, between well-established knowledge and elements of new tangible cultures" (Losasso 2019, p. 55).

Within contemporary and future scenarios, one can recognize, in the orientation towards sustainability, an effective bonding agent for the urban-rural continuum, bearing in mind that, since very remote times and with continuity, the reference to the natural elements of the specific place was considered essential for the built environment in its links to agricultural activities. This reference, which is the basis for the general bioclimatic approach to architecture (Olgyay 1963), is summarized in its attention, inter alia, to the following elements and aspects: exposure to the solar path and to the dominant air movements; weather conditions; configuration and nature of the soil; vegetation; circularity of flows; passive solutions for well-being. For this reason, identifying and acting on the common basis of agricultural and construction activities today acquires a particular and incisive meaning, regarding the search for greater concreteness in the orientation towards sustainability.

Lewis Mumford, asserting that the city of the dead would be born before the city of the living, linked the beginnings of construction with those of cultivation, fixing them both in the Mesolithic, an era of transition from nomadism to permanent settlements (Mumford 1961, Chap. 1). With this assertion, he indicated the suggestive hypothesis according to which burial places would attract relatives, allowing one to better observe the effects of seasonal cycles on the growth of spontaneous vegetation and creating the conditions for developing the long-term vision and awareness of the link between cause and effect. From these attitudes came the rooted settlement in a specific place, which marked the epochal shift from the cave to the hut, from hunting to breeding, from gathering of spontaneous berries to sowing. Mumford's hypothesis helps to underline the fact that this transformation probably did not have utilitarian origins, but might have arisen from profound spiritual needs, since endearment to one's relatives would have brought about settlement, a premise and a condition for agriculture and architecture, in which the improvement of man's living conditions on earth certainly had its beginnings. Furthermore, certain well-known considerations by Martin Heidegger confirm the relationship between agriculture and construction on the basis of the profound cultural affinities between Building, Dwelling, Thinking. In fact, in the German language the word bauen, which indicates the act of building as erecting something new, also means cultivating, taking care of something that already exists (Heidegger 1954) (Fig. 7.2).

Therefore, after actually settling in a place, one might recognize in a continuous care, a second common basis of agriculture and architecture, the relevance of which is obvious, for the former in its immediacy and, for the latter, a more drawn-out time-span. A more prosaic relationship between the early development of agriculture and architecture has been argued on a technical level on the basis of archaeological evidence, regarding the use of permanent construction materials that have accompanied the progressive domestication of humans, animals and plants (Wilson 2007); for example, the beginnings of arboriculture were related, in particular, to the improvement in metallurgical and textile production techniques, which contributed to the formation of human settlements (Fuller and Stevens 2019).

In addition to the previous interpretations, a technological reading of the relationships between agricultural and construction activities may be even more incisive in demonstrating the role of the bioclimatic approach in the urban-rural continuum. In

Fig. 7.2 The archetypal primitive hut (Rykwert 1972) suggests a sort of primordial bioclimatic basis of architecture. On the left, Lauger, M. A., 1755, Essai sur l'architecture, Frontispiece of the 1st English edition, ed. by S. Wale [Illustration in the public domain, courtesy of Open Library, openlibrary.org]. On the right, image modified by Viollet-Le Duc, E., 1875, Histoire de l'habitation depuis les temps préhistorique jusqu'à nos jours, Bibliotèque d'éducation et de ricreation, Paris, pag. 6. Av. at https://gallica.bnf.fr/ark:/12148/bpt6k6258576d.texteImage

terms of product technology, the results of the two activities (mainly food for the former and the built environment for the latter) are indispensable for the survival of human beings and they both reflect, at community and individual level, their sedimented cultural identity (roots of tradition) and evolutionary (modernity and—also innovative—trends). Product technology can also be traced back to the use, in construction, of cultivated produce or agricultural processing waste, frequent in the past and still the object of increasing interest in the field of bio-based solutions for building materials and components (Sandak et al. 2019), aimed at providing an alternative to building materials that erode non-renewable natural resources (Violano 2018). As far as process technology is concerned, one could emphasize the fact that it underlies both agriculture and construction. In fact, they are both, primarily, planned activities, evaluating the objectives to be achieved in relation to the available resources and operational constraints; then they are designed, defining all the aspects involved in the actual activities; then carried out, implementing them as soon as they have been planned and designed; finally practiced, during the consequent operating and management phase.

The procedural dimension is present in all forms of agriculture and construction, from the most complex to the simplest and most direct. For this reason, one can easily find parallelism between the production processes of these two activities in certain aspects, including: (1) the similar evolution in the transition from pre-industrial tradition to industrialization, which has frayed the connection with the site factor, with which both activities were related, whilst leveraging their awareness of natural aspects (Fig. 7.3); (2) the contemporary trends in agriculture and construction, increasingly focused on environmental sustainability, through short supply chains and a circular economy; (3) the permanence of the artisan dimension

Fig. 7.3 Agricultural production has also lost many of the links with the site specificities in the transition from the pre-industrial tradition to industrialization. At the top, the old millstone space; on the left, millstone and oil mill dating back to the nineteenth century; on the right, a machine for pressing grapes in a rural building from the 1970s. Fontanarossa estate in Cerda (Sicily). Photo by courtesy of M. Botti and A. Lo Bello, 2010

of many processes, in which the manual component and the direct transmission of the rules of the art continue to prevail; (4) the coexistence of activities, sometimes conflicting or in contradiction, which are carried out in parallel, on the local and global level, in different ways; (5) the growing awareness of the incisive impact of agriculture and construction on individual and collective health; (6) the consequent importance of the regulatory sphere, from which issues such as monitoring, certification and qualification derive, which increasingly involve operational practices in the agricultural and construction field, with the exception of those falling within the amateur and hobbyism fields.

The urban-rural continuum was identified in the first half of the twentieth century as a research topic (especially focusing on the identifying elements of the two opposite poles) in the context of urban sociology studies (Dewey 1960). In Italy during the 1970s, the urban-rural continuum, in which pieces of the chipped city and of "de-ruralized" countryside coexist, was identified as a significant phenomenon in the evolution of the landscape and human settlements (Assunto 1973; Samonà 1976); it is still an important test-bench for contemporary research regarding the architectural design (Sciascia 2013) and territorial planning (Schroeder et al. 2015). By placing the emphasis on the bioclimatic approach, the perspective shifts from an urban-centric vision to a focus on the rural environment, due to a character that is connected more closely to the natural identitary elements of the sites. Obviously, this does not mean seeking the irremediable lost virginity of Eden, with the ambition of ruralizing the whole territory, giving into bucolic chimera and ignoring the need for differentiating settlements with distinct vocations. In fact, the reference to natural elements is a necessary, but not sufficient, condition for fully understanding the bioclimatic approach as a tool of the sustainable design.

The principle of drawing inspiration from natural self-regulation systems is justified by the hypothesis of biophilia as innate leanings in human beings towards other living organisms (Beatley 2011, pp. 45–46), and to the consequent need for continuous contact with nature (Wilson 1984); this takes place by naturalizing the built environment through increasing its plant component. This is connected to the numerous design experiments on the theme of Urban and Building Integrated Vegetation, as a tool for improving air quality and mitigating the heat island phenomenon (Akbari et al. 2016). The concept of biomimetics, in which the reference to living organisms becomes an even more persuasive model for artifacts, inspiring morphologies and processes (Chiesa 2010; Gruber 2011), summarizes more recent developments in the bioclimatic approach. However, by only focusing on the natural factor, paradoxically we remain anchored to the obsolete contrast "city vs country", now surpassed by the urban-rural continuum. This phenomenon, however, suggests as preferable a bioclimatic approach that includes anthropic and artificial elements, based on the understanding of the complex and dynamic intertwining between natural and human systems, as already asserted back in the 1960s (Olgyay 1963). The bioclimatic approach can thus contribute to defining an analytical and design methodology aimed at positively governing the connections between urban and rural, ensuring food production and water management, with reference to the multiple dimension of sustainability through "Rural Design" (Thorbeck and Troughton 2016).

7.2 Rural Built Environment, Between Tacit and Explicit Knowledge

The rural built environment includes the results of the building interventions aimed at accommodating or facilitating agricultural activities (cultivation and breeding). A close relationship with the natural elements features this kind of built environment, because the individual buildings (intended for the storage or processing of products, the housing of animals or the dwellings of owners and workers), are interrelated with the orographic and hydro-graphic connotations of a wider environmental system: access roads and country roads, fences, rows of trees, hydraulic arrangements (canals and artificial lakes), retaining walls, etc. (Germanà 1999). In its multi-scalar, tangible (physical and material) and intangible (social, cultural, economic) dimensions, the rural built environment can be observed holistically, referring both to the building elements and materials and to the agricultural landscape, i.e. the form that man, in the course and for the purposes of his agricultural productive activities, consciously and systematically gives to the natural landscape ["quella forma che l'uomo, nel corso e ai fini delle sue attività produttive agricole, coscientemente e sistematicamente imprime al paesaggio naturale"] (Sereni 1961, p. 29).

A sort of preconception recurs in many approaches to the rural built environment, namely the fact that it is mainly characterized by being contrasted with urban settlements. In the analytical-cognitive and synthetic-design methodologies, this preconception has often produced a form of reductionism, pressing to circumscribe the rural in a "picturesque" dimension (Fig. 7.4), out of time and far from a precise place, towards which contemporaneity has often felt contrasting feelings: nostalgia (of the good old days) and repulsion (for isolation and backwardness).

The consolidation of the phenomenon of the urban-rural continuum in contemporary and future scenarios, is assisting in a distancing from this contrast, by seeking new approaches and methodologies. In fact, the reductionist tendency to consider the rural construction as "non-urban" and far from contemporary, seems to distinguish most of the studies devoted to the topic, and from which the following stages have been highlighted, with reference to Italy and especially to Sicily (Germanà 1999, par. II.2), summarily reported in chronological order:

(1) the years following the unification of Italy (1861), during which the rural buildings became emblematic of the very poor conditions of peasant life and of social inequalities; (2) the decades around 1900, when anthropological-ethnographic studies highlighted the documentary value of the traditional built rural environment that already appeared endangered; (3) the period between the two World Wars, when the rural world became a political priority of the regime, while the architectural world recognized a character of essential functionality in the morphology of the rural buildings (Pagano 1935), which in the same period also became a design theme for new settlements, such as for example in the context of the "restocking" strategies of the Sicilian rural areas; (4) the second post-war period, when human geography studies scrutinized the typological aspects of rural houses throughout Italy, linking them to the evolution of the dominant crop forms (a period during which the dream of

Fig. 7.4 The rural built environment is often idealized in a picturesque scenery. Sicilian cottage. Coloured print in P. De Wint, W. Light, Sicilian Scenery, 1823, London, anastatic reprint

"agrarian reform" was nourished in Sicily, which left visible traces in the hinterland, with the construction of rural villages, today mostly left derelict); (5) the last decades of the twentieth century, in which a clearer bifurcation between contemporary rural buildings and those inherited from tradition arose; on the one hand, the rural built environment relinquished the prerequisites for an important national issue, worthy of cognitive and planning commitment, while new constructions connected to agricultural activities were adopting industrial models, and abandoning all links with local tradition and site specificities (Fig. 7.5); on the other, the awareness of the landscape dimension of the rural building was growing, together with the conservational demand for traditional rural architecture, endangered by the abandonment due to "de-ruralization" or by interventions rather disrespectful of their identity (Fig. 7.6).

There has often been a misunderstanding that has led to confusing the traditional rural built environment with the "vernacular" one, due to the accentuation of certain recurring characteristics (coherence with the specific microclimatic conditions and preference for the available on-site building materials; marked adherence to needs; continuity with construction traditions and contribution to the cultural identity of communities; ICOMOS 1999), and this can be traced back to the same reductionist tendency. This misunderstanding threatens to blur the correlation to agricultural practices as the main and essential peculiarity of the rural built environment; the fact that

Fig. 7.5 Additions to a nineteenth-century "masseria" carried out in the 1970s due to the new needs of agricultural production. Fontanarossa estate in Cerda (Sicily). Ph. by M.L.G. 1987 (on the left) and 2003 (on the right)

Fig. 7.6 The art of getting by has very often justified interventions that have used inexpensive building materials, unsuitable for maintaining the original identity. Baite Cisterna Val Sassina, Barzio (northern Italy). Photo by courtesy of A. Doda, 2020

"Non-Pedigreed Architecture" unifies "vernacular, anonymous, spontaneous, indigenous, rural" (Rudofsky 1964, Preface: page not numbered), may have contributed to increasing the lack of distinction between rural and vernacular.

Similar ambiguity can be dispelled by emphasizing the intangible and procedural aspects of the rural built environment, which—as well as any artifact—is always a cultural expression representative of whoever has produced it (on an individual and social level) (Germanà 2005, p. 460). This clarification is pertinent, considering both the existing (pre-industrial and subsequent) buildings and new buildings. Whenever connected to agricultural practices limited to family sustenance and whenever the agricultural produce is not intended for sale, the rural built environment can be traced back to the vernacular category (Guillaud 2014, p. 33). On the contrary, when connected to more complex activities, with an economic impact and supported by the agronomic sciences ("primary" productive sector), the rural built environment—as a whole—takes on a character that differs from the free-range vernacular, through being "scholarly" or "learned". This is because it goes beyond local specificity and derives from a more advanced design intention, based on dedicated technological knowledge and precise cultural references (traditionally, the "Art of building"; today, technical and professional knowledge).

Both on the analytical level (aiming to study the existing traditional or contemporary rural buildings), and on the design level (aiming to define new construction interventions connected to the agricultural activities, from scratch or through recovery) it is essential to distinguish between the vernacular and scholarly natures of the rural buildings, understanding to what extent, depending on the case, they might contrast or coexist. This is necessary, because it helps to understand that the rural building, seen as the result of a production process, may express different epistemological bases. In fact, if the vernacular nature prevails, it can be said that the production process was guided by "tacit knowledge" (Polanyi 1966, Chap. 1), implicitly transmitted through direct experience, within traditional organizational models and linked to specific local contexts. By comparison, if the scholarly nature of the rural building prevails, the signs of an "explicit knowledge", codified and transmitted in formalized forms, will be identified as the consequence of an effort of objectification, simplification and sharing (Nonaka 1994, p. 16). Regardless of the field of application, understanding how far you can (or should) refer to implicit or rather explicit knowledge, is also important for dealing with the operational aspects of the bioclimatic approach (see the following par. 7.4).

7.3 Bioclimatic Approach to the Traditional and Contemporary Rural Built Environment

The bioclimatic approach is one of the pillars of sustainable architectural design and has been connected to energy aspects, given that—by favouring the use of climatic resources such as sun and wind and exploiting passive systems that leverage only natural physical phenomena—allows one to achieve the well-being of the users of the built environment, minimizing or reducing the mechanical equipment (see for example the "Three-Tier Design Approach" in Lechner 2015IV, p. 9).

In addition to the centrality of the specific place, there are two interrelated conceptions at the basis of this approach: the built environment seen as a "system", namely a series of mutual interacting intertwined elements (Morin 1977), to be considered "open", given that the system interweaves dynamic relationships with other systems (the users and the surrounding environment); the built environment seen as an "organism", organized and structured as a constantly evolving living being (Laborit 1971).

These conceptual foundations are particularly consistent with the field of application of the rural built environment, due to: its adherence to place, its multi-scale identity and its connection to agricultural practices, which are intrinsically focused on living organisms. For this reason, the bioclimatic approach can be applied profitably, adapting to the possible articulations of the theme, starting from the distinction between the existing rural built environment and the new one.

The rural architecture inherited from the pre-industrial tradition has offered numerous insights for focusing the potential of the bioclimatic approach, providing various examples of ante litteram sustainability, despite its deriving from a time when the very idea of sustainability was intrinsically foreign (Germanà 2005; Niglio 2018). In cases where the rural building has a vernacular nature, numerous studies have shown that there are multiple and multidimensional antecedents of today's orientation towards sustainability (see, for example, the summary of the "Wheel of environmental, socio-cultural and socio-economic sustainable principles" in Carlos et al. 2014, p. 27). The recurring insistence on environmental and bioclimatic characteristics has been indicated as one of the most frequent "shortcomings" in the numerous studies on the vernacular construction, with: the "technological bias", which has led to the loss of the link between tangible and intangible dimensions; the "romanticized approach", which prompted insistence on the banal contrast with contemporary constructions; the "essentialist representation", with the tendency to generalize conclusions regarding circumscribed realities (Vellinga 2014, pp. 4–6).

The bioclimatic approach is an irreplaceable cognitive tool, also in cases where scholarly nature prevails in the rural buildings. In fact, the works of the treatise writers that aimed to provide specific indications on how to design and to build rural constructions and their pertinences, demonstrate that "explicit knowledge" includes numerous references to principles that we would define today as sustainability, which at the time were merely considered to be common sense. The treatises from ancient times contain numerous indications on the best orientation for the various parts of the rural buildings, based on their specific destination (distinguishing between houses, stables, barns, warehouses, etc.), geared towards healthy conditions for humans and animals and for the best conservation of products (for example, cellars in the north and olive oil warehouses in the south) (Micheli 2016). The treaties in the Renaissance period move in the same vein, focusing greater attention to the hierarchization of spaces and emphasizing a different morphological characterization for rural buildings from that in the urban area (Germanà 1999, par. II.1).

In the texts on rural buildings produced after the advent of the capitalist model in agricultural production, attention to bioclimatic factors continues to be present. Given the more recent nature of these literary sources, their interest increases thanks to the possibility of connecting them to contemporary rural buildings and the subsequent ones still existing. For example, this comparison was debated with regard to the "masseria", a form of rural building widespread in the internal areas of western Sicily especially after the Unification of Italy (1861), linked to wide-scale cultivation and bearing testimony to a decisive historical phase in the agrarian history of the island (Germanà 2005). In the numerous nineteenth-century Sicilian "masserie" (one hundred and fifteen of which have been counted in the Madonie area alone;

D'Amore 2020), some characteristic elements occur that can today be appreciated for their bioclimatic aspects, starting from the articulation of the spaces around one or more courtyards, basically square in shape and often paved with cobblestones. This peculiar spatial layout (in addition to facilitating control of the productive activities and reducing construction costs), offers the possibility of greater protection from external (natural or anthropic) agents, consenting utilization of the spaces and taking into account the shady conditions during daylight hours (Fig. 7.7). The courtyard of the "masseria" can be assimilated to the empty spaces (atrium, patio) around which there revolve typologies recurring in the Mediterranean area since ancient times (Rajapaksha et al. 2018); the consequent, beneficial effects on one's well-being have been amply demonstrated, in terms of the bioclimatic action of air movement activated by the differential exposure to solar irradiation and by day-night alternation (Sartogo and Calderaro 2008, p. 124).

Contemporary technical literature highlights other elements of the "masseria" connected to the bioclimatic approach, which shift the focus from the vernacular nature of the rural built environment to the scholarly one. In 1871 the Società

Fig. 7.7 Daylight and shadow within the courtyard of the "masseria". Fontanarossa estate in Cerda (Sicily). Photo by courtesy of M. Botti, 2010

di Acclimatazione e di Agricoltura di Sicilia [Sicilian Acclimatization and Agriculture Society] made an announcement for the best practical manual that would indicate to the owners the cheapest construction of rural factories needed to build a complete "masseria", taking into account the conditions of the various districts of Sicily ["miglior manuale pratico che additerà ai proprietari la costruzione più economica delle fabbriche rurali indispensabili a comporre una completa masseria, avuto riguardo alle condizioni delle varie contrade della Sicilia"] (Basile 1873, pp. 1–2). The winning treatise was the one presented by Giovanni La Mensa (La Mensa 1880), who together with others (Tirrito 1881) outlined a series of suggestions in which it is possible to identify certain ideas of ante litteram sustainability (Fig. 7.8). Among these, those that we would today call bioclimatic are listed below: (1) regarding the criteria for the location within the agricultural holdings, the Sicilian manualists highlight proximity to water sources (springs and wells) and salubrity (recommending ventilated sites, well exposed to solar irradiation, protected from the danger of floods and, in the case of proximity to places infested with malaria, safeguarded by a higher altitude and the interposition, not too close to the buildings, of plant barriers—in the suggested form of poplars, linden, mulberry and eucalyptus); (2) as regards the distribution of the spaces according to their specific function, the orientation criteria follow the advice of the ancient treatises, suggesting, for example, locating the warehouses in basements facing east, not subject to sudden temperature changes, which would compromise the conservation of products; (3) regarding indoor environmental conditions, with reference to both humans and animals and the conservation of agricultural products, Sicilian manualists propose solutions appropriate to specific needs and always geared towards low cost and ease of management: in order to remove the moisture from the ground, both in existing buildings and in new ones, it is suggested that a perimeter trench should be dug, deeper than the foundations, sloping down towards a suitable disposal site; (4) in order to obtain adequate ventilation in the buildings, it is suggested that multiple-exposure openings should be provided, to procure controllable air-movement; for the ventilation of the stables, where strong draughts could harm the animals, activation of "cross ventilation" is proposed, with windows opening on two opposite sides (to be opened alternately during the seasons) and vents to be placed lower down (at least two for each window) and also ventilation shafts (La Mensa p. 55 and foll.; Tirrito, p. 41 and foll.).

The bioclimatic approach to the traditional rural built environment offers an important opportunity for a more complete understanding of the reasons that guided original construction; in fact, it shows that the recurring characteristics are not accidental, in aspects attributable both to tacit and to explicit knowledge; it contributes to the environmental performance of this form of architectural heritage, while simultaneously assuming the significance of testimony to the technological culture of the past.

The theme of the recovery of traditional rural buildings in recent decades has been studied in various and wide-ranging aspects, including for example: the need for the definition of specific intervention criteria (Germanà 1999, par. V.1); the formulation of design approaches that can be shared within homogeneous territories, based on knowledge of local construction and architectural aspects (Musso and Franco

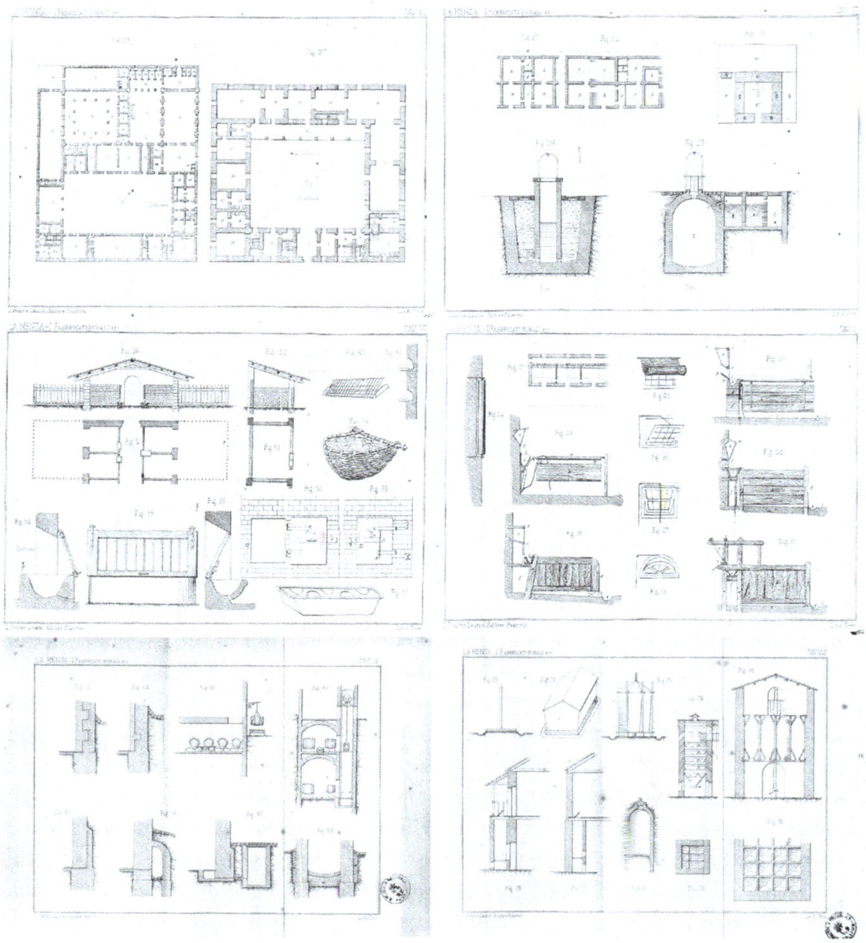

Fig. 7.8 Some of the tables accompanying the treatise on the ideal Sicilian "masseria" by Giovanni la Mensa (La Mensa 1880)

2014, p. 533); the effectiveness of regional programming (Bosia and Savio 2014); opportunities for enhancing homogeneous typologies on a large scale integrated with cultural and experiential tourism (D'Amore 2020). Therefore, the bioclimatic aspects are part of a much broader framework of issues to be taken into account in interventions on traditional rural buildings, which are largely associated with reuse, due to the functional obsolescence determined by the evolution of agronomic techniques.

The attention to these aspects does not only concern interventions on rural buildings, but should inspire any building intervention, starting from a preliminary and accurate analysis of the site, with the following precautions: to avoid waterproofing neighbouring land; to encourage all forms of integrated vegetation; to attribute a correct orientation of the functions; to leverage passive solutions for well-being; to aim to restrict energy consumption with insulating envelopes and shields; to use renewable energy sources; to reduce the use of water during construction and utilization phases; to prefer local materials and durable and maintainable solutions. For the rural built environment of preindustrial origin specifically, the interventions should also bear in mind the way in which the constructive culture that produced them, both in implicit and explicit form, took into account the bioclimatic aspects in updating them, rather than deleting them because of careless design choices (Germanà 1999, 2008).

The buildings, considered as a whole, attributable to agricultural activities, in addition to the rural buildings inherited from tradition, include constructions built in recent decades, reflecting the superimposition of the industrial model on agriculture, to the extent of justifying the statement that there no longer exists, if it ever existed, a 'rural world', but only an economic and social sector not different from the 'urban' one if not for the distribution of the population in the territory, the particular technology and the specific working environment ["non esiste più, se mai è esistito, un 'mondo rurale', ma soltanto un settore economico e sociale non diverso da quello 'urbano' se non per la distribuzione della popolazione nel territorio, la particolare tecnologia e lo specifico ambiente di lavoro"] (Daneo 1972).

With continued reference to Sicily, this type of built environment is consistent with almost all the buildings constructed in the second half of the twentieth century, because it reflects building production standardized on the various levels of poor architectural quality. As containers without any contextual connection and devoid of cultural identity, these largely prefabricated buildings epitomize significant general problems, in terms of: energy inefficiency; presence of dangerous building materials (asbestos); conditions of advanced degradation due to lack of maintenance, caused by a devastating mix of managerial and financial difficulties. These considerations can be applied to the entire country, and there was no shortage of alarm bells regarding the fact that Italian rural buildings were becoming an undifferentiated amalgam, indifferent to regional traditions, which would soon create a widespread qualitative emergency (Assunto 1973; Maldonado 1977).

In order to resolve this lack of quality, today there is no shortage of technical solutions, supported by an adequate methodological and operational background (redevelopment and maintenance processes, with particular attention to the energy aspects of retro-fitting and integration of renewable energy source systems). What are actually missing, in general are the social, economic and financial preconditions indispensable for a productive and sustainable agricultural reality. The size and seriousness of the issue are summarized by a single data-item, indicative of the radical reorganization of the primary sector: the number of agricultural workers, as a percentage of the whole active Italian population, went from 70% in 1871 to less than 10% in 1994 (Fanfani 2004, p. 135).

The contemporary economic trends involving the primary sector are associated with the theme of newly-built rural constructions (considering only productive buildings and excluding extra urban residences or accommodation facilities), for which there is no shortage of well-known examples, such as the Dominus Winery in Napa

Valley, California (http://dominusestate.com/the-estate/architecture). Completed in 1997 on a project by the architects Jacques Herzog and Pierre De Meuron, the building, after many years, has maintained a seminal value in testifying to the high aesthetic and environmental quality that can be achieved thanks to an architectural design that is attentive to the specificity of the place; it has become iconic owing to the solution of the facade with gabions filled with basalt rocks from the nearby canyon. This technical solution, seen from the outside, contributes to successful insertion into the landscape; seen from the inside, it filters the light and the view towards the surrounding vineyards, as well as contributing to protection from external temperatures. Here it should be stressed that the project is consistent with a corporate philosophy entirely oriented towards sustainability, through a focus on biological aspects, a philosophy that can be applied in contexts beyond the single building: in the control of the natural habitat of the stream that crosses the property; in the adoption of dry-farming techniques; in using "cover crops" to prevent soil erosion and encourage biodiversity, avoiding the use of pesticides; in the predisposition of places for the nesting of birds of prey to oppose the rodents that damage the vineyards (Fig. 7.9).

Within the limits of its particularity, the Dominus Winery is a significant example of the general trend towards sustainability that can be recognized in the most recent rural productive buildings coherently with the actual agricultural practices, but which cannot be considered a specific identifying characteristic. In fact, the use of renewable energy sources with integrated systems, passive solutions for heating, cooling and ventilation, a preference for durable materials and prefabricated elements, are technical solutions that have taken hold globally in almost all new buildings, regardless of function, in the last ten years. The choice of frugal, exposed materials, which might be connected to a "rural" dimension, in search of morphological and functional essentiality, can also be found in numerous contemporary examples, regardless of place and intended use.

This consideration could suggest that the autonomous development of the rural building, which has distanced it from urban models, making it a category in its own right (Gambi 1964, p. 448), has now come to an end; in addition, it could demonstrate how sterile the question is if there is still a "typology" of rural production building, within a more general "typological question", made problematic by the ever more rapid changes in the forms of use of the built environment (Koch 2014). However, widening our scope from building typology to the role of the rural building in the contemporary landscape, a distinct trace of identity can still be found in buildings playing host to agricultural activities, in which, often in conflictual manner, traditional and more recent buildings come together. For example, precisely by leveraging this common identity, a unitary project was formulated with guidelines geared towards quality as regards salvage operations, the replacement and enlargement of rural buildings in a large area of northern Italy hit by the earthquake in 2012; one criterion among several determined the positioning of buildings with regard to the

Fig. 7.9 The Dominus winery in California is a seminal example of a holistic approach to sustainability, which unites winery design (by Jacques Herzog and Pierre De Meuron) and vineyard management. Images from http://dominusestate.com/the-estate/architecture

orientation of the sun and winds, underlining the importance of this criterion both for sustainability and for adhering to the natural order of the landscape (Marangoni ed. 2013, p. 37).

Although the degree of denaturalization of some agricultural activities is dramatically high (for example, in greenhouses or intensive farming), the need to orientate agriculture towards sustainability is now widely shared by technicians and consumers. The bioclimatic approach will, therefore, play an increasingly central role, adding an item to the evolving picture of the affinities between agricultural and construction activities, which will find tangible expression precisely in the rural built environment taken as a whole.

7.4 Future Scenarios for the Urban-Rural Continuum: The Bioclimatic Bonding Agent

A useful reflection regarding the operational consequences of the bioclimatic approach to architectural design may derive from focusing attention on the rural built environment, taking into account its multidimensional and multi-scale features and—above all—its possible, vernacular or scholarly, character. In fact, in the bioclimatic approach reference can be made to two distinct epistemological bases: one definable as a "vernacular basis" because it is intuitive, empirical, bordering on obviousness and is taken for granted (there is no designer that ignores the theoretical indications as regards the best orientation, except to disregard them at the first design difficulty). The other is definable as a "scholarly basis", which is reflected in scientific methods formalized decades ago, recently supported by more and more qualified tools (for instance, Chiesa and Grosso 2019); this, however, risks remaining confined within a niche of hyper-specialism, without affecting wider professional and administrative practice. In particular, knowledge of the environmental components of the site, as a precondition of each design action, is often rejected, so as to justify the expression "knowledge gap", as an obstacle that prevents a wide practical application of the bioclimatic approach (Germanà 2016, p. 276). Due to the interruption in the pre-industrial tradition, uniquely empirical knowledge has now been lost; therefore it is necessary to apply greater leverage on the scholarly dimension, overcoming the barriers between experts and users that prevents any fully technological practice (Pacey 1983) and implementing appropriate actions in the field of education, training and dissemination.

Despite all forecasts, it is not at all certain that the twenty-first century will be the "city century"; instead, it is more likely that it will be the "urban-rural continuum century". Leaving behind a vision of the "empty world", a container of unconditional human actions, and finally applying, in practice, the "responsibility principle", invoked theoretically for decades for technological processes of any nature (Jonas 1979), we have understood that in a sustainable built environment, "built capital" must integrate with "social capital" and "natural capital" (Elmqvist et al. 2013). It is no coincidence that an architect-theorist such as Rem Koolhaas said that the countryside is today the place where future civilization is taking shape in the most radical way and where the pressing questions posed by global issues such as climate change, migration and conservation of nature may find answers (Koolhaas 2020). This affirmation is deeply rooted in the thoughts of a designer who, after years of disinterest, recognized an increasing new interest in the countryside: "a world formerly dictated by the seasons and the organisation of agriculture is now a toxic mix of genetic experiment, science, industrial nostalgia, seasonal immigration, territorial buying sprees, massive subsidies, incidental inhabitation, tax incentives, investment, political turmoil, in other words more volatile than the most accelerated city. The countryside is an amalgamation of tendencies that are outside our overview and outside our awareness. Our current obsession with only the city is highly irresponsible because you cannot understand the city without understanding the countryside" (Koolhaas 2014).

The urban-rural continuum, with an elusive physicality that goes beyond the previously consolidated categories of thought, condenses the transformations that are taking place in the contemporary world, offering a test bench for all fields of knowledge and for all technological disciplines, as demonstrated by the effort to define a set of integrated "Guiding Principles" for urban-rural linkages (UN-Habitat 2019). Much more than the material aspects, it is the intangible ones that connote this intermediate and contradictory place (such as the effects of digitalization) so decisively. For this reason, a common paradigm such as sustainability may serve to address its contradictions and, in particular, the bioclimatic approach might be considered as both a sort of theoretical and operational bonding agent, because it contributes to creating a common methodology, suitable for tackling many diverse contemporary challenges.

Acknowledgements I dedicate this chapter to the memory and example of my beloved father Virgilio, a great agronomist expert in the Sicilian rural world.

References

Akbari H et al (2016) Local climate change and urban heat island mitigation techniques—the state of the art. J Civ Eng Manage 22(1):1–16. https://doi.org/10.3846/13923730.2015.1111934

Assunto R (1973) Il Paesaggio e l'estetica. Giannini, Napoli

Basile M (1873) Il caseggiato delle aziende rurali. Trattato agronomico con mescolanze morali e politiche. Tipografia D'Amico, Messina

Beatley T (2011) Biophilic cities: integrating nature into urban design and planning. Island Press, Washington

Bosia D, Savio L (2014) Actions and tools for the conservation and valorisation of rural architecture and landscape. Techne. J Technol Architect Environ 7:87–93. https://doi.org/10.13128/Techne-14536

Carlos GD et al (2014) Research method and operative approach. In: Correia M Dipasquale L Mecca S (eds) VERSUS. Heritage for tomorrow. Vernacular Knowledge for Sustainable Architecture. FUP, Firenze, pp 23–31. Available at https://issuu.com/dida-unifi/docs/versus_fup_google

Chiesa G (2010) Biomimetica: tecnologia e innovazione per l'architettura, English edition (2017) Biomimetics technology and innovation for architecture. CELID, Torino

Chiesa G, Grosso M (2019) A parametric tool for assessing optimal location of buildings according to environmental criteria. In: Sayigh A (ed) Sustainable building for a cleaner environment. Innovative renewable energy. Springer, Cham, pp 115–129. https://doi.org/10.1007/978-3-319-94595-8_11

D'Amore A (2020) Conoscere per conservare e valorizzare. Le masserie quale elemento di sviluppo del territorio delle Madonie. Ph.D. Thesis Università di Palermo

Daneo C (1972) Agricoltura e sviluppo capitalistico in Italia. Einaudi, Torino

Dewey R (1960) The rural-urban continuum: real but relatively unimportant. Am J Sociol 66:60–66. Available at www.jstor.org/stable/2773223

Dymitrow M (2018) Rural/urban: laying bare the controversy. Geographia Polonica 91:375–397. https://doi.org/10.7163/GPol.0126

Dymitrow M (2019) The concept of 'Rural' as a psychosocial process: From concept attainment to concept unlearning. Quaestiones Geographicae 38:15–28. https://doi.org/10.2478/quageo-2019-0036

Elmqvist T, Redman CL, Barthel S, Costanza R (2013) History of urbanization and the missing ecology. In: Elmqvist T et al (eds) Urbanization, biodiversity and ecosystem services: challenges and opportunities: a global assessment. Springer, Dordrecht Heidelberg New York London, pp 13–30. https://doi.org/10.1007/978-94-007-7088-1_2

Fanfani T (2004) La ricostruzione in Italia nel secondo dopoguerra. Provvedimenti e linee guida per la ripresa dell'agricoltura. Rivista di Storia dell'Agricoltura XLIV: 125:154. Available at http://rsa.storiaagricoltura.it/scheda.asp?IDF=119&IDS=6&IDP=1

Fuller DQ, Stevens CJ (2019) Between domestication and civilization: the role of agriculture and arboriculture in the emergence of the first urban societies. Veg Hist Archaeobotany 28:263–282. https://doi.org/10.1007/s00334-019-00727-4

Gallo P, Casazza C, Sala M (2016) Performances and potential of a productive urban green infrastructure. Techne J Technol Architect Environ 11:104–112. https://doi.org/10.13128/techne-18408

Gambi L (1964) Per una storia dell'abitazione rurale in Italia. Rivista storica italiana LXXV:427–454

Germanà ML (1999) L'architettura rurale tradizionale in Sicilia: conservazione e recupero. Publisicula, Palermo. Available at https://www.researchgate.net/publication/297484956

Germanà ML (2005) La sostenibilità inconsapevole del costruito rurale tradizionale: l'esempio della masseria siciliana. In: Mecca S, Biondi B (eds) Architectural heritage and sustainable development of small and medium cities in South Mediterranean Regions. Results and strategies of research and cooperation. ETS. Pisa pp 459–467. Available at http://hdl.handle.net/10447/3207

Germanà ML (2008) Ruolo delle tradizioni costruttive nella salvaguardia del costruito rurale storico. In: Dell'Acqua A, Degli Esposti V, Mochi G (eds) Linguaggio edilizio e sapere costruttivo. Edicom, Monfalcone (GO), pp 733–745. Available at http://hdl.handle.net/10447/35612

Germanà ML (2016) Criteri per la sostenibilità ambientale degli interventi nel centro storico di Modica. In: Trombino G (ed) Modica. Contributi per il recupero e la riqualificazione del centro storico. 40due Edizioni, Palermo, pp 273–292. Available at https://iris.unipa.it/retrieve/handle/10447/201137/342884/Germana_Modica.pdf

Gruber P (2011) Biomimetics in architecture. Architecture of life and buildings. Springer, Verlag/Wien

Guillaud H (2014) Defining vernacular architecture. In: Correia M, Dipasquale L, Mecca S (eds) VERSUS. Heritage for tomorrow. Vernacular knowledge for sustainable architecture. FUP, Firenze, p 33. Available at https://issuu.com/dida-unifi/docs/versus_fup_google

Heidegger M (1954) Vorträge und Aufsätze. Neske, Pfullingen. English transl. Hofstadter A. Building dwelling thinking. In: Krell DF (ed) (1993) Martin heidegger: basic writings, revised and expanded edition. Routledge, London, pp 217–65

ICOMOS (1999) Charter on the built vernacular heritage. Available at http://wp.ciav.icomos.org/wp-content/uploads/2019/04/charter-vernacular-en.pdf

Jonas H (1979) Das Prinzip Verantwortung. Insel Verlag, Frankfurt am Main. Eng. transl. 1984 The imperative of responsibility: in search of ethics for the technological age. University of Chicago Press, Chicago
Koch D (2014) The typological question and the formal basis of architecture. J Space Syntax 5:168–189. Available at http://joss.bartlett.ucl.ac.uk/journal/index.php/joss/article/view/208/pdf
Koolhaas R (2020) Countryside, the future. Guggenheim Museum Exhibition catalogue and video. Available at https://www.guggenheim.org/exhibition/countryside
Koolhaas R (2014) Countryside architecture. ICON 35. Available at https://www.iconeye.com/architecture/features/item/11031-rem-koolhaas-in-the-country
La Mensa G (1880) Trattato sui fabbricati rurali e relativi alle industrie agricole che concorrono a formare una completa fattoria. Luigi Pedone Lauriel Editore, Palermo
Laborit H (1971) L'homme et la ville. Flammarion, Paris
Lechner N (2015IV) Heating, cooling, lighting. Sustainable design methods for architects. Wiley, Inc., Hoboken New Jersey
Losasso M (2019) Culture, project and environment for the development of rural and suburban territories. In: Fanzini D, Tartaglia A, Riva R (eds) Project challenges: sustainable development and urban resilience. Maggioli, Santarcangelo di Romagna (RN)
Maldonado T (1977) Agricoltura: un appuntamento mancato. Casabella 426:9
Marangoni B (ed) (2013) Paesaggi da ricostruire. Linee guida per la tutela, valorizzazione, ricostruzione del paesaggio rurale nella bassa pianura emiliana. Regione Emilia Romagna, Bologna. Available at https://territorio.regione.emilia-romagna.it/paesaggio/pubblicazioni/paesaggi-da-ricostruire
Micheli B (2016) Architettura solare del passato. I consigli di Vitruvio. Av. at https://docplayer.it/11490297-I-consigli-di-vitruvio.html
Morin E (1977) Le Mèthode. I. La Nature de la Nature. Édition de Seuil, Paris
Mumford L (1961) The city in history: its origins, its transformations, and its prospects. Harcourt, San Diego
Musso SF, Franco G (2014) "Guidelines" for sustainable rehabilitation of the rural architecture. In: Mileto C et al (eds) Vernacular architecture, towards a sustainable future. CRC Press Taylor & Francis Group, London, pp 531–536
NCSU North Carolina State University (2007) Mayday 23: world population becomes more urban than rural. ScienceDaily. Available at www.sciencedaily.com/releases/2007/05/070525000642.htm
Niglio M (2018) Past as future in adaptive buildings. Climatic adaptation in ancient constructions. Adv Mater Res 1149:1–9. https://doi.org/10.4028/www.scientific.net/AMR.1149.1
Nonaka I (1994) A dynamic theory of organizational knowledge creation. Organ Sci 5:4–37. Available at https://www.jstor.org/stable/2635068?seq=1
Olgyay V (1963) Design with climate: bioclimatic approach to architectural regionalism. Princeton University, New Jersey
Pacey A (1983) The culture of technology. MIT Press, Cambridge
Pagano G (1935) Documenti di architettura rurale. Casabella 95:18–25. Eng. translation by Sabadino M (2010) Journal of Architectural Education ACSA:92–98
Polanyi M (1966) The tacit knowledge dimension. Doubleday & Comp, New York
Rajapaksha I, Fiorito F, Lazer E, Sartogo F (2018) Exploring thermal comfort in the context of historical conservation. A study of the vernacular architecture of Pompeii. Architectural Sci Rev 61:4–14. https://doi.org/10.1080/00038628.2017.1405790
Rudofsky B (1964) Architecture without architects. An introduction to non-pedigreed architecture. Publ. The Museum of Modern Art, New York
Rykwert J (1972) On Adam's house in Paradise: The idea of the primitive hut in architectural history, The Museum of Modern Art New York
Samonà G (1976) La città in estensione. Stampatori tipolitografi associati, Palermo
Sandak A, Sandak J, Brzezicki M, Kutnar A (2019) Bio-based building skin. Springer Open, Singapore. 10.1007/978-981-13-3747-5_1

Sartogo F, Calderaro V (2008) The Mediterranean: a cool temperate climate. In: Hyde R (ed) Bioclimatic housing. Innovative design for warm climates. Earthscan, London, pp 117–147

Schroeder J, Carta M, Ferretti M, Lino B (eds) (2016) Territories. Rural-urban strategies. Jovis Verlag, Berlin. Available at https://www.jovis.de/en/books/product/territories-rural-urban-strategies.html

Sciascia A (2013) The rough edges of the "expanding town" projects between Palermo and Partinico. FAMagazine 23:42–47. https://doi.org/10.12838/issn.20390491/n23.2013/5

Sereni E (1961), (second edition 1982) Storia del paesaggio agrario italiano. Laterza, Bari. English edition: 2014, Princeton University Press, Princeton

Thorbeck D, Troughton J (2016) Connecting urban and rural futures through rural design. In: Maheshwari B et al (eds) Balanced urban development: options and strategies for liveable cities. Water science and technology library 72, pp 45–55, https://doi.org/10.1007/978-3-319-28112-4_4

Tirrito R (1881) Norme pratiche di architettura rurale per l'agronomo. Tipografia Lo Casto, Palermo

UN United Nations (2015) The 2030 agenda sustainable development goals, goal 11: make cities inclusive, safe, resilient and sustainable. Available at https://www.un.org/sustainabledevelopment/cities/

UN United Nations (2017) Quito declaration on sustainable cities and human settlements for all, in "New Urban Agenda". Available at http://habitat3.org/the-new-urban-agenda

UN-Habitat United Nations Human Settlements Programme (2019) Urban-rural linkages: guiding principles. Framework for action to advance integrated territorial development. Available at https://urbanrurallinkages.files.wordpress.com/2020/03/url-gp-framework-for-action_english.pdf

Vellinga M (2014) Vernacular architecture and sustainability: two or three lessons. In: Mileto C et al (eds) Vernacular architecture, Towards a sustainable future. CRC Press Taylor & Francis Group, London, pp 3–8

Violano A (2018) Beyond materials: the experimentation of biobased grown materials from mycelia. Techne J Technol Architecture Environ 16:299–307. https://doi.org/10.13128/Techne-23029

Wilson EO (1984) Biophilia. The human bond with other species. Harvard University Press, Harvard

Wilson PJ (2007) Agriculture or architecture? The beginnings of domestication. In: Cassidy R, Mullin M (eds) Where the wild things are now. Domestication reconsidered. Bloomsbury Academic Berg, Oxford, pp 101–122. https://doi.org/10.5040/9781474215954

Chapter 8
Pulsations in Technology Innovation

Michela Toni

Abstract I'm honoured and pleased by the invitation to narrate firsthand some of my design experiments in "bioarchitecture", a field I have been involved in since my post-graduate experiences until today with in-depth studies. Definitely, this article gives me an opportunity to revisit my past and recent works and share them with young researchers or any professionals willing to know more about the approaches pursued by the scholars of the Schools of Architecture in Italy in the past and still underway. I'm honoured by the opportunity of comparing my views against others' on the vision of the architecture project in order to provide fresh inputs. I'm pleased because it gives me the opportunity to make my work known with the expectation that the conditions to build synergies and make either a practical or theoretical contribution to the community I live in, will be accomplished. In particular, research topics are outlined on building typologies, processes, construction and environmental systems with prime focus on the design of a human living environment and new solutions to counteract the urban and landscape decay compatibly with available resources.

Keywords Scale-ups · Local resources · Light technologies · Quality processes

8.1 Introduction

To introduce my personal research approaches to the theme under this publication, I need to dwell a bit on the cultural period when a new sensitivity emerged towards the environment, and no definition of "bioclimatic design" existed yet in the vocabulary of the architect. Let's me start with a picture of myself as a student at the then—Faculty of Architecture in Florence when strolling around the old town in the late 1970s. I got captured by the bookshop windows and by the enthralling image of the Sun in some book covers. The authors were dealing for the most with issues related to the principles of thermodynamics. The only book available to me was dealing with the "solar house" concept; it was co-authored by Giovanni Del Signore

M. Toni (✉)
Università degli Studi di Ferrara, Ferrara, Italy
e-mail: michela.toni@gmail.com; michela.toni@unife.it

G. Chiesa (ed.), *Bioclimatic Approaches in Urban and Building Design*, PoliTO Springer Series, https://doi.org/10.1007/978-3-030-59328-5_8

and Andrea Ponsi, a physician and an architect respectively. The latter—also thanks to this book—won a scholarship to explore this issue in the United States. As a matter of fact, whilst the Italian architecture was evolving around different paradigms, instead overseas—and in some European countries—many progressive designers were radically experimenting how to build a house by using natural resources.

While researching for my dissertation, Giovanni Del Signore welcomed me at his lab-home in Arcetri, on the hills of Florence, and he showed me the pioneering solar collectors he had engineered himself. What impressed me most was his very approach to self-construction as a freedom tool, and how he meant the sun energy as a democratic, free, easy-to-get source vs petroleum seen as a wealth in the hands of a few powerful companies.

According to this stimulating experience, my research work throughout the years was mainly to keep away from both a technicity, efficiency-centered approach on one side, and a strictly ecological vision, on the other side. To put it simply, my aim has always been that of designing a living environment where people can freely develop their needs at best or to find new solutions to urban and territorial decay in tune with a shared use of the resources.

Again, in the late 1970s but from a quite different cultural perspective, Società Metallurgica Italiana (SMI), a leading manufacturer of added-value solar collectors with copper absorbent plate, was active in Florence. I visited the SMI' site in the Tuscan Apennines but unfortunately I was able to get only but a few information about how they were manufacturing their solar collectors. As I learned afterward, SMI was primarily a weapon manufacturer, so secrecy at its sites was a must. Anyway, there are people who make a difference, and I want to share a precious experience I had as a young eager-to-learn student. So, at SMI headquarters in Florence, one of their heads, *ingegner* Porta, taught me how to use a portable card computer, a true novelty, that I could use to calculate the heat load efficiency of the solar collectors I had included in a building project for my dissertation, that I defended in a.y. 1978–79. This meeting was an opportunity to advance my knowledge outside the university setting, then quite behind in this field.

Beyond the technological innovations in the solar field, I was looking for data about new architectural design experiments—which information was very difficult to find.

Opportunities for discussion and learning arose while attending a symposium—*Piccolo e bello*—held by Università IUAV di Venezia (Venice 1979), the *Expo Solare* (Genoa 1979) and the *Brighton Exhibition* (Brighton 1980) (Fig. 8.1). There, you might concurrently mix with radical—and marginal as such—circles trusting a global strategy was the right one to challenge the energy and environment issues on one side, and high-tech systems manufacturers ready to advance the building industry, on the other side: two totally different approaches, and the challenge was to take into account both of them at the same time.[1]

In those years, the energy issue and associated economic and political implications emerged clearly with the oil crisis after the 1973 Arab–Israeli War.

The Italian Government tried to counteract with an efficiency-centered strategy and enacted Law 373 on energy saving in 1976. Also, the UNI Standard #4184 titling

Fig. 8.1 A small craft with solar cell propulsion, that flew across the English Channel. *Brighton Exhibition*, 1980. Picture by Michela Toni

Collettori solari, criteri e raccomandazioni—to say, criteria and recommendations for solar collectors—was published in 1979. This was the first standard to set the requirements for solar collectors in Italy.

Overseas, instead, there was no certainty any longer as to whether a technical approach could actually bring efficient actions: from an international perspective, the Vietnam war and its disasters on human beings and the environment; from the perspective of the common man, the severe effects of daily pollution on the environment—the two poles of awareness of the global environmental issue (Toni, 2012e).

In such a milieu, as a senior, then as an architect and teacher, I was logically driven to investigate the design evolutions which can be generated at different scales when focusing on environmental issues. These design evolutions then resulted into research themes. As conveyed by the title of this publication, let's call them *pulsations in technology innovation.* Pulsations that cannot be set chronologically because they are closely intertwined, as I discovered myself when trying to summarize their key concepts and narrate them as a whole.

8.1.1 Scale-Ups

From the early researches on the design implications associated with the use of renewable resources in building (sun heat and light, natural ventilation, vegetated soil, rain water), to the next researches aiming at improving the sound performance, the interactions among the multiple morphologic, technical and environmental issues are reorganized and investigated at different design scales against the variables which determine the living quality from a sensory and perceptive perspective.

The relationship between technologies and building types when the design takes root from environmental factors with least energy waste, was the very core of my dissertation work titled *Implicazioni tecnologiche nella progettazione in riferimento a fonti energetiche alternative*, Toni, M., supervisor Prof. Zaffagnini, M., 1979. The winner of a national prize, *Il sole e l'habitat*,[2] (AAVV 1980a, 1980b; Toni 1980a), this research dealt with public housing buildings including terraced houses, tenements with communal balconies and line-type buildings in the framework of Italian Law 457/1978.

In this project, the microclimate takes advantage of a multipurpose balcony-glasshouse space with vegetated soil to absorb the rainwater. Also, passive solar devices (glasshouses) and active solar devices (collectors) in the building envelope and criteria tailored to optimize the thermodynamic qualities by exploiting the free energy (glass surface, lighting and cooking equipment, thermal inertia of floors and glasshouse soil, and dwellers' activities) create innovations in building types (Figs. 8.2 and 8.3).

The brick and concrete envelope ensures a high thermal inertia, which contributes to wellbeing.

For more than a decade, as a collaborator in the research team of Edinricerche, Bologna, I investigated how conventional technologies could improve wellbeing and control the energy. The outcomes of this work have been disclosed in exhibitions or conferences and can be found in collective publications[3] (AAVV 1981a, b, 1982; Toni 1980b, 1981, 1987).

New ideas were experimented in a promising design competition period. In particular, we won the *Concorso di idee per la nuova sede municipale di Casalecchio di Reno in provincia di Bologna*, a competition to bring in fresh ideas for the new headquarters of the Town Hall and refurbishment of the old town in Casalecchio di Reno, in the Province of Bologna (Edinricerche and Tecnicoop) (1981). Here, the team's attention to the social environment and a new approach to sustainability—such as the facades for solar flow control resulting into an excellent natural illumination and improved management economies—was the winning idea (Fig. 8.4).

As a matter of fact, between the late 1980s and the early 1990s a cultural milieu was established seemingly ready to a leap in quality in the building field, in particular, when a subsidized housing trial program (the so-called *Programma di Sperimentazione di Edilizia Agevolata*) was created pursuant to Art. 4, Law 94 dated March 25, 1982—with the Emilia Romagna Region in charge of coordinating the local activities.

Through Edinricerche, a number of projects (including operational feasibility plans (*Piani Operativi di Fattibilità*, POF) and final design project (*Progetti Definitivi di Intervento*, PID) were created with the idea of proposing innovation elements acceptable to the small to medium-sized companies—at that time, they were responsible for a high percentage of building projects—having the right building and organization capabilities as a key player to make the sector to evolve. Innovations in

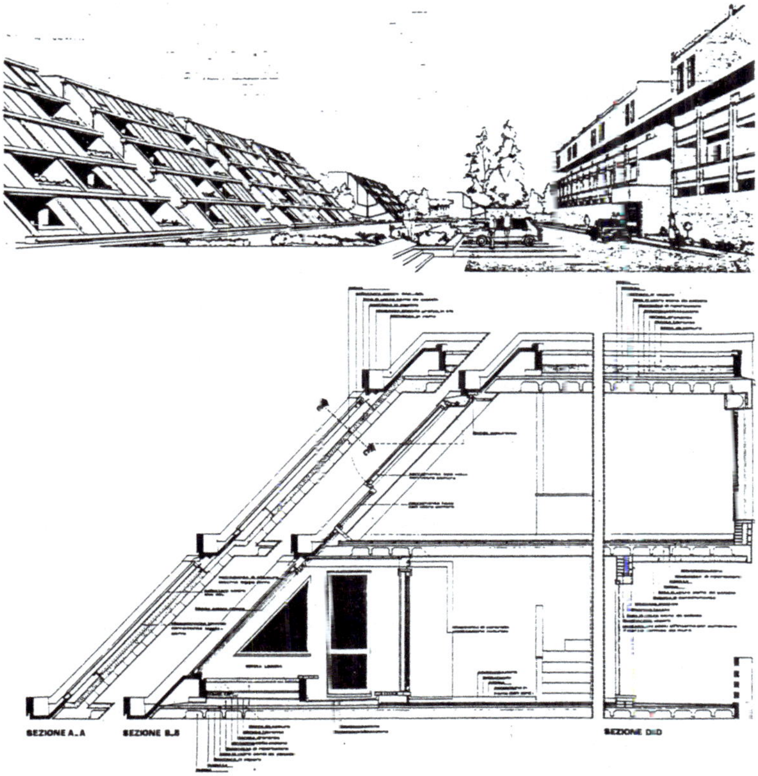

Fig. 8.2 Balcony-glasshouse space (Toni, M., Zaffagnini, M.)

building types to meet new requirements for a flexible use of the housings, were investigated against the close relation between thermal loads and daylight (Fig. 8.5).

According to this climate, the research doctorate—the highest academic degree—was established in Italy in 1980 (Law 28 of February 21, 1980) and definitely opened up new avenues for the research. While attending the very first cycle of Architecture Technologies in the years 1984 to 1987, by leveraging on my previous work, I joined a research project focusing on the environmental quality of the architecture design at different scales (from the technical piece to more aggregate scales, such as basic space units, housing, residential unit, settlement). The effort of this commitment resulted into my dissertation titled *Qualità involucro—Contributo attorno ai problemi del comfort abitativo* (Toni 1987). My goal was mainly a methodological approach based on a responsive design, far from a determinist, specifically efficiency-centered approach.

The issues associated with the wellbeing design under multiple sides—acoustic (A), thermogeometric (T), and visual (V)—have been investigated specifically with reference to the residential building. It was demonstrated that set quality levels can

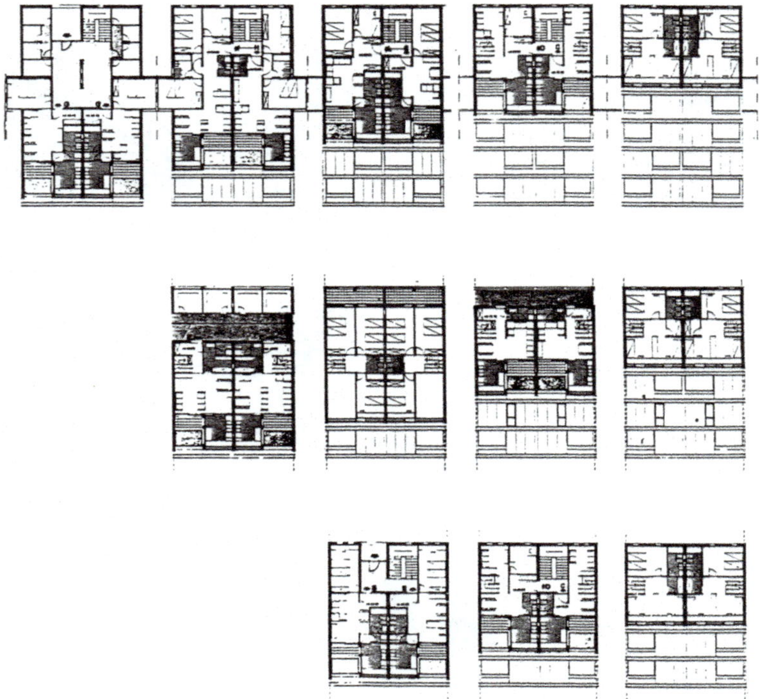

Fig. 8.3 Building types (Toni, M., Zaffagnini, M.)

be attained by using *design variables* (based on calculations and/or metaproject diagrams) concerning both the morphology and the materials at different scales (Fig. 8.6). Anyway, the approach complexity enables to appreciate that the same specific wellbeing goal can be met through different solutions. Besides it, a crossed analysis of the design variables belonging to the three above-mentioned classes (A, T, V) shows the concurrency of some of them in different classes, in such a way that conflicting situations emerge and decisions are to be taken (Fig. 8.7). In this connection, a method to assess the interference is envisaged, which is absolutely necessary to determine priorities for a quality-centered design (Toni 1990).

As proven, the challenges related to acoustic wellbeing cannot be disregarded when investigating an architecture centered on environmental factors and wellbeing alike. Nevertheless, until the late 1980s, only little information was available to the designer on how to attain an acoustically comfortable environment, except for the technical data related to plants or installations. It was then very interesting to study how impactful the physical and technical principles were on the knowledge for a proper architectural design, taking into account both the technological solutions of the construction elements and the morphologic characteristics of the spaces[4] (Fig. 8.8). A later stage of this research was about the protection from acoustic pollution, through a contribution to the ministerial working group entrusted

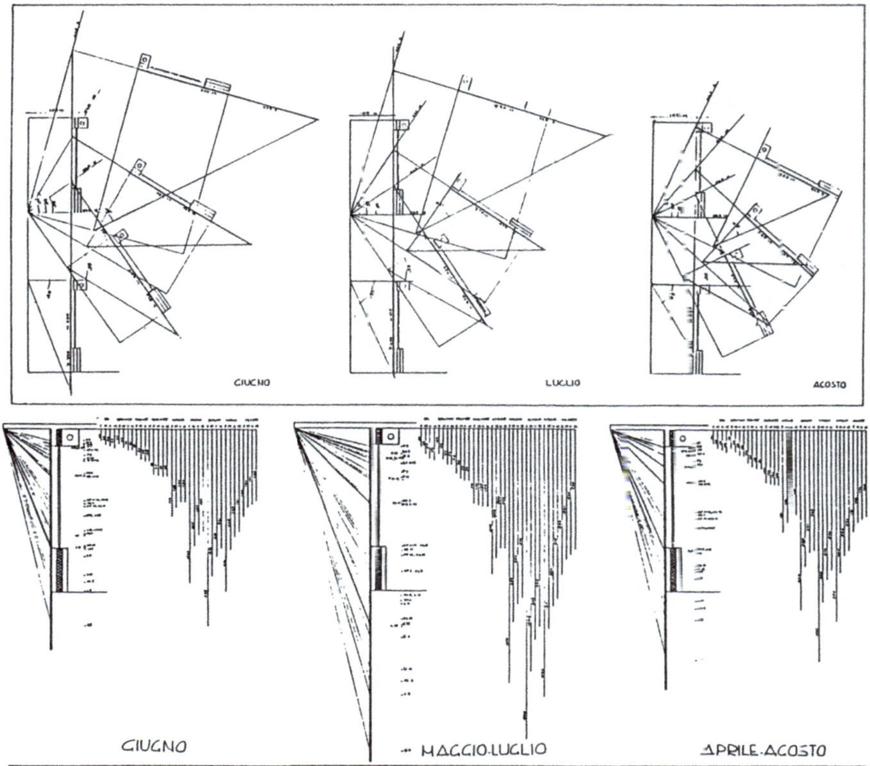

Fig. 8.4 Shielding system study (Toni, M.)

to prepare the 1990s rules which started raising awareness towards the environment and human health (DPCM/Council of Ministers Presidential Decree for workspace health promotion dated March 1, 1991; Framework Law No. 447 concerning acoustic pollution of October 26, 1995) (Toni 1989, 1996).

Improved noise proofing and reception of sound were the subject of a research intended to provide background information for the architecture design. It investigated how to obtain a range of performances by using different materials in sound proofing systems (monolithic or layered elements); and sound-absorbent systems (porous, flexible, resonating elements or mixed systems) (Fig. 8.9) (Toni 1997).

The sound-related issues in theatre design were the theme of an interesting follow-up research (Fig. 8.10) (Toni 1998).

Among the many effects on my teaching activity, an experiment about the acoustic performances of an open-air theatre was implemented through a 1:50 model of professional sound apparatus. The aim was to highlight the focus of acoustic waves and reverberations (Fig. 8.11).

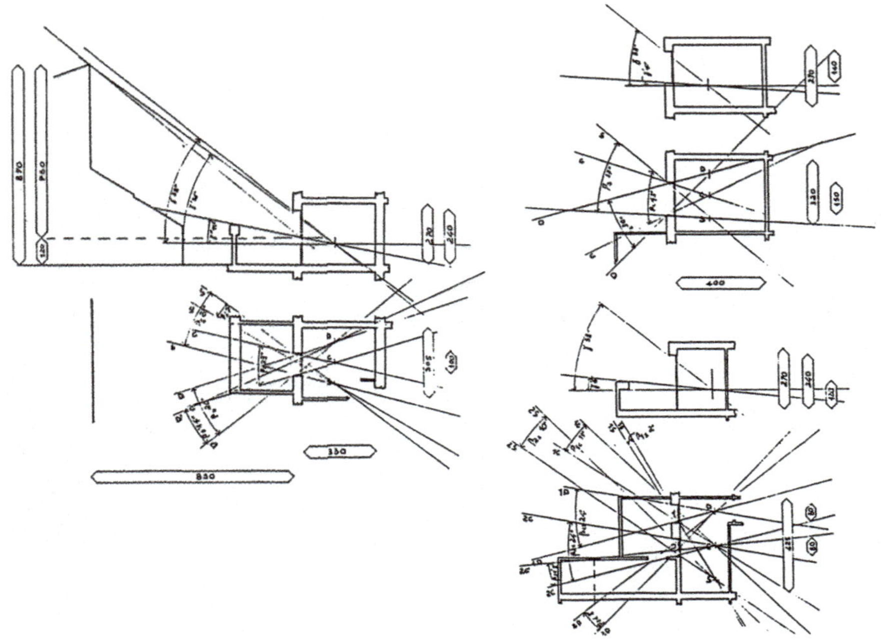

Fig. 8.5 Control of average daylight factor in building type innovations designed for API (the Association of small and medium-sized enterprises) (Toni 1990)

8.1.2 Local Resources

The processes supporting the use of building materials—in particular, natural materials such as materials obtained from plants, reclaimed stones or recycled paper—found locally, were investigated to understand their real potentials in low-impact projects. At a wider extent, the use of wood, soil and plants for restoration of natural environments through naturalistic engineering techniques, was investigated.

The use of materials obtained from plants is a fascinating opportunity for architects to leverage on natural resources, as they demonstrate technically effective for wellbeing. The recovered or recycled materials (mainly, stones or cellulose) are also interesting resources: they are zero impact (to say, available sustainably) as recycling saves energy by eliminating the need to make materials from scratch.

This specific area of research was focusing on the techniques used in the past against potential new models in today's applications.

According to the technical data collected about their actual productions at field companies and operators, it was demonstrated that some natural materials (specifically, reeds) have characteristics useful to the microclimate quality (heat insulation, permeability, sound proofing, healthiness etc.) (Figure 8.12) or to create low-impact constructions and small structures (reversibility, recyclability, biodegradability etc.). This approach also enables for installing yards compatible with requirements for

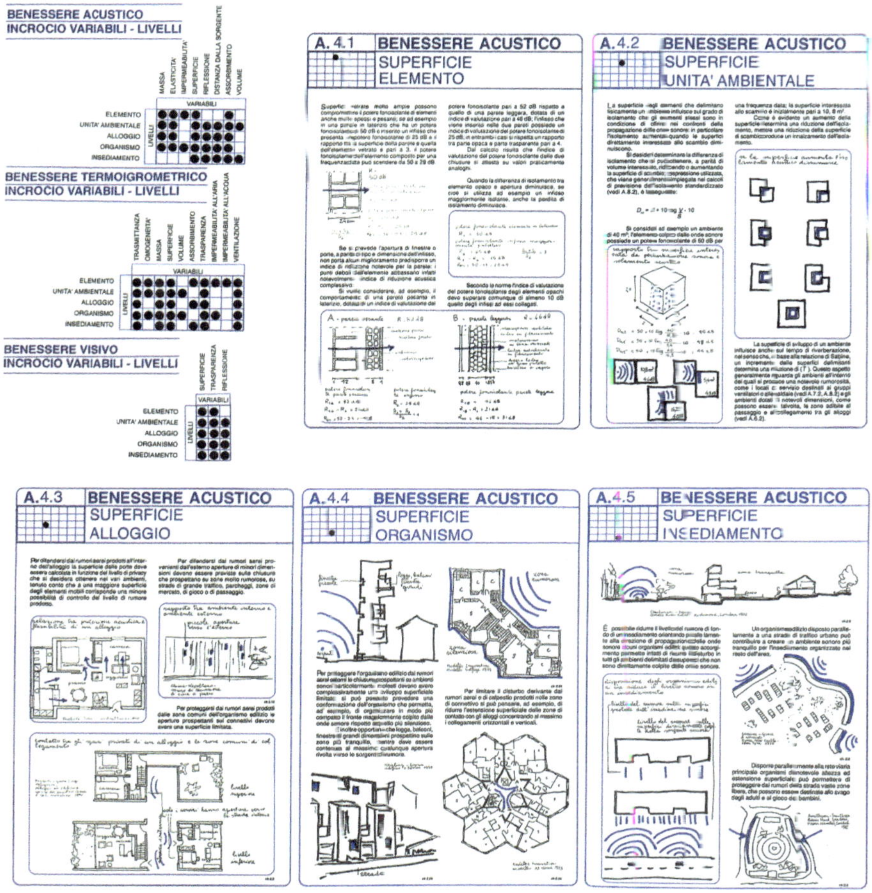

Fig. 8.6 *Design variables—survey levels (or scale) grids* for the executive project. A sample of acoustic wellbeing worksheets, with an in-depth review of the *Surface* variable for the five survey levels (Toni, M.)

environmental protection of natural sites or urban places. In this connection, different types of use for high performance structures are envisaged (Fig. 8.13).

Very interesting outcomes highlight also the close relationship between the available materials and the protection of the environment they come from (not only a natural environment, but also a historical, economic, and social environment, including heritage, know-how, technical abilities, and territory-related structures). Specifically, to obtain vegetal materials, agricultural techniques which can promote biodiversity and comply with the time and mode of collection according to natural cycles, are needed. To have available recovered stones, quarry and lab waste must

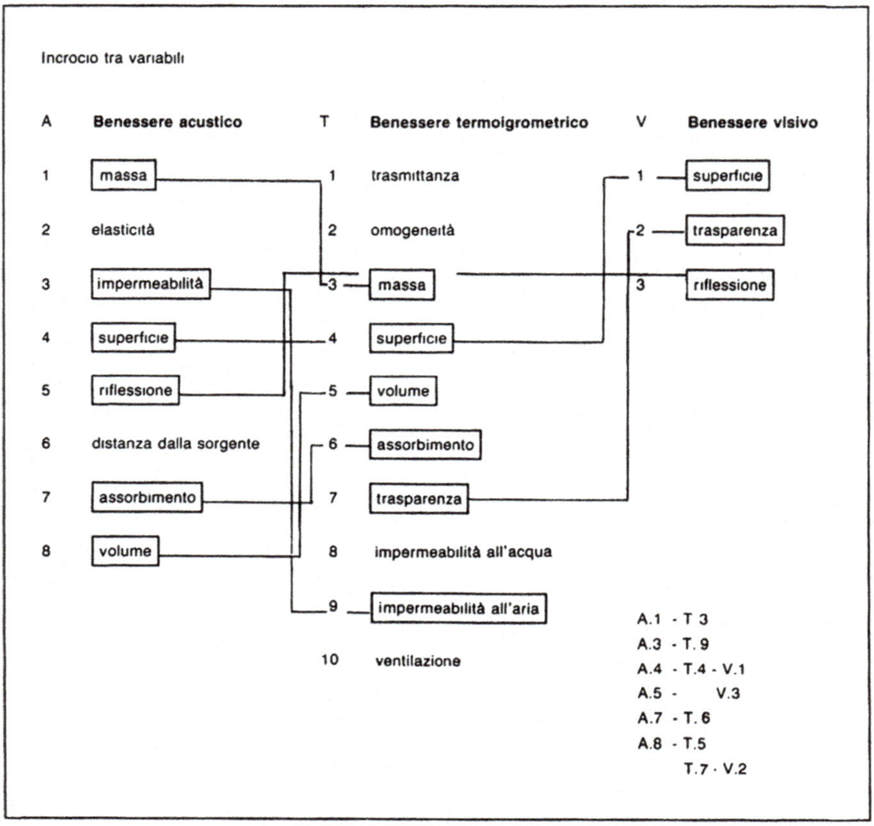

Fig. 8.7 Crossing variables (Toni, M.)

be used wisely based on upstream quarrying activity where raw material and local abilities are used in a discerning manner (Toni 2011, 2012d, 2014b, 2015, 2016a, b, c).

This research activity was combined with design experiments in cooperation with students from multiple academic years. The aim was to investigate how to implement actions compatible with the environment, especially areas of great natural interest[5] (Figs. 8.14 and 8.15).

As activities of key importance for the local economy (such as marble quarrying or fishing) are carried out in some of these areas, then the research emphasizes on how the enhancement of the natural resources through actions in line with complex local dynamics, can prove interesting opportunities to strengthen and develop work activities (Toni 2012b, c).

A specific research project was developed on the use of natural materials for restoration of environmental instabilities in mountain areas, mainly due to poor main-

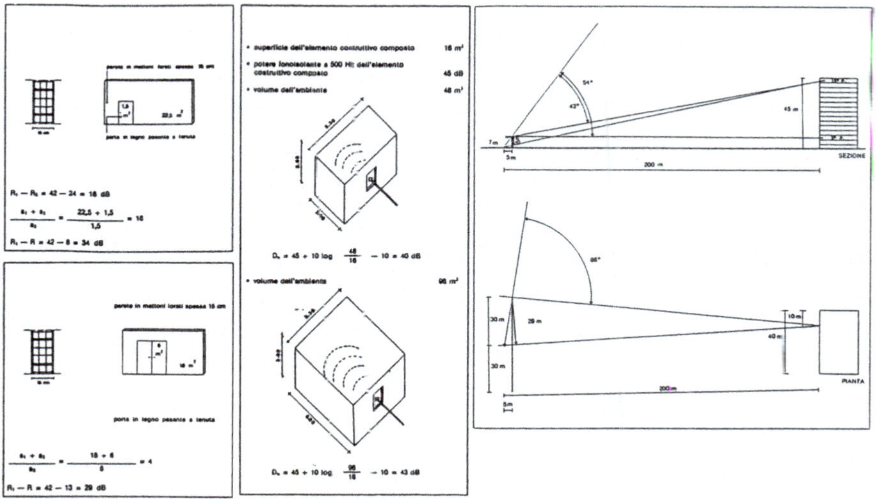

Fig. 8.8 Noise control for improved wellbeing: Impact of a door on sound proofing; change of sound insulation of a construction element depending on the room volume; noise level reduced by a screen (Toni, M)

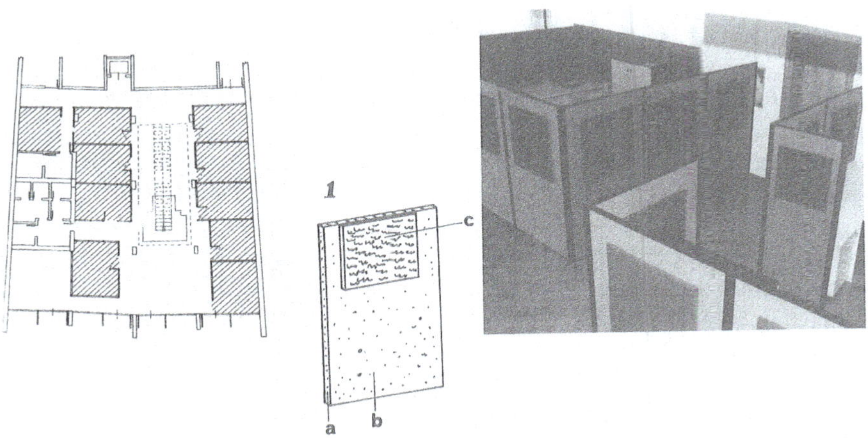

Fig. 8.9 Mobile screens for controlling the reverberation time and reducing the background noise to improve the voice reception in large-size conference rooms (Design for the Gran Sasso National Laboratory, Toni, M., 1994)

tenance of the territory and crop abandonment—all worsened by road infrastructures created without consolidating the slopes or performing regular works on the creek banks.

Project types based on natural engineering, to be made through poor materials, for the most available in situ (wood, stones, soil and plants) are taken into account.

Fig. 8.10 Open air theatre: the set is made up of arches covered with vegetation (*2002 Tercas architecture competition, 15th edition, Design of Palazzo Cicada, Atri, and arrangement of the archaeological site with open air theatre*, Capuani, D., Toni, M., 2002)

This approach marks a sharp change in design strategy in comparison with an indiscriminate use of concrete, as it helps the restoration of the natural abilities to drain the rainwater, covering with vegetation and support sides and slopes, which reflect the natural balance conditions of a territory, both under a functional and sensorial perspective (Fig. 8.16).

The case study of Upper Versilia and Garfagnana—with so many projects developed following floods in late 1900s—was the focus of this research.

The study mainly investigated the impact of works aimed at consolidating the banks of the streams from erosion and reducing the speed of the waters; restoring the slope of sides that prevents the formation of landslides and breakaways (Fig. 8.17); regulating and channelling rainwater against run-off. On some simple typologies, work experience with students was carried out (Fig. 8.18).

Research shows that these works allow the installation of sites with minimal impact on the environment, as they use light equipment; furthermore, they promote the employment, since local labour can specialize without too much difficulty (Toni 2007b).

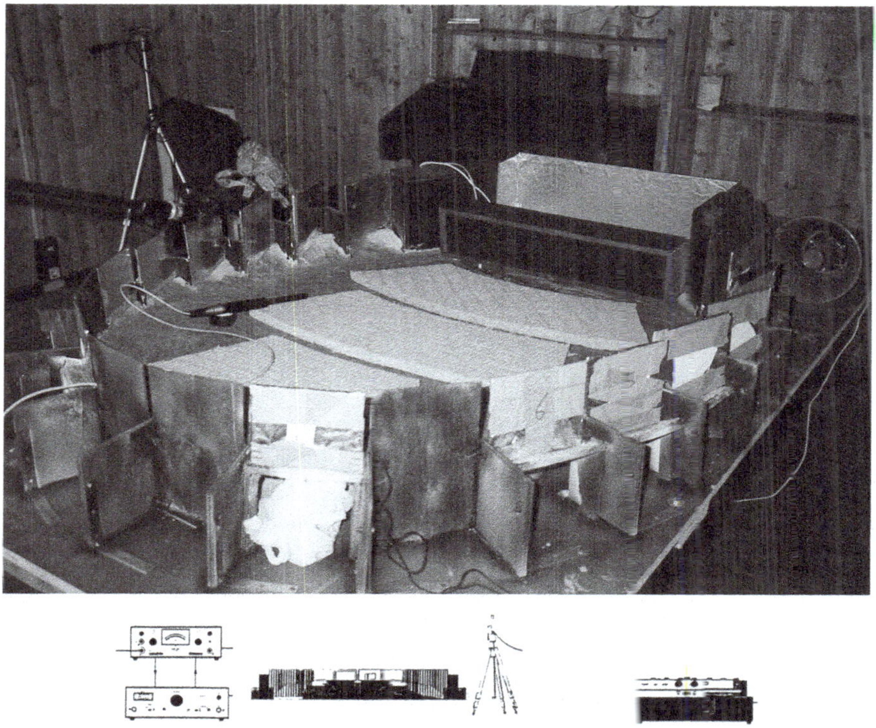

Fig. 8.11 Acoustic trials on a 1000 people theatre mockup (sound assessment, Toni, M.; theatre design, dissertation work by Ballanti, C., Bonetti, A., and Gamberini, A., 1999)

Fig. 8.12 A school building, heat insulated by using reed panels (pictures by Leo Bodner)

Fig. 8.13 A structure made of reeds (dissertation work by Ciociola, A., Palmisano, P., and Vischi, G., supervisor Toni, M., co-examiners Capuani, D., and Iacono, G., 2009)

8.1.3 Light Technologies

Technologies used in construction in a still marginal way, such as Fiber Reinforced Plastics (FRP), demonstrated to be proactively used also in the field of bioclimatic architecture, despite the apparent contradiction of being formed with synthetic materials.

The research deals with production outputs (mainly FRP pultruded profiles) which at the time still represented a potential market for innovation, being mainly used in sectors other than construction (aeronautics, aerospace, shipbuilding); but they can turn into a real market under the pressure of a new attention to specific problems, including environmental ones.

Information was obtained from manufacturers of semi-finished products based in Italy and Europe, participation in the Jec Composites Show in Paris, study trip to Maunsell, in Kent, a leading company in the design and production of FRP (1999); moreover, visiting projects in England and Scotland (1999).

The research focused on those characteristics that make composites different from other materials, so they are irreplaceable for specific applications in which environmental and safety-related aspects of people in particular contexts also come into play.

Chemical resistance against aggressive industrial and atmospheric agents, salt, rust which saves materials for restoration and hours of work in repeated maintenance cycles.

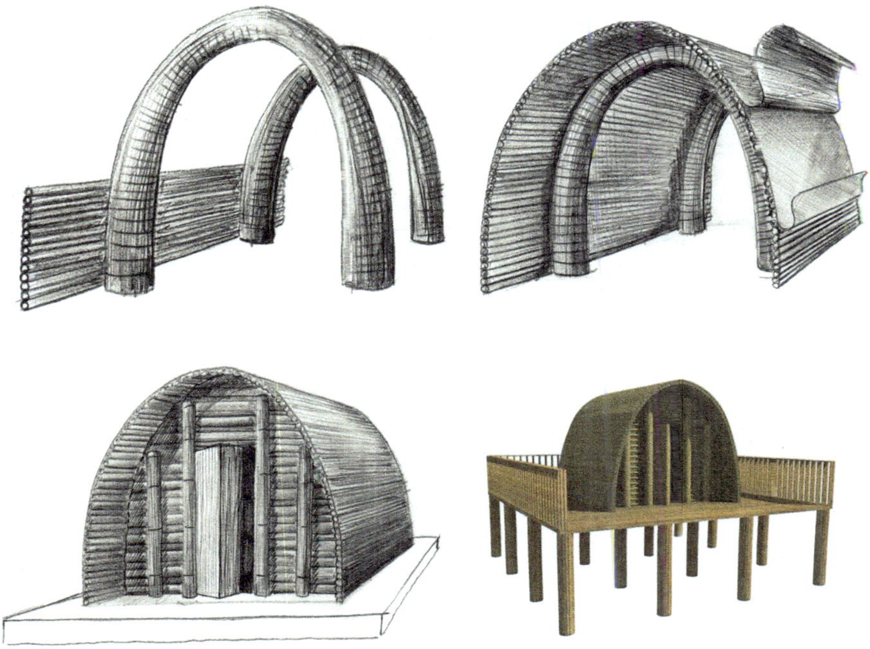

Fig. 8.14 Temporary huts built of reed bundles for a sustainable tourism project. They reappraise the conventional houses in the Tigris and Euphrates wetlands (from a dissertation work by Beltrami, S., and Cherubini, V., supervisor Toni, M., 2009)

Low conductivity for buildings without thermal bridges.

Radar transparency, respect for natural electromagnetic fields and electrical insulation for applications in which sources of danger to humans can be determined (Fig. 8.19).

Lightness that allows you to create mobile structures; pre-assemble parts of structures in the workshop, transport them to the construction site and assemble them in a few hours, reducing the impact of the construction site (for noise, dust, functional interference); disassemble a product without leaving traces on the environment if the parts are dry mounted; and burden the foundations with less loads.

Ability to deform without breaking, offering the possibility of enduring seismic actions with greater safety; improving the passive safety of people against impact (for example, in road fences).

Possibility to colour the paste material, carry out skin treatments in a bath and obtain translucent materials with savings on maintenance and interesting eye-appealing characteristics.

Evaluations of the impact produced on the environment by FRP products over a certain number of years have shown that in some applications the eco-balance can also be advantageous in comparison with natural materials such as wood.

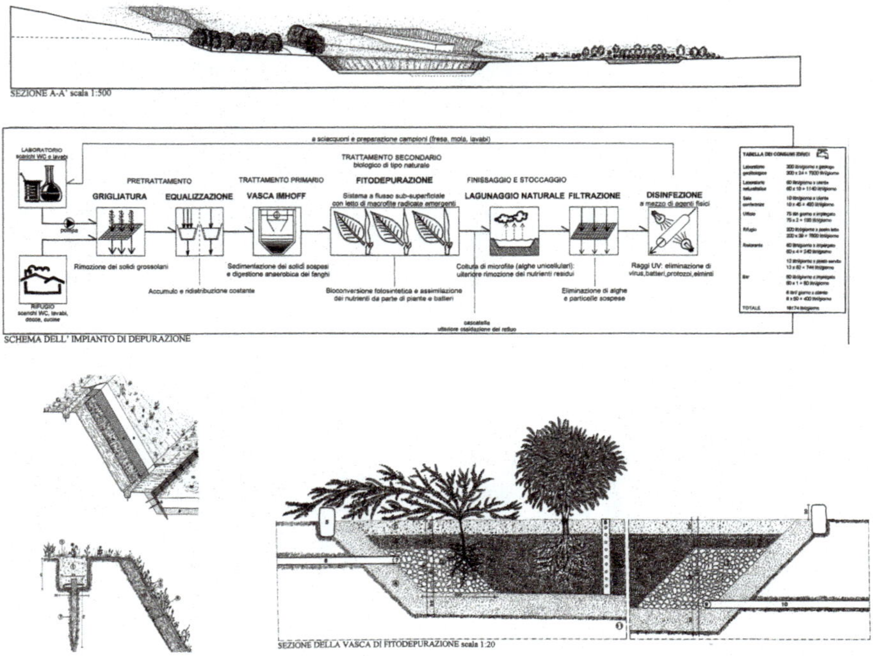

Fig. 8.15 Project of a phytopurification plant (from the dissertation work of Zaccherini, S., supervisor Toni, M., co-examiners Capuani, D., and Iacono, G., 2010)

Fig. 8.16 Contrast to the impoverishment of the places with restoration of unevenness, creation of fertile ground and growth of vegetation for the enhancement of the various environmental components (Toni, M.)

Fig. 8.17 Osiers for soil containment and vegetation reconstitution (Biancalion, F.)

There is also the possibility of modifying the performance during production (choice of resins and fibres) to reduce critical issues (fire behaviour, moisture absorption, and UV resistance) (Toni 2002).

During the research, the construction possibilities were investigated by checking the chance of assembling FRP semi-finished products manufactured by companies on the European market, connecting them dry or by gluing them depending on the need of the construction system (Fig. 8.20).

To form a critical vision of the architectural language that can be developed with the use of composites in resins and fibres, the research activity has allowed to discover unpublished examples of achievements in the several European countries (United Kingdom, Denmark, Finland, Spain, France, Germany, Italy, and Switzerland).

It has been discovered that pedestrian bridges and small driveways are the most documented construction typology in current applications, also because they can be positively included into the environment and a construction site can be easily installed with reduced working times and minimal impact with the places (Fig. 8.21). Furthermore, particularly innovative cases are facade systems with high thermal insulation in variously coloured composite panels which determine an interesting visual impact (Toni 2012a).

Starting from here, the research allows to develop reflections on the possibility of obtaining an architecture consistent with the characteristics of this material and constructions that are coherent with the contexts (Toni 2005, 2007a). Confronting the use of FRP in new projects proves, therefore, an interesting stimulus (Fig. 8.22).

Fig. 8.18 Construction site with students involved in the construction of a channel for water management (Regional Park of the Apuan Alps, 2005) (picture by Michela Toni)

Fig. 8.19 Reinforcement meshes for concrete made of AR glass fibres to replace the current reinforcement meshes of steel—interesting in the field of bio-architecture (pictures by Fiberline Composites)

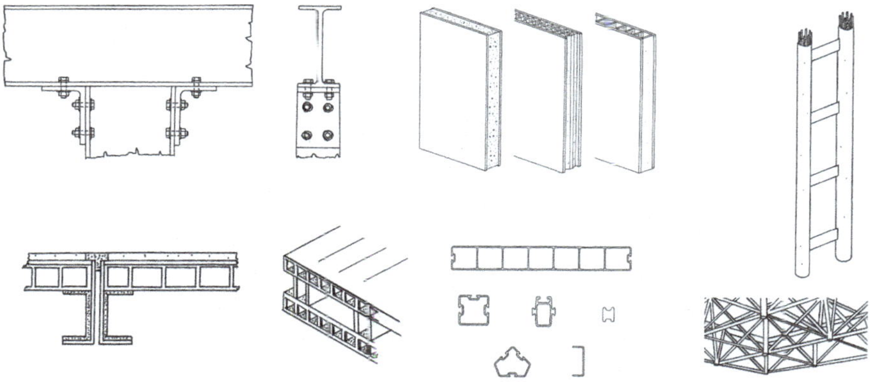

Fig. 8.20 Construction systems with FRP semi-finished products (Toni, M.)

Fig. 8.21 Pedestrian bridge made entirely of FRP, integrated in a golf course in Aberfeldy, Scotland (picture by Michela Toni)

The experience also had a strong impact on an intense project research work with the students, as they demonstrated that this technology can be proposed both for natural sites and historical towns (Fig. 8.23), with the advantage of being able to completely disassemble the works without leaving traces to meet new needs.

8.1.4 *Quality Processes*

The theme of the relationship between technology and the territory in which it is used is addressed to deepen the contribution that certain visions of construction practice can make to the social and environmental ecosystems of the different places. The search for a link between technical skills and culture of the place they are used in is

Fig. 8.22 A technological mix in Viareggio: FRP for a mobile bridge that must be light, and for access ramps; marble, the historical material of the nearby Apuan Alps, for the two towers that make up small squares at high altitude for stopping people during the walkway motion. (*Competition for the design of a mobile cycle-pedestrian walkway for crossing the Burlamacca Canal in the Municipality of Viareggio*, Toni, M., 2001)

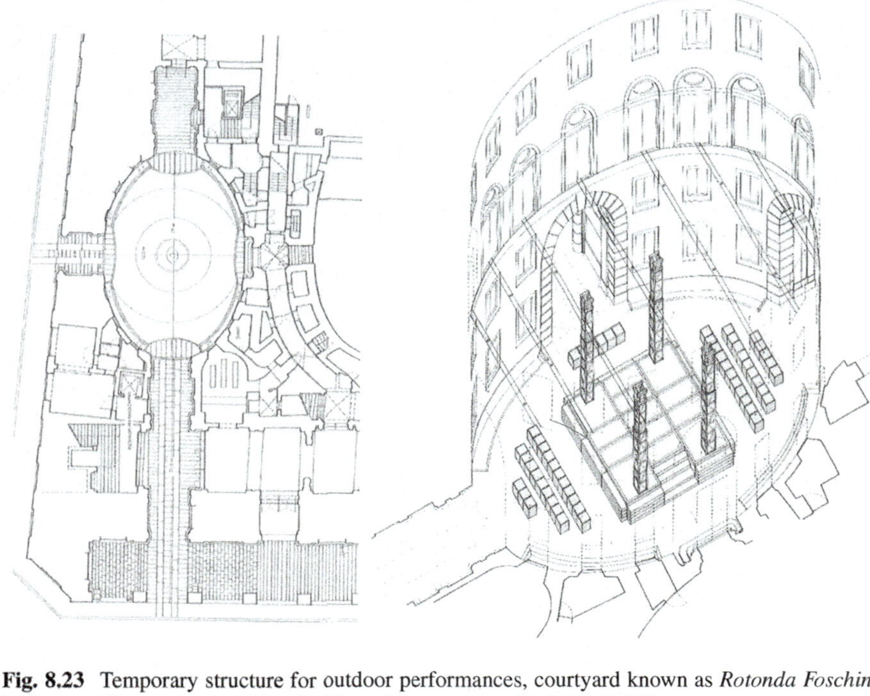

Fig. 8.23 Temporary structure for outdoor performances, courtyard known as *Rotonda Foschini*, in Ferrara (dissertation work by D'Allara, C., supervisor Toni, M. (1990)

chosen as a key to understanding what are the conditions that allow you to develop innovative processes in advanced contexts. The aim is to understand how to obtain the success of similar approaches in other contexts and on what factors to act beyond the mere technical aspect.

Focusing on energy control and environmental care, a first-line observatory for construction in Italy has been identified in the Autonomous Province of Bolzano. It was, therefore, obvious to study the reasons for their quality process.

Thanks to years of attendance, direct experience has been gained with numerous construction solutions that take into account the enhancement of natural resources (Figs. 8.24 and 8.25). In addition to the concrete results, however, what particularly struck, was the strong integration between the various subjects involved in the process of construction (administrators, public and private customers, designers, construction companies, manufacturers, etc.), which highlighted a territory where all the community players interact to achieve a common goal[6] (Toni 2013, 2014a).

In fact, the research work has helped to highlight the strong link between technical skills and social development, making it clear that even if centered on shared values, such as the environment, a way of building cannot be exported dogmatically and used successfully outside of a process of interrelation among the components that make up the material and intellectual territory in which it must be adopted. In this

Fig. 8.24 Pfitscher winery in Montagna (Bolzano), first company in the sector with *CasaClima Wine* seal of quality, based on: eco-compatibility and energy efficiency of buildings; use of local materials; architecture modelled on the topography of the land; rainwater recovery; reduction of the soil waterproofing level; rational use of water and control of waste production and pollution of the winemaking, packaging and product transport process, arch. German Gabaliri (picture by Michela Toni)

Fig. 8.25 Geothermal plant with the soil as an energy source, Georg Felderer, Energytech (picture by Georg Felderer)

sense, South Tyrol can make a contribution to think about which land management models can be suitable for other contexts, essential in a historical phase such as the one in which we find ourselves to build wealth and ties to look beyond the crisis.

As a further study, the case of IPES (Istituto per l'Edilizia Sociale della Provincia Autonoma di Bolzano), the largest builder in South Tyrol, has been committed for years to create high housing quality buildings, with particular attention to the energy and social sharing sphere; recently (2014–2018), IPES also participated in European programs for the construction of sustainable urban areas[7] (Toni and Coccagna 2018).

The goal of the research activity was to study the working method and tools developed by the Institute to obtain excellent results.

Data collected through field analysis made it possible to learn about innovative technological solutions regarding buildings, systems, and external arrangements resulting from IPES know-how. With regard to the method used, IPES entrusted the design to external professionals and engaged its in-house technicians in the works management in order to get new ideas and the best quality for money. With regard to the tools, IPES has developed a regulatory framework to facilitate the designers and builders' choices by correlating the required quality levels (on energy consumption, CO_2 emissions, safety, usability, durability, maintainability and other variables) and the related costs that you are willing to bear.

The research showed that the outcomes are supported by a fruitful management model of the design and construction process. As a matter of fact, wide-range environmental advantages along with economic advantages for the builders and economic and social benefits for the dwellers, are obtained (Fig. 8.26) (Toni and De Fonzo 2014). Therefore, sharing this method with other subjects involved in the

Fig. 8.26 Full-scale models of parts of buildings to be built, delivered to IPES by construction companies on the occasion of tenders to demonstrate the technical capacity to build them (picture by Michela Toni)

building activity could contribute to experiment new approaches and improve the environmental quality at multiple scales.

A final research work I feel the need to share is the creation of an application resulting from the belief that an architecture aware of the climatic factors and the many issues of the local context cannot ignore people who live there, not only because it is based on their needs, but also because it is conditioned by their behaviour.

In concrete terms, the creation of the ASA (Sustainable Approach to Living) application aims to actively develop the involvement of the inhabitants for the use of a building, which leads to better liveability and reduced management and pollution related costs (Toni 2019).

The field of action is that of Public Residential Building, where different players converge. The research deals with the process of knowledge that involves each of them (designers, technicians of housing agencies and construction companies, tenants) and brings in new actors in the role of tenant trainers: the students from the last year of upper technical schools with the assistance of their teachers. Through the shared work of students and tenants supported by the ASA App, the latter learn how to manage their home based on factors that change over time (activities, climatic factors and more), to achieve savings and well-being (Figs. 8.27 and 8.28).

Research is underway; in fact, we are currently looking for parties interested to support the experimentation of the project on the basis of the economic and social advantages it promises for the entities that manage the buildings and the inhabitants.

And it is significant to end the story on the different stages of the work that has committed me from the beginning of my personal journey on this theme, with a research that awaits further developments, because the activity that has been—and is—one with life goes ahead at its own pace.

Notes

1. At that time, the solar energy world was quite a small world! At the *Brighton Exhibition*, I met by chance several technical heads from SMI, and Mario Grosso,

Fig. 8.27 Process outline: Family household using the ASA app on iPad to identify the rooms where each member is engaged in his/her routine activities (Toni, M.)

Fig. 8.28 An ASA customization of the link between the heat generated by each occupant and the activity he/she is engaged in inside a room (heat is expressed as Kcal/h) (Toni, M.)

at the time a young researcher at Politecnico di Torino, Turin, whose research and teaching work was an incentive for this publication.

2. *Il sole e l'habitat*, a competition published by the then Ministry of Industry (the Ministry of the Environment was established later) and INARCH, the Italian Institute for Architecture, in 1979. According to the competition announcement, 70% of the thermal load for heating and supply of sanitary warm water in public housing buildings and school buildings had to be provided by active and passive solar devices.
3. Sponsored by ANDIL-ASSOLATERIZI (the Italian Association of Brickwork Contractors) and CPTC (the Society of Concrete Roof Tile Manufacturers).
4. For the account of ANCE (the Italian Association of Private Construction Contractors).
5. Specifically, within the course *Progettazione Ambientale* (academic years 1995–1999) and *Laboratorio di Sintesi Finale di Progettazione Ambientale* (academic years 1999–2012) I taught as regular professor at the University of Ferrara's Department of Architecture.
6. The research experience continues today. In the 2012–2013 academic year, a sabbatical was dedicated to structuring the vast study material collected between 2006 and 2012; the result is a history of the human, social and cultural environment of that area (Toni (25).
7. In particular, the Sinfonia (Smart Initiative of Cities Fully Committed to Invest in Advanced Large-scaled Energy Solutions) project, co-financed by the European Union on the basis of the 7th Framework Program, through the "Energy—Smart Cities & Communities 2013" call, with which Bolzano, together with Innsbruck, is preparing to become the first smart city on the continent.

References

AAVV (1980a) *Il sole e l'habitat—162 progetti per l'impiego di energia solare nell'edilizia residenziale e scolastica*. Edizioni Kappa, Roma

AAVV (1980b), Concorso nazionale "Il Sole e l'habitat" per l'impiego di energia solare nell'edilizia scolastica e residenziale. In *L'Architettura. Cronache e storia*, n. 292, febbraio 1980

AAVV (1981a) *Il tavellone—Impieghi e prospettive di sviluppo di un semilavorato per l'edilizia*. Laterconsult, Roma

AAVV (1981b) *Il laterizio tra passato e futuro—Mostra organizzata dalle Sezioni Merceologiche dell'ANDIL. Associazione Nazionale degli Industriali dei Laterizi*, Laterconsult, Roma

AAVV (1982) *Il laterizio e la qualità dell'abitare*. Laterconsult Roma

Toni M (1980a) Il collettore solare come componente edilizio. In: AA, *Il nodo energia*. Edizioni Ente Fiere di Bologna, Bologna

Toni M (1980b) *Tetti ed energia*, Vecom, Verona 1980, SAIE 1980

Toni M (1981) *Il sistema tetto—Accessori e dispositivi di completamento della copertura*. Vecom, Verona 1981, SAIE 1981

Toni M (1987) La qualità ambientale, La qualità tecnologica: la resistenza al fuoco. In: AAVV (ed) *Rosso Mattone*. Luigi Parma, Bologna

Toni M (1989) *L'isolamento acustica nell'edilizia—Guida alla progettazione del benessere acustico.* Edilstampa, Roma

Toni M (1990) *Qualità involucro—Contributo attorno ai problemi del comfort abitativo.* Pitagora Editrice, Bologna

Toni M (1996) *L'isolamento acustico nell'edilizia—Guida alla progettazione del benessere acustico—2° edizione interamente rivista e aggiornata con la legge n. 447 del 26 ottobre 1995.* Edilstampa, Roma, aprile 1996

Toni M (1997) *Interventi, materiali e strumenti per l'isolamento acustico degli edifici.* Maggioli, Rimini

Toni M (1998) Il teatro Verdi: un 'idea di flessibilità. In: *Ingegneri Ferraresi*, periodico dell'Ordine degli Ingegneri della Provincia di Ferrara, numero zero 1998

Toni M (2002) Ipotesi sull'utilizzo dei materiali compositi in edilizia civile. In: Toni M, Bassi C (eds) *Il filo di Arianna—Idee su Ferrara.* Alinea, Firenze

Toni M (2005) *FRP Architettura. Costruire con materie plastiche rinforzate con fibre.* Alinea, Firenze

Toni M (2007a) *FRP Architecture. Building by fiber-reinforced plastics.* Alinea, Firenze

Toni M (2007b) *Rinforzare sostenibile—Idee sull'utilizzo dell'ingegneria naturalistica attraverso esperienze in Alta Versilia e in Garfagnana.* Alinea, Firenze

Toni M (2011) *Un materiale e il suo ambiente—Utilizzo della canna palustre nelle costruzioni.* Alinea, Firenze

Toni M (2012a) *Colorata e leggera—Architettura in materie plastiche rinforzate con fibre.* Alinea, Firenze

Toni M (2012b) *Foce di Pianza—Ipotesi per un laboratorio di progettazione ambientale.* All'Insegna del Giglio, Borgo San Lorenzo

Toni M (2012c) *Goro lavoro natura Gorino—Salvaguardare l'ambiente, consolidare le opportunità di lavoro.* All'Insegna del Giglio, Borgo San Lorenzo

Toni M (2012d) *Un ambiente di marmo.* All'Insegna del Giglio, Borgo San Lorenzo

Toni M (2012e) *Sole e non solo. L'architettura prende coscienza.* Alinea, Firenze

Toni M (2013) *Sudtirolo Architettura—Le differenze di un territorio attraverso costruzioni realizzate nel rispetto della normativa CasaClima (Südtirol Architektur—Die Eigenheiten eines Gebietes am Beispiel von Gebäuden mit KlimaHaus Standard).* Edicom, Monfalcone

Toni M (2014a) Il territorio culturale di un'eccellenza tecnologica—Una ricerca sul campo in Sudtirolo (The cultural territory of a technological excellence—a field research in south tyrol). In: *TECHNE*, vol 08, No 7

Toni M (2014b) Isolare con la carta—Alte prestazioni dalla natura (Insulating with Paper. High performances from nature). In: Alfonso Acocella (a cura di) *Paper Design.* Altralinea, Firenze

Toni M (2015) Produzione edilizia in continuità con il territorio (Interdipendence between building resources and territory). In: AAVV *Sostenibilità ambientale, economia circolare e produzione edilizia. La ricerca scientifica nel Settore delle Costruzioni nell'era delle nuove sfide ambientali e digitali (Environmental sustainability, circular economy and building production. The scientific research in the constrution Industry in the age of new environmental and digital challanges).* Maggioli, Santarcangelo di Romagna

Toni M (2016a) Architettura a chilometro zero ad Amsterdam (A zero-mile architecture solution in Amsterdam). In: Toni M (a cura di) *Tecnologie e territorialità. Riflessioni su nuove visioni di sostenibilità (Technologies and territorial milieu. Addressing new visions of sustainability).* Moderna, La Spezia

Toni M (2016b) Costruire con materiali naturali locali (Building with Local, Natural, Materials). In: Toni M (a cura di) *Tecnologie e territorialità. Riflessioni su nuove visioni di sostenibilità (Technologies and Territorial Milieu. Addressing New Visions of Sustainability).* Moderna, La Spezia

Toni M (2016c) Tecnologie e territorialità (Technologies and Territorial Milieu). In: Toni M (a cura di) *Tecnologie e territorialità. Riflessioni su nuove visioni di sostenibilità (Technologies and Territorial Milieu. Addressing New Visions of Sustainability).* Moderna, La Spezia

Toni M (2019) The ASA (Approccio Sostenibile dell'Abitare) App. A sustainable housing approach tool to raise tenants' awareness in the public housing sector. In: Daniotti B, Imperadori M, Lupica Spagnolo, S (eds) *SBE 19 Milan—resilient built environment for sustainable mediterranean countries*, IOP conference series. Earth and environmental science, vol 296. IOP Publishing
Toni M, Coccagna M (2018) The role of IPES social housing in the EU Sinfonia Project for a "Bolzano Smart City". In: *TECHNE Special Issue*. 01, 2018
Toni M, De Fonzo M (2014) Tools for enhanced sustainability plans in urban areas. In: 40th IAHS world congress on housing, *sustainable housing construction*, Funchal, 16–19 dicembre 2014, IteCons, Coimbra

Chapter 9
Bioclimatic Design and Environmental Impact

Adriano Magliocco

Abstract The paper focuses on the relationship between bioclimatic design and environmental impact in the evolution of what is called a "solar city". Since the ultimate goal of a sustainable project should be to reduce the overall environmental impact, it is appropriate to reflect on the relationship between building and site, considering that, from the settlement point of view, it cannot be said that many of the most famous interventions, from the 80s to today, are to be considered successful.

Keywords Settlements · Solar city · Environmental impact

9.1 Introduction: Bioclimatic Design and Solar Technologies, a Misunderstanding

When it comes to bioclimatic design, we cannot fail to refer to the text that has trained all the scholars of this design approach, i.e.: "Design with Climate. Bioclimatic approach to architectural regionalism" by Olgyay (1963). Olgyay explicitly proposes an inter-disciplinary approach, in which biology, meteorology and engineering skills are applied to building design. The aim of bioclimatic design is to make buildings manageable from the point of view of hygrothermal comfort, using strategies and construction technologies that take into account the local climatic characteristics, reducing the intervention of heating and air conditioning systems as much as possible. In the 1960s, there was a tendency to make the environmental problem coincide with the problem of the fossil fuels. The issue was how to reduce oil and gas consumption by carefully designing these aspects. Olgyay spoke of "architectural regionalism", highlighting how, in pre-industrial times, vernacular construction was clearly characterised, in forms and construction methods, by available resources, climate and, consequently, socio-economic structure. It was clear that he did not want to re-propose the building types of ancient peoples but to indicate the need to employ adaptation principles that reduced the consumption of resources with limited

A. Magliocco (✉)
DAD, Università di Genova, Genoa, Italy
e-mail: adriano.magliocco@unige.it

G. Chiesa (ed.), *Bioclimatic Approaches in Urban and Building Design*, PoliTO Springer Series, https://doi.org/10.1007/978-3-030-59328-5_9

availability (even if the reference to vernacular construction was taken as a starting point by many "bio-architects" also from a lexical point of view).

This important text guided the entire first season of design experiments inspired by the bioclimatic approach and, in particular, by "solar" architecture, often in a "didactic" way, being oriented towards experimenting with technological solutions. A great misunderstanding in the transfer of this approach to professional practice was to match the bioclimatic approach to the use of passive solar technologies, which led to a new "international movement" of greenhouses and roof pitches facing south. In the 1990s, solar greenhouses also flourished in climatic zones where what is gained in winter is abundantly thwarted by the risk of overheating in summer. This misunderstanding, while being understandable given that solar energy is a much better distributed and manageable resource even with minimal equipment compared to other free sources (wind, etc.), has often shifted bioclimatic design away from more complex concept of "designing with the climate" simplifying it into "designing with the sun".

If we review those experiments, particularly those on an urban scale, it is evident that it had been forgotten that a human settlement was born not only to repair man from bad weather (Olgyay speaks several times of "shelter") but to create a favourable social environment, with interactions influenced partly by spatial organisation (but probably spatial organisation is a mirror of the inhabitants' relational modalities). Living is more than just an energy problem, although this cannot be ignored, and a city is not a set of buildings, but a "place of opportunities". But perhaps we are at the end of the city and distinguishing it from differently organized territory (Benevolo and Erbani 2011) is increasingly difficult.

9.2 Solar Cities

The concept of "solar city"—a city that owes its energy supply mainly to local production using solar energy—has changed slightly over time, shifting gradually from prevalently passive to active energy, particularly photovoltaic (PV). The first interventions experimented with different technologies for the exploitation of solar energy but calling these agglomerations cities is extremely misleading. The name city in the English-speaking world is a large town of considerable importance. This term can hardly be used with reference to settlements that are mainly residential and peripheral. The city was born as a meeting place, for commercial exchange and social interaction; the city was born around communication routes, be they roads or rivers. The first generation of solar cities, on the other hand, seems to refer more to the garden city of Ebenezer (1898) than to the Roman civitas. In 1991, Jeffrey Ross Cook, a Canadian-born professor and researcher known for his studies on bioclimatic construction and Native American vernacular architecture, was "Searching for the bioclimatic city" (Cook 1991). It is clearly from bioclimatic design that the first concept of Solar City was born.

Solar City Linz is one of the first constructions of settlements guided by the reduction of energy consumption. Designed in the late 90 s and built at the beginning of the new millennium, it was considered a reference model, having achieved energy values that were very low for the time, thanks to the bioclimatic strategies implemented. It can be considered a success story from several points of view. Despite its structure, it cannot be considered a case of urban sprawl. It is rather small and is located in the immediate outskirts of the city, on former agricultural land intended for industry, which hit recession, resulting in the area's reassignment for residential expansion (Castelli 2010). Its settlement layout is typical of the US and more recent northern European peripheral expansions. It has their mono-functional destination and configuration, dictated more by the desire to experiment with design strategies and technological solutions than to constitute a real urban fabric. Possibly also due to its name, it was an initial model of "solar city" which, if replicated on a large scale, configures a settlement mode (as has actually happened in many places), potentially heralding impacts due to the extension of infrastructure and consumption due to transport. A slightly more "radical" version can be found in Denmark's ecovillages (Magliocco 2004), where passive solar solutions and increasingly mature energy production technologies (solar thermal vacuum, micro-wind, etc.) combine with morphological references taken from the first experiments of the American counterculture—remember the "zomes" of Drop City? (Sadler 2006)—in small peripheral urban areas made up of one-two-storey houses.

Williams (2016) illustrates how experiments conducted on the construction of entire low-energy neighbourhoods—with additional features aimed at reducing the impact of human activities such as water consumption, waste management etc.—limited themselves to innovating technologies and reference standards (on energy consumption but also on other environmental aspects) without, however, making a transformation in the approach to the development model. Nonetheless, they have created replicated models, even in countries other than those in which they were developed. In the three examples brought by Williams—BedZed in England, Hammarby in Sweden and Vauban in Germany—it is evident that, between the beginning and end of the first decade of the 2000s, a process of technological updating developed. Within this process, active and passive solar technologies work side by side and become more and more present, applying a model of solar city in transition. "These projects incorporated several innovative concepts which had become popular within Swedish government (national and local) over a 30 year period—systems thinking, holistic solutions, circular metabolism, integrated infrastructure planning and low carbon development—but had not yet materialised".

Regarding BedZed, a project that has had great echo in the media and is considered the first zero-energy district, Chance (2009) from Bioregional (the non-profit company that promoted the operation). analysed it during the seven-year follow-up and wrote "…BedZED residents live and work outside of the boundary of the site, so most of their lifestyle choices are affected by the infrastructure, products and services that enable (or frustrate) their ability to adopt sustainable lifestyle choices". The ecological footprint calculated in 2009 (Hodge and Haltrect 2009) was 4.67 ha per inhabitant, compared to an average 5.32 ha per inhabitant in the suburb of Sutton, where BedZed is located. This is due to the weight of transport (most of the inhabitants, at least that year, were professionals) and the failure of the biomass central heating system. In the file published by Bioregional, from which Chance took the data, it is interesting to see how, in the face of a reduction in energy consumption for space heating, the environmental weight of transport increased, despite a reduction in the parking spaces that should have discouraged inhabitants from owning their own car, preferring a car sharing system. The neighbourhood looks pleasant and the evaluation revealed an appreciation by the inhabitants, also in terms of social cohesion. The architecture of the buildings mediates between tradition (thanks to the use of exposed brick) and contemporaneity. The defect is one shared by all these interventions: life in the suburbs needs new infrastructural networks and bioclimatic buildings need inhabitants who are willing to learn how to manage their home. It isn't merely a matter of using a remote control to turn a system on and off. However, the countries that began addressing environmental problem a few years ago refer specifically to projects like this (Xue and Meng 2013).

Both these experiments and the resulting rules and regulations mainly affect new buildings, having only a marginal influence on extensive renovations. Williams (2017) also reminds us that, in the three examples, a cultural change was also hoped for (especially in the cases of Hammarby and Vauban, with participatory forum processes etc.), shared and supported by public bodies. This change, however, is not always transferable to other social and political contexts. The three cases in question were built on a brown field—on land to be redeveloped—making one wonder if it is possible to take them as a model, imagining a total renewal of the existing buildings and a substantial transformation of the urban fabric and the image of the consolidated city, or imagining a continuous expansion in the face of uncertain demographic trends. In both cases, the risk is the shift of the impact from energy consumption during operation to the production and construction phases (in relation to installation equipment and construction methods. See, for example, Goggins et al. 2016).

9.3 The Environmental Impact of the Construction Sector

The bioclimatic project was created in order to re-propose an "intelligent" approach to construction, in an era in which a globalising approach to construction design—which seemed to ignore place and climate due to the availability of fossil fuels at low prices—began to seem absurd. And if the focus was on controlling the consumption of energy resources, which seemed to be the main environmental impact, it was necessary to ask ourselves today how interests in this area have changed. The technologies used for the exploitation of energy sources have evolved and their use extended. The dependence of the price of oil on issues related to international politics and competition among producers is increasingly evident and there is no longer any talk of exhaustion of the availability of raw materials. The environmental crisis seems to have shifted to aspects more closely related to the phenomenon of climate change, such as the availability of food and water, and environmental quality in a broad sense, in relation to anthropic activities: effect of the heat island phenomenon, particulates, ozone, etc.

To have a sufficiently shared vision in the technical-scientific world of what the environmental impact indicators of civil construction are, we can refer to the environmental certification systems of buildings and, more specifically, the evaluation systems for settlements (Neighbourhood Sustainable Assessment NSA). Numerous parameters are considered in the current vision of controlling the environmental impact of settlement processes. They are linked particularly to the site identification. The buildability of a site is usually defined in municipal urban plans, so the characteristics of the areas should be assessed in Strategic Environmental Assessment processes (SEA—Directive 2001/42/EC). It would appear that the scientific communities that deal with the environmental certification systems of buildings and settlements also consider evaluation a posteriori, probably considering that the construction of a site, despite the obligation (at least in Europe) to apply the SEA, is often influenced by political reasons rather than environmental protection. Although these systems are developed by large groups of scholars and have evolved over time, they tend to offer a rather standardised vision, often outside of local specifications, exporting a specific cultural model. In this regard, and referring above all to the well-known LEED system, Claire Doussard (2017) writes about these evaluation systems. “Some of these systems aim to be universal through the use of standards, while paradoxically contributing to the certification of projects in culturally diverse contexts”. Nonetheless, these systems offer a rather broad overview of the points of attention of a sustainable project that goes well beyond issues related to bioclimatic design and energy saving.

According to the LEED evaluation system, the most widespread in the world, in the edition “for Neighbourhood Development” (2018), while this certification system identifies many alternative solutions to obtain a good score, attention to the choice of site is high and there are many unavoidable prerequisites. The “Smart Location and Linkage” parameter, for example, assigns a score of 1 or 2 points to a new settlement on a brownfield, instead of a maximum of 10 points obtainable from a combination of settlement situations aimed at the requalification of abandoned areas within the urban body or immediately adjacent to it. Up to 3 points are assigned if the site is able to provide “Housing and Job Proximity” in order to limit the need for mobility. Proximity to public access systems—“Access to Quality Transit”—assigns up to 7 points. Respect for natural habitats is also very important. There are criteria for the organisation of the building fabric so that it can also be travelled safely on foot, criteria for density, as well as criteria relating to energy consumption, the use of water, and comfort in confined and open spaces.

Referring to the Italian cultural and technical sphere, the areas of assessment are not very different. Obviously, the weight that is credited to the individual parameters may be different. The ITACA interregional protocol has now become a Reference Practice of the UNI body i.e. UNI/PdR 13.1: 2019 and this makes it a reference also for private construction. This system—more easily applicable by designers than LEED which, due to its characteristics, requires the use of certified experts—similarly pays attention both to settlement criteria and to energy consumption criteria, but also to the choice of materials and the definition of conditions of hygrothermal comfort.

Another factor to consider as an impact parameter is the loss of the urban fabric in its characteristics as a socio-economic relational system, closely correlated with the settlement structure. In the transition from constructive experimentation to the construction of a settlement, there should be a process which, besides considering the aforementioned biological and meteorological aspects, also considers relational and perceptive aspects between the elements and consequently between the hosted organisms: the inhabitants. In the project for the Ecolonia district—in the municipality of Alphen aan den Rijn (Netherlands), in the early 1990s—Lucien Kroll, who organised the structure to then entrust the construction of the buildings to several designers, proposed a dense settlement, not by elevation but by proximity between the building bodies (Bakker and Thijssen 1991; Magliocco 1996). Kroll renounces "solar gain" in many buildings by adopting other hygrothermal control strategies, with the aim of creating a "human" fabric, in which it is easy to meet the neighbours and possibly develop relationships with them. The criticism is not so much of solar architecture as peripheral urbanisation (which is what Ecolonia was considered at the time of its realisation), organised only through hierarchical road layouts, in which the car-garage-home model defines the living space as a complete refuge, in which to shelter overnight. These settlements organised according to the principle of "solar access" promote a functionalist approach (in the worst sense of the term), combining building hygiene and low density. However, it is to this model that many of the "sustainable" design experiments refer.

9.4 Solar Cities Today, Goodbye to the Bioclimatic Model?

It is possible today to notice a sort of dissociation between the new solar city models and the bioclimatic approach. The passive house model promoted by the research centre founded in 1996 by Dr. Wolfgang Feist (The Passive House Institute -PHI) has defined new consumption standards and construction methods. In Italy, CasaClima certification (a brand managed since 2002 by the South Tyrolean Energy Agency) defines very similar criteria. The European Directive on Nearly Zero-Energy Buildings (NZEB, Directive 2010/31/EU) in a certain sense has validated—albeit not explicitly—a constructive approach that makes hyperinsulation the main strategy for reducing energy consumption. The bioclimatic approach requires great multi-inter disciplinary skills to be applied in different climates; the new reference model, on he other hand, is based on a conservative approach (hyperinsulation), impermeability to air and the use of small heat pump systems with heat recovery for air exchange. It was initially thought that it could be applied anywhere, varying only the transmittance of the opaque and transparent elements and the power of the heat pump. Consumption is so low that it allows the feeding of the heat pump with small photovoltaic systems. However, a resource consumption monitoring system that does not take the entire production process into account cannot detect its real performance. An approach that promotes efficiency—of technologies, of building performance—instead of considering the effectiveness of solutions and strategies, does not necessarily lead to important results.

All this has led to a new business, photovoltaics, first pushed and supported with various forms of incentives and then placed in the hands of large industries. Byrne et al. (2016) talks about the Solar City strategy as an operation to maximise production from solar sources through photovoltaic systems which, without waiting for private initiatives (with all the pertinent constraints and difficulties) could contribute strongly to the needs of large cities as well, using infrastructure and publicly owned real estate as support surfaces. The calculation he proposes, for six case studies—Amsterdam, London, Munich, New York City, Seoul, and Tokyo—obviously ignores architectural results. Consequently the model of the solar city changes, becoming increasingly a mere technical-economic issue. Currently, the term seems to be used to describe large new generation agglomerations, like Chinese or American cities (Hosseini 2019), in which the "solar" adjectivation is linked mainly to the production of solar thermal and, above all, photovoltaic energy. The development of the solar city concept is now strongly influenced by industrial development, particularly in large-scale production in Asian countries, when most of the active solar systems (SWH and PV) distributed throughout the world come from. One of the cases that can be cited is the city of Dezhou in China, which has become the seat of a large industrial district for the production of solar systems, initially created on the initiative of the Himin industry. In 2005, the Dezhou government implemented a Solar City Strategy and introduced a series of preferential policies and plans to provide comprehensive technological and financial support to the solar industry. Dezhou boosted its international reputation when it hosted the International Solar City Congress in 2010 (Yu and Gibbs 2017). This can be seen from a positive perspective, as in the aforementioned paper, because "green" entrepreneurs are able to boost processes that would otherwise take much longer to develop due to various types of resistance, or to a negative perspective, as urban development is subjected exclusively to the interests of industry. Both aspects are true. At building scale, the Solar Valley Micro-E International Hotel is an obvious expression of this combination: each person evaluates the results.

9.5 Conclusions

The bioclimatic approach to design has been an important step in defining contemporary building models, demonstrating that it is possible to severely limit energy consumption in the civil sector starting with construction methods, while maintaining a strong link with the place. The simplification of the bioclimatic approach in a simple exploitation of solar energy has led to the contradiction of a new globalising model, that of the solar city, which is less and less a habitat for man and more and more a machine for energy production. This model is essentially unable to dialogue with the consolidated urban fabric (under penalty of its distortion) and is strongly focused on the definition of urban agglomerations where construction is influenced by the goal of maximising the use of solar energy. The risk is that the control of energy consumption may be unable to coexist with the other requirements of what, even today, in the European city, we consider as human space.

References

Bakker FE, Thijssen AP (1991), An environmental housing design for Ecolonia, Alphen aan den Rijn, The Netherlands. In: Architecture and urban space: proceedings of the Ninth International PLEA Conference, Seville, Spain, September 24–27, 1991. Kluwer, Dordrecht, pp 359–364

Benevolo L, Erbani F (2011) La fine della città. Intervista a cura di Francesco Erbani, Laterza, Bari

Byrne J, Taminiau J, Kim KN, Seo J, Lee J (2016) A solar city strategy applied to six municipalities: integrating market, finance, and policy factors for infrastructure-scale photovoltaic development in Amsterdam, London, Munich, New York, Seoul, and Tokyo. WIREs Energy Environ 5:68–88. https://doi.org/10.1002/wene.182

Castelli G (2010) Solar city a Linz: qualità, bassi consumi, regia pubblica. Urbanistica n. 141–2010:58

Chance T (2009) Towards sustainable residential communities; the Beddington Zero Energy Development (BedZED) and beyond. Environ Urbaanization 21(2):527–544. https://doi.org/10.1177/0956247809339007

Cook J (1991), Searching for he bioclimatic city. In: Proceedings paper 9th international passive-and-low-energy-architecture (PLEA) conference, Seville 24–27, 1991. Kluwer Academic Publ

Doussard C (2017) Assessment of sustainable neighbourhoods: from standards to cultural practices. Int J Sustain Dev Plan. https://doi.org/10.2495/sdp-v12-n3-368-378

Ebenezer H (1898) Tomorrow, a peaceful path to real reform. Swan Sonnenschein & Co, London

Goggins J, Moran P, Armstrong A, Hajdukiewicz M (2016) Lifecycle environmental and economic performance of nearly zeroenergy buildings (NZEB) in Ireland. Energy Build 116(2016):622–637

Hodge J, Haltrecht J (2009) BedZED seven years on. The impact of the UK's best known eco-village and its residents. BioRegional Development Group, Published July 2009

Hosseini SE (2019) Development of solar energy towards solar city Utopia. Energy Sources Part A Recovery Utilization Environ Effects. 10.1080/15567036.2019.1576803

LEED v4 for NEIGHBORHOOD DEVELOPMENT Updated July 2, 2018

Magliocco A (1996) Olanda: politica ambientale e politica residenziale. In: Raiteri R (ed) Progettare la residenza tendenze innovative, pp 25–116. Maggioli Editore, Rimini

Magliocco A (2004) L'ambiente in testa. Edilizia sostenibile in Danimarca. Costruire n. 253, giugno 2004, pp 56–60

Olgyay V (1963) Design with climate. Bioclimatic approach to architectural regionalism. Princeton University Press, Princeton

PRASSI DI RIFERIMENTO UNI/PdR 13.1:2019 Sostenibilità ambientale nelle costruzioni - Strumenti operativi per la valutazione della sostenibilità - Edifici residenziali

Sadler S (2006) Drop city revisited. J Architectural Educ 59:5–14. https://doi.org/10.1111/j.1531-314X.2006.00029.x

Williams J (2016) Can low carbon city experiments transform the development regime? Futures. Elsevier. http://dx.doi.org/10.1016/j

Williams J (2017) Lost in translation: translating low carbon experiments into new spatial contexts viewed through the mobile-transitions lens. J Cleaner Prod. Elsevier. http://dx.doi.org/10.1016/j.jclepro.2017.03.236

Xue YW, Meng CL (2013) Energy conservation technology in BedZED and the enlightenment for China. In: Advanced materials research (Volumes 690–693), pp 839–842. https://doi.org/10.4028/www.scientific.net/amr.690-693.839

Yu Z, Gibbs D (2017) Unravelling the role of green entrepreneurs in urban sustainability transitions: a case study of China's Solar City. Urban Studies, pp 1–17. Urban Studies Journal Limited. https://doi.org/10.1177/0042098019888144

Part III
Environmental Building Design Concepts and Methods

Chapter 10
Climate, Place, Ecology

Maria Cristina Forlani

Abstract During the last century there has been an accelerated transformation in architectural studies and practices, which brought about structural and technological innovation. But the contradictions of this design approach are obvious. Environmental emergencies and energy problems are, therefore, the urgencies to be faced in the field of architectural design for a responsible approach to the sustainability of the sector. To overcome the current crisis (both economic and environmental) we need to review the cultural model of development and, particularly, of living. We need to be aware of the natural limits of growth and rethink the territory and the exploitation of local resources (tangible and intangible). The goal is the innovation of the model according to the parameters of ecology.

Keywords Climate change · Mental ecology · Design change

10.1 Introduction

During the last century there has been an accelerated transformation in architectural studies and practices, which brought about structural and technological innovation. Eliminating the weight of structures and creating increasingly light and "dematerialised" systems, while ignoring the problems of heating and cooling houses, which are now entrusted to increasingly sophisticated and efficient systems, the designers have been freed from a a series of design difficulties. This has given them the illusion of extreme liberty in the definition of "design" subject only to the laws of formal composition.

[1]This refers to the environmental impacts and energy consumption for the extraction of raw materials, the production of materials and components, and difficulties with management and disposal when the products reach the end of their life.

M. C. Forlani (✉)
Department of Architecture, Università Degli Studi "G. D'Annunzio" Di Chieti-Pescara, Chieti-Pescara, Italy
e-mail: cforlani@unich.it

G. Chiesa (ed.), *Bioclimatic Approaches in Urban and Building Design*, PoliTO Springer Series, https://doi.org/10.1007/978-3-030-59328-5_10

As a result, materials and systems considered obsolete and/or uneconomical have been abandoned: heavy and bulky structures have been replaced by light skeletons capable of guaranteeing structural efficiency with minimum use of materials and extensive availability in the management of wall-voids and planimetric spaces; the range of new systems on offer has also contributed to the possibility of minimising thicknesses and even allowing complete openings.

But the contradictions of this design approach are obvious. The structural "challenges" and those of artificial comfort should not be the only yardsticks of design: the actual "price", in terms of energy and environment, is too high.[1]

The most recent challenge, that of energy saving, can be riddled with ambiguous implications too. We cannot focus only on the amount of energy directly required by HVAC systems; we also have to consider the energy needed to obtain the most advanced materials (embedded energy). Then we must also take into account the energy required for the ever-increasing volume of transport, caused by centralised and de-territorialised industrial production.[2]

The architect's job involves many responsibilities, from the consumption of land, energy and materials to the production of waste and emissions at every stage of the building process. Just think of the fact that over 35% of energy consumption takes place in the civil sector and 40% of the waste produced is related to construction and demolition (maintenance). It should also be remembered that about 20% of pollution come from the construction sector.

Environmental emergencies and energy problems are, therefore, the urgencies to be faced in the field of architectural design for a responsible approach to the sustainability of the sector.

10.2 Climate and Place, Reflections on the Design of Architecture from the Recommendations of the Author of the First Architectural Treatise

The very first treatise on architecture known to us included knowledge of the climatic situation among the basic recommendations for good construction (Vitruvio, De Architectura, Book VI).

The acquisition of climatic parameters in architectural design is, therefore, an ancient question and closely linked to formal and material choices. Starting from the evolution of systems for controlling indoor comfort, however, the reference in design to geographical and climatic issues has gradually become of secondary importance.

---

[2]As early as 1972 the O.C.S.E. (Organizzazione per la Cooperazione e lo Sviluppo Economici) noted the need to improve material policy at individual country level, pointing out once again that each country must take into account a set of strictly local conditions which differ enormously from country to country. (L'Observateur de l'O.C.D.E., n. 59, August 1972).

The close connection between climatic environment and building product remains evident in all so-called spontaneous or traditional architecture. Architectural production from the second half of the twentieth century, however highlights the possibility of conditioning a building without having to refer to the characteristics of the place. This releases us from the need to abide by encoded schemes, giving us more and more design freedom. The project becomes detached from the 'place' and can be developed elsewhere, far away from the construction site and the local culture.

It was only in the aftermath of the energy crisis of the 1970s that catastrophic predictions regarding the scarcity of energy resources, which, had, until then, been the subject of ever-increasing consumption, brought attention back to the climate issue. But the energy problem did not reverse the trend. The consumerist and energy-intensive development model did not change. Instead, technological solutions to the problem were sought. Attempts were made, for example, at reducing the consumption of energy for heating (the 'peak' at the time) by limiting dispersion in buildings, improving wall and window insulation (see Law 373 of 1976).

Only marginal attention was focused on renewable energy sources, such as solar power, and on reconsidering the relationship between building and climate

In actual fact, climate data had already been re-introduced into design as a basis for appropriate choices as early as the 1950s: the Olgyay brothers (Olgyay 1963), for example, had proposed a scientific systematisation of traditional empirical knowledge, and the classification of climate types had already been carried out a few years earlier by climatologist W. Koppen.[3]

Essentially, some references from which it was possible to begin rebuilding and updating that successful relationship (between climate and local resources) which had provided such interesting and unique "products", and which could have helped to make design more sustainable, had emerged.

So-called 'spontaneous architecture', based on local resources and capable of using climate to improve environmental wellbeing (Rudofsky 1964), was brought to the attention of scholars. This led them to rediscover a number of extremely interesting ways of responding to climatic peculiarities with ingenious building systems. The work on the design of sustainable air conditioning systems carried out by nineteenth century technicians was, however, less well known. Banham (Banham 1969) offers some excellent examples but a targeted study could surprise us with *"their perhaps interesting topicality [...] as experiences of "protoclimatisation" which [...] are now proposed as tangible and operational suggestions for those who really want to create sustainable air conditioning [...] in order to recover a line of experience brutally interrupted by the typical progressive enthusiasm regarding techniques which are now revealing all their limits and which must be reviewed and reassembled with some urgency".* (Cetica 2004)

In the technical literature of the 19th and the early decades of the twentieth centuries, a great deal of information was produced in relation to the "conditioning"

[3]In 1936, meteorologist W.P. Koppen published the most complete version of his classification of climates.

of buildings using design methods which considered the basic principles of environmental technical physics. Convective motions were used to configure heating and ventilation systems; a basic layout, consisting of heaters and/or fireplaces, combustion exhaust pipes and cavities in walls and/or underfloor ducts, were applied in many variations. But with the aim of ensuring that no energy was lost, indications were given to use even those convective motions activated by light fittings with surprising brilliance. In addition to walls, floors and ceilings, other construction elements, such as architraves, cornices, attics and lanterns, were involved too.[4]

Though little publicised, the research has continued on from the above manuals and has never stopped. The definition of the last "pan de verre" (in which the analysis of an important element of the building envelope, the window, reveals notations of specific interest with respect to the sustainability issues of the design-constructive proposals) by Le Corbusier, is an example of particular relevance and strength. Le Corbusier himself, in a letter to Alazard (his old assistant who lived at rue de Sèvres 35, who had moved on to become manager of glassworks), wrote *"In India, I came face to face with the problem of breathing and, as a consequence, ventilation inside buildings. I classified the two functions of pan de verre: aerating and illuminating. I separated the two functions, obtaining efficiency and economy [...] installing fixed glass panels (which could not be opened) to let in the light and hollow struts to support them, the vertical opening in the struts allowed me to adjust and regulate the ventilation as I liked, from floor to ceiling, adjusting it between the narrowest vertical air blade to a strip 17 cm wide"*. (Le Corbusier 1948)

Designing in harmony with the climate to make conscious use of the resources available and to favour an internal comfort without major burdens on the consumption of non-renewable energy requires knowledge of the physical phenomena and the dynamics that regulate the interaction between building and climate. Some basic principles on heat transmission, the rules of air stratification, ventilation, evaporation and condensation must be acquired. Location and orientation, as well as the behaviour of shape, must also be taken into account.

"Bioclimatic" is the name given to "climatically responsible" products, which first came on the scene after the recessions in the 1970s, their main aim being to save energy and introduction of renewable energies. They are characterised by solar capture systems and components designed to store energy. Due to the decision not to use power generation systems, bioclimatic architecture is also referred to as "passive" and a harbinger of changes in appearance or new formal aspects.

Lastly, the acquisition of sensitivity towards climatic issues makes it possible to make appropriate choices among the various passive technologies currently codified,

---

[4]See the numerous texts which, from the 17th century to the early 20th century, contain detailed information on heating and ventilation in buildings. The most significant texts can be found in the Biblioteca dell'Ecole Nationale des Ponts e Chausseés (Bosc E., Traite complet, historique et pratique, du chauffage et de la ventilation des habitations particulieres et des edificies publics, Paris, Morel 1875) and the National Library (Beaurienne A., Le chauffage economique dans l'industrie, Paris 1922) of Paris; in the library of the British Museum (Ritchie R., A treatise on ventilation natural and artificial, 1862) and of London's Victoria and Albert Museum (Shuffrey L. A., The English fire-place, London 1912).

but also to configure new ones, as a precise result of the acquisition of a design method which based technical and formal solutions on knowledge.

> To correctly arrange buildings, one must first consider the regions and latitude where they are to be built. It cannot be assumed that a single architectural model will be suitable for all regions [...]. The drawbacks of nature must be remedied by technique. In the other regions, efforts must be made to improve environmental conditions, using appropriate criteria for different latitudes.[...] as the various regions are characterised by different climatic conditions [...], I have not hesitated to indicate a different type of building to suit their varying characteristics.[...]. (Florian 1978)

We want to propose a way to approach design with awareness of the area's cultural roots, paying attention to environmental issues and the needs of users, supplying "sustainable" development in the true sense of the term.

Design must pay close attention to every phase of the process due to the implications of each one on the environment. It must tend to embody the "design" in the difficult balance between the choices that involve the entire "package of requirements" without favouring one aspect (structural, aesthetic, functional, etc.) over the others.

The systemic approach and the assessment of the requirements-performance ratio—attentive to environmental impact—when makes design decisions becomes a way of guaranteeing the best results in any transformation activity.

To overcome the current crisis (both economic and environmental) we need to review the cultural model of development and, particularly, of living. We need to be aware of the natural limits of growth and rethink the territory and the exploitation of local resources (tangible and intangible). We have to reflect on the new complexities triggered by the paradigm of sustainability (consumption, emissions, waste) and on the introduction of further analysis of requirements (approach to environmental protection). Basically, we need to need to move from a linear to a cyclic method in which the concept of "assessment" emerges as a fundamental.

Designing from a life cycle viewpoint requires particular attention to be paid to the choice of materials, and consequently construction systems, to the use of climatic resources and to the quality of comfort. Such attention requires constant reference to the local situation, the broadest and most in-depth knowledge of the "place", as already introduced by the "appropriate technologies" and now cited and updated as a strategy to respond to the demands of sustainability.

The work method is based on interdisciplinary collaboration in order to foster the excellence available on the territory (at cultural and productive level). The aim is to provide the client with adequate support in order to qualify the demand and make it conform to the 'real needs' by spreading and sharing the culture of sustainability and of new forms of economy and living, which must be flexible and adaptable to new requirements.

The goal is the innovation of the model according to the parameters of ecology. In this regard, the reference to Odum's study (Odum 1983) seems fundamental: the closure of cycles through the recovery, regeneration and renewal of the existing material and spatial heritage.

10.3 Ecology as a New Design Paradigm in View of Environmental Problems

The events of the early 1970s suddenly highlighted a process triggered a few centuries earlier and deliberately ignored. Back in the nineteenth century, the results of innovation in the fields of agriculture and industry and the radical transformation of society, its attitudes and culture, were already clear to scholars and scientists working in the disciplinary fields related to the environmental system. It was then that the new science called ecology began to take shape.[5]

The environment is the result of the deep and continuous interaction between natural and anthropic components. In every territory, the transformations determined within the "short term" of history and the changes that have occurred over the "long term" of nature visibly overlap. Nature and culture form an interacting whole, with each action of one determining a response by the other, on the basis of which new choices and new behaviours are determined.

An in-depth knowledge of the environment is the basis for a different and innovative management of the territory, in which economic needs, wellbeing/living and environmental protection are compatible and inseparable goals. In this way, the environment becomes a field of investigation and action not only for scientific disciplines but also for humanities. The complexity inherent in the new concept of the environment makes it essential to have different skills and reading methods in order to create a system in which nature and culture can interact through positive feedback in view of new and profitable balances.

In the construction sector, the need to address the problems inherent in sustainable development in scientific terms and the goal of creating balanced situations of sustainability, without postponing them in time and space (with "remedies" such as emission trading) leads to a method based on the knowledge of the level of local sustainability to indicate which strategies to undertake in the construction sector.

Alongside the increasingly sophisticated refinement of 'bioclimatic systems' and the attention paid to the energetic and ecological efficiency[6] of buildings which—with good reason—have influenced research in our sector, one of the fields of interest of research that must be addressed is the control of the choice of materials (from procurement to transformation and recycling) in relation to performance requirements (environmental, technical and structural) and territorial resources.[7]

---

[5]The term "ecology"—"Ökologie"—was first used by Ernst Haeckel in 1866.

[6]Interventions aimed at the best energy performance can but be closely linked to those of eco-sustainability. Energy savings cannot be calculated solely in terms of management, but must take into account the extraction of materials, the transformation of components and the disposal of the product. In short, a Life Cycle assessment must be carried out. This must consider "the potential impact on the natural environment, human health or the depletion of natural resources, determined by the interactions between technosphere and ecosphere".

[7]From the perspective of sustainable development, the three environmental, social and economic spheres of 'sustainability' must be taken into account. For any intervention, therefore, not only the general quality that design for construction must guarantee, but also the possibility of generating

The aim should be to orient the activities that have an effect on a territory towards a better use of resources, while trying to maintain or improve the system's "state of health". The tools required to understand and assess these domains are complex and multiple, just like the concept of sustainability. This makes it essential to use interdisciplinary teams of scholars.

The attitude of balance in the design of construction-architecture, which seems to have to be taken for granted and entrusted to natural common sense, is actually characterised by particular complexity when it becomes unavoidable to fall into a rich system of interactions, where economic, social and cultural factors still have a considerable influence. As a result, the need to adopt a new behaviour that coincides with that outlined in sustainable "action" emerges.

Sustainability challenges the specific aspects of the more recent "creation of architecture" and consequently even current indications for design. Knowledge of the history of construction and its theories can therefore be a key to interpretation and a tool for operating in substantial continuity which can, at the same time, take on the meaning of sustainable action.

10.4 Research Topics

The main topics of study have been essentially suggested by the observation of an inappropriate and environmentally critical development, even in cases of good intentions at local government level. All this has led to reflections on the need to provide active collaboration, in close connection with government agencies, in order to produce guidelines for operation and to coordinate sharable and effective control tools.

It must be said that an extensive base of previous studies had been focused, over the years, on certain topics which subsequently became the subject of specific regional laws[8] aimed at recuperating and safeguarding particular heritages of the "genius loci" considered capable of typifying and strongly conveying the image of the area to tourists.

The critical points of this action were seen in the consequences, which were not in line with the intentions of the government bodies. The aforesaid laws were intended to provide funds for very delicate actions in particularly fragile and vulnerable areas (coastal, rural and mountain territories) and referred to elements relating to a constructive tradition that had been abandoned and forgotten.

new entrepreneurship and innovations capable of determining social satisfaction and sustainable economy at local level must be indicated.

[8]Particular reference is made to the following Regional Laws: n. 93/1994 Provisions for the recovery and enhancement of the Trabocchi of the Teatina Coast; n.17/1997 Provisions for the recovery and enhancement of the huts in tholos and raw earth houses. In recent years, guidelines have been developed for sustainable interventions aimed at the recovery and appreciation of this heritage with a strong innovative connotation in terms of image and management.

This made it hard to intervene because there was a widespread lack of technical skills capable of ensuring compatible and non-destructive actions. There was also a lack of widespread culture with regard to sustainable recovery and reuse.

It was considered essential to strongly link research to didactics, i.e. the training of new technicians whose preparation was based not only on critical local issues but also on environmental issues, with a view to sustainable development. More recently, the matter of the emergency which focuses on the need—with regard to recovery—to deal seriously with the problem of sustainability, not only at an environmental level but also with economic and social implications, has been the subject of attention.

This was one of the reasons why the topics chosen prioritised the regional situation in line with what emerges, in greater depth, from research into sustainability.

10.4.1 Experimentations[9] with Zero Ground Consumption: From Emergency to an Opportunity to Promote Innovative Models

The characteristics of the Italian territory (which has 8000 municipalities) must make us reflect on the effectiveness of current development policies, aimed at prioritising large urban areas. Italy's metropolitan areas, comprising about 1700 municipalities distributed in predominantly flat areas (under 30% of the territory), include less than half of the resident population, with about 25 million inhabitants.

This means that human settlement is strongly differentiated and unbalanced: more than 70% of the territory, represented by hilly-mountainous areas, includes the remaining population distributed in medium and small towns and villages. It therefore seems necessary to reconsider this large part of the territory which, in the face of interesting cultural and environmental peculiarities, is prone to abandonment and depopulation. These are due both to the difficulties of competition with big cities (conflicts of scale) and to continuous emergencies, further accentuated by the continuation of the socio-economic crisis.

These territories, beginning with those in central Italy affected by the earthquake, where reconstruction processes already have to be triggered, can be ideal places for experimenting with new development proposals based on the value of local identities and cultural capital (resource/place-based development).

The basis for development lies in the understanding of the critical issues (particularly environmental issues) that will have to be analysed and addressed in order to avoid the failure of reconstruction and regeneration.

---

[9]Two experiences, considered significant with respect to the problems highlighted, are presented here: the recovery of a village aiming at development through creative workshops and the post-earthquake reconstruction to test the possibility of new development models based on economic autonomy.

The aim is to take action to reduce emissions of pollutants (which cause climate change), starting with the prudent use of resources, and to prepare for serious events/emergencies.

The proposal is based, therefore, on outlining a clear and inclusive environmental policy capable of managing the territory in an interdisciplinary way, providing the elements to “know” and “decide”.

10.4.2 Promoting Small Towns and Promoting Sustainable Development

There are two fundamental aims, the first aimed at shaping a local economy capable of supporting basic needs; the second aimed at building a cultural system capable of bringing the territory to life and generating appeal. Both are to be assessed in terms of the environmental and socio-economic impacts and repercussions of the actins carried out, and in relation to their ability to activate endogenous “energies” capable of enhancing the quality of life and producing local wealth in a sustainable way.

2003 “WHAT RECOVERY?” to promote the village of Castelbasso (TE).

A starting point to ensure the maintenance of the pre-existing heritage must also be established on the basis of certain reflections to configure the conditions without which it is impossible to imagine the continuity of the use, life and heritage in question. The village hosts a summer event (art exhibitions and musical performances) among the ruins of the buildings and in the open spaces between them. The proposal is to focus on strengthening the event by promoting a permanent cultural workshop, open all year round, both for planning and for further events to be held indoors during the winter.

We propose, therefore, a scenario beyond the survival or maintenance of a status quo: a development capable of updating the life/use of pre-existing heritage in a continuous way.

The main aims that embody the specific actions that characterise the project can be summed up in a performance improvement in environmental comfort and in the reversibility of the interventions in order to make the structure flexible with respect to evolving needs.

The recovery of existing buildings identifies actions aimed at improving the performance of said buildings. It identifies the peculiarities inherent in their construction system (of the conservative type) and their development, introducing new technical elements that are minimally invasive and have a “light” figurative weight, in order to avoid the loss of formal identity and, above all, to respect the behaviour of traditional buildings (Fig. 10.1).

New products propose a strong innovation, with the desire to test systems capable of providing specific performance and also to define a functional integration with the pre-existing heritage. The choice of a real system of “arrangements” (a technology mediated by temporary works is proposed) places the emphasis on reversibility. This

Fig. 10.1 Exemplification of functional recovery: a loggia used as a greenhouse for solar gain (Forlani et al. (2008), in Recupero e riqualificazione del borgo di Castelbasso (TE), Alinea editrice, Florence)

seems fundamental—in order to guarantee the experimentation—and necessary, at the same time, to leave the "field" free from "marks" at a later date (when needs change and new goals are set) (Fig. 10.2).

a b

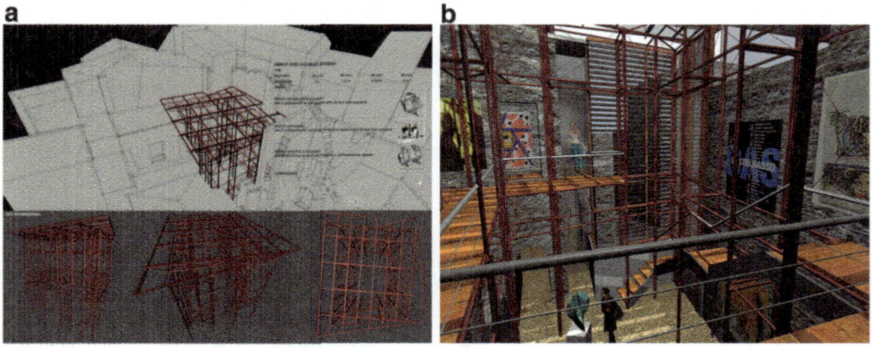

Fig. 10.2 **a, b** New spaces in courtyards structured by provisional systems (Forlani et al. (2008), in Recupero e riqualificazione del borgo di Castelbasso (TE), Alinea editrice, Florence)

2009 "WHAT DEVELOPMENT"? for the 'reconstruction' of Caporciano (AQ), from the village to the territory.

In the aftermath of the emergency caused by the earthquake in the area of L'Aquila, the opportunity arose for an incredible building site, a workshop for testing new development models, was presented: "the guidelines are aimed at ensuring the socio-economic recovery and redevelopment of the town: the strengthening and promotion of local territorial systems, with the identification of homogeneous areas based on strategic sectors of intervention: the redevelopment of environmental and historical-cultural networks; the rationalisation of territorial and urban mobility; the dissemination, penetration and efficiency of infrastructure networks, services and centres".[10]

While the main aim was the socio-economic recovery of the territory (environmental protection, tourism, local production and production of energy, agriculture, industrial ecology), specifically, attempts were made at promoting a new development model by exploiting local resources in order to foster the emergence of a system based on local economies.[11]

Reflections have been developed to outline a future scenario starting from an administratively feasible hypothesis based on the constitution of supra-municipal districts.[12]

The strategies were identified starting from the main problems that emerged from the analysis of the territory:

- to create new local employment to prevent population depletion;
- to develop high quality settlements at low cost to generate "appeal".

In order to support the actions for the implementation of the proposals, possible funding opportunities have been identified in the Structural Funds approved by the European Commission[13]; in particular, funding opportunities have been found in the documents for the planning of spending and territorial development at regional level (DOCUP):

- industrial symbiosis, such as the organisation of local resources in industrial ecology (rubble, agriculture, breeding, energy_PCR FESR line I research, innovation and competitiveness)
- reuse of the built environment (functional mix) and transformation of existing constructions in the form of environmental requalification (new controlled

[10]Decree Law n. 39 of 28 April 2009, the "strategic nature" of the reconstruction plans (art. 14, paragraph 5).

[11]In Leipzig, the need to develop a local economy and a local labour market was the focus of the 2007 "Charter on Sustainable European Cities".

[12]The mountain community of 'Campo Imperatore-Altopiano di Navelli' was considered: 17 municipalities with a total of 9600 inhabitants, from a minimum of 90 to a maximum of 1800.

[13]The European Regional Development Fund (ERDF), which promotes public and private investment, as well as the reduction of regional imbalances in the European Union; The European Social Fund (ESF), which supports policies to promote and improve employment, quality and productivity at work; The European Agricultural Fund for Rural Development (EAFRD), which contributes to the promotion of sustainable rural development in the European Union.

spaces/superfetation), energy efficiency (bioclimatic prostheses with anti-seismic value_POR FESR line II energy)

- regeneration of urban areas and proposals for new housing, (one-parent groups, temporary use for work_POR FESR line IV territorial development/IV 2.1. Promotion of promotion of reception and renovation of buildings) (Fig. 10.3).

The new housing proposal aims to attract a new population and to propose a transformation with high quality technological-environmental products; it also embodies

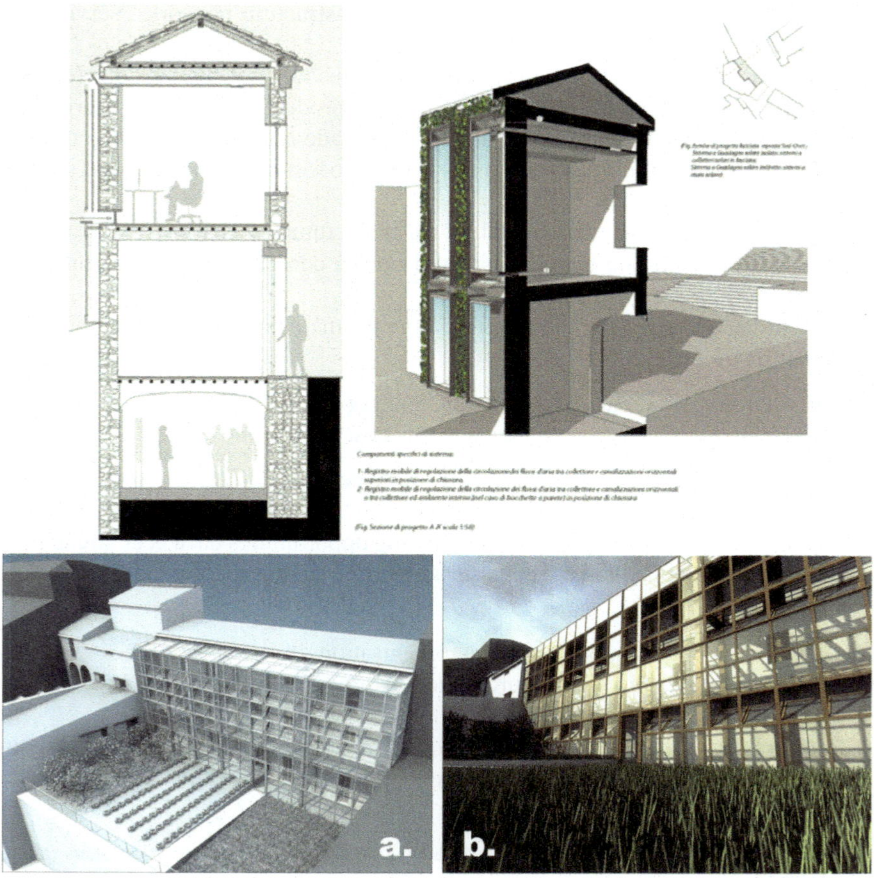

Fig. 10.3 Reuse with functional mix (collective uses, housing, offices) and improvement of comfort with added bioclimatic systems (Comune di Caporciano (AQ), Reconstruction Plan, scientific manager prof. M. Cristina Forlani (architect); presentation of the preliminary phase, tables of the exhibition held in the Abbey of Bominaco, August 2011); **a**, **b** innovation of the aspect: "controlled superfetation" like anti-seismic bioclimatic prostheses (Comune di Caporciano (AQ), Reconstruction Plan, scientific manager prof. M. Cristina Forlani (architect); presentation of the preliminary phase, tables of the exhibition held in the Abbey of Bominaco, August 2011)

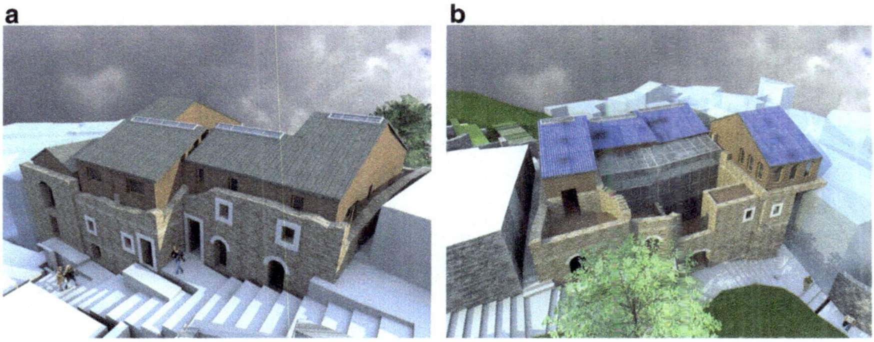

Fig. 10.4 **a**, **b** Maintenance of "memory"': the recovery of ruins and the "grafting" of new technologies from local resources

the surpassing of reconstruction "the way it was" with the proposal of a particular conservation of the historical-constructive "memory".

Thermo-hygrometric comfort and natural lighting were the guiding elements of the project, which took into account the use of climatic resources to supplement the heat input, facilitate ventilation and improve natural lighting by means of bioclimatic devices, in particular a leaning greenhouse, containing the horizontal and vertical distribution system, and a system for transporting light, from the reconstructed roofs, to the rooms on the ground floor (Fig. 10.4).

In terms of environmental sustainability, priority was given to reducing emissions, waste production and, particularly, the consumption of resources (energy, water and building materials). It goes without saying that, at the same time, considerable economic savings were also made, notably in the management of the complex:

- as far as energy saving is concerned, solar heating panels were used for domestic hot water and heating, and photovoltaic panels were used for the production of electricity. The solar heating system covers 70% of domestic hot water and heating requirements. The remaining 30% is entrusted to a gas boiler with provision for the use of local plant biomass. The photovoltaic system covers over 50% of the domestic electricity requirements, along with solutions aimed at increasing the use of natural light (light transport);
- as far as drinking water savings are concerned, a rainwater collection and grey water management system is envisaged. Rainwater recovery and grey water management make it possible to minimise the consumption of drinking water and to have an even greater quantity of it (given the local climate) to be used outdoors, for the irrigation of green areas, for example;
- for building materials, reference was made to natural and renewable resources with low incorporated energy. Technological packages made by companies located as close as possible ("zero km") were chosen for the envelope, with hopes of local production in the area of use in the near future, as indicated in the Strategic Development Plan;

- the containment of polluting emissions is ascertained by the lowering in production of CO_2 mainly as a result of the use of renewable energy and certified materials as well as the choice, in a first phase, of more effective and efficient systems, than those used traditionally;
- to reduce waste production, the decision was made to take into account mainly construction and demolition waste, which represents one of the major sources (about 40% of all waste production). It was with this in mind that the ruins, which would otherwise have constituted (in the case of demolition for new construction) a large amount of rubble with the consequent environmental impact and landfill problems, were recuperated.

The decision to recuperate and renovate the existing building also ensures the limitation of land consumption necessary for the development of new projects. Most of the costs of recovering and consolidating the existing parts are also offset by the absence of costs for the demolition of the ruins and the storage of the rubble with the addition of the costs and environmental impacts of transport to the landfill.

Reference

Banham R (1969) The architecture of the well-tempered environment. Architectural Press, London, GB
Bevilacqua P (1996) Ambiente, economia, risorse in Italia, Donzelli, Roma, IT
Bevilacqua P (2006) La terra è finita, breve storia dell'ambiente, Editori La- terza, Roma-Bari, IT
Bonomi A, De Rita G (1998) Manifesto per lo sviluppo locale, Bollati Boringhieri, Torino, IT
Butera F (2007) Dalla caverna alla casa ecologica, Edizioni Ambiente, Milano, IT
Cetica PA (2004) L'architettura dei muri intelligenti, Angelo Pontecorvo Editore, Firenze, IT. p 8
Commoner B (1990) Making peace with the planet. Pantheon, New York, USA
Florian G (1978) interpretazione a cura di, Vitruvio Pollione - Dell'architettura, Giardini Editori e Stampatori, Pisa, IT. Libro VI, I, Influenze del clima sull'architettura pp 109–111
Forlani MC (2010) Cultura tecnologica e progetto sostenibile, Alinea editrice, Firenze, IT
Forlani MC (2011) "Uso sostenibile delle risorse e scelta dei materiali a basso impatto", in Tucci, F., Efficienza ecologica ed energetica in architettura, Alinea, Firenze, IT
Forlani MC (2018) Ricostruire la città, Edicom Edizioni, Monfalcone (GO), IT
Jackson T (2017) Prosperità senza crescita, Edizioni Ambiente, S.Giuliano Milanese (MI), IT
Koppen W (1931) Grundriss der klimakunde, Walter de Gruyter & Co., berlino-Lipsia
Latouche S (2006) Le pari de la décroissance, Librairie Anthéme Fayard, Paris, FR
Le Corbusier (1948) Le Modulor, 1974 per l'edizione italiana Gabriele Mazzotta editore, Milano, p 183 e segg
Odum EP (1983) Basic ecology, saunders. Philadelphia, USA
Olgyay V, Olgyay A (1963) Design with climate. Princeton University Press, USA
Magnaghi A (2010) Progetto locale. Verso la coscienza di luogo, Bollati Boringhieri, Torino, IT
Norberg-Schulz C (1979) Genius loci, Gruppo Editoriale Electa, Milano, IT
Pulselli RM (2010) Integrative emergy evaluation and geographic information systems for monitoring resource use in the Abruzzo region (Italy). J Environ Manage 91:2349–2357
Rudofsky B. (1964) Architecture without architects. MoMA Press NY, USA
Tiezzi E (1992) Tempi storici Tempi biologici. Garzanti, Milano, IT

Chapter 11
The Ecological Culture of the Project: A Critical Vision

Andrea Giachetta and Maria Canepa

Abstract The last decades witnessed an unexpected spread of the concept of sustainability. This phenomenon has also involved the urban and architectural design field for its obvious and known environmental impact. This cultural process, undoubtedly positive, is leaving a trace in our time more than others. It is leading to a growing awareness of the environmental problem, to an ever-wider diffusion of new design approaches and new technologies and to a significant reduction in energy consumption and polluting emissions. However, this same cultural process has exponentially accelerated over the last decades, partly because of emergency circumstances, causing an extreme simplification of its content and a subsequent rise in false beliefs. An example is the belief that the direction towards environmental culture is going is unambiguous and commonly shared. The following essay is a reflection on this false belief and its possible consequences.

Keywords Sustainability · Ecological design · Cultural ecology · Environment

11.1 Introduction

This essay provides an unusual reflection on the current ecological culture in architectural projects. Until only a few decades ago, ecological culture was a niche, under development reality, albeit based on incontrovertible principles and solid foundations. It was prerogative of a small number of isolated researchers who worked bravely, often with makeshift means.

Today, the situation has almost reversed. The culture of sustainability, even in the urban and architectural design, represents a wide and largely funded field that employs many people at all levels and with the most diverse roles. The project's ecological culture is the pride of whom contributed to its incredibly fast growth. However, this process, so quickly developed, shows today at a closer look some small cracks.

A. Giachetta (✉) · M. Canepa
Departments of Architecture and Design, University of Genova, Genova, Italy
e-mail: andrea.giachetta@unige.it

G. Chiesa (ed.), *Bioclimatic Approaches in Urban and Building Design*, PoliTO Springer Series, https://doi.org/10.1007/978-3-030-59328-5_11

This essay aims to be a food for thought about what is happening by providing the reader with test on opening questions about sustainability in a general and applied to urban and architectural design. The reader is invited to analyse issues and wonder what position he would take with respect to them. In the end, those who have the patience to analyse the different cases proposed will understand the goal of this unusual exercise.

11.2 Case Study N. 1: Environmental Assessment Tools

Environmental assessment tools of buildings are voluntary control procedures managed by various different bodies. They allow the assessment of buildings' sustainability through the definition of parameters related to energy and environmental aspects. In particular, they evaluate site sustainability, water management, energy management and consumption, environmental indoor quality, material quality, waste management and service management (Magliocco and Canepa 2015).

The main goal of environmental assessment tools is to validate design and construction processes aimed to crate high-quality buildings with the lowest consumption of resources possible. This comes with additional costs that, however, can be justified by the greater value acquired by the certified buildings on the real estate market. Another reason underpinning these extra expenses might be the superior energy and environmental performance, and subsequent less maintenance and management costs, of certified buildings.

Nevertheless, environmental certifications do not go against the realization of new buildings, despite the current need of reducing land use. In Italy the offer of the real estate market, especially in the residential sector, definitely exceeds the demand. Sustainability should take this into account by promoting the recovery of existing buildings rather than the creation of new ones.

In view of these facts, two considerations can be made:

(A) Environmental assessment tools certify construction industry's quality. Certified building are more valuable compared the uncertified ones, since they are more sustainable and cost-effective. Of particular interest is the unregulated usage of material resources and its repercussions on construction and product choices, and technological solutions. A related issue that must be taken into account in an architectural project is indoor pollution caused by building materials.

(B) The application of the environmental assessment tools is not necessarily synonymous of greater sustainability; it is necessary to make a broader reasoning. In most European cities there is, in fact, no real need to build new buildings, but rather attribute greater energy performance to the old ones. But this is not always easy. Greater efforts must be made to manage cities as a whole rather than individual buildings. Energy savings can only be valid in areas where the minimum levels of energy performance are very low. Only this condition would determine a significant difference in terms of resource consumption during the operating phase between a certified and an uncertified building. In Europe, the legislation that defines the minimum levels of energy consumption is increasingly restrictive. Energy certification consists mainly in the evaluation of the energy performances; this trend however has to change in the near future. Materials sustainability in terms indoor environmental quality must be favoured, since for example the same energy performance required by directives (thermal transmittance values) can be achieved with different materials that, however, can be more or less energy-demanding in the production phase.

11.3 Case Study N. 2: The Research Centre

A few years ago, to one of the authors of this essay, still not fully engaged in university, was offered a very special professional assignment. It involved the design of a research centre on renewable energy in an abandoned former quarry of a tourist town on the Ligurian coast. The client, who was also the owner of the site, was an entrepreneur from Milan who worked and still works in the field of renewables with great passion and with a close-knit group of technicians, dealing in particular with large wind turbines and solar thermodynamic energy. His idea was to build the centre behind a small town whose main resource is seaside tourism, in a panoramic site of great charm, fairly well connected, even if it lacks the services and opportunities present in a metropolitan area. This idea was founded on the belief that the beauty of the place would have contributed to attracting resources and scholars also from abroad and that the relative isolation would have allowed the formation of a very united community by creating a working atmosphere suitable for cutting-edge research in the sector.

The intervention involved the construction of buildings (with laboratories, classrooms, exhibition spaces, conference centre, canteen), accommodation for researchers and open spaces equipped as test fields. A wind turbine the core business of the sponsoring company, would have represented both a symbolic identification signal and an important means to guarantee economic resources to run the centre. In the entrepreneur's intentions, the area should have become a veritable open-air museum of renewables with near zero energy buildings. The designers involved (architects, plant engineers, structural engineers, geologists, naturalists, etc.) were all very passionate about this initiative that seemed to offer to a region like Liguria a big opportunity, with innovative, poetic and somewhat utopian features, that recalled an entrepreneurial spirit of the past. The idea was obviously presented to the municipality that strongly supported the initiative.

However, when the project (please refer to Longiardi et al. 2010; Giachetta 2011 for further information) was presented to the Region committee, albeit with the consensus of the president, it was hampered by the landscape and environmental protection offices. The technicians, while fully appreciating the initiative, expressed concerns. The project area was indeed an abandoned former quarry, but over the year it had undergone a process of re-naturalization (also thanks to the work of Genoese university departments) and has been repopulated by some protected species (chiropters and animal species listed in Annex II of Directive 92/43/CEE, in particular Rana dalmatina, Pelodytes punctatus, Bufo Viridis, Hyla Meridionalis, as well as species listed in Annex I of the Birds Directive 79/409/CEE).

The site was included in a Provincial Protected Area and a Site of Community Importance. The intervention, even if designed with the involvement of herpetologists, chiropterologists and ornithologists, could have put these species at risk. Despite the good intentions of the promoter, the size and nature of the research centre would in fact have led to a construction site with a significant environmental impact. All the regional technicians objected to the construction of the wind turbine, although it was considered indispensable by the client. Birdlife would have been put at risk by the wind turbine. Furthermore, while acknowledging the particularity of the case, the regional technicians did not want to grant a derogation for the construction of the wind turbine in an area so close to the sea, naturalistically delicate and protected because this would have created, according to them, a dangerous precedent. Finally, the regional technicians were not inclined to grant the possibility of creating volumes also for housing purposes (although clearly linked to the permanence of researchers) in an area already widely disfigured by the massive and uncontrolled construction of holiday houses that took place over the past decades.

Due to these obstacles the process of approving the project became complicated and the entrepreneur desisted from continuing the project.

Both regional technicians and the entrepreneur tried to promote sustainable territorial development in different but absolutely legitimate fashions. In regards to what briefly described here, readers are asked the following question: which of the following two opinions do you agree with?

(A) The entrepreneur had a good idea. However, his project would endanger the environment. It is true that the area where he intends to operate is an abandoned former quarry but, has it has become an area of recognized naturalistic importance. Allowing this project to start would create a dangerous precedent. If the entrepreneur really wants to operate in Liguria, he could do it in former industrial areas that really need to be redeveloped (there are many!).

(B) The entrepreneur had a good idea that could attract economic resources and employ researchers in the field of renewables in Liguria, a region that has suffered the economic crisis more than others; we must support him and understand his intention to operate in a beautiful place, or rather to requalify it. The choice of the centre location is important to attract people and funding, also from abroad. Why deny him this chance? (Figs. 11.1 and 11.2).

Fig. 11.1 Design render of the centre

Fig. 11.2 Presentation of the project to the former President of Regione Liguria

11.4 Case Study N. 3: The "Big" Wind Turbines Farms

We have already mentioned the problem of wind turbines that is a sensitive topic, especially in Italy. Energy production from renewable sources is what most would consider the best solution for environmental sustainability; at the same time, they support actions to protect fauna and flora and guide or, if necessary, limite the anthropization processes of the territory. Today, however, it happens more and more often that these actions, although virtuous, are not compatible with each other (Giachetta 2010, pp. 17–25). Some regional regulations have issued criteria and guidelines for the installation of wind turbines, limiting the construction of wind farms in a more or less severe fashion related to birdlife, or more generally, naturalistic and landscaping values.

These norms have often been contested by municipal and provincial administrations, professional associations and groups of citizens, since they are considered too penalizing for wind turbines or, on the contrary, not enough to safeguard the local ecosystem and the landscape. Hence, debates between opposing factions such as industry experts and well-known environmental associations have started.

Liguria has always pursued an ambiguous wind turbines policy. Environmental Energy Plan envisages installing and administrating new turbines (both by politicians and non) and theoretically promoting this form of sustainable energy production. Nonetheless, it contains significant restrictions on building new large-scale wind farms. Just by looking at the cartography of the "Areas unsuitable for the construction of wind farms" on the regional cartographic portal (https://geoportale.regione.liguria.it—last web access December 2019), one can appreciate the difficulties to overcome to build wind farms in Liguria. It is clear that this resistance is not only the result of what is called NIMBY syndrome (Not In My Back Yard), but it can be traced back to a desire to preserve one of the Italy's orographically more complex and naturalistically rich landscapes, with a coastal strip characterized by avifauna's corridors and one of the most wooded hinterlands.

This summary provides you with sufficient elements to reason about this open issue. Do you agree with the current policy that allows the installation of wind turbines, but rigorously limits it in so far as it has landscape/environmental impacts (on birdlife in particular), or would you support a greater "opening" towards wind energy despite the risk of landscape/environment impacts?

(A) I agree with the current policy.
(B) I support a a greater "opening" towards wind turbines (Fig. 11.3).

11.5 Case Study N. 4: The Bioclimatic Guidelines

A few years ago, one of the authors of this essay was in charge of drafting the Article 11 bis "Guidelines for bioclimatic architecture and green building" of the Plan Implementation Rules Territorial Coordination for the Metropolitan area of Savona. The aim of these guidelines, also due to the nature and specific skills of the administration, was to stimulate the Municipalities of the Metropolitan area to include rules in their Urban Plans and encourage a change in the attitude of contractors and designers, by pushing them to work with solutions respectful of the environment and its resources.

Article 11 bis suggested the adoption of solutions such as: the reduction of urbanization costs for new buildings or the increase in the admissible constructible volumes

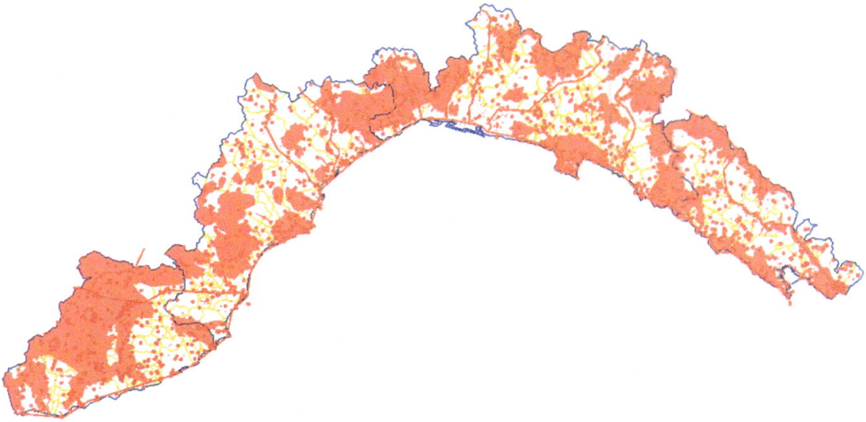

Fig. 11.3 From Regione Liguria. In red the unsuitable areas for the construction of wind farms (https://srvcarto.regione.liguria.it/geoviewer2/pages/apps/geoportale/index.html?id=1235 (December 2019)

as reward measures to be implemented as the application of sustainable intervention strategies. Eventually the article was approved and some Municipalities took it into account in the following years for the preparation of their urban plans. To make the article effective, after listening to the opinions of the competent commissions, the approval of the Metropolitan area Council was required. On the day of the vote, some resistance was expected from political groups that normally are not very interested in environmental issues, but there was no opposition on their part. Some unexpected criticisms, not such as to jeopardize the vote, were instead made, on that and other occasions, by representatives of environmental associations that considered the rules based on incentives too "soft".

In their opinion, for an environmental policy to be effective, it was necessary to impose and not only suggest new and more virtuous behaviors; moreover, the application incentive rules could be dangerous, if not subject to careful and constant checks. For example, granting volumetric incentives to those who use bioclimatic technologies such as solar greenhouse can lead some unscrupulous entrepreneurs to define as passive solar system volumes that are not at all and exploit economic or volumetric incentives (causing a double damage to the community).

To conclude, the adoption of reward or mandatory measures is a fundamental issue that biased the success and failure of many environmental policies, also at national and European level. It is a too complex subject to be treated in this essay thoroughly, but the example reported opens up to another question for readers. Which of the two statements do you feel closest?

(A) Regarding the building sector, better environmental policies (energy saving, reduction of polluting emissions, etc.) are those that focus on specific rules and regulations that oblige operators, in the case of new buildings or redevelopment interventions of a certain entity, to adopt minimum measures to protect the environment and its resources and the most advanced systems and technologies; the incentive policy (especially in Italy) could economically support those who do not even deserve it!

(B) Regarding the building sector, better environmental policies are those that guarantee environmental safety but, at the same time, aim to promote economic incentives, tax reduction and increase in building permits. There is no point in having rules that citizens and operators involved would break as considered inadequate (Fig. 11.4).

Fig. 11.4 Bioclimatic guidelines cover

11.6 Case Study N. 5: The Historical Centre

Considering the norms on the adoption of sustainable building technologies, especially in Italy, there is at least one other point of discussion that involves two opposing sides. On one hand, one argues that the application of technologies for sustainability must be carried out at any cost, if we really want to achieve a result in a reasonable time; on the other hand, there are those who argue, legitimately, that the protection of the built environment of historical value must also be recognized among sustainable practices, even when it ends up constituting a constraint on the application of elements, innovative building components and systems in the energy and environmental field. For instance, the installation of photovoltaic panels might be a sustainable practice that however could have a significant impact on historical centres.

Until now this issue has been addressed with rather naive attempts of regulatory mediation (Giachetta, 2013, pp. 76–77) or by developing systems for renewable energy that can be camouflaged, almost always significantly more expensive and not exactly performative. Thus, photovoltaic and thermal solar panels have been developed in the form of slate-like or coloured elements, like brick, or in the form of roof tiles in order to be used even in historic centres.

Do attempts like PV tiles shown in the Fig. 11.5 make sense? Can they somehow represent a possible solution for interventions in historic urban fabrics?

(A) Yes, in some cases they can be just mediation!
(B) No, it is necessary—if necessary—to learn how to manage the historical pre-existences in a more courageous fashion!

Fig. 11.5 PV tiles

11.7 Case Study N. 6: The Biofuels Debate

Biofuels are sustainable fuels of natural origin that can replace fossil fuels. Bioethanol and biodiesel are two examples of biofuels obtained from sugar can and vegetable oils, respectively.

Some poor or developing countries, such as Brazil, have largely invested on the production of these renewable fuels to boost their economy and become more independent from richer countries. On the downside, this massive production is affecting their agriculture and environment because it requires large portions of agricultural land, sometimes obtained with uncontrolled deforestation, and allows the use of modified organisms and pesticides that would otherwise not be used for nutritional purposes.

We are aware of the fact that this problem would need more in-depth analyses, however we ask you the following question: do you agree with the large use of biofuels even in countries that lack of strict regulations?

(A) Yes, absolutely, the large use of biofuels can significantly reduce the use of polluting and non-renewable fossil sources and can severely limit dependence on oil-producing countries. This is especially important for poor or developing countries that have no other resources apart from their territory. The advantages that might arise exceed the environmental and health hazards. In any case, we do not have the right to limit their energy production from renewable sources.

(B) I have some concerns. The massive use of this type of fuels can affect biodiversity by promoting monocultures and less rigid controls on the use of GMOs and pesticides.

I think that biofuels production should be much better regulated internationally with stricter controls and limitation policies. Sugar cane, for example, is one of the most extreme monocultures and in some countries, such as Mauritius and Barbados, it occupies more than half of the cultivated land causing food and environmental problems. The spread of palm oil crops in Southeast Asia and West Africa has resulted in uncontrolled deforestation and environmental disasters.

11.8 Conclusions

At this point, you will be disappointed if you think that it is possible to define your 'environmental' profile based on the answer provided. This essay wanted to show that there is no single thought on environmental sustainability and a considered reflection is necessary to overcome a homologating and simplified culture.

One should bear in mind that the answers provided by the readers do not result from more or less correct attitudes towards the environment, but simply from different and equally legitimate visions.

In the future, the environmental-friendly management of our life style might be less biased by the backward cultures. Larger economical resources will be allocated to support a more environmentally conscious vision of the world and fossil fuel lobbies will gradually lose power.

It is very likely that the most important challenge for the near future will consist not so much in convincing the last irreducible anti-environmentalists (who hopefully will no longer occupy positions of power as unfortunately still happens today), but in finding common ways of working for a shared environmental vision.

Almost no one seems to be aware of the complexity behind the fulfilment of a sustainable future. Often, in fact, we take for granted concepts such as 'sustainable design' as if they were recognized and accepted by the majority of public administrators, citizens, teachers, entrepreneurs, professionals and only a small portion were against them.

Upon closer examination, however, when trying to face real issues, unexpected problems often emerge and not even the experts—sometimes especially them—seem to really know what they are dealing with.

Are we so sure to know what it really means to have a more correct attitude towards the environment in term of training, economy, design activities that we carry out, plan to work, relax, travel, live in our buildings, our cities, the territory we occupy, our planet?

As highlighted by the previous questions, more and more often we should ask ourselves what sustainability really is: it means to equip buildings with technical instruments for the production of renewable sources or to protect the integrity of historic buildings and the landscape from human interference? To what extent we can exploit the territory to obtain renewable energy? Should we improve the existing buildings or build new ones? Should we improve the production of objects that we consider essential or try to produce less (Latouche 2004, 2007)? Do we aim at environmental sustainability or sustainable development?

And, furthermore, are bottom-up or top-down policies more needed for sustainability? Who will be the real protagonists of the sustainable future: all citizens or multinationals (Sukhdev 2012)? What ecological approaches will be sustainable in the so-called developing countries? And will these same adjectives make sense in a fairer and more sustainable world? Is it possible to think of organic food for everyone or is it just a green alternative for wealthy people? How will we deal with waste: convert it into energy, recycle it, stop producing it? Is naturalistic tourism sustainable? Will we have to increase or limit the influx of people into sensitive areas? Does it make sense to lash out – even in a hysterical fashion - against some types of products that may be harmful for the environment, such as palm oil, and replace them with lower yields and perhaps even more destructive? The fact is that the opinions of the environmentalists are many and equally reasonable (or almost). Moreover, there exist different recognizable cultural attitudes: sustainability, conservationism, degrowth, deep ecology, strong, weak anthropocentrism, ecocentrism, biocentrism, just to name some of them. But what does it mean today to have, for example, an eco-centric or biocentric position regarding the management and planning of our production, tourism, housing and life activities?

It might be not possible to answer, but raising the question seems indispensable to understand how important it is. A critical approach must be adopted in order to guarantee the existence of a sustainable project in a future.

These criteria should be also included in the trainings of future generations. We are still at the dawn of only a six-year-old environmental culture, but the time to change our behaviors might have come. We should settle for easy solutions. For a truly sustainable approach to the project to be possible, we can no longer rely only on experts, on the intuition of genius which will save us; the complexity this approach require an increasing and closer confrontation with the natural, historical, cultural and social environment and, above all, with the different opinions and skills needed.

The management of ecological complexity might require a joint effort, with greater skills, greater operational humility, more open-mindedness, more listening skills and a slightly less boundless self-love.

References

Giachetta A (2013) La gabbia del progetto ecologico. Carocci, Roma

Giachetta A. (2011), "Project for a renewable energy research centre", in Techne, Journal of Technology for Architecture and Environment, vol.1, Firenze University Press, Firenze (pp.144–147).

Giachetta A (2010) Il progetto ecologico oggi: visioni contrapposte. Alinea, Firenze

Latouche S (2007) Petit traité de la décroissance sereine. Mille et une Nuits, Paris

Latouche S (2004) Survivre au développement. Mille et une Nuits, Paris

Longiardi B, Casadei B, Grassi A, Camana S (2010) Ortofabbrica. 1° Contest di creatività sostenibile, Guaraldi, Rimini, p 85

Magliocco A, Canepa M (2015) Tools for sustainability not for the market: a critical analysis of environmental assessment tools. In: Khan A, Allacker K (eds) Architecture and sustainability: critical perspectives for integrated design. ACCO, Leuven, pp177–180.

Sukhdev P (2012) Corporation 2020: transforming business for tomorrow's world. Island Press, Washington

Chapter 12
Interaction Between Anthropic and Natural Phenomena in Urban Contexts Re-equilibration

Rossella Franchino

Abstract The protection of the urban contexts from the increasing anthropization development necessarily requires the adoption of appropriate and adequate strategies for the environmental re-balancing that are crucial especially when the environmental components: water, air and soil have been compromised by previous anthropic activities. Achieving a renewed quality of the environmental components is the necessary prerequisite for reuse and reconfiguration interventions. With such premises this paper focuses on the application of natural resources in the re-balancing interventions highlighting the significant environmental benefits arising from the use of principles of nature as a management model. All this in order to define the possible interferences between the natural and anthropic aspects that underlie the sustainability of urbanization and intervene on urban development with appropriate approach in order to find an alternative to the model that was previously used.

Keywords *Green + Blue* strategies · Environmental systems · Urban ecosystem · Sustainable cities

12.1 Introduction

In recent decades, environmental issues have taken priority among the main concerns of contemporary man. They have significantly affected the study, implementation and management of new technologies, both innovative as well as simply evolutionary. For this reason, the analysis of convenience, oriented only by the need to save resources, has become a compatibility analysis that has added to the aim of saving that of mitigating the environmental impact, both seen in their impact on the remaining availability for future generations and the ability to enjoy a habitat not completely degraded, despite having undergone repeated transformations. In this

R. Franchino (✉)
Department of Architecture and Industrial Design, Università Degli Studi Della Campania "Luigi Vanvitelli", Aversa (CE), Italy
e-mail: rossella.franchino@unicampania.it

G. Chiesa (ed.), *Bioclimatic Approaches in Urban and Building Design*, PoliTO Springer Series, https://doi.org/10.1007/978-3-030-59328-5_12

regard, the identification of strategies for environmental re-balancing is crucial especially when the environmental components: water, air and soil have been compromised by previous anthropic activities. Achieving a renewed quality of the environmental components is the necessary prerequisite for reuse and reconfiguration interventions.

Starting from these premises, the present work focuses in particular on the environmental conditions of urban areas because *"cities are a key nexus of the relationship between people and nature and are huge centers of demand for ecosystem services and also generate extremely large environmental impacts"* (Elmqvist et al. 2015).

All this also with a view to achieving the sustainable development goal "*11-Cities and sustainable communities*" of the 2030 United Nations Agenda (United Nations 2015).

"*Managing urban areas has become one of the most important development challenges of the twenty-first century. Our success or failure in building sustainable cities will be a major factor in the success of the post-2015 UN development agenda*" (Wilmoth 2014). This statement should make us reflect on the future of urban areas, especially in view of the growth of the world population and its urbanization. A further consideration, on which the future of urban areas strictly depends, is also that of the correct use of natural resources. It is very important, therefore, to verify the control of transformations so that urban areas can be able to regulate themselves from the point of view of the use of environmental resources as urbanization has a significant impact on them.

To intervene on urban development in order to find an alternative to the model that was imposed during the last century, a sustainable approach is essential required that places the territory as a fabric in which the border between the artificial and natural environments is not detected and in which each process is controlled so that its impact induced is the minimum possible.

This in order to define the possible interferences between the natural and anthropic aspects that underlie the sustainability of urbanization, the land use control and the territorial fragmentation in line with the strategies set by the European Union to stop the loss of biodiversity and degradation of ecosystems in Europe (Communication from the Commission to the European Parliament 2011).

The conservation of biodiversity (Padovani et al. 2009) and the control of the state of natural ecosystems are substantial aspects in the identification of sustainable redevelopment interventions. All this in order to control the environmental fragmentation (Battisti and Romano, 2007) which constitutes a serious threat to the ecological diversity.

Currently, one of the most significant issues in the urban contexts rebalancing is related to the environmental renewal of open spaces (Bassani 2011; Marcus and Francis 1998; Wolley 2003). The open spaces assume, in fact, a strategic function for their role as nodal elements within the city both in terms of fruition and ecological-environmental. Their renewal constitutes an added value for the urban fabric, especially in the presence of unfavourable environmental conditions or when the eco-systemic balance is compromised precisely because they play a significant role in activating the rebalancing processes.

12.2 The Environmental Compatibility

In the field of environmental protection, the search of reversible practices with particular attention to the resource management issues in recovery interventions of the territory is a prerequisite for the implementation of the intergenerational equity principle.

For this reason, any intervention that interest the urban territory transformations in the context of sustainability must necessarily be structured with particular attention to environmental compatibility.

The territory as a whole, which consists of a set of important architectural and functional features that are surrounded and interrelated with the environmental matrices: air, water and soil, should therefore be configured by technological interventions validated within a more general environmental compatibility debate. The overall aim of studying environmental compatibility also specialises in a number of specific objectives, which essentially consist of the ability to detect natural and anthropic environmental aspects as well as contain the transformations within the context of secure sustainability.

The territory should, therefore, be conceived as an organism with a dynamic equilibrium achieved through the technological control of complex functions. The compatible and sustainable territory should be configured as an *urban ecosystem*, with high ecological and environmental standards in which each process is controlled so that its impact, and consequently the induced irreversible degradation, is the minimum possible in relation to the process constraints (Franchino and Frettoloso 2017).

With these premises it is therefore important to be able to consider the territory as a real ecosystem (Adler and Tanner 2013) (Aitkenhead-Peterson and Volder 2010) by entering into the relationships between anthropic and natural phenomena with the aim of identifying strategies and means to rebalancing them. In order for this to happen, it must be taken into account that human activities with their complex interrelated structures and relationships determine different levels of footprint in the environment. For contain the ineradicable footprint as much as possible, it is necessary to structure the recovery interventions of degraded areas from previous anthropogenic activities on a case-by-case basis. This is possible increasing the carrying capacity defined as the ability to absorb and control the anthropic phenomena with a sustainable impact on the ecosystem.

12.3 Green + Blue Corridors

In order In order to appropriately structure eco-oriented interventions aimed at the environmental re-equilibration of urbanized contexts, this paper focuses on the contribution of natural resources, *blue* + *green*, with the aim of using the principles of nature as a model of sustainable management by stimulating the potential intrinsic nature of these resources.

Since the natural hydrological cycle of water is strongly influenced by urbanization in order to structure the interventions of transformation and rebalancing of the urban environment in an appropriate and sustainable way, the control of rainwater flows is of fundamental importance.

In the urban contexts the continuous increase of impervious surfaces often leads to considerable problems of erosion and flooding disturbing the equilibrium between precipitation, evaporation, surface runoff and groundwater supply. It is, therefore, evident that a change of attitude in the management of water resources in urbanized areas must accommodate the natural cycle.

The actions to be put in place for to achieve this goal including the re-waterproofing of the ground to allow for the infiltration of rainwater and recover it in order to reuse it.

The natural water cycle in rural areas, with extensive permeable surfaces and the presence of vegetation and surface water is characterized by high evaporation, high infiltration and low runoff.

The urban water cycle is characterized by less evaporation, less infiltration and an appreciable increase in runoff due to the waterproofing of surfaces. This means that the water must be quickly collected and discharged into the sewer system and therefore does not recharge the aquifer.

This involves the use of infrastructures that are costly in economic, environmental and landscape terms.

Especially in view of current and future potential climate changes, using passive systems that utilize nature's own ability can be particularly effective not only environmentally, but also economically since it allows to limit the use of expensive solutions.

This objective can be achieved by adding or replacing where possible traditional infrastructures with the so-called green infrastructures, which allow water management through rainwater collection strategies carried out with natural systems that consent the creation of a self-regulating basin. These natural systems are even more effective if, in addition to managing the water cycle also controlling the structure of the urban green. With a view to exploring the interrelationships between anthropic and natural phenomena a fundamental role is played by urban green which, in some current examples, is no longer just a plant material, but which it performs eco-technological functions absolutely innovative compared to those traditionally. The use of urban green according to the aforementioned premises can allow us to aspire to look at the urban territory as a real ecosystem by going into the evaluations of the interrelationships between human and natural phenomena and provide tools and methods for rebalancing them.

This interconnection allows to realize blue + green corridors that cross the urban environment, connect the natural systems and at the same time contribute the improvement of the ecological-environmental characteristics of the area.

The infrastructure systems that enable the achievement of the above objectives are referred to as green and blue infrastructures and can ensure a sustainable future for urban areas as well as generate multiple ecological-environmental, social and economic benefits.

The collection of rainwater as a function of its subsequent reuse can be done either by simple collection systems that either take advantage of the gradients of soil and permeable surfaces (passive systems) or with more complex water collection systems including impervious surfaces such as roofs of houses, paved surfaces, etc. (active systems) (Kinkade Levario 2007).

In the United States of America, there are various examples of the use of green and blue infrastructures according to these principles. It is worth mentioning the experiences of Seattle and Olympia (Washington), Portland and Wilsonville (Oregon), Philadelphia (Pennsylvania), Emeryville, Santa Monica and San Jose' (California), Chicago (Illinois), Alachua County (Florida), Stafford County (Virginia) and Lenexa (Kansas) (EPA 2010).

12.4 Applications to Case Studies

The contribution from the conversion of urban open spaces is particularly important, since they constitute nodal elements capable of performing the delicate function of linking the urban systems with the surrounding natural ones, while assuming a strategic role in the transformation of anthropized areas. All this only if the reconversion is outlined as a tangible re-naturalization, with it being possible to activate those processes that use the principles of nature as a model of sustainable management and stimulate the intrinsic potentialities of natural and undeveloped resources due to intense anthropization (Franchino and Frettoloso 2017).

For a better definition of the concepts presented below are discussed two proposals for the redevelopment of urban open spaces that are interesting case studies because referring to areas with characteristics of highly urbanized and their transformation consists in a real intervention of urban contexts re-equilibration.

The proposed interventions include a re-functionalization and re-naturalization of two urban open spaces in Torre del Greco (NA) (Figs. 12.1, 12.2, 12.3 and 12.4) and three urban open spaces in Aversa (CE) (Figs. 12.5, 12.6, 12.7, 12.8 and 12.9).

The two case studies share common characteristics. The open spaces taken into consideration are located in the Campania Region in Italy and refer to areas with massive urbanization. These are urban open spaces in the inhabited centre currently without a specific destination of use, which in some cases are also in degraded conditions.

For both case studies, the redevelopment intervention focuses on the use of *green + blue* eco-oriented technology strategies.

The peculiarity of the areas studied, characterized by intense urbanization, necessarily requires to intervene through the use of technological strategies with minimum environmental impact, low energy consumption and reduced surface consumption.

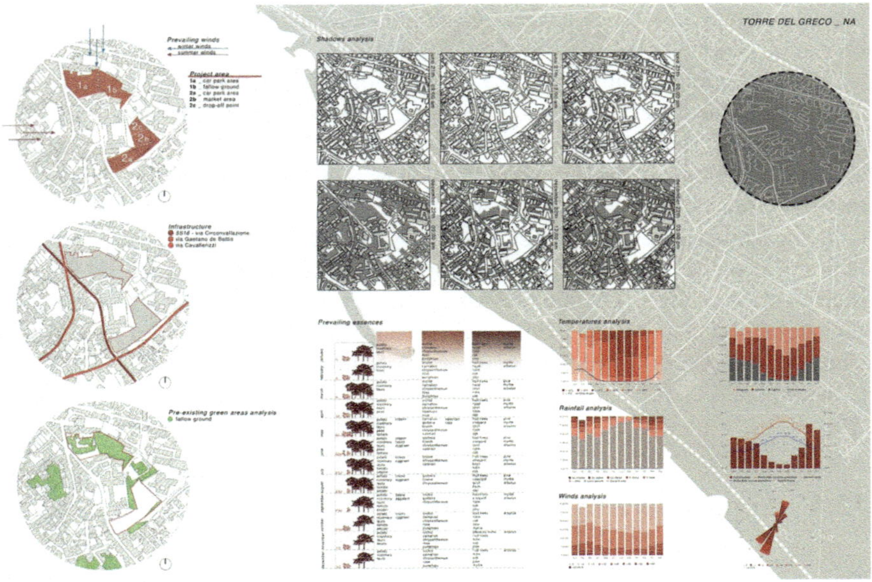

Fig. 12.1 Redevelopment of urban open spaces in Torre del Greco (NA)—Italia_ initial analysis (*credits* Iorio Federica, Palumbo Angela)

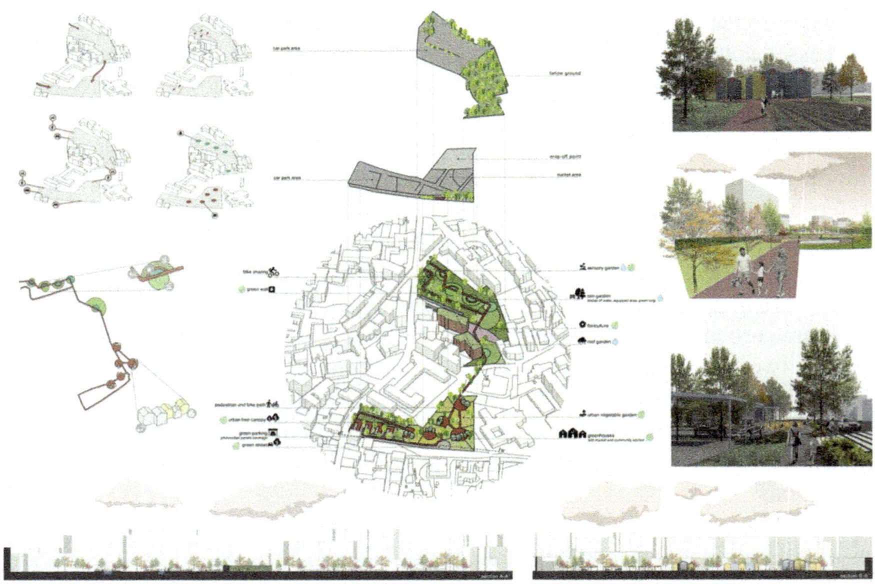

Fig. 12.2 Redevelopment of urban open spaces in Torre del Greco (NA)—Italia_ concept (*credits* Iorio Federica, Palumbo Angela)

Fig. 12.3 Redevelopment of urban open spaces in Torre del Greco (NA) —Italia _ proposal for area 1 (*credits* Iorio Federica)

Fig. 12.4 Redevelopment of urban open spaces in Torre del Greco (NA)—Italia _ proposal for area 2 (*credits* Palumbo Angela)

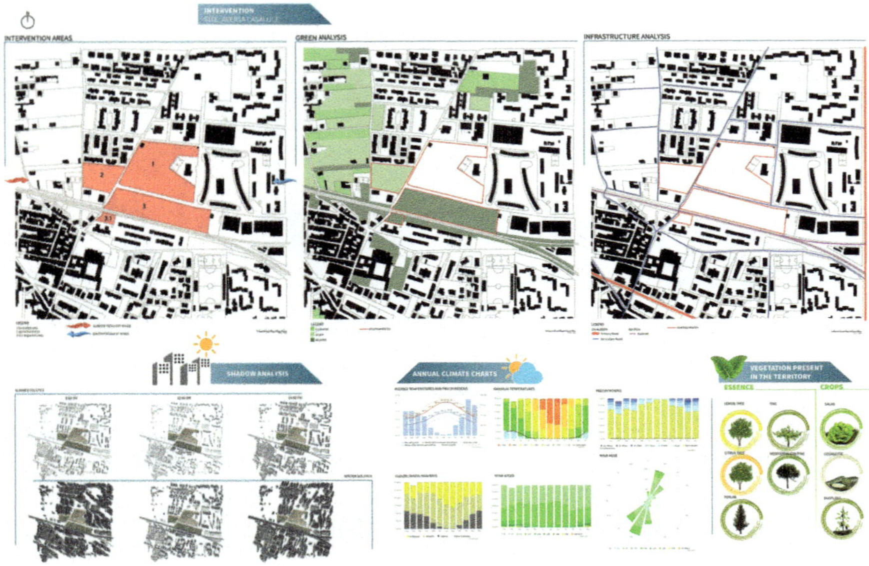

Fig. 12.5 Redevelopment of urban open spaces in Aversa (CE)—Italia_ initial analysis (*credits* Ciarmiello Pasquale, Piscitelli Paolo, Tammaro Giandomenico)

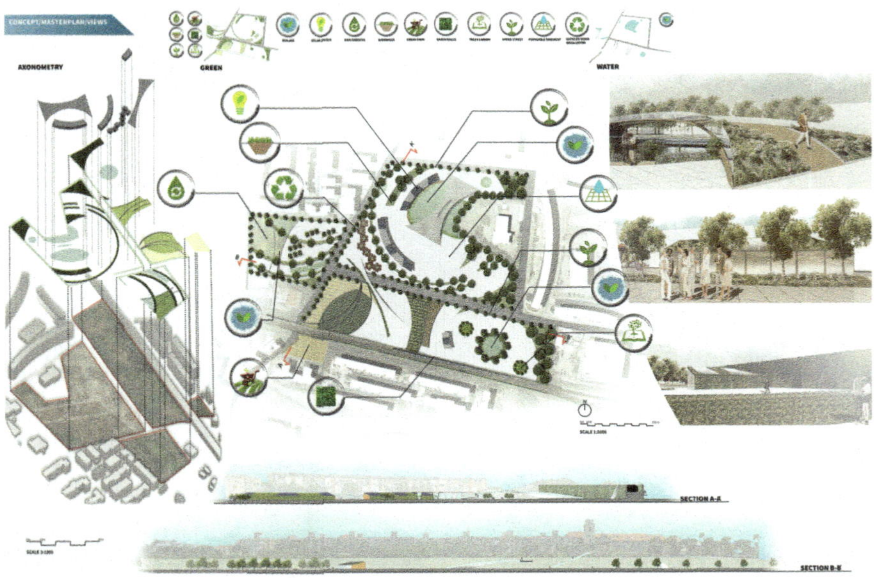

Fig. 12.6 Redevelopment of urban open spaces in Aversa (CE)—Italia_ concept (*credits* Ciarmiello Pasquale, Piscitelli Paolo, Tammaro Giandomenico)

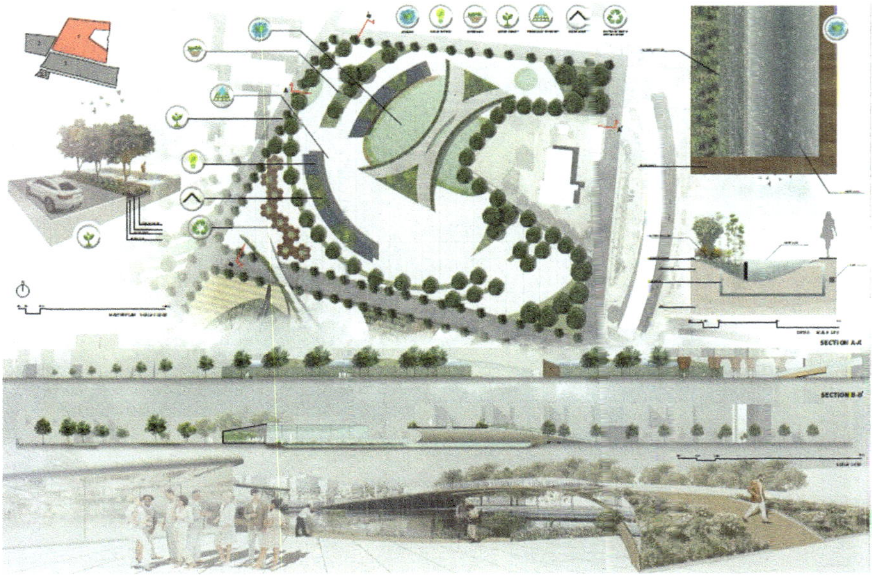

Fig. 12.7 Redevelopment of urban open spaces in Aversa (CE)—Italia _ proposal for area 1 (*credits* Tammaro Giandomenico)

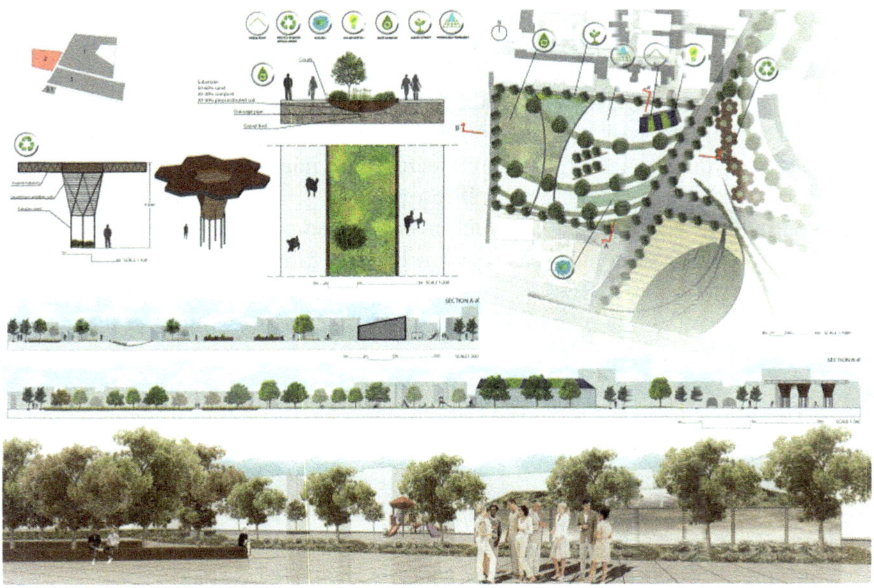

Fig. 12.8 Redevelopment of urban open spaces in Aversa (CE)—Italia _ proposal for area 2 (*credits* Ciarmiello Pasquale)

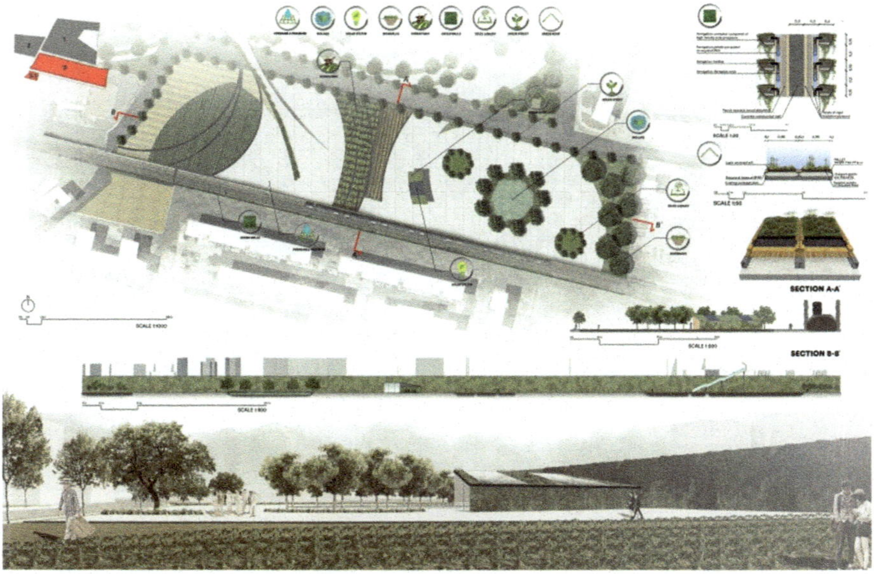

Fig. 12.9 Redevelopment of urban open spaces in Aversa (CE)—Italia _ proposal for area 3 (*credits* Piscitelli Paolo)

The organization of the technology strategies was particularly delicate and necessitated to structure the interventions with demonstration systems that were self-sufficient as possible and not connected to large distribution and treatment networks, while also using low energy consumption as well as energy and water recovery technologies according to the off-grid architecture that manages the energy, gas, water and wastewater needs by using natural resources.

In particular, the proposed interventions are based on the use of passive systems with *green* + *blue* strategies that perform the mitigation function with the aim to conserve the natural resources as well as connect the environmental and landscape values of the areas for an overall development of the territory. In the context of blue strategies, preference has been given in particular to those that allow the control of rainwater management and to restore as much as possible the natural hydrological cycle strongly altered by urbanization. Specific attention is also given to passive systems for the recovery of rainwater such as permeable pavements, green streets, bioswales and rain gardens.

The green streets are positioned in particular on the sides of the roads to collect rainwater from the roadway, while the rain gardens are placed at strategic points near the buildings to collect rainwater from roofs or any paved surfaces.

Among the *green* strategies, particular importance is given to the use of green roofs, green walls, and urban farming solutions, both indoor and outdoor, as a productive redevelopment tool.

The use of urban farming (Bit 2014; Fox 2011) is the main redevelopment strategy adopted both because it also allows to encourage the productive activities of the territory and because it contributes to the improvement of the ecological-environmental conditions of the areas. In this regard the proposed interventions focused on the use of urban farming both indoors in the form of greenhouse with hydroponic cultivation and outdoor with the application of synergic agriculture realized with pallets above ground.

L'urban farming, already successfully tested in some cities, can represent an important strategy of open spaces redevelopment from an ecological-environmental point of view. The cultivated areas, infact, can be networked with the other green spaces that may be present and create a real connection with the surrounding natural and rural territory.

12.5 Conclusions

Since the environmental transformations are dynamic and essentially depend on how moving systems fit and possibly alter static systems, the structure of the intervention to rebalance the highly urbanized contexts is, therefore, complex and is based on an integrated and coordinated set of interventions in order to achieve the protection of the environmental conditions of the territory from the impact of development and anthropization and at the same time qualify life, protect health, increase the safety of the inhabitants.

With such premises this paper has studied the application of natural resources in the sustainable management of urban transformation and has shown, through appropriately structured case studies, the significant environmental benefits arising from the use of principles of nature as a management model. Using nature's own ability to control the negative effects of urban transformation and focusing on technological green + blue strategies is, among other things, not only environmentally, but also economically convenient.

References

Adler RF, Tanner CJ (2013) Urban ecosystems ecological principles for the built environments. Cambridge University Press

Aitkenhead-Peterson J, Volder A (ed) (2010) Urban ecosystem ecology. Agronomy Monogr 55

Bassani R (2011) Spazio aperto e dinamica urbana, Maggioli Editore

Battisti C, Romano B (2007) Frammentazione e connettività. Dall'analisi ecologica alla pianificazione ambientale. Editore Città Studi

Bit E (ed) (2014) Come costruire la città verde. Dalla riqualificazione edilizia all'urban farming, Sistemi editoriali

Cahill TH (2012) Low impact development and sustainable stormwater management. Wiley

Comunicazione della Commissione al Parlamento europeo, al Consiglio, al Comitato economico e sociale e al Comitato delle regioni: La nostra assicurazione sulla vita, il nostro capitale naturale: strategia dell'UE sulla biodiversità fino al 2020 [COM(2011) 244 def. del 3 maggio 2011]. Available at: https://eur-lex.europa.eu/legal-content/IT/TXT/?uri=LEGISSUM:ev0029

Elmqvist T, Setala H, Handel SN, Van Der Ploeg A, Aronson J, Blignaut JN, Gomez-Baggethun E, Nowak DJ, Kronenbert J, De Groot R (2015) Benefits of restoring ecosystem services in urban areas. In: Curr Opin Environ Sustain 14:101–108

Fox T (2011) Urban farming—sustainable city living in your backyard, in your community, and in the world. Hobby Farm Press

Franchino R, Frettoloso C (2017) Open spaces as dynamic urban environments. Edicom edizioni

Kinkade Levario H (2007) Design for water. New Society Publishers, Canada

Marcus CC, Francis C (eds) (1998) People places—design guidelines for urban open spaces, 2nd edn. Wiley

Padovani LM, Carrabba P, Di Giovanni B, Mauro F (2009) Biodiversità - Risorse per lo sviluppo - Focus 2009 Sviluppo Sostenibile. ENEA Agenzia nazionale per le nuove tecnologie, l'energia e lo sviluppo economico sostenibile

United Nations (2015) Sustainable development goals. Available at https://sustainabledevelopment.un.org/, last access June 2020

EPA, United States Environmental Protection Agency EPA-841-F-10–004/August 2010—Green Infrastructure cases studies—Municipal Policies for Managing Stormwater with Green Infrastructure. Available at nepis.epa.gov/Exe/ZyPURL.cgi?Dockey=P100FTEM.TXT, last access June 2020

Wilmoth J (2014) World's population increasingly urban with more than half living in urban areas. Department of Economic and Social Affairs—United Nations. Available at https://www.un.org/en/development/desa/news/population/world-urbanization-prospects.html, last access June 2020

Wolley H (2004) Urban open spaces. Taylor & Francis

Chapter 13
Advanced Sustainable Design (ASD) for Resilient Scenarios

Consuelo Nava

13.1 Introduction

In the era of "rampant post-production", a time that could be seen as "the fastest technological revolution in history", defining a new paradigm with radical and advanced connotations means taking even more responsibility to nurture that required disruptive visionary shift; especially in the areas where different aspects of living resist this said change.

Entrusting the project with this task insinuates a need to closely compare old disciplinary standards with new mechanisms of knowledge. In such a performative and creative scenario, it is also necessary to rewrite the variable categories of the project, to redefine the concept of "context", "landscape and environment", "city and architecture", "technologies and techniques". The "hyper-sustainability" filter conjoins itself with the instrumental capacities of "enabling technologies" to implement "change-friendly" processes. Such procedures become even more necessary, when overcoming the more adaptive approach, to guide towards a circularity of tactics and actions that reveal capable and innovative strategies, therefore bringing about a new quality to the project (Nava 2019a, b).

The proposed treatment pursues the objective of the "new quality of the project", which addresses the particular practices of "advanced sustainable design for resilient scenarios", thus illustrating two trajectories. The first investigates the themes of new paradigms and operational methodologies of design; the second proposes and experiments with multi-level information practices, related to design/drawing in sections for the transforming advanced design of the building environment. Both sections express theoretical and methodological references, while also illustrating experimental examples.

C. Nava (✉)
Università Degli Studi Mediterranea Di Reggio Calabria, Calabria, Italy
e-mail: consuelo.nava@unirc.it

G. Chiesa (ed.), *Bioclimatic Approaches in Urban and Building Design*, PoliTO Springer Series, https://doi.org/10.1007/978-3-030-59328-5_13

ASD by Paradigms and Operational Design Methodologies

13.2 The Definition of "Advanced Sustainable Design": The Mapping of Topics

The quality of the production is obtainable from the building's environmental transformative conception and realization processes. According to sustainable strategies, this quality must aim to have a higher performance rate, throughout all its components, than those related to the benefits and satisfaction of the project's requirements. The service attributes must find "an added quality" on which to express their capabilities whereby, on a multiple-level relationship scenario, any technological system that enables an effective process for operations responds to the needs expressed. Product technologies are required to have their own qualitative rating scale that depend upon the physical and material characteristics, durability, efficiency, configuration, operation, etc.; process technologies must refer to the question of the system's capacity for innovation and integration, operation and management, competitiveness and improvement (incremental or radical), adaptation to the evolution of sociocultural models and for the ability to create new meanings within the evolved design processes. Moreover, high performances are comparable in terms of evaluation, in terms of a shorter or longer response time respectively. For this reason, their function is to act as the caliber of interventions, both for transition and resilience scenarios.

The multilevel relationship scenario produces a new advanced operative design model which activates "agile" processes. Between circular and resilient systems, such an activation targets all scales of the construction environment (cities, buildings, products) while simultaneously referencing the 2030 objectives of JTM (Just Transition Mechanism) and the SDGs (Sustainable Development Goals). Managing and keeping data resources available and KET's as sensitive devices triggers a level of continuous advancement in the long-term, this perspective is capable of pushing the process of success through into "management innovation" (Celi 2005). The management tools operate on applying areas of soft- (for the process) and hard- (for the product) technologies, in the physical field as well as the digital fields, in order to finalize an advanced design with high performance results (Fig. 13.1).

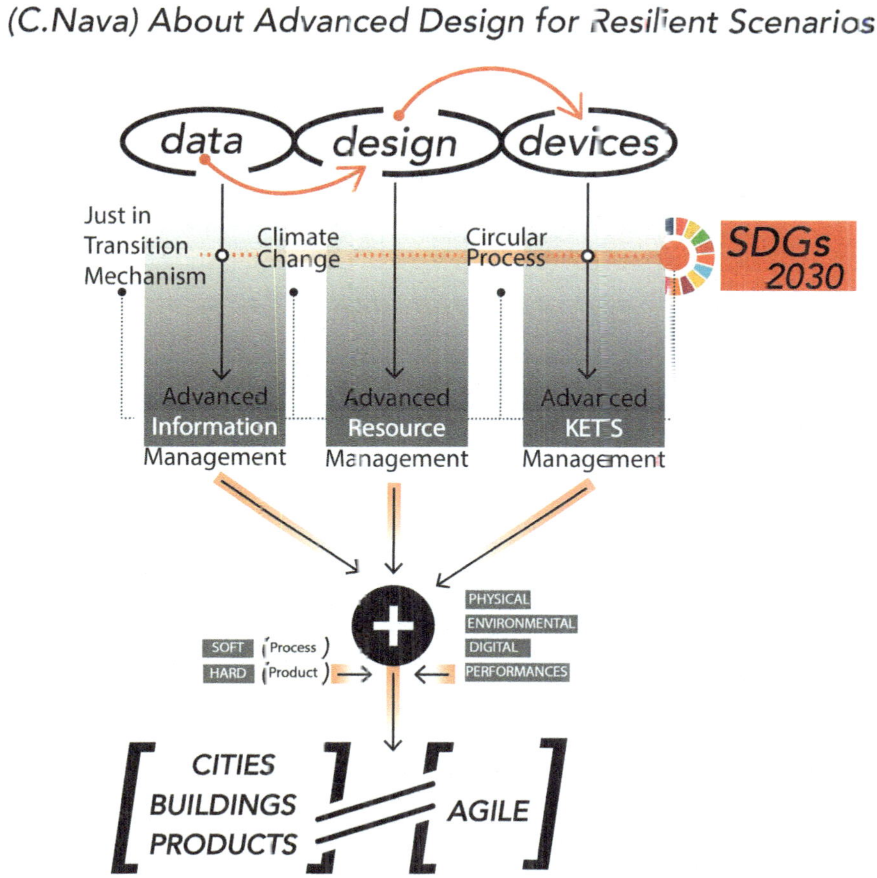

Fig. 13.1 Scheme on innovative management of agile processes. *Source* Research in progress, C. Nava, University Mediterranea of Reggio Calabria, 2019

13.2.1 An Application Example on the Proposed Model: Innovation Management in Circular Processes About "Frontier Industrial Components for Sustainable Construction

The experimentation model illustrated in Fig. 13.2, presents the advanced design process in an operational form (Celi 2005). The figure reports a frontier industrial component structuring the framework that explains a relationship between key enabling technologies, circular economy, and sustainable design.

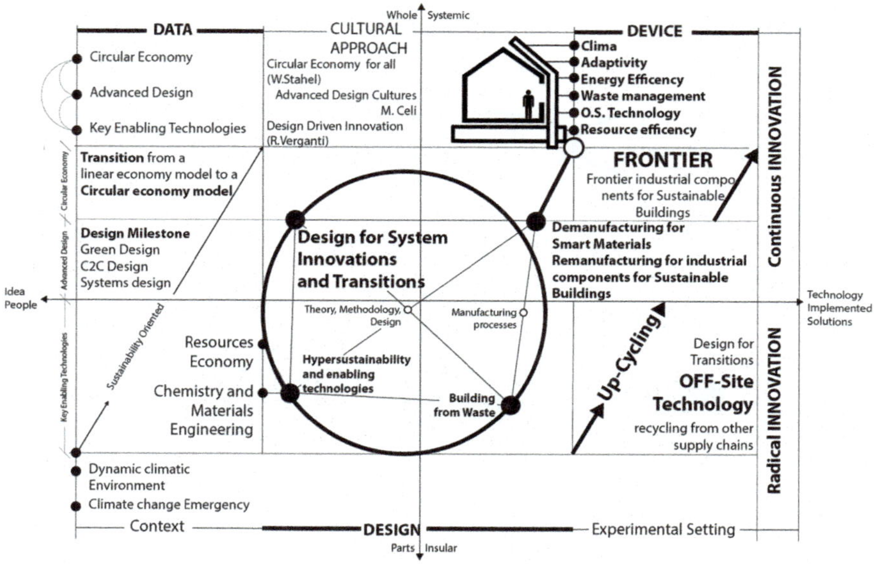

Fig. 13.2 Experimentation model of advanced design. *Source* Research in progress, Domenico Lucanto, University Mediterranea of Reggio Calabria, 2020

The phrase "frontier industrial components for sustainable construction" signifies casing systems, which not only concern the skin of the building or its structural system, but the entire technological-environmental system. The concept references a complete array of energetic-environmental requirements and performances, paying particular attention to the morphology and constitution of the materials for the components and their production cycle. The frontier may be classified as a "sensitive device".

The innovative nature of the advanced design operation is defined by the circular process. It is a matter of applying the themes of radical innovation (Verganti 2009), starting with the definition of physical-environmental contexts and moving to reduce upcycling on all the frontier industrial components.

By specifying the performance requirements and the technological qualities that the device must have, all characteristics are created with a form of continuous innovation, thereby qualifying the systemic approach towards overcoming the aspects strictly related to its technological evolution.

Off-site technologies allow sustainability-oriented processes to be industrialized by integrating enabling technologies associated with the upcycling of waste or waste materials in the testing phase.

These have a highly developed ecological profile because they come from the supply chains of materials and resources whose scraps or residues can also be traced in sectors that are not dedicated to general building or construction.

Such an upcycling-oriented experiment takes on additional significance when considering the digital processes for the design, from de-manufacturing to the re-manufacturing of frontier industrialized components for Sustainable Construction (Smart Strategy—S3/EU).

13.3 Design Methodologies for Resilient Scenarios: Climate Change + Circular Process

The advanced design methodology aims to identify operational steps of both the design and theoretical work, with a view to composing resilient scenarios. Such situations may include emerging environmental issues, the effects of climate change and the application of circular models; for material resources with energetic and environmental flows the importance lies in responding with a high level of performance. "Total design" tools and devices are reported to be able to govern from new scales such as urban, building and prototype.

For a varying context that changes the configurations at all points of the calibrated system via an interpretation of the transformations, it is necessary to work with a project culture that matches the aspects of an organizational-management "design". A structure that operates matrices through normative, interpretative, generative models. In other words, at the same time as the expression, technical and graphic information of the project states exchange and transmission; transferring contents and solutions to different points.

The sustainable project is always, by its very nature, a "total design", able to find moments of compatibility with other structural dimensions of the environment. Such as those more relevant to the landscape and the device's efficiency system, even including its installations. In regimes of hyper sustainability, including short-medium- and long term, and encompassing the verification and measurement of economic feasibility and the evaluation of viable strategies, there is no longer a real need to distinguish between a sustainable approach to the project and a conventional way of thinking about cities, buildings, and innovative living service products. This is a historical moment in consciousness that presents the only possible way to limit the consumption of exhaustible resources, to renew those of the local and super-local biophysical system. Thinking univocally about the same system's capacity to produce a comprehensive quality of life, in physical, material and performance aspects, for the environment and other non-urban areas. The question of "total quality" is answered by the definition of "total design"; "using advanced techniques involving preplanning. One must know what has to be done before doing it, so that one can organise the work properly. The key to a building operation ought to be what one might call the total design—which foresees and plans every stage in the operation. This total design must nowadays be the work of many people, each contributing his particular knowledge, and this team should be controlled by a strong leader, now an executive team with power to decide priorities and reconcile conflicting claims" (Ove Arup 2012).

This goes beyond the meaning of integrated design, by involving sectors of applied skills and contexts to propose a transversal awareness of controlling the design at all levels, from its programming intentions (definition of a master plan) to its figurative and technical declinations (design and detail). The design process can be defined and observed in the explanation and methodology of "sustainable total design" whereby the procedures act according to a self-informative method. The domains of data, design, simulations and mapping are assumed, produced and tested; these variable settings are closely reported on following the scenarios that communicate the visions (Nava 2019a, b).

Step 1 Circular and Resilient Cities, Ecologically Adaptive.

(a) Changeable design scenarios with programs, actions and visions of '20/'30/'50 and refers to the 2030 Agenda [SdG 3,6,7,9,10,11,12,13,14,15].
(b) Plan an ecologically adaptive and resilient area; climate, water, soil, and air.
(c) Setting up physical functions and spaces (confined and not) through the CITY Mining principle—mining city (resources from circular cycles and circular flow operations—energy, water, waste, and materials).
(d) Mobility and accessibility with low impact and high efficiency.
(e) Spaces and networks of social interaction and exchange (public, public–private, sharing, co-managed spaces).
(f) Design and organize functions and spaces of autonomous districts and/or ecological neighborhoods
(g) Complete the urban fabric program through BUILDING DENSIFICATION and/or GREENING and NEW PERMEABILITY actions (productive nature, soils and network, mitigation) (Figs. 13.3 and 13.4).

Step 2 Advanced Sustainable Building—Housing Agile Building

(a) New user for flexible types—spaces, sections and program layout and operation for buildings and complexes of varying size and structure over time.
(b) Organic Architecture (inside and outside)—form/function/climatically "adaptive" structure.
(c) Process, innovation, construction site in "agile" mode (digital, physical, environmental): offsite and upcycling.

Step 3 Advanced model prototype

(a) to Advanced technological systems (adaptive and cybernetic) and improved processes for components and materials (hybridization and eco-design).
(b) Spatial and functional models (building-system-component relationships, alternatives in open platforms).
(c) Realization and design processes "for manufacturing" and "for testing" (Figs. 13.5 and 13.6).

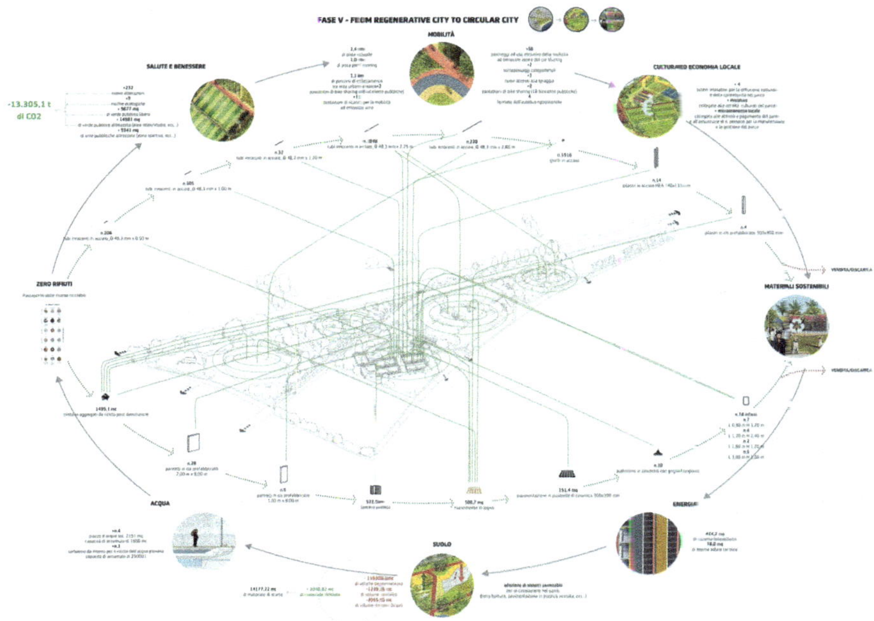

Fig. 13.3 Experimentation model of advanced circular design. *Source* Research of Thesis C. Nava with G.Lico, University Mediterranea of Reggio Calabria, 2019

13.4 Physical—Environmental—Digital Performances: Agile Cities/building

The topic of "agile buildings in resilient circular cities" finds its paradigms not only in the ability to express sustainable innovation on the urban and building scale, but even more so in the time of the fourth industrial revolution. Thus, being faced with the chance to respond to the general public's need for a change in the quality of life with opportunities that infrastructures, processes, projects and products must be able to express.

"Being agile" means responding to these changes with a high performance standard. Adapting to such shifting needs in increasingly "circular" systems can trigger radically innovative procedures throughout the entire project, including the processes, as well as the product itself. Enabling technologies become the "service devices" for this "performance", they themselves adapt to the contexts in which they are called to govern the process and/or the project in order to produce "quality" throughout its usage, operation, as well as the configuration of urban systems and structural architecture. The city—under the pressure of these recent years due to the environmental and immigration problems/issues—turned its weakness into a privileged context and a resilient environment where it's possible to activate these processes and to insert

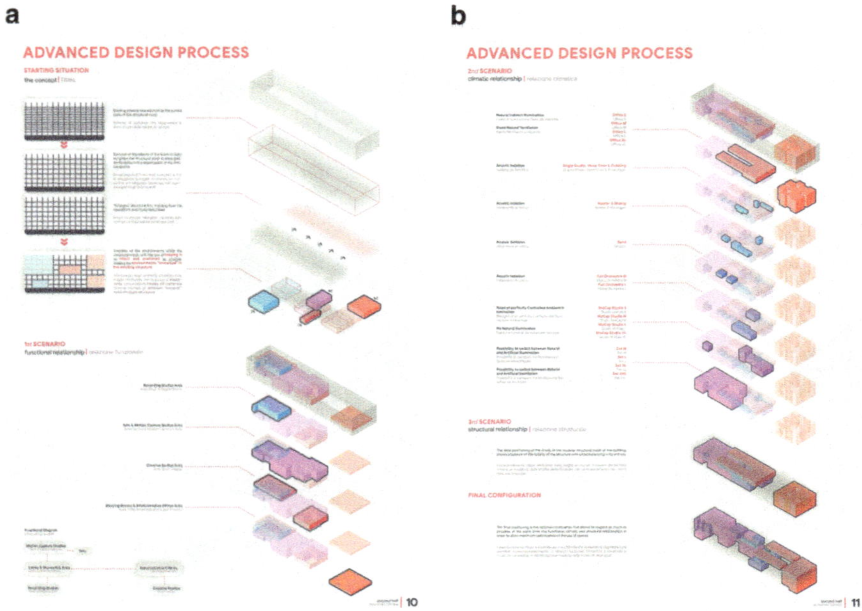

Fig. 13.4 **a** Experimentation model of advanced process design—the high environmental quality building_starting and 1nd scenario. *Source* Research of Thesis, C. Nava with C.Mazzei, University Mediterranea of Reggio Calabria, 2020, **b** experimentation model of advanced process design—the high environmental quality building_2nd/3nd scenario and final configuration. *Source* Research of Thesis, C. Nava with C.Mazzei, University Mediterranea of Reggio Calabria, 2020

Fig. 13.5 Experimentation prototyping and area test of advanced circular design—Resilient Green Parking. *Source* PVCupcycling, C. Nava with PMopenlab, 2019

three different levels of innovation in favour of the users/citizens. Thus, a sustainable and innovative, high-performance project is achieved through:

- The control of physical components for city buildings: throughout 2020/30/50 it will be important for existing surroundings to become "regenerative systems" and to respond to new structures without the need for large economic investments. By implementing circular economy policies, which include metabolism, densification, compensation, resilience, recycling, etc. exemplary scenario procedures will be able to take place for energy transition structures, such as user-producer technologies, renewables, smart grid systems…

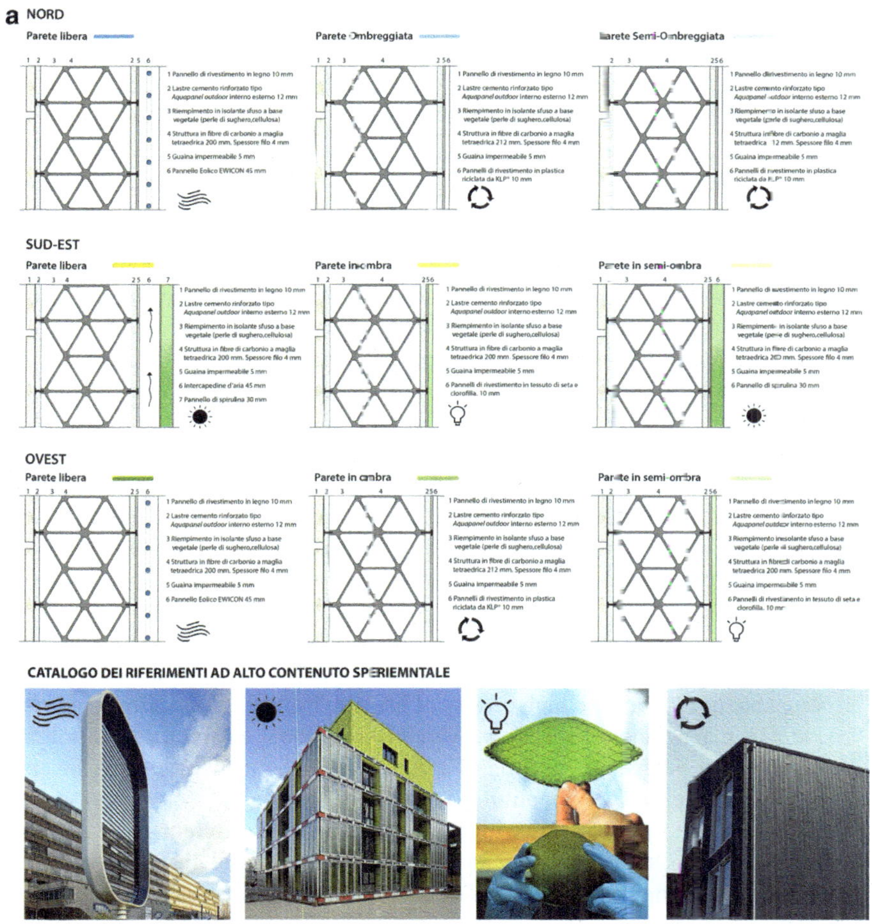

Fig. 13.6 **a** Experimentation design of advanced agile design—structure and adaptive skin. *Source* Research of thesis C. Nava with G.Marano, 2020, **b** experimentation prototyping of advanced agile design—structure 3D printing. *Source* Research of thesis C. Nava with G.Marano, 2020

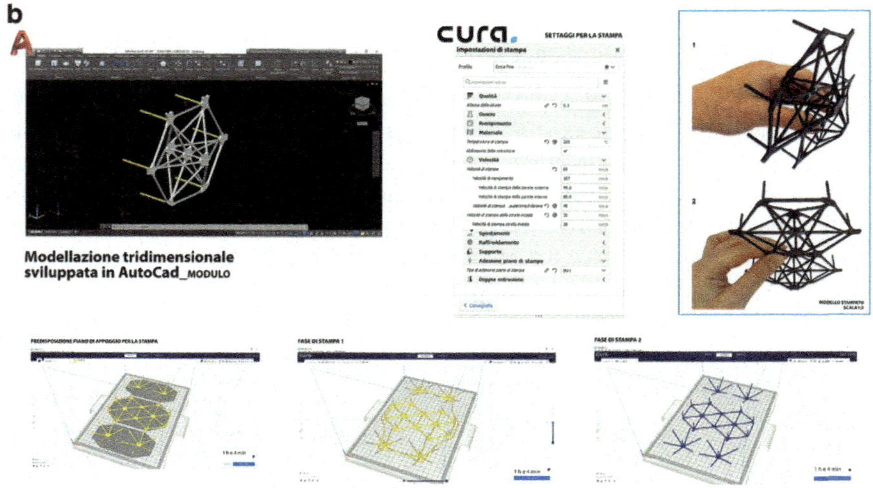

Fig. 13.6 (continued)

- The integration of digital devices for city buildings: useful for informing the project of possible sites and functional structures on a large urban scale, with maps presenting the correspondence between the location and the appropriate resources. Digitalisation provides the ability to predict, control, monitor and facilitate the use of urban services, as well as repurposing confined spaces from passive into active. Ultimately utilizing advanced "smart" technologies in order to control the project and facilitate the creation of highly adaptive systems. Such additive manufacturing, that is used for scale prototypes, includes censoring the integration of traditional systems and materials with innovative systems and materials in newer hybrid models. This controls the building's envelope performance in the design and operative phase through using the information method with applications and devices.
- The environmental factors of the project's buildings: by controlling each product's life cycle process, the building's influential standard of spaces and constructions is capable of producing a high quality of life. Environmental systems require high levels of management and safety, which brings about the reduction of climate-changing emissions, the creation of spaces in a resilient environment, the ability to control levels of "compatibility" and "compensation" between systems at all scales of the project and with the involvement of different natural/artificial systems. This is about choosing an urban building "design" that is capable of re-defining its configuration in terms of energy, material innovation and recycling, hybrid systems, which lends itself to be the most highly responsive performing operational model in its context.

ASD by the Multilevel-Information in Section

13.5 Sustainability Multilevel

The measurement of interventions on particularly sensitive contexts, such as those related to natural or fragile ecosystems, is always measurable through an interscalar process, which must control actions, transformations and impacts and additionally evaluate several reference measures. The project and the programme that seek to pursue such methods are always named as "multidimensional." Any transformation of the building environment or of an ecosystem with a stable natural value produces both direct impacts and indirect impacts, equally measurable at various times. Through enabling technologies and dedicated devices the design process instructions never have a hierarchical process of operation, but take information and data to produce them as a system to support the design. Using the logic of an open platform the system is capable of triggering reversible and evaluable activities in itinere.

The object's dimensions within the intervention already belong to three very different configuration scenarios; the resources involved remain as the context of reference. Having included all transformative actions (incl. infrastructural) the need to penetrate the processes as receptors or transformation systems throughout the life cycle is clear, with relationships and operations that accommodate their own characteristics and behaviors. Controlling the design in a multi-scalar key allows the ideation or definition phase of the models/prototypes to be active, conditioning a controlled response focussed towards getting high performance results. Therefore it could be said that according to the strategies of hyper sustainability the project may face a scale of problems and their consequences, to which however, the definition and performance quality of the project must correspond with an inter-scalar response (Nava 2012).

13.6 Design and Information for the Hypertext Project

The research project matured into one dedicated to sustainability during years of didactic and experiential exercises that proved, in terms of a textual approach and thanks to its integration of open structures and hierarchical characters, to have soared into hypertextual. Adopting this perspective for the project in question means considering several aspects of design and technical information. The first concerns content, meaning of choices intended for transference; the other involves method and expression, with which these choices are transferred. But hypertextual dimension traces such capabilities to an expression of contemporary plurality, in which logos and techne are expressed through both technical and graphic information when the logic of a technological design system is fully realised. Such a discourse is entrusted absolutely to the descriptive and typological role, which is assumed by the section from either the horizontal or vertical plane, in the advancement of the project transferred through the manual or digital drawing tool.

This discourse has been made possible thanks to the basic principles of this section's design, which has most certainly been made with reference to the scale of representation, that is conventionally becoming the very tool for unfolding the complexity of the project. This is how representative communication transfers the measured logics of the typological, morphological and constructive characters of the sustainable project. The investigation therefore goes on to find some issues with the contemporary project and on the innovation of its process. For example whenever the project becomes design for manufacturing could it ever be possible to address the issues of urban, environmental design at the scale of the construction components, the detail and its specifications, as far as the architecture has assumed the value of an instrumental code. Another theme may contain questions such as how much the material must be understood to be as equally important in all areas; matter of form, expression of resources, structural component of each system, from the territory to the landscape, in construction and in how much this must be read through multiple scales of representation and more information, what ways of organizing the available data and reporting on those produced with the project, communicated through the hypertext of the section representation conceived in its three dimensions.

The section is a studied method of tentative design in which the experimental investigation of the design phases is transferred and thereby entrusts the spellings that are used to simulate visionary plans, operating models, alternatives, possibilities and performance capabilities. A representation that almost always declines on environmental aspects, such as the project's natural material and resource usage, tests their ability to produce site, building, technological components and usage patterns over time. A tool that is used to investigate the different combinations of the project's physical and structural components, as well as give notice of the spatiality and function organization, which other types of representation rely upon. The section's hypertext capacity becomes an operative method to try for possible solutions, to choose plans for the different sections and their most significant paths, which are deducted from the description of the most dedicated data. It is the instrument which entrusts possible changes to decisions and thus their consequential choices throughout the history of the project. Sometimes the sections themselves leave room for a more open project, one that is more participated in, not only by the aspect's contributions that produce the different qualities in an integrated way, but as representations of those visions that over time can accommodate other solutions, without losing the reasons of the project itself.

The section takes the form of a layered representative system. The content lies in the meaning it wants to convey, which is why it can be peeled without decomposing its structure, just like "an onion", peel until you find the reason and the origin of expression.

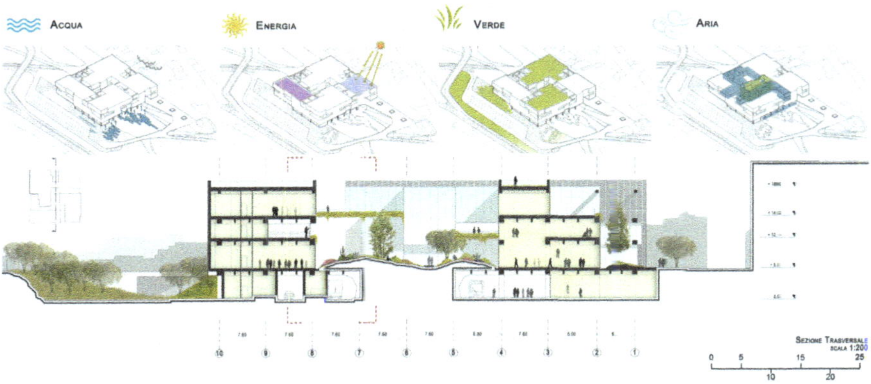

Fig. 13.7 Experimentation of advanced agile design—resources. *Source* Research of thesis C. Nava with A.Franco, 2013

A diary of sorts, in which comments and sketches verify the space in the different representations, mixing up the reading keys of the graphic scales, and then reconstructing the ways in which the information supports the final decisions for the project. It is because of this that the horizontal and vertical planes of the sections often generate from the same point, at the very start of the process, right from the heart of its concept, with proposals that are in fact impossible, but which assist, in short, to fix the limit and the resource of the project itself.

The section is a useful code for the project that becomes its model. It is the most innovative interpretation of a representation that is capable of transferring technological information components that model usage patterns over time (Fig. 13.7).

13.6.1 Territory and Urban Section—Re-morphing

The section on territorial-urban scale describes the morphology of the environment, which includes the natural, the physical and the anthropic, such transformations in the project are capable of expressing strategic levels and the impact the implemented changes may have on sites.

Relevant information on the infrastructure of the territory and urban contexts is detected, translated and in turn, declined, through studying the changes in the soils' morphology, as well as the relationship between whether to build or not to build, and the metabolic organization of the flow of resources—water, energy, green, air and any flows or routes used.

The section is built as a result of data organization and its graphic representation that is then transferred to the project, this highlights the systems that are progressively completing the strategic vision of urban-territorial transformation, through:

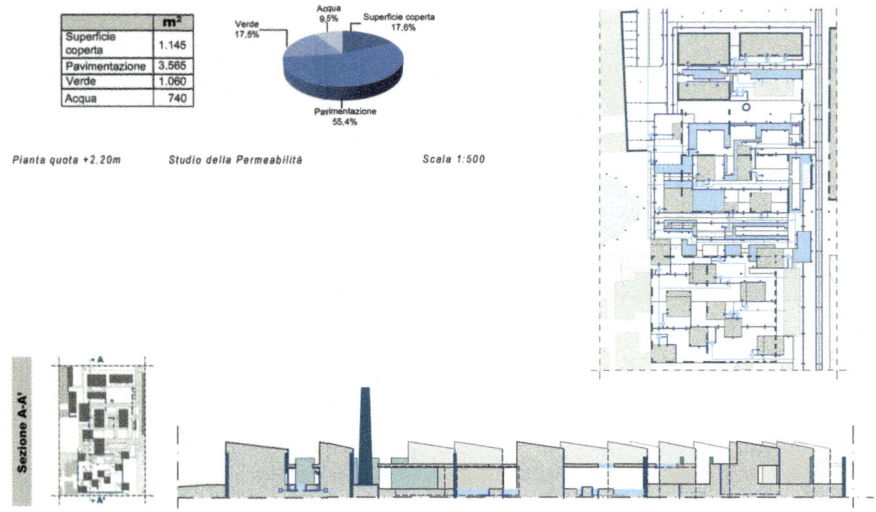

Fig. 13.8 Experimentation of advanced agile design—warer management. *Source* Research of thesis C. Nava with R.Viviana, 2012

- the possibility of a program of actions that will be carried out in several time phases;
- the analytical-environmental study on the settlement scale and the identification of the contexts affected by the transformations;
- horizontal and vertical re-morphing, which reads the changes in the territorial and urban skyline;
- the visions of the physical and functional change between the zero state of the project and its subsequent possible alternatives, to measure its feasibility and models-intensity of use;
- the reconnection between the project area and contiguous areas in a single physical—configurative system, which re-establishes its recognizability and attributes values to the new identity of the territorial-urban system;
- environmental accounting illustrated as a data system responding to typological alternatives, with reference to the framework of resources to be activated, recovered, compensated, dedicated (quantitative and qualitative data of the new circular cycle of resources) (Fig. 13.8).

13.6.2 Landscape + Environment Section—Re-habitat

The section on the environment and landscape scale describes the stability of the physical environmental systems and the quality of the landscapes. It represents the terrain's ability to withstand the transformative impacts when it comes into contact with the contexts concerned.

Information is detected, translated and declined by referencing the flows of resources, such as structuring components and possible configurations of "new units" not only territorially, but also at an urban and micro-urban level.

The section is built through the organization of data and the representation of that data when it is transferred to the project. It highlights the morphological, material and performance characteristics of the inhabited systems of the landscape and its environment, in their ability to be valued and recovered, redeveloped and recycled, to raise the level of usability and therefore maintenance. This is achieved through:

- referencing master plans and functional layouts that can be capable of describing the areas of intervention in qualitative and quantitative terms for landscape/environment features;
- the reactivation of environmental systems through the circular cycle of natural resources (green, water, climate, soil, species, etc.) as poles catalysts of the smart-grid environment/landscape;
- the implementation of the preliminary design at both the contextual and building levels, in order to reconstitute the landscape units via environmental compatibility actions in different seasonal structures and/or for different usage scenarios;
- the choice of materials and components in the reconstituted units from controlled production processes, with low environmental impact and/or from recycling, comes with high durability and maintainability;
- the study of the systems' life cycle and their operating processes, with reference to the new components of the project, both natural and artificial (Fig. 13.9).

Fig. 13.9 Experimentation prototyping of advanced agile design—structure 3D printing. *Source* Research of thesis C. Nava with S.Vartolo, 2012

13.6.3 Context + Building + Construction Section—Re-building

The section that synchronically describes the combined relationship of the context + building + construction. It illustrates the rules and methods with which the transformation of the construction environment is carried out. This is achieved by using settlement technologies and fabrication typologies that are typical of an innovative design process, from the very conception to the programming of the green construction site.

Information is collected, translated and addressed with reference to environmental and physical data. Preceding and succeeding the transformation of the data, it reduces the proximity of the scale to the context and evaluates quality at the initial site. This indicates the types of environmental and technological requirements, whilst bringing into question the feasibility of the work and its environmental and economic sustainability. The project answers this question thanks to the many layers in the body of the building and its characteristics, which are determined by the construction procedures. The ability to meet the contemporary demands of the sustainable project with the integration of innovative features during the construction processes depends on the complexity. It requires identifying them as connotative conditions of the typological, morphological and material quality of the different types of buildings.

The section is built through the organization of data and the representation of that data when it is transferred to the project. It highlights:

- simulations with outlines and references of plans, conceptualising the relationship between confined spaces and open spaces;
- in-depth analysis to assess whether the technicalities are appropriate for the building envelope, definition of the technological systems and the material and component qualities;
- constructive-technical logistics of the frontier systems with detailed implementation layouts;
- the abacus of the components and how to use them;
- selection and integration of the designed construction system, using the profiles compatibility;
- to work out the engineering of the project through the creation of a prototype (material and digital) (Fig. 13.10).

13.6.4 Context + Building + Resources Section—Re-generative

The section that describes the circular relationship between the context + building + resources, illustrates the operating models over time and at different climatic conditions. This evaluates what the best choices are for obtaining high performance levels that can allow the buildings to be seen as having an "agile system". Highlighting that resources are being used in an increasingly "circular" and "regenerative" context.

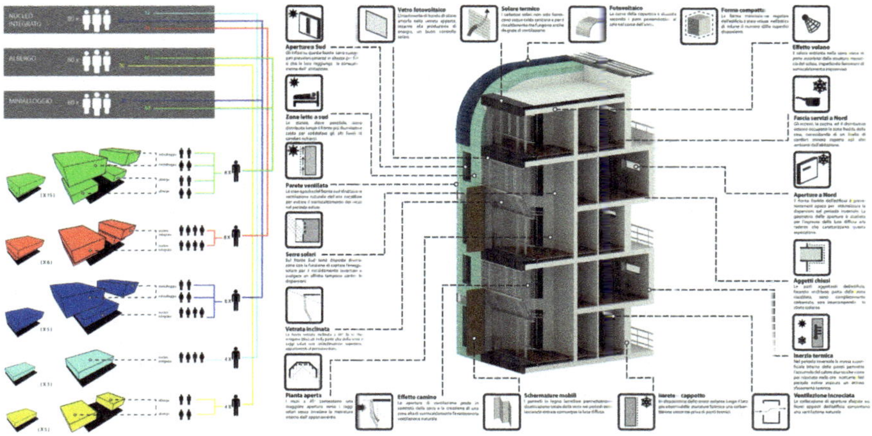

Fig. 13.10 Experimentation of advanced agile design—environmental tools. *Source* Research of thesis C. Nava with F.Megna, 2012

The information is collected, translated and condensed taking into consideration the dynamic behavior of the building's function that has been inserted in a defined context. Thus illustrating an integrated way in which passive reactivity can include the comfort and safety levels of the confined spaces as well as be influenced by the open space systems. The proposal of possible alternatives requires a definition and deepening of even the most innovative elements, through the detailed study of morphological, typological and material characteristics alongside the synthetic simulations of performance. The integration of the building's form, structure and technologies, when viewed as an open and dynamic response to environmental energetic behavior, brings about changes in the whole scenario affecting the different types of usage by specific users. The section is built through the organization of data and their representation when transferred to the project, thus highlighting:

- environmental and energetic simulations with reference to the regulatory domains at the context and building scale;
- bio-climatic simulations of hybrid behavior (active–passive) in the different performances;
- operating levels and technology of the exposed facades;
- indication and study of integrated technological systems (systems, solar technologies, innovative high-performance systems);
- environmental energy profiles of the components;
- useful life cycle of the building and its technological systems and integrated regenerative systems;
- implementation of data after the project's sustainability assessment in specially designed systems;

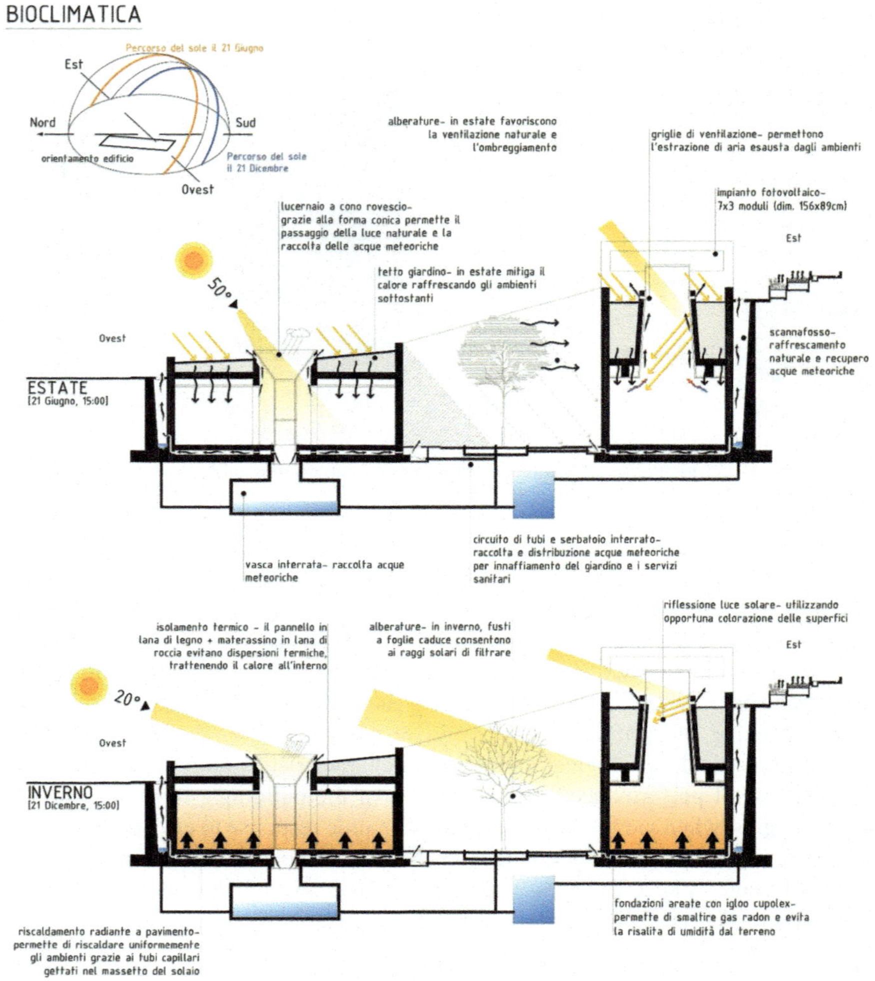

Fig. 13.11 Experimentation of advanced agile design—bioclimatic performances. *Source* Research of thesis C. Nava with G.Mangano, 2013

- programming and designing of a green construction site that seeks to reduce the use of natural resources (soil, water, etc.), in order to have energy effective productivity, the kind that is necessary in these processes (Fig. 13.11).

13.7 Conclusions

We adopt the term "design" rather than project, in order to favour a particular approach. As a concept related to innovation, a design-driven formula, which provides the necessary space for interpretation and communication, in the projected life context, between those who must bear the fruit of the project in its production and those who must use and consume them in post-production. The collective laboratory for each process is defined as a "design discourse", in which those who operate the project always assume the role of interpreter/researcher and give various contributions. "Companies that produce design-driven innovations place great value on their interactions with the interpreter network. These companies know that they are immersed in a collective research laboratory through which companies, designers, artists and schools are carrying out their exploration. These researchers are involved, explicitly and implicitly, in an ongoing dialogue: they exchange insights, interpretations and proposals in the form of studies, conversations prototypes, images and products. They test the robustness of their assumptions and share their visions. This process of 360-degree research into the possible meanings of things is design discourse" (Verganti 2009).

For this reason and as discussed in further ongoing lessons, the declination of the project is addressed whilst considering the characterizations and reasons of the "design form". The definition of the equivalent adjective announces its purpose: impact design, high performance design, cybernetic design. It works to trigger a projective process, which can translate procedures of "total design" and is comparable to proposed sustainable inter-scalar operations that have the ability to control the same definitions of the project (Nava 2012). With the aim of configuring a possible commentary on the quality produced in architectural landscapes, a conscious use of technologies can enable an understanding of the functioning life-cycle and the relational energetic conditions among all the biophysical and socio-economic systems that react to a certain "level of turbulence". This is indicative of all building environmental dimensions involved in fast, innovative, complex projects.

The "total sustainable design" is therefore charged with a methodical task of knowing how to effectively manage innovation through a generative, multi-level design model (from strategic planning to the open design of the system/object). To govern the ways in which the activities of the processes affect the technical and performance aspects and to express all levels of useful knowledge about the operating systems involved. Such a definition allows the project to assume an "exploratory", multidisciplinary, non-sectoral or specialized character, but even more a narrative and dedicated role to illustrate paradigmatic issues of interest, of which a possible definition guided by innovation must be found. An approach that although it preserves that projective character of the project, by axiom it leaves room for a fertile state of design thinking, even without its need to achieve a transformation of the building environment. "On the one hand, the project, precisely as an anticipation, refers as a temporal horizon to the future: that is, the project is inseparably linked to the affirmation of the idea of the future. On the other hand, the design activity must necessarily confront the category of the possible. This clarification puts the project away from deterministic interpretations, but at the same time leaves room for the possibility of thinking about design activity outside of an effective desire to transform the present" (Campioli 1997).

References

Campioli A (1997) Le culture del progetto. In: Nardi G (a cura) Aspettando il progetto. FrancoAngeli, Milano

Celi M (2005) Advanced design cultures. Long-term perspective and continuous innovation, Springer

Nava C (2012) Edifici Sostenibili. Particolari costruttivi DEI, Roma

Nava C (2019a) Ipersostenibilità e Tecnologie Abilitanti. Teoria Metodo e Progetto. Aracne, Roma

Nava C (2019b) Sezioni Sostenibili. Design e Informazione per il progetto ipertesto Aracne, Roma

Verganti R (2009) Design-driven innovation. Rizzoli, Milano

Chapter 14
Building's Operational Versus Embodied Energy: Needs and Barriers for a More Reliable Environmental Impact Balance

Ernesto Antonini

Abstract To comply with EU directives, which require new buildings to be nZEB by 2021 (2010/31/EC) and all buildings to be nZEB by 2050 (2018/844/EC), the construction industry has mainly focused on the containment of energy consumption during use, but scientific literature highlights that techniques employed to achieve higher operational energy efficiency lead to an increase of embodied energy, embodied carbon and other environmental impacts, especially those related to the manufacturing of building materials. The more stringent energy standards have thus reduced the impacts of the building operational phase, but raising those relating to materials and components which are required to allow the buildings achieving these benefits. Several evidences of this trend are provided by applying the Life Cycle Assessment (LCA): although that method is not currently addressed by the EPBD, the key role that the embedded impacts play on the building environmental balance is confirmed. Though a precise calculation of embedded impacts is difficult to perform, due to the many factors involved, the estimation of their share compared to the operating impacts is crucial to determine an effective building energy and environmental balance. Moving toward this target, the Sustainable Building Rating Systems (SBRS) are progressively integrating LCA-based criteria within their protocols. This should provide a simplified and standardized approach thus helping their application in building sector, as preconized by the literature. However, several discrepancies among the different SBSRs in use worldwide emerge in both method's application and interpretation.

Keywords Energy efficiency · NZEB · Embodied energy · LCA · Sustainable building · Rating system

---

E. Antonini (✉)
Department of Architecture, Università di Bologna, Bologna, Italy
e-mail: ernesto.antonini@unibo.it

G. Chiesa (ed.), *Bioclimatic Approaches in Urban and Building Design*, PoliTO Springer Series, https://doi.org/10.1007/978-3-030-59328-5_14

14.1 Moving Towards the EPBD Goals

Climate change, acidification, eutrophication, stratospheric ozone depletion and photochemical ozone creation as well as water consumption, waste production and resource shortage, are just a few major environmental issues that World is currently facing. These phenomena underline the urgent need of mitigation and control over the side effects arising from human activities (Canadell et al. 2009).

The building sector has a significant weight over the total resource depletion and on the related emissions released, as buildings generate impacts on the environment during their entire life cycle: from construction to disposal (Cabeza et al. 2014).

In fact, construction is one of the least environmental sustainable economic sectors globally, since the production and operation of buildings and infrastructures are estimated to account for around 50% of greenhouse gas emissions, 40% of water pollution and 50% of waste disposal in landfills, in addition to absorbing 50% of energy, 50% of drinking water and 60% (by bulk) of materials from the ecosystem (Edwards 2014). This means that about half of non-renewable-resources depleted across the planet are used in the construction industry (Doan et al. 2017), thus making it one of the main resource consumer (Edwards 2014).

Commercial and residential buildings account for 38.7% of the total energy consumption in Europe (Allouhi et al. 2015), which is among the most resource-hungry economies in the world, also because it is largely depending from raw materials and energy imports. (European Commission 2011).

The use of materials, a relevant share of which are employed in construction, has grown enormously at global scale within the last century, and an intense further increase is estimated over the next 30 years. (Pachego-Torgal et al. 2013). Resource depletion and use are also strictly related to pollutant emissions and waste production, resulting in harmful impacts for the environment and ecosystems.

Tables 14.1, and 14.2 provide some significant data.

In recent decades, this has led international organizations to promote—and many countries to adopt—more stringent mitigation actions and regulatory standards in

Table. 14.1 Global resources used in building sector

Estimate of global resources used in buildings	(%)
Energy	45–50
Water	50
Materials (for buildings and infrastructures)	60
Agricultural soil	80
Timber products	60 (90% of hardwoods)
Coral reef destruction	50 (indirect)
Rainforest destruction	25 (indirect)

Source Politi (2019), reworked from Edwards (2014)

Table 14.2 Global pollution attributed to buildings

Estimate of global pollution that can be attributed to buildings	(%)
Air quality	23
Global warming gasses	50
Drinking water	40
Landfill waste	50
Ozone thinning	50

Source Politi (2019), reworked from Edwards (2014)

order to control the negative effects resulting on the environment from the building activities (Nilsson et al. 2012).

The founding initiatives date back to the end of the 1980s with the Brundtland Report (WCED 1987), which has been followed by other important events embracing sustainable development: the Kyoto Protocol (1997) on global warming mitigation, the Lisbon Strategy (2000–2010) and Europe 2020 (2010–2020) including improvement actions for greenhouse gas emissions and energy efficiency in EU, the more recent Agenda 2030, which established a set of Sustainable Development Goals at the UN Sustainable Development Summit in 2015 (continuing the Agenda 21 action plan launched in 1992 in Rio de Janeiro) and the 2015 United Nations Climate Change Conference (COP 21) at which the Paris Agreement, a global agreement on climate change, was negotiated.

In order to achieve a significant reduction in GHG emissions and energy demand, the EU has introduced ambitious and progressively stricter requirements for energy performance of buildings, till the 2018/844/EC Directive, which prescribe all buildings comply with the Net Zero Energy Buildings (nZEBs) standard by 2050, while the new buildings must already match the nZEB standard by 2021. According to the Directive, nZEBs are buildings whose very low energy demand should be extensively covered by renewable sources, produced on-site or nearby.

According to the EeBGuide Guidance Document (Wittstock et al. 2012), the energy consumption and related emissions of new and refurbished buildings has been significantly shorted within the European context in the last 20 years, thanks to the enhanced relevance that energy efficiency of buildings has gained.

Nevertheless, the sole improvement in efficiency has been shown to be not a mean effective enough to reduce the total energy needs, because of the considerable share of energy included in the upstream and downstream processes of building materials production. Several Authors support this fact, pointing out that energy involved in off-site production of building components accounts for over 75% of the total embodied energy in buildings (Ding 2004).

In fact, the envelope insulation and airtightness -as well as the adoption and possibly integration of both mechanical ventilation systems (CMV) and devices allowing the exploitation of renewable energy sources (RES)—play a crucial role among the design strategies that can be applied to reduce the building in-use energy demand till the achievement of a "near-zero-energy" standard (Stephan et al. 2013).

This means that buildings need an increasing quantity of high-performance materials and equipment, from which it is highly likely to expect a substantial rising in energy demand for product manufacturing, maintenance and disposal.

However, standards and regulation target mostly focus on the energy required to operate and use the building, i.e. Operational Energy (OE), which is the "energy required for keeping indoor comfort conditions and providing ordinary maintenance of the buildings thus including energy for HVAC (heating, ventilation and air conditioning), domestic hot water, lighting, and appliances" (Cabeza et al. 2014).

Conversely, little importance is given to the energy expended to make dwellings achieve their performances and maintain them over time, i.e. Embodied Energy (EE), which can be defined as "the energy content of all the materials used in the building and technical installations, and energy incurred at the time of new construction and renovation of the building" (Cabeza et al. 2014).

While there is broad accordance about the definition of Operational Energy, not all the authors agree about definition of Embodied Energy. While the above mentioned Cabeza' definition doesn't include the energy spent during the end of life phase which he pointed out as Demolition Energy, other Authors discussed this approach, giving a more comprehensive characterization of Embodied Energy, as the energy "sequestered in building materials during all processes of production, on-site construction, and final demolition and disposal" (Dixit et al. 2010).

Although precisely tracking the boundaries is essential for computation, both definitions converge in considering as embodied the share of energy "hidden" within the materials which are needed by the building process, thus focusing on these elements as they are crucial to assess the on duty behaviour of the building.

Since the manufacturing of building materials and products is responsible for 20% of the world's fuel consumption, Embodied Energy become a critical issue to consider when national and global strategic plans for energy are designed, and regulations as well (Dixit et al. 2010).

14.2 Operational Versus Embodied Share

The available literature provides much evidence of the growing relevance of EE in relation to the OE reduction, especially when OE levels are significantly lowered, like in high-energy-efficient buildings (Azari and Abbasabadi 2018)

By analysing the case studies of high-energy-performance buildings found in 20 journal articles, Optis and Wild (2010) observed that the EE varies between 2 and 51% of the Total Life Cycle Energy, while OE varies between 49 and 98%.

Looking for possible differences in the behaviour of traditional and low-energy buildings, Sartori and Hestnes (2007) surveyed 60 LCAs of buildings from 9 countries and proved a linear relationship between OE and Total Energy demand. The EE of low-energy buildings, however, spanned from 9 to 46% of the Total Energy demand, while it ranged between 2 and 38% in traditional ones.

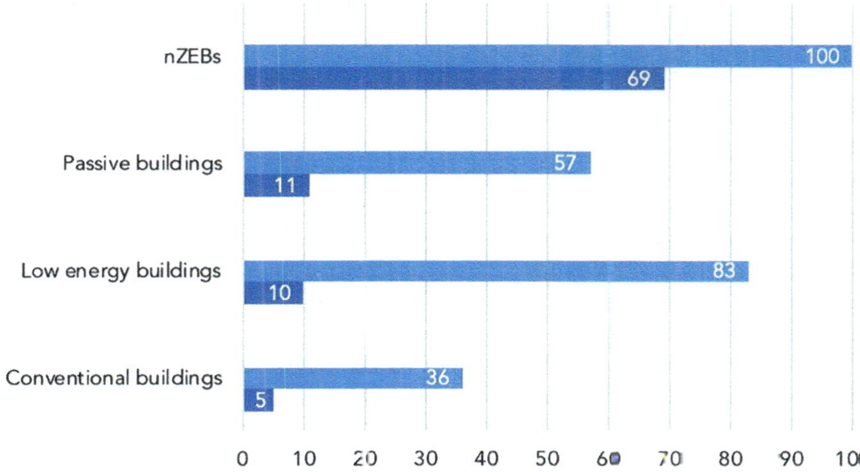

Fig. 14.1 Embodied energy share (%) in different types of buildings. *Source* Foliti (2019), reworked from Chastas et al. (2017)

The study on low-energy houses conducted by Thormark (2002) shows that Total Energy demand over a 50-year Life Cycle (energy for building production, maintenance and operation) is higher in buildings with lower OE levels.

After examining the total life cycle energy of 90 residential buildings, Chastas et al. (2017), determined that the share of EE varies between 26 and 57% in low-energy buildings, while in more performative passive buildings it spans from 11 to 33% of the energy consumption within the whole life cycle. Figure 14.1 shows the study outcomes.

This evidence confirms the results of several previous LCA surveys, showing that operational energy demand represents less than 40% of the total energy consumption of a passive house, when a conventional 50 years lifespan is considered (Stephan et al. 2013),

A cradle-to-grave LCA study (Proietti et al. 2013) carried-out in 2013 on an Italian Passivhaus outlined that the embodied impact share of envelope and installations ranges between 48 and 61% in terms of Gross Energy Requirements (GER), when excluding from computation the contributions of renewable sources.

Aiming at providing a more comprehensive assessment of the impacts, other studies refer to Embodied Carbon (EC) or Global Warming Potential (GWP), as the relative measure of how much heat a GHG captures in the atmosphere, compared to the amount of heat trapped by the same mass of CO_2. Therefore, GWP quantifies the emissions associated with the manufacture and use of a product or service (Anderson and Thornback 2012).

Consequently, according to an LCA performed on different conventional Italian buildings (Asdrubali et al. 2013) within the whole GHG impact, the GWP share of the operational stage accounts for approximately 77% in detached houses and 85%

in office buildings, while 13–22% of GWP belongs to materials and components and around 1.3–2.0% to the end-of-life phase.

A similar breakdown is documented by Anderson and Thornback (2012), who stated that Operational Carbon (OC) is usually the primary cause for GHG emissions in existing buildings, with a share of about 75–90% of the total emission, while the share of Embodied Carbon ranges between 10–25%.

Engin and Frances (2010) demonstrated that Embodied Carbon can account between 10 and 50% of the Total Carbon emissions for 60 years building life span.

Eaton and Amato (2005) showed that embodied carbon in UK buildings can reach 37–43% for 60 years life-cycle carbon emissions, while Smith (2008) estimated this share to be up to 80%.

Performing a life cycle carbon analysis on a PassivHaus, Din and Brotas (2016) demonstrated that buildings with a long lifespan have a more impactful GWP during the initial phases (cradle to gate) than all other building life stages.

As the GHG gas emissions are closely related to energy, EE and GWP indexes—although not equivalent—can be compared effectively. Especially because the improvement of building energy efficiency is leading the overall growth of the relative share of CO_2 embedded emissions, being their magnitude not negligible compared to the Total Life Cycle emissions (Ibn-Mohammed et al. 2013).

Even though it is not the only characteristic to consider, energy is a good indicator to obtain a broad picture of how much each phase of a building life weighs in terms of its impact on the environment.

Figure 14.2 summarize the data about Embodied Energy and GWP, collected from the above mentioned literature.

14.3 Data Interpretation and Consequences

The literature outcomes confirm that the OE reduction always increases the EE, as well as other environmental impact indexes. Although this trend was foreseeable, OE and EE correlation ratios show large variations instead, due to the heterogeneous casuistry and the differences in methods adopted to assess each case, from which a reliable reference value cannot be derived. However, the trends confirmed by the data allow at least two conclusions.

Firstly, a building energy balance focusing on just the energy demand for operation leads to some relevant elements being neglected, significantly affecting the perception of its overall environmental behaviour, especially when effective measures have been adopted to reduce the building OE share.

Secondly, the challenge to reduce the building-in-use energy demand till the achievement of a "near-zero-energy" standard seems to lead to a substantial increase of building EE and GWP, due to both the intensive use of high-performance materials and the broad application of RES devices, which contribution to EE can equal and even surpass those related to the basic core of the building (Chastas et al. 2016).

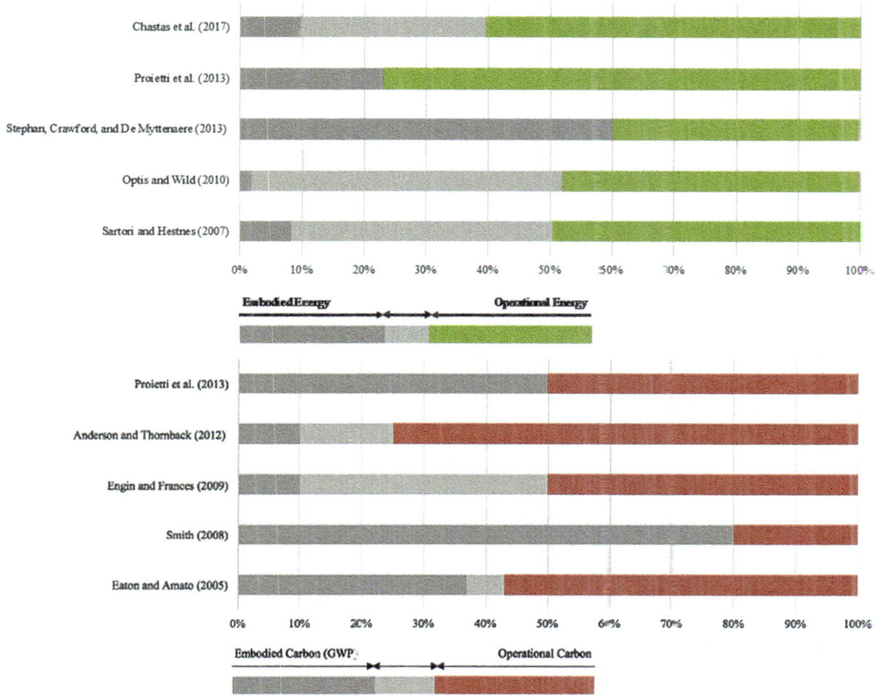

Fig. 14.2 Summary of data found in literature: embodied energy data (top); GWP data (bottom). *Source* Politi (2019)

As underlined by various literature sources, a boomerang effect may occur when the reduction of the operation energy leads to increase both the embodied energy and associated impacts of the building materials and components, thus intensifying consumption rather than decrease it (Sartori and Hestnes 2007; Chastas et al. 2017).

To avoid running into the "Jevons paradox" (Palumbo and Politi 2018) an optimal level of energy efficiency must be identified (Copiello 2017), which is those "capable of "repaying" the larger use of resources in non-operational phases which is needed to achieve lower consumption during the life service cycle." (Palumbo et al. 2019) (Fig. 14.3).

Therefore, since low energy buildings are expected to spread in next future, greater attention should be paid to EE of materials and installation devices, because of their relevant contribution to the environmental balance. For this reason, design strategies will have to include a careful selection of products able to provide maximum operation performances to the building, with the minimum embodied impacts. This challenge requires to measure the sustainability levels, energy needs and other environmental impacts within the entire building life cycle, through methods suitable to allow the comparison between alternative technical options

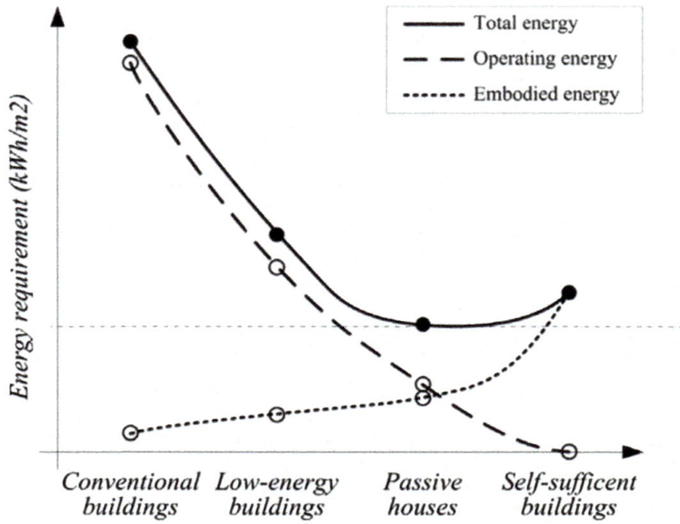

Fig. 14.3 Comparison of operational energy and embodied energy for different types of buildings. *Source* Copiello (2017), reworked by Palumbo et al. (2019)

14.4 Moving Toward a Building Whole Environmental Balance

LCA is considered the most appropriate scientific tool for this purpose (Proietti et al. 2013), as it provides an analytical, reliable and life-cycle-spanning quantification (Sartori and Hestnes 2007) of EE and OE effects on GWP (Goggins et al. 2016).

Both OE and EE must be quantified to draft the energy balance but, while the OE can be estimated by simulating the energy behaviour of the building, through an EPBD-compliant dynamic simulation software, the adoption of LCA analysis for the computation of Embodied aspects requires some cautions.

In fact, the assessment must mainly focus on building materials, which endorse a prominent role in shaping the overall balance of energy and GHG emissions, thus significantly affecting the OE and EE shares, as reported by the literature.

By analysing the strengths and the weaknesses of LCA within the building sector, some Authors however identified several barriers affecting the assessment, especially when the object of the study is a material (Rønnin and Brekke 2014). The main detected obstacles concern both the inventory stage and the data interpretation stage of the assessment process. (Politi et al. 2018) Regarding inventory, the more relevant handicaps rely to the process complexity, the difficulty to collect all the significant data and the scarcity of harmonisation about the outcome comparison. While the arbitrariness of the results, the scarcity of cooperation between product manufacturers and potential customers, difficulties in understanding and applying LCA results and low link with the energy certification are those mainly affecting the interpretation stage. (Bribián et al. 2009).

Additionally, adopting a life cycle approach requires a complex prevision of the service life of building materials, which depends on many variables, such as the user' life patterns, the maintenance cycle, the climatic conditions and the workmanships performed during design and construction. The typical wide life-cycle span of the buildings, as well as the huge number of processes and players involved within the building process, can seriously jeopardize the consistency and reliability of these elements, thus the assessment effectiveness.

A further barrier is the deep dependence on the availability and reliability of robust data sources to feed Life Cicle Inventory (LCI) as well as the quality of the provided data, as they are strictly depending on the production geographic location, process, plants (Ding 2014).

Thus, the sources from which the products and processes emission values are taken strongly affect LCA outcomes, besides being influenced by climatic conditions, context features and building characteristics (Optis and Wild 2010).

This means that the assessment can suffer from inaccuracy and unreliability of the methodologies, as well as from the great variations and lack of homogeneity in EE available databases (Dixit et al. 2010). In addition, the insufficient documentation used to quantify the EE could make ineffective the data obtained by the LCA in order to compare different scenarios (Optis and Wild 2010).

These complexity and uncertainty lead in the most frequently criticisms addressed to the LCA, which stress on the complicated calculation and high cost required to perform the assessment, especially when it applies to the whole building processes.

14.5 Integration of the Life Cycle Approach Within Sustainable Building Rating Systems

Due to these serious constraints, instead of implementing LCA for providing a comprehensive evaluation of buildings related impacts, the construction sector has been led to resort to Sustainable Building Rating Systems (SBRS) aiming at having easier and faster to apply, though less accurate, evaluations. These protocols express as a "score" the performances gap between the benchmarks set by the systems and the outcomes of the assessment carried-out on a building, scanning a multitude of environmental related aspects (Say and Wood, 2008). LEED in the US, BREEAM in the UK, DGNB in Germany, Green Star in Australia and the internationally developed Green Building Challenge Schemes, are just a few examples of the several protocols which are utilized worldwide today. Table 14.3 reports the number of operations performed by some most used SBRS, based on the information issued by each of them (the data refer to 2018).

Most of these protocols shape the overall rating by including some indicators that assess the embodied energy of building materials and components, sometimes

Table 14.3 Number of operations performed by some most used SBRS, based on the information issued by each of them

Origin	RS name	Certification issued	Construction typology			
			New		Existing	
			Residential	Non-Res	As Built	Refurbishment
USA	LEED	78,379	Yes	Yes	Yes	Yes
U.K.	BREEAM	7,746	Yes	Yes	Yes	Yes
France	HQE	266.000	Yes	Yes	Yes	N.A
Germany	DGNB	>1160[a]	No[c]	Yes	Yes	N.A
Italy	ITACA	619	Yes	Yes	Yes	N.A
Italy	CASACLIMA	1689[b]	Yes	N.A	Yes	N.A
Japan	CASBEE	>450	Yes	Yes	Yes	Yes
Canada	GREEN GLOBES	816	No	Yes	Yes	N.A
Qatar	QSAS	> 128	Yes	Yes	Yes	N.A

Source Reworked from Politi and Antonini (2017)
[a]including both certificated and pre-certificated projects
[b]618 of these in Bolzano/Bozen Province
[c]under realization
N.A.: Not available data

referring to LCA, or to LCA-related labelling protocols. Integrating these indicators involve significant benefits, as improving the crediting of environmental performance, as well as an easier and cost-effective assessment process (Trusty and Horst 2002).

Thus, according to the classification proposed by Ding (2014), SBRS can be considered as "third level" LCA tools, since those operating at the sole material level belong to the first level, while the second level includes the design decision- making tools, able to evaluate the whole building environmental performance, elaborating inputs like building geometry and presenting aggregated impacts as results.

Type III Ecolabels, or Environmental Product Declarations (EPD), are the available source from which the building material environmental features can be retrieved for SBRS purposes, since they provide LCA compliant written reports including both qualitative and quantitative information about the product "using predetermined parameters and, where relevant, additional environmental information" (ISO 14025 2006).

Nevertheless, the SBRS currently in use are still paying a very few importance to the LCA criteria in terms of credit potential and relative weight. According to a study carried out on the four most used international Rating Systems (LEED, DGNB, BREEAM, Green Star), the average share of LCA and EPD related criteria weight is 8.63% over the total, spanning from 13.5% of DGNB to 3.7% of LEED. (Politi et al. 2018).

Table 14.4 The credit potential and relative weight, given to the LCA and EPD related criteria by the analysed rating system

Rating systems information	LCA + EPD categories data				LCA criteria data			
Name/version /year/country of origin	LCA related categories	N. of criteria within the category	Category credits potential LCA related	Category relative wight (%)	LCA related criteria within the category	Criterion credits achievable	Relative weight (%)	Total LCA weight (%)
DGNB/Int. CORE14 Office2014/GER	Effects on the global and local environment (ENV 10)	3	110	12.4	Env.1.1 life cycle impact assessment	70	7.9	13.5
	Resource consumption and waste generation (ENV 20)	3	90	10.2	Env.2.1 life cycleassessment-primary energy	50	5.6	
BREEAM/Int. New Constr./2016/UK	Materials	6	14	12.5	Mat 01 life cycle mpacts	0.5	5	4.5
					Mat 02 hard landscaping	/	/	
LEED V.4/BD + C (New Constructions)/2017/ USA	Materials and resources	5 + 2	13	11.8	Building life-cycle impact reduction	3/5	2.7	2.7
GREEN STAR/Design and As Built v.1.1/2015/AUS	Materials	4	14	14	Life cycle impacts	7	7	7

Source Reworked from Politi and Antonini (2017)

Table 14.4 summarizes the credit potential and relative weight, given to the LCA and EPD related criteria by the analysed rating systems.

Looking in more detail at how different SBRS incorporate LCA elements, some similarities emerge as well as several differences.

In general, the comparison shows that the investigated SBRS adopt a similar strategy towards the LCA framework, including the number of LCA modules, impact categories, building elements to comprehend in the assessment, reference functional units and reference lifespan. Regarding the sharing extend for the LCA modules, all the investigated SBRS converge on 100% sharing for Raw Material Supply (A1), Transport (A2), Manufacturing (A3) and Disposal (C4); 83% for Maintenance (B2), Replacement (B4), Demolition (C1), Transport (C2) and Water processing (C3). Instead, greater discrepancy was reported for the remaining modules.

With respect to impact categories: 100% sharing was reported for Climate Change (GWP), Stratospheric Ozone Depletion (ODP), Eutrophication (EP), Photochemical Ozone Creation (POCP) and Acidification (AP); 67% for Fossil Fuel Depletion (ADP), while minor agreement was reported for the other impact categories. This outcome seems to confirm the directions of EN 15804:2012 and EN 15978:2011, which define seven impact categories for buildings LCA, namely GWP, ODP, AP, EP, POCP, ADP-non-fossil, ADP-fossil.

Further analogies concern the building elements considered by the majority of GBRSs, which are: ground slabs, floor slabs, external envelope, windows and doors (100%), footings, foundation and roof assemblies (83%), while sharing to a lesser extent was reported for the other building elements. Again, the four considered SBRSs include EPD-specific credits, generally rating the number or the percentage of building components with EPD, while other SBRSs include EPD within the LCA criteria as a possible source of data to perform the analysis.

Regarding the rating methods, two different approaches can be observed: the majority of the analyzed SBRSs rate the LCA impact reduction with respect to a reference building that must be specifically designed according to protocol requirements (usually according to national regulations), while others one centred on predefined impact benchmarks. In particular:

- LEED (USA): the building to be used as a reference must reflect the project building with regard to: size, function, orientation and operating energy performance. The significative differences between the baseline and the project building are represented by the structural typology and the technological configurations such as the external envelope and technical installations which must, however, be compliant with the local minimum regulations for energy and thermal performance. LEED requires a (at least) 10% improvement for (at least) three indicators, of which GWP is mandatory. In addition, all other environmental indicators must not increase more than 5%. It is therefore not clear what can be considered a good result in absolute terms, as the improvement achieved depends on the initial conditions, which are conditioned by the designer's choices.

- Green Star (Australia) enables two pathways for the LCA analysis interpretation. The first one is similar to the one proposed by LEED, in which the performance of the project building has to be compared to a standard practice reference building which is a hypothetical standard contemporary construction with the same structural requirements, scale, function, location, tenant requirements, materials, aesthetics, site conditions (including underlying geology), planning constraints, orientation and construction season. The reference building, in addition, must be compliant with the national codes and regulations on materials and energy performance. The second pathway requires the project building to be compared with an actual reference building, constructed in the last five years that is similar to the usage, construction and operation of the project building. The building must match the project building at least in terms of the structural requirements, scale, function, location and site conditions. Unlike LEED, which only permits one level of improvement (10%) with respect to the reference building, assigning only 3/110 points, Green Star awards up to 6/100 credits (+ 1 extra point) based on increasing the cumulative impact reduction from 30 to 130%.
- DGNB (Germany) is one of the few GBRSs setting predefined benchmarks for the lifecycle impacts of buildings. The threshold values were identified through research on the German building stock, promoted by the German Federal Ministry of Transport, Building and Urban Development (BMVBS) and they were updated to align with new national regulations (Ganassali et al. 2016). Similarly, Active House also defines specific benchmarks, but it is not clear how they are determined.
- BREEAM requires the use of a specifically developed tool ("Mat.01 Calculator") to rate a variety of LCA analysis aspects.
- One of the first consequences emerging from this study is the impossibility of comparing the LCA outcomes from different GBRSs as the calculation and rating methods adopted differ greatly from one system to another. With respect to threshold values, two main approaches were identified: one is based on the comparison with a reference building, while the other one relies on predefined benchmarks. In the first case the thresholds vary from project to project and credits are awarded depending on the relative improvement achieved with respect to one or more design alternatives which, however, still depend on the designer's choices. This method, therefore, cannot provide an objective measure of the severity of the actual impacts deriving from the project. The approach specified by DGNB, conversely, aims to provide objective values to refer to in order to understand the relevance of the environmental impacts over the building's life cycle.

14.6 Conclusions

Despite its relevant and increasing weight (especially for high energy performance buildings), the embodied energy contribution to the building's environmental balance remains difficult to determine, mainly due to the lack of common protocols for the life cycle assessment (Din and Brotas 2016).

The assessment made using sustainability rating tools (SBRS) as well as the application of an LCA analysis represent valid means for the design of new energy efficient buildings as well as for retrofitting of the existing ones towards an EPBD target. This kind of outcome, in addition, makes it easy to spot the relationship between different building performances. For example, indoor comfort (air quality, thermal and hygrometric comfort) can be related to the environmental performance of materials. The same goes for the energy performance, which can benefit from a proper and conscious choice of new materials, technical installations and also the reuse of upcycled building components. Low-energy buildings, such as passive houses, should take into account both environmental profiles and energy efficiency, and therefore the application of a method able to evaluate both these aspects seems a good opportunity not only for new projects but also for in-depth renovations.

Despite some evidences found in literature are confirmed regarding the simplified and standardized approach provided by SBRS in integrating of LCA issues, several discrepancies are also underlined in performing the assessment and in the interpretation of the results among different SBRSs.

References

Allouhi A, El Fouih Y, Kousksou T, Jamil A, Zeraouli Y, Mourad Y (2015) Energy consumption and efficiency in buildings: current status and future trends. J Clean Prod 109(2015):118–130. https://doi.org/10.1016/j.jclepro.2015.05.139

Anderson J, Thornback J (2012) A guide to understanding the embodied impacts of construction products. Construction Products Association, London

Asdrubali F, Baldassarri C, Fthenakisb V (2013) Life cycle analysis in the construction sector: guiding the optimization of conventional Italian buildings. Energ Build 64(2013):73–89. https://doi.org/10.1016/j.enbuild.2013.04.018

Azari R, Abbasabadi N (2018) Embodied energy of buildings: a review of data, methods, challenges, and research. Energ Build 168(2018):225–235. https://doi.org/10.1016/j.enbuild.2018.03.003

Bribián IZ, Usón AA, Scarpellini S (2009) Life cycle assessment in buildings: state-of-the-art and simplified LCA methodology as a complement for building certification. Build Environ 44(12):2510–2520. https://doi.org/10.1016/j.buildenv.2009.05.001

Cabeza LF, Rincón L, Vilariño V, Pérez G, Castell A (2014) Life cycle assessment (LCA) and life cycle energy analysis (LCEA) of buildings and the building sector: a review. Renew Sust 29-C: 394–416. https://doi.org/10.1016/j.rser.2013.08.037

Canadell JG, Ciais P, Dhakal S, Le Quéré C, Patwardhan A, Raupach MR (2009) The Human perturbation of the carbon cycle: the global carbon cycle II; United Nations Environment Programme (Scientific Committee on Problems of the Environment), New York, NY

Chastas P, Theodosiou T, Bikas D (2016) Embodied energy in residential buildings-towards the nearly zero energy building: a literature review. Build Environ 105(2016):267–282. https://doi.org/10.1016/j.buildenv.2016.05.040

Chastas P, Theodosiou T, Bikas D, Kontoleon K (2017) Embodied energy and nearly zero energy buildings: a review in residential buildings. Procedia Environ Sci 38(2017):554–561. https://doi.org/10.1016/j.proenv.2017.03.123

Copiello S (2017) Building energy efficiency: a research branch made of paradoxes. Renew. Sustain. Energy Rev 69:1064–1076. https://doi.org/10.1016/j.rser.2016.09.094

Din A, Brotas L (2016) Exploration of life cycle data calculation: lessons from a passivhaus case study. Energ Build 118(2016):82–92 https://doi.org/10.1016/j.enbuild.2016.02.032
Ding G (2004) The development of a multi-criteria approach for the measurement of sustainable performance for built projects and facilities. Ph.D. Thesis, University of Technology, Sydney
Ding G (2014) Life cycle assessment (LCA) of sustainable building materials: an overview. In: Pacheco-Torgal F, Cabeza LF, Labrincha D, de Magalhiies A (eds) Eco-efficient construction and building materials. Life Cycle Assessment (LCA), Eco- Labelling and Case Studies, Woodhead Publishing, Philadelphia, pp 38–62
Dixit MK, Fernández-Solís JL, Lavy S, Culp CH (2010) Identification of parameters for embodied energy measurement: a literature review. Energ Build 42(8):1238–1247. https://doi.org/10.1016/j.enbuild.2010.02.016
Doan DT, Ghaffarianhoseini A, Naismith N, Zhang Z, Ghaffarianhoseini A, Tookey J (2017) A critical comparison of green building rating systems. Build Environ 123(2017):243–260. https://doi.org/10.1016/j.buildenv.2017.07.007
Eaton KJ, Amato A (2005) A comparative environmental life cycle assessment of modern office buildings. The Steel Construction Institute, London
Edwards B (2014) Rough guide to sustainability: a design primer. RIBA Publishing, London
Engin A, Frances Y (2010) Zero carbon isn't really zero: why embodied carbon in materials can't be ignored. Available at https://www.ci.net/articles/archive/zerocarbon/. Accessed 28/12/2019
EU Commission (2011) Roadmap to a resource efficient Europe (COM(2011) 571). EC, Brussels
Ganassali S, Lavagna M, Campioli A (2016) Valutazione LCA all'interno dei protocolli ambientali multicriteri per il settore delle costruzioni. In: Dominici Loprieno A, Scalbi S, Righi S (eds) Life cycle thinking, sostenibilità ed economia circolare, ENEA. Roma, pp 415–422
Goggins J, Moran P, Armstrong A, Hajdukiewicz M (2016) Lifecycle environmental and economic performance of nearly zero energy buildings (NZEB) in Ireland. Energ Build 116(2016):622–637. https://doi.org/10.1016/j.enbuild.2016.01.016)
Ibn-Mohammed T, Greenough R, Taylor S, Ozawa-Meida L, Acquaye A (2013) Operational versus embodied emissions in buildings. A review of current trends. Energ Build 66:232–245. https://doi.org/10.1016/j.enbuild.2013.07.026
ISO 14025 (2006) Environmental labels and declarations—Type III environmental declarations—principles and procedures
Nilsson M, Zamparutti T, Petersen JE, Nykvist B, Rudberg P, McGuinn J (2012) Understanding policy coherence: analytical framework and examples of sector-environment policy interactions in the EU. Environ Policy Governance 22–6(2012):395–423. https://doi.org/10.1002/eet.1589
Optis M, Wild P (2010) Inadequate documentation in published life cycle energy reports on buildings. Int J Life Cycle Assess 15(2010):644–651. https://doi.org/10.1007/s11367-010-0203-4
Pacheco-Torgal F, Cabeza LF, Labrincha J, de Magalhaes A (eds) (2013) Eco-efficient construction and building materials, life cycle assessment (LCA), eco-labelling and case studies. Woodhead Publishing, Series in Civil and Structural Engineering Book, Cambridge UK, pp 1–10
Palumbo E, Politi S (2018) Improving building envelope efficiency: interaction between embedded energy and operational energy. TECHNE-J Technol Archit Environ 16(2018):247–257. https://doi.org/10.13128/Techne-23039
Palumbo E, Ciulla G, D'Amico A, Traverso M, Antonini E (2019) Energy and environmental profile of building trade-off between energy retrofit and production of materials: a case study. In: Proceedings of the 14th conference on sustainable development of energy, water and environment systems (SDEWES). Dubrovnik, KR, 1–6 October 2019
Politi S, Antonini E (2017) Buildings hidden energy and environmental consequences. Sustain Mediter Constr J 6:19–23
Politi S, Antonini E, Wilkinson SJ (2018) Overview of building LCA from the sustainability rating tools perspective. In: ZEMCH 2018 international conference, Melbourne, Australia, 29 Jan–1st Feb 2018, pp 237–250
Politi S (2019) Buildings embodied impacts over the life cycle: an essential assessment framework for the early design phase. PhD Thesis University of Bologna

Proietti S, Sdringola P, Desideria U, Zepparelli F, Masciarelli F, Castellani F (2013) Life cycle assessment of a passive house in a seismic temperate zone. Energ Build 64(2013):463–472. https://doi.org/10.1016/j.enbuild.2013.05.013

Rønning A, Brekke A (2014) Life cycle assessment (LCA) of the building sector: strengths and weaknesses. In: Pacheco-Torgal F, Cabeza LF, Labrincha J, de Magalhiies A (eds) Eco-efficient construction and building materials, life cycle assessment (LCA). Eco-Labelling and Case Studies, Woodhead Publishing, Sawston UK, pp 63–83. https://doi.org/10.1533/9780857097729.1.63

Say C, Wood A (2008) Sustainable rating systems around the world. CTBUH J II:18–29

Sartori I, Hestnes AG (2007) Energy use in the life cycle of conventional and low-energy buildings: a review article. Energ Build 39(2007):249–257. https://doi.org/10.1016/j.enbuild.2006.07.001

Smith BP (2008) Whole life carbon foot printing. Struct Eng Mag 86–6(2008):15–16

Stephan A, Crawford RH, de Myttenaere K (2013) A comprehensive assessment of the life cycle energy demand of passive houses. Appl Energy 112(2013):23–34. https://doi.org/10.1016/j.apenergy.2013.05.076

Thormark C (2002) A low energy building in a life cycle—its embodied energy, energy need for operation and recycling potential. Build Environ 37(2002):429–435. https://doi.org/10.1016/S0360-1323(01)00033-6

Trusty WB, Horst S (2002) Integrating LCA tools in green building rating systems. In: The Austin papers: best of the 2002 international green building conference. Building Green Inc, Brattleboro VT, pp 53–57

Wittstock B, Gantner J et al (2012) EeBGuide guidance document. Part B: BUILDINGS. Operational guidance for life cycle assessment studies of the energy-efficient buildings initiative. Available at https://www.eebguide.eu/eebblog/?page_id=659. Accessed 09/01/2020

WCED-World Commission on Environment and Development (1987) Our common future. Oxford University Press, Oxford

Part IV
Urban Environmental Design

Chapter 15
Bioclimatic Architecture and Urban Morphology. Studies on Intermediate Urban Open Spaces

Alessandra Battisti

Abstract The contribution follows the European guidelines for the adaptation of the built environment to climate change, illustrated at the COP21 (2015) and collected in the European Climate Adaptation Platform. The text concerns requalification process of the urban open spaces by connecting social and technological features characterizing the urban realm and the related environmental-climatic performance. Through an in-depth study of materials and possible combinations of innovative materials, urban vegetation, water elements and urban shades, the current knowledge of the complex issue of outdoor comfort and urban microclimate is being extended to provide innovative social and functional solutions. The paper comprises the relationship between energy use and urban morphology, studies on intermediate urban open spaces, on environmental and bioclimatic comfort, the interactions between biophysical and microclimatic factors, ultimately the tests of innovative technologies.

15.1 Introduction

The 2015 Paris COP 21, and those that followed, until the last one in Madrid, forcefully underscored that the struggle against climate change is one of the environmental issues most debated on the global level in recent years, and that an innovative vision for cities has never been so important as in our century. In fact, currently more than half of the world's population lives in cities, and this migratory trend should continue in the years to come: by 2050, more than two thirds of the world's population will be concentrated in urban settlements. In the desire to stimulate a new vision of the environmental government of cities, at the United Nations in September 2019, UN Secretary General Antonio Guterres addressed all the world's leaders, asking them to participate in the meeting in New York on 20 September with concrete, realistic plans for increasing their national contributions for 2020 (precisely the commitments made during the 2015 Paris Agreement to cut greenhouse gases) in line with the objective of reducing greenhouse emissions by 45% during the next decade, to achieve

A. Battisti (✉)
PDTA Department, Sapienza University of Rome, Rome, Italy
e-mail: alessandra.battisti@uniroma1.it

G. Chiesa (ed.), *Bioclimatic Approaches in Urban and Building Design*, PoliTO Springer Series, https://doi.org/10.1007/978-3-030-59328-5_15

zero net emissions for 2050. This position was already drastically hoped for in the IPCC's 2018 report that indicated that the risks—for the human race, the biosphere, and economies—were higher than previously stated, and that encouraged actions aimed at reducing global temperatures (IPCC 2018), as well as reducing green house gas concentrations (UNFCCC 2019). These actions were polarized on two levels: the first regarding the planet's major ecosystems (rain forests, major river systems, etc.); the second focusing on urban areas, since these are the ones most seriously stricken by the effects of climate change, showing a high level of vulnerability and of environmental risk.

It is thus brought to light that with the gradual phenomenon of urbanization taking place, it will in the future be cities to determine the possible achievement of a radical turning point, capable of dealing with climate change and inclusive economic growth able to create jobs, without making foolish use of non-renewable resources, on the path shown by environmental sustainability and resilience. This is why the Sustainable Development Goal (SDG) 11, "Make cities and human settlements inclusive, safe, resilient and sustainable," holds such importance at the moment. In particular, objective 11.B states that by 2020 we might see substantial growth in the number of cities and urban settlements that have adopted and implemented integrated policies and plans towards inclusion, the efficiency of resources, mitigation of and adaptation to climate change, and resilience to disasters, and that have developed and implemented, in line with the Sendai framework for disaster risk reduction 2015–2030, a disaster risk management that is holistic on all levels. There are, however, still many challenges to be faced in order to achieve SDG 11. It will be necessary to conceive necessary innovative solutions to help inhabitants behave in a more aware fashion (consider Switzerland's 2000W citizen), and to design models of city life in which the inhabitants are involved in positively influencing their life style and can interact actively with their community.

The increasingly frequent heat waves triggered by climate change focus attention on the important relationship between land cover and temperature, particularly in urban areas with a greater presence of artificial surfaces, often asphalted and built in concrete or with fewer planted areas. Accompanied by the reduction in urban green areas, these are all characteristics that cause open spaces to absorb heat in the summer, and do not permit the land's adequate transpiration and evaporation, thus resulting in the phenomenon of urban heat islands. In this perspective, the strategies of mitigation and adaptation to resilience of public spaces are seen as the most important ones to realize in the short term and, as in many other cases, the identified objectives must be tangible, concrete, and measurable so as to be able to make a significant impact on changing course and having the power to make a difference.

According to Landsberg (1983), the first scholar to use the term "heat island" was Manley in 1958, in his discussion on the effect of London's artificial heat (Manley 1958). The term has now taken on the meaning of a phenomenon taking place in urban environments, characterized by higher summer temperatures—varying from 2 to 12 °C—than the surrounding rural areas (US EPA 2009a, b; Pearlmutter 2000), attributable to the built urban environment and presenting particular risks for the urban population (Oke 1982). Moreover, experts often divide heat islands into three different types:

1. Surface heat islands: the infrared radiation emitted and reflected by surfaces, which allow the portions of building where the surfaces are hottest to be identified, must be measured.
2. Canopy-level heat islands: which exist in the air stratum where human activities are performed, from the ground to beneath the tree canopy and building roofs.
3. Urban and boundary layer heat islands: which start from the level of building roofs and tree canopies, extending to the point where the urban landscapes no longer influence the atmosphere. This region generally extends no more than a mile from the surface (US EPA 2009a).

Urban heat islands at the surface and canopy level, more commonly studied in architecture as cited in this document, are those in which the built environment and its artificialization has the most pronounced effect because air flow and energy exchanges are controlled for micro-scale, and in which the site's specific characteristics are extremely important (Golden 2004; Santamouris 2007).

In Europe, massive urbanization today sees the intensification of the phenomenon, which contributes to the increased thermal discomfort of the city's inhabitants, and therefore to increases in energy-intensive behaviour and practices to combat the phenomenon (intense use of artificial ventilation systems). Some cities like Paris and London, after intense heat waves, recorded exponentially increased mortality (Besancenot 2002) and adopted strategies to mitigate urban heat islands, seeking to react with initiatives making it possible to protect the population by increasing its capacity to adapt to these extreme levels of the phenomenon. Therefore, based on existing knowledge, several questions were raised from the standpoint of urban planning, relating to the mitigation and adaptation strategies to combat the risks of UHIs and of climate change. In particular, in line with the growing importance of mitigation and adaptation measures, and in light of the concept of energy requalification, especially of consolidated and historic fabrics, the identification, study, and design of high-performance, innovative systems is one of the main fields of investigation, and a significant contribution that technology can provide to combat the negative effects of the UHI and the risks due to climate change in highly manmade settings. In this sense, technological research never ceases to suggest alternative, innovative, and creative paths, and to offer new components and materials for designing; it is the architect's task to grasp their possibilities in power, in order to trigger that process of acting, operating, and working, in which to express one's will to change course towards a resilient planning of public space (Erell et al. 2011; Santamouris and Asimakopoulos 2001).

15.2 Mitigation and Adaptation Strategies Toward Urban Resilience

In recent years, parallel to the sustainable development and climate change debate, there has been an increasing interest in the concept of 'climate resilience' and more generally 'urban resilience' (Hassler and Kohler 2014). The terms 'resilience' compromises different levels of interest and provides insights into complex socio-ecological systems and their sustainable management. Urban resilience refers to the ability of an urban system and all its constituent socio-ecological and socio-technical networks, across temporal and spatial scales (Battisti 2015), to maintain or rapidly return to desired functions in the face of a disturbance, to adapt to change, and to quickly transform systems that limit current or future adaptive capacity (Meerow et al. 2016). As underlined by the OECD, a resilient city has the ability to absorb, recover and prepare for future shocks, may it be economic, environmental, social and institutional (OECD 2018). The concept of a city resilient to climate change, comprises two main aspects or actions: mitigation and adaptation strategies. While mitigation tries to reduce the impact that can lead to higher energy consumption and emissions (Santamouris 2014), adaptation aims to decrease the other harmful effects of climate change and to prepare the built environment to climate emergency (EEA et al. 2016). Several strategies may belong to both categories, since they represent a complementary set of actions to counteract climate change. In order to simplify the consistent and transdisciplinary body of researches, it is possible to divide the current prevalent studies in three fields of analysis, studies focused on: (i) smart materials, such as high reflective materials (e.g. cool materials, cool colored materials), (ii) on green infrastructure, such as parks, green roofs, green walls, garden systems, pervious pavements and trees in general, and (iii) blue infrastructures, in particular water mist cooling systems. Among these three fields of study effective mitigation strategies are: the increase of vegetated surfaces (such as grass, green roof and green walls) and green spaces and the raise of the urban materials' albedo (that is the ratio of global radiation reflected to the global radiation received by a surface) (Shashua-Bar and Hoffman 2000); nevertheless the increase of green spaces may be included, together with the rain garden systems, to the adaptation measures.

15.2.1 Cooling Materials

Among the geo-engineering based solutions used for mitigating climate change phenomena in public space, the most relevant and consolidated category are high reflective materials, also referred to as 'cool materials' for their capacity to present low surface temperatures, followed by phase change materials (PCM) and thermochromic materials. As underlined by Pisello in her comprehensive review (Pisello 2017), the cool materials class is large and comprises: the first generation, including natural materials with high reflectivity, the second generation, based on the development of artificial white materials design to have a high albedo equal or higher than 0.85, the third generation of cool materials, known as "cool colored materials", comprehending colored coatings that present a high reflectivity in the infrared spectrum of solar radiation (NIR), thus appearing like traditional dark materials but with low surface temperature, the fourth generation, based on nanotechnological additives like thermochromic paints or PCM doped cool materials. As observed by Yang et al. (2015) previous researches on high reflective materials were focused primarily on their effects on the outdoor temperatures thus focusing only on microclimate parameters such as air temperature and surface temperature variations and in that regards a consistent review has been made by Santamouris and Hakbari et al. (Santamouris 2014; Akbari et al. 2016) and recently by Pisello (2017); nevertheless recently studies have highlighted the importance to comprise also other parameters such as mean radiant temperature (Tmrt) and above all comfort analysis (in terms of PET or UTCI), since it appears that 'cool materials' have different side effects on outdoor thermal comfort depending on the surface of application, on their exposure to solar radiation and on the combination with other elements (Yang et al. 2015). In general, increasing reflectivity facilitates reduction of daytime surface temperature during the day, whereas during nighttime, albedo of materials becomes ineffective due to the absence of solar radiation, however since more radiation is reflected during the day, less heat is stored in reflective materials compared to conventional ones thus leading to small reductions of night-time surface temperatures. Cool materials can be applied on roof (i.e. cool roof), wall (i.e. cool wall) and pavements (i.e. cool pavements). A field study conducted in Athens, in particular a transformation of a traditional pavement in an urban park in 'cool pavements' (i.e. 'Flisvos Project) (Santamouris et al. 2012), showed that reflective pavements could reduce surface temperature in an urban park up to 7.6 °C under non-shaded condition. However it is fundamental to evaluate how much of the reflected radiation is absorbed by the surrounding surfaces of the built environment such as walls or by pedestrian themselves, as several studies have underlined (Yang et al. 2015; Erell et al. 2014; Taleghani and Berardi 2018; Chatzidimitriou and Yannas 2015). For example, a study conducted by Chatzidimitriou and Yannas (2015) highlighted that from pedestrian comfort's viewpoint, achieving lower surface temperatures by use of reflective materials does not compensate for the higher amounts of reflected radiation released in the urban environment and directly affecting pedestrian comfort. In another study made by Erell et al. (2014) the authors underlined that use of cool colored materials, may increase albedo substantially without altering visual appearance; however, because the human body is non-selective in its absorption of solar energy, the effect of albedo modification on thermal sensation could be negative. On the other hand, combination of cool materials and trees showed positive effects both on outdoor temperature and on thermal comfort as well, as demonstrated in a study of Shahidan et al. (2012) and of Laureti et al. (2018). Regarding cool roofs measures, recent literature contributions showed that cooling need reduction due to cool roofs corresponds to 2–44%, with an average benefit around 20% (Kolokotsa et al. 2012; Synnefa et al. 2006). Cool roofs present higher heat island mitigation potential when the albedo is equal or higher than

0.7. However, it must be underlined that most cool roofing materials lose a fraction of their initial reflectance by weathering and aging. Regarding thermochromic materials for the building envelope, this type of innovative materials can change their optical and thermal properties in a dynamic way, responding to their environment, changing the color and the solar properties—moving from darker to lighter tones as temperature rises. This change is reversible, which means that as temperature decreases, the material turn back to its original color. Karlessi et al. (2009) have developed 11 thermochromic coatings for building envelopes using thermochromic pigments into an appropriate binder system and investigated their performance in comparison with similarly colored highly reflective (cool) and common coatings. The results demonstrated that the surface temperatures of thermochromic samples were lower than the temperatures of color-matched cool and common coatings by a maximum of 7 °C and 11 °C respectively. However, the application of thermochromic materials in building and urban structures presents problems in terms of durability and stability even for short outdoor exposure and thus, further investigations are required (Bretz and Akbari 1997).

15.2.2 Green Infrastructures

Regarding the second type of mitigation and adaptation measures in the city in between spaces as well, 'green infrastructures' are considered to be one of the most effective strategy for urban cooling—primarily through shading and the reduction of ground surface temperatures, and in some cases through evapotranspiration as well (Saaroni et al. 2018). Defined as a «strategically planned network of natural and semi-natural areas with other environmental features designed and managed to deliver a wide range of ecosystem services» (EU-DGE 2014), due to their cooling potential and their mitigation effects for the outdoor and indoor (Saaroni et al. 2018), they comprise: the increase of vegetated surfaces, such as parks, trees, and green roofs and green walls. As underlined by Kumar (EU and UN-Habitat 2016), studies show that green infrastructures can positively contribute to reducing the impacts of climate change by, for example, reducing heat stress, improving air quality and decreasing flood risks. If we start from the big to the small scale of GI, parks with large grassy areas have been found to be effective cooling agents during the night (e.g. Shahidan et al. 2012), but during the day they can potentially be warmer than the surroundings, especially if not adequately irrigated (Saaroni et al. 2015). This temperature difference between urban green spaces and their surrounding built-up terrain, referred to as the 'park cool island' effect (PCI) (Spronken-Smith and Oke 1998), display a relevant diversity in the effectiveness in their surroundings areas and throughout the day, therefore it presents some controversial observations. Bowler et al. (2010) underline how most of the studies dealt with "micro-scale" PCIs, and that their thermally mitigating effect is localized to the immediate vicinity of the vegetated area and may underestimate the cooling effect beyond the green area. In a recent study, Ketterer and

Matzarakis proposed three green renovation scenarios for the environmental renovation of a consolidated urban area in Stuttgart (Ketterer and Matzarakis 2014), results showed that air temperature is highest in the asphalt scenario and lowest in the area with trees, whereas grass and park presented the same air temperature. Considering thermal comfort, a green surface reduces PET by up to 7 K in the neigbhouring streets and areas, especially on the lee side, and it rises up to 58 °C above sealed surfaces with low albedo, high solar irradiation and low wind speed. In green areas, PET ranges between 18 and 28 °C in shaded and not more than 35 °C in unshaded areas. These results are in line with another recent study conducted by Lee et al. (2016) in Friburg, where trees display a mean PET variation of 3 K and grassland of 1 K. Thus, it appears that the shading factor is the most dominant in the comfort perception, and it may vary considerably if it is from a building, a vegetated element (e.g. tree) or an artificial element (e.g. umbrella). In a recent study conducted in Canada, building shade was found to be the most effective cooling strategy, followed by the tree strategy and the umbrella strategy (Lee et al. 2018). Nevertheless, green spaces offer different benefits related to clean air, natural ventilation and psychological benefits that must also be considered. If we consider trees individually, they display a higher performance on thermal comfort and heat mitigation than grass, because of their geometry and their related higher shading effect on one hand, on the evapotranspiration processes and on the other the capacity, due to the foliage density, to modify and interact with the natural ventilation; thus, tree canopies can create a microclimate beneath them (Bowler et al. 2010). Among all the green infrastructures, trees have proved to be a feasible form for the mitigation of heatwaves, in fact trees dissipate heat by long-wave radiation, convection of heat into the air, and transpiration (water loss from leaves). In particular, as underlined by Upreti et al. (2017), the shading effect of urban trees reduces the net energy absorption thus modifying the urban energy balance and cooling the urban canopy and boundary layers by reducing the sensible heat (Roy et al. 2012; Rahaman et al. 2015), they also influence the indoor energy consumption for cooling (Akbari et al. 2001). The results from the study of Upreti et al. (2017) show that shading from trees reduces the amount of heat absorbed by roads and walls during the day significantly, about 6 °C nearby and 1–4 °C for the metropolitan area, and during the nighttime shows a cooling effect of UHI of 1.5 °C. It is evident that trees show a stronger impact on the surface temperature than on the air temperature, for surface heating of engineered materials is stronger than atmospheric heating by sensible heat. Moreover, trees represent a good solution for water absorption and evaporation, since they absorb water from the soil beneath the pavement and transpire to the atmosphere, creating a pathway for water evaporation through also impermeable layers (Fini et al. 2017), such as asphalt. However, there is a recent controversial side-effect due to climate change and air pollution on the plant behavior and thus on parks efficiency. A recent collaborative study with plant biologists and climate scientists (Kala et al. 2016), has pointed out how under certain extreme outdoor conditions, such as summer heatwaves and high air pollution rate, parks not always ‘behave’ as they should, in fact they decrease their cooling capacity and may not improve outdoor thermal comfort. Other problems and hazards associated to the extensive use of trees within cities are related to a reduction of the solar access and wind speed, carbon pollution through landscape and tree management practices, maintenance problems and pollen diffusion, as underlined by Roy et al. (2012). Thus, in evaluating and designing green spaces it is fundamental to consider all the risks derived from the initial environmental conditions and consequentially to design according to local characteristics. Finally, regarding systems of ‘superficial greenery’, green roofs represent an interesting measure because they proved to have positive mitigation effects both on indoor as well on the outdoor (GGCA and R20 2012), they can reduce the risks of heat waves, catch run-off water and reduce the need for indoor cooling. By preventing incoming solar radiation from reaching the building structure below and by the evapotranspiration process of plants, green roofs guarantee a mitigation of the ambient outdoor temperatures. Their thermal performance depends on: direct shading of the roof, evaporative cooling from the plants and the growing medium, additional insulation values from both the plants and the growing medium, and the thermal mass effect of the growing medium (Liu and Baskaran 2003). As underlined by Santamouris (2014), the specific energy benefits related to green roofs depend on the local climate, the green roof design itself and on the specific building characteristics. Lazzarin et al. (2005) showed that the average absorbed solar radiation by the greenery is close to 23%. Regarding the indoor, a research conducted in south Italy reveals that green roofs (without any insulation material) are cooler by about 12 °C in comparison with conventional roofs, according to the average surface temperature measurements in summer. Even in winter season, the temperature difference between conventional and greens roofs is remarkable by almost 4 °C. From this point of view, regarding the indoor, buildings with green roofs consume less energy in comparison with conventional roofs (Synnefa et al. 2006). Regarding the outdoor there have been many studies which investigated air temperatures variations; in a review of technologies using reflective roofs and green roofs for UHI mitigation (Santamouris 2014) the conclusion was that green roofs applied on a city scale may reduce the average ambient temperature of the city by a value between 0.3 and 3 °C. In a recent study conducted by Berardi (2016) the results indicate that increasing the leaf area index (LAI) would lead to an increased cooling effect of the air temperature up to 0.4 °C during the day at pedestrian level, while a more significant temperature reduction would be obtained only at the rooftop level. In New York City, afternoon temperatures were found to be reduced by up to 0.6 °C with the adoption of green roofs (Columbia University 2006), while in Hong Kong, air temperatures were reduced by up to 0.7 °C at 2:00 pm with green roof applications—see also Smith and Roebber (2011) for Chicago. Another interesting strategy belonging to GI and to the adaptation measures, are rain gardens also referred to as Sustainable Drainage Systems (SuDS). The main purpose of these measures is to reduce the impact of development by restoring the natural water cycle and promoting in situ management of storm-water runoff through infiltration and evapotranspiration. Sustainable Drainage Systems (SuDS), as rain gardens, bring back water to the surface, encouraging an interactive relationship with local residents and helping to improve urban landscape, they focus on the use of new paradigms, acting at the root point of the problem that is to say at the points of contact between rain water and the city: rooftops, private plots, gardens, squares, streets, etc. Because of their capacities, such as preventing disasters caused by floods, reducing non-point source pollution, providing urban green spaces and improving urban landscapes, they are gradually becoming important not only as adaptation measure but as mitigation as well. Sustainable Drainage Systems (SuDS), include also another significant strategy: permeable pavements (e.g. interlocking concrete pavers) and porous pavements (e.g. porous concrete). These are a hybrid strategy because of their multiple effects on soil, atmosphere and indirectly on human comfort. Studies on porous pavements showed how this technology allowed both higher infiltration and evaporation of water than asphalt and thus, they contribute to the mitigation of high outdoor temperatures, both of air and surfaces. In a recent study conducted by Fini et al. in Italy (i.e. Como) (2017) four different types of pavements have been tested and the authors have observed that a reduction of evaporative cooling from soil paved with permeable and impermeable pavements contributed to significant soil warming: at 20 cm depth, soils under concrete pavers and asphalt were 4 and 5 °C warmer than soil covered by porous pavements and unpaved soils, respectively. Thus, enhancing evaporation from paved soil by the use of porous pavements may contribute to mitigating urban heat islands (Asaeda et al. 1996).

15.2.3 Blue Strategies

They include ‘urban blue-space’ that refers to all substantial bodies of static or dynamic surface water found in urban areas. It is important to underline that the influence of water bodies on urban climate have been studied under different point of views, ranging from the absence of water and evapotranspiration from vegetation (Oke 1987). At the same time, water has been studied for its influence on outdoor comfort (Akbari 1992) as well as for its impact on the energy consumption of surrounding buildings (Santamouris et al. 2001). In general water bodies (such as lake, pond, river) show several advantages as a natural cooling technique because of their thermal and optical properties (Santamouris and Kolokotsa 2016). However, among the ‘blue strategies’, the most promising techniques are those based on the evaporative cooling. If we consider the minimal intervention measures for historical context, which is the most interesting and complex area in the Mediterranean cities, the most used are spray fountains and water mist cooling systems. In particular, direct cooling with mist is expected to relieve thermal stress through two main routes: reducing the air temperature and light wetting of the skin with evaporation, both effects promoted by the forced convection from the fan. In a study conducted by Montazeri et al. (2017) on the cooling effects of water spray in Rotterdam they observed as a first result a maximum air temperature reduction of 5 °C and UTCI (Universal thermal Comfort Index, McGregor 2012) reduction of 7 °C at 1.75 m height, secondly, the thermal comfort at pedestrian height is improved up to 5 m away from spray system. A second study conducted by Farnham et al. (2015) in Japan in 2015, shows a range of interesting impacts on human comfort as well as air temperature. The results showed that mist fan systems have potential to yield both physiological and psychological relief from thermal stress even in relatively humid climates. In fact, the mist and fan combination can easily yield localized cooling rates over 200 W/m^2 and the registered skin temperature was significantly reduced almost instantaneously by the mist (about 1 °C for each droplet), the comfort survey showed a decrease of three steps on a nine-step thermal sensation scale, and a three-step increase in a seven-step thermal comfort scale. In addition, since misting is a visible form of cooling, it can be use freely by the pedestrians: those seeking relief can easily self-regulate their cooling level by choosing where to enter the mist, while those who do not wish to risk wetting can easily avoid it.

15.3 Conclusions

Many obstacles are still encountered on various levels in the development of designs incorporating innovative materials and technologies in the regeneration processes of urban basins. On the one hand, there is a lamentable dearth of national regulations of reference. On the other, over time, serious problems have emerged, connected with the maintenance of many of the investigated technologies. This factor requires a reinterpretation, in a broader sense, of tolerance of error and malfunction, especially as regards experimentation still in progress on innovative materials and components. Lastly, there has been a widespread lack of communication, and therefore of understanding, by most people, of the effectiveness and efficiency of innovative technologies within the urban basin, and of the performance that materials can attain in terms of the inhabitants' thermal/hygrometric comfort, as well as their health and well-being.

While the practical consequences of these operations are still neglected, the disciplinary theoretical ones appear relevant: the technological reason for the design must not be limited to generalizing the observation, or to merely ordering what can be analyzed in the past and present. It must, however, project itself into the future, offering itself as a vision. Planning born from anticipating the reason, combined with knowledge of what is produced that is most innovative and evolutionary, can make the inferences on the situation that are capable of changing course for the phenomena in progress (climate change), especially when important issues like the survival of the human race are in play.

References

Akbari H (1992) United States Environmental Protection Agency, Climate Change Division, Lawrence Berkeley Laboratory, United States; Department of Energy Cooling our communities: a guidebook on tree planting and light-colored surfacing. U.S. Environmental Protection Agency, Office of Policy Analysis, Climate Change Division: for sale by the U.S. G.P.O., Supt. of Docs.: Washington, DC. ISBN 978-0-16-036034-3

Akbari H, Cartalis C, Santamouris M, Synnefa A, Kolokotsa D, Muscio A, Pisello AL, Rossi F, Wong N.H, Zinzi M (2016) Local climate change and urban heat island mitigation techniques—the state of the art. J Civ Eng Manag 22:1–16. https://doi.org/10.3846/13923730.2015.1111934

Akbari H, Pomerantz M, Taha H (2001) Cool surfaces and shade trees to reduce energy use and improve air quality in urban areas. Sol Energy 70:295–310. https://doi.org/10.1016/s0038-092x(00)00089-x

Asaeda T, Ca VT, Wake A (1996) Heat storage of pavement and its effect on the lower atmosphere. Atmos Environ 30:413–427

Battisti A (2015) Die energetische Requalifizierung als Chance für die Regeneration des städtischen historischen Erbes und die soziale Kohäsion| La riqualificazione energetica come opportunità di rigenerazione del patrimonio storico urbano e di coesione sociale. In: Battisti A, Endres E, Santucci D, Tucci F (eds) Energie Bedrohung oder Chance für die europäische Stadtlandschaft?|Energia Occasione o Minaccia per il Paesaggio urbano europeo? TUM Technische Universität München, München|Monaco, pp 20–41. ISBN: 9783941370685

Berardi U (2016) The outdoor microclimate benefits and energy saving resulting from green roofs retrofits. ENB Energy Build 121:217–229

Besancenot JP (2002) Vagues de chaleur et mortalité dans les grandes agglomérations urbaines. Environ Risques Santé 1(2002):229–240

Bowler DE, Buyung-Ali L, Knight TM, Pullin AS (2010) Urban greening to cool towns and cities: a systematic review of the empirical evidence. Landsc Urban Plan 97:147

Bretz SE, Akbari H (1997) Long-term performance of high albedo roof coatings. Energy Build 25(2):159–167

Chatzidimitriou A, Yannas S (2015) Microclimate development in open urban spaces: the influence of form and materials. ENB Energy Build 108:156–174. https://doi.org/10.1016/j.enbuild.2015.08.048

Columbia University (2006) Center for Climate Systems Research Mitigating New York City's heat island with urban forestry, living roofs, and light surfaces: New York City regional heat island initiative final report. New York State Energy Research and Development Authority, Albany, NY

EEA et al (2016) Europæiske Miljøagentur Urban adaptation to climate change in Europe 2016: transforming cities in a changing climate. European Environment Agency, Copenhagen. ISBN 978-92-9213-741-0

Erell E, Pearlmutter D, Boneh D, Kutiel PB (2014) Effect of high-albedo materials on pedestrian heat stress in urban street canyons. UCLIM Urban Clim Part 2(10):367–386. https://doi.org/10.1016/j.uclim.2013.10.005

Erell E, Pearlmutter D, Williamson TJ (2011) Urban microclimate: designing the spaces between buildings. Earthscan, London, Washington, DC. ISBN 978-1-84407-467-9

EU-DGE (2014) Building a green infrastructure for Europe (WorldCat.org). Available online at: https://www.worldcat.org/title/building-a-green-infrastructure-for-europe/oclc/875969246&referer=brief_results. Accessed 4 Aug 2018

EU, UN-Habitat (2016) United Nations Human Settlements Programme. European Commission, Directorate-General for Regional and Urban Policy The State of European cities 2016 cities leading the way to a better future. Publications Office, Luxembourg. ISBN 978-92-79-64259-3

Farnham C, Emura K, Mizuno T (2015) Evaluation of cooling effects: outdoor water mist fan. Build Res Inf Build Res Inf 43:334–345

Fini A, Frangi P, Mori J, Donzelli D, Ferrini F (2017) Nature based solutions to mitigate soil sealing in urban areas: results from a 4-year study comparing permeable, porous, and impermeable pavements. YENRS Environ Res 156:443–454

GCCA, R20 (2012) A practical guide to cool roofs and cool pavements on CoolRoofToolkit.org. Available online at: https://www.coolrooftoolkit.org/read-the-guide/. Accessed 22 Jul 2018

Golden J (2004) The built environment induced urban heat island effect in rapidly urbanizing arid regions—a sustainable urban engineering complexity. Environ Sci 1(4):321–349

Hassler U, Kohler N (2014) Resilience in the built environment. Build Res Inf Build Res Inf 42:119–129

IPCC (2018) Intergovernmental Panel for Climate Change, Global warming of 1.5 °C An IPCC Special Report on the impacts of global warming of 1.5 °C above pre-industrial levels and related global greenhouse gas emission pathways, in the context of strengthening the global response to the threat of climate change, sustainable development, and efforts to eradicate poverty. IPCC, Geneva. ISBN 978-92-9169-151-7

Kala J, De Kauwe MG, Pitman AJ, Medlyn BE, Wang Y-P, Lorenz R, Perkins-Kirkpatrick SE (2016) Impact of the representation of stomatal conductance on model projections of heatwave intensity. Sci Rep 6:23418

Karlessi T, Santamouris M, Apostolakis K, Synnefa A, Livada I (2009) Development and testing of thermochromic coatings for buildings and urban structures. Sol Energy Phoenix Ariz Then N Y 83:538–551. https://doi.org/10.1016/j.solener.2008.10.005

Ketterer C, Matzarakis A (2014) Human-biometeorological assessment of heat stress reduction by replanning measures in Stuttgart, Germany. Landsc Urban Plan 122:78–88. https://doi.org/10.1016/j.landurbplan.2013.11.003

Kolokotsa D, Maravelaki-Kalaitzaki P, Papantoniou S, Vangeloglou E, Saliari M, Karlessi T, Santamouris M (2012) Development and analysis of mineral based coatings for buildings and urban structures. SE Sol Energy 86:1648–1659. https://doi.org/10.1016/j.solener.2012.02.032

Landsberg HE (1983) The urban climate. Academic, New York, ISBN 978-0-12-435960-4

Laureti F, Martinelli L, Battisti A (2018) Assessment and mitigation strategies to counteract overheating in urban historical areas in Rome. Clim Clim 2018:6. https://doi.org/10.3390/cli6010018

Lazzarin RM, Castellotti F, Busato F (2005) Experimental measurements and numerical modelling of a green roof. Energy Build 37:1260–1267

Lee H, Mayer H, Chen L (2016) Contribution of trees and grasslands to the mitigation of human heat stress in a residential district of Freiburg, Southwest Germany. Land Landsc Urban Plan 148:37–50

Lee I, Voogt JA, Gillespie TJ (2018) Analysis and comparison of shading strategies to increase human thermal comfort in urban areas. Atmosphere 9(3):91. Available online at: http://www.mdpi.com/2073-4433/9/3/91. Accessed 22 Jul 2018

Liu K, Baskaran B (2003) Thermal performance of green roofs through field evaluation

Manley G (1958) On the frequency of snowfall in metropolitan England. QJRMS 84(359):70–72

McGregor GR (2012) Special issue: Universal Thermal Comfort Index (UTCI). Int J Biometeorol 2012:56

Meerow S, Newell JP, Stults M (2016) Defining urban resilience: a review. Landsc Urban Plan 147:38–49

Montazeri H, Toparlar Y, Blocken B, Hensen JLM (2017) Simulating the cooling effects of water spray systems in urban landscapes: a computational fluid dynamics study in Rotterdam, The Netherlands. Landsc Urban Plan 159:85–100

OECD (2018) Resilient cities. Available online at: http://www.oecd.org/cfe/regional-policy/resilient-cities.htm. Accessed 21 Jul 2018

Oke TR (1982) The energetic basis of the urban heat island. Q J R Meteorol Soc 108:1–24

Oke TR (1987) Boundary layer climates. Routledge, London, England; New York. ISBN 978-0-415-04319-9

Pearlmutter D (2000) Patterns of sustainability in desert architecture. Arid Lands Newslett 47:1–12. Available at: http://ag.arizona.edu/oals/ALN/aln47/pearlmutter.html. Accessed 12 May 2019

Pisello AL (2017) State of the art on the development of cool coatings for buildings and cities. Sol Energy 144:660–680. https://doi.org/10.1016/j.solener.2017.01.068

Rahman MA, Armson D, Ennos AR (2015) A comparison of the growth and cooling effectiveness of five commonly planted urban tree species. Urban Ecosyst Urban Ecosyst 18:371–389

Roy S, Byrne J, Pickering CA (2012) Systematic quantitative review of urban tree benefits, costs, and assessment methods across cities in different climatic zones. Urban Forest Urban Green 11:351–363. https://doi.org/10.1016/j.ufug.2012.06.006

Saaroni H, Amorim J, Hiemstra J, Pearlmutter D (2018) Urban Green Infrastructure as a tool for urban heat mitigation: survey of research methodologies and findings across different climatic regions. UCLIM Urban Clim 24:94–110. https://doi.org/10.1016/j.uclim.2018.02.001

Saaroni H, Pealmutter D, Hutuka T (2015) Human-biometeorological conditions and thermal perception in a Mediterranean coastal park. Int J Biometeorol 59:1347–1365. Available online at: https://link.springer.com/article/10.1007%2Fs00484-014-0944-z. Accessed 22 Jul 2018

Santamouris M (2007) Heat Island research in Europe—the state of the art. Adv Build Energy Res (ABER) 1(1):123–150

Santamouris M (2014) Cooling the cities—a review of reflective and green roof mitigation technologies to fight heat island and improve comfort in urban environments. Sol Energy 103:682–703. https://doi.org/10.1016/j.solener.2012.07.003

Santamouris M, Asimakopoulos DN (2001) Energy and climate in the urban built environment. James X James, London. ISBN 978-1-873936-90-0

Santamouris M, Gaitani N, Spanou A, Saliari M, Giannopoulou K, Vasilakopoulou K, Kardomateas T (2012) Using cool paving materials to improve microclimate of urban areas—design realization and results of the flisvos project. BAE Build Environ 53:128–136. https://doi.org/10.1016/j.buildenv.2012.01.022

Santamouris M, Kolokotsa D (2016) Urban climate mitigation techniques. Routledge-Earthscan, s.l. ISBN 978-0-415-71213-2

Santamouris M, Papanikolaou N, Livada I, Koronakis I, Georgakis C, Argiriou A, Assimakopoulos DN (2001) On the impact of urban climate on the energy consumption of buildings. Sol Energy 70:201–216. https://doi.org/10.1016/S0038-092X(00)00095-5

Shahidan MF, Jones PJ, Gwilliam J, Salleh E (2012) An evaluation of outdoor and building environment cooling achieved through combination modification of trees with ground materials. Build Environ 58:245–257. https://doi.org/10.1016/j.buildenv.2012.07.012

Shashua-Bar L, Hoffman ME (2000) Vegetation as a climatic component in the design of an urban street: an empirical model for predicting the cooling effect of urban green areas with trees. Energy Build 31:221–235. https://doi.org/10.1016/S0378-7788(99)00018-3

Smith KR, Roebber PJ (2011) Green roof mitigation potential for a proxy future climate scenario in Chicago, Illinois. J Appl Meteorol Clim 50:507–522

Spronken-Smith RA, Oke TR (1998) The thermal regime of urban parks in two cities with different summer climates. Int J Remote Sens 19:2085–2104

Synnefa A, Santamouris M, Livada I (2006) A study of the thermal performance of reflective coatings for the urban environment. SE Sol Energy 80:968–981. https://doi.org/10.1016/j.solener.2005.08.005

Taleghani M, Berardi U (2018) The effect of pavement characteristics on pedestrians' thermal comfort in Toronto. Urban Clim 24:449–459. https://doi.org/10.1016/j.uclim.2017.05.007

UNFCCC (2019) Yearbook of global climate action 2019. ISBN 978-92-9219-186-3

Upreti R, Wang Z-H, Yang J (2017) Radiative shading effect of urban trees on cooling the regional built environment. UFUG Urban Forest Urban Green 26:18–24

US EPA (2009a) Reducing urban heat islands: compendium of strategies. Available online at: http://www.epa.gov/heatisld/resources/compendium.htm. Accessed 12 May 2019

US EPA (2009b) Smart growth and urban heat islands. Available online at: http://www.epa.gov/heatisld/resources/pdf/smartgrowthheatislands.pdf. Accessed 12 May 2019

Yang J, Wang Z-H, Kaloush KE (2015) Environmental impacts of reflective materials: is high albedo a 'silver bullet' for mitigating urban heat island? Renew Sustain Energy Rev 47:830–843. https://doi.org/10.1016/j.rser.2015.03.092

Chapter 16
Environmental Design Challenges for Multi-beneficial Urban Open Spaces

Renata Valente

Abstract Pluvial flooding is an increasing plague among the current urban problems. Green streets reconvert urban public connection space for multi-benefit uses such as preventing flooding, improving air and rainwater quality, increasing clean water infiltrated into groundwater, reducing the effects of urban heat island, providing open spaces, recreational habitats and green pedestrian links. An interscalar research project applies good practices of sustainable rainwater road management through nature-based solutions on an axis of strategic connection in the Terra di Lavoro plain (Campania, Southern Italy), offering original technical protocols. Main points of reflection are the central role of local public governments as drivers of integrated projects, the interdisciplinary vision and the meaning value of the interconnected green stormwater infrastructures (GSI) network.

Keywords Interdisciplinarity · Green streets · Implementation tools

16.1 Environmental Concerns and Social Roles

Third decade of the third millennium: the environmental issue is apparently on political and social agendas around the globe. Discussed from the level of dissemination to the in-depth and controversial scientific predictive analysis, it is intertwined with transversal political visions. The theoretical reflection on the topic, since Leopold (1949–68) to Carson (1962), Bateson (1972), Lovelock (1979), Morin (1987), has always underlined its subversive dimension, also inviting to the new alliance of knowledge, both humanistic and scientific (Prigogine and Stengers 1981). Based on such a complex vision, the environmental project integrates cultural, social and perceptive components, considered as important as the technical ones (Bocchi 1985). Citizens perceive the same subversive dimension according to their cultural level and social habits; the awareness of the risks to which one is exposed corresponds to the

R. Valente (✉)
Università degli Studi della Campania Luigi Vanvitelli, Aversa (CE), Italy
e-mail: renata.valente@unicampania.it

G. Chiesa (ed.), *Bioclimatic Approaches in Urban and Building Design*, PoliTO Springer Series, https://doi.org/10.1007/978-3-030-59328-5_16

need for a change in lifestyle paradigm and for related projects towards the adaptation of the built environment.

As a result, the administrations in charge of managing the transformations of the territory express attention to environmental problems in relation to political and cultural guidelines. However, observing the international virtuous examples shows that it is possible to impress an environmental operational effectiveness depending on the technical-scientific preparation and updating of the officers. The single designer's action is limited if not framed and interconnected in a systemic vision of context. Therefore, the academy's task is crucial to provide contents and training experiences enabling the construction of operational frameworks where each contribution is functional to the general system to transform places and the elements composing them. Every action aimed at a complex project of integrated management has not only considerable technical value, but above all the political significance of the propensity towards equity of resources, application of environmental strategies and induction of positive interconnected effects, mutually reinforcing. On the other hand, designers must be aware and prepared to provide cultural and operational guidance, as Chris Reed urges.

> We can't simply wait around for clients or patrons to invent projects for us. We need to take a more proactive, entrepreneurial role in showing potential clients the extraordinary range of possible projects and the various tools that will help realize them. We need to break down traditional disciplinary distinctions and professional versus academic divides. And we also need to tap into broad networks of colleagues – such as applied ecologists, housing specialists, anthropologists, community organizers, economic advisors, local foundation officers and management consultants. And we must integrate research and funding initiatives that can help propel projects forward, and positively test the limits of knowledge. (C. Reed in Riano 2011)

To effectively achieve this dynamic interdisciplinary collaboration, it is also necessary to overcome the interest in different ways of considering innovation. Each traditionally scientific environment looks forward achieving cutting edge results in its own field, while it does not satisfactorily assess the collegial consequences of integration. This can undermine the common discourse, the expended energies and the effects; only a deep conviction in the value of shared outcomes, useful first for the community, can overcome the narcissism that is in the specific and sectorial finding.

16.2 A Collegial Case Study

Pluvial flooding is a plague among the current urban problems: the frequency with which it occurs, the power of the water and the amount of damage have increased. Moreover, they are among the outcomes that most directly and physically affect citizens causing significant social impacts. We know that, in addition to meteorological and climatic causes, the increase in soil consumption and surface sealing are determining factors, but these phenomena are proceeding at a relentless pace, although in recent times the environmental project culture evolved towards an eco-systemic

Fig. 16.1 Example of a planter operating diagram (from Sandpoint City Council, Sandpoint Urban Renewal Agency (SURA), Downtown Streets Plan and Design Guide, City of Sandpoint, p. 44, December 2012)

approach for sustainable rainwater management projects. In fact, as an alternative to traditional systems of rapid channelling into underground sewerage infrastructure, surface areas can serve to collect, filter, retain, purify and possibly infiltrate the water itself (Fig. 16.1).

A green street reconverts urban public connection space for multi-benefit uses such as preventing flooding, improving air and rainwater quality, increasing clean water infiltrated into groundwater, reducing the effects of urban heat island, providing open spaces, recreational habitats and green pedestrian links. Best practices are the satisfactory experiences of North American cities such as Portland, Seattle, Washington, Philadelphia, Boston, San Francisco, Los Angeles (USA) and Toronto (Canada), although previous historical references are the German experiments in environmental redevelopment of the Ruhr brownfields area and the city of Berlin. The international examples show how the application to relatively recent urban realities are more suitable than those to historical tissues, where the use of places has sedimented morphologies and habits. However, technicians and administrations are not always prepared to approach the problem in a sustainable way, nor basic data, updated tools and economic resources are always available. Therefore, it happens that further conditions of environmental injustice increase and self-feed. In the light of these considerations, the current phase of research on these aspects develops a significant depth when it declines theoretical concepts, integrated data readings and interdisciplinary mapping to deduce context requirements and site-specific solution systems. After a decade of applications, one of the research main topic is to evaluate the benefits related to different systems, places, climates and habits. Thank to periodical reports, researchers and public officers drive interesting considerations about air quality improvement, increased greenness, energy savings, greenhouse gas reduction, community cohesion, environmental equity (Valente et al. 2019).

From these assumptions an interscalar project model experiments the application of good practices of sustainable rainwater road management through nature based solutions (European Commission 2015; Valente 2017) on an axis of strategic connection in the Terra di Lavoro plain (Campania, Southern Italy).[1] This allows to trigger adaptive transformation processes and to rethink the best management practices used elsewhere. A peripheral university offers a contribution to the requalification of the surrounding territory by operating at a very close scale, considering difficulties and differences compared to most cases of literature, studying the feasibility of a green street with symbolic value for an unqualified urban area. As always, the boundaries of municipalities and provinces do not follow geographical limits of the catchment area and therefore the studies involve several local governments, intercepting different types and qualities of urbanization, in a territory with strong tendency to flooding that worsens with the increase in land consumption and climate change.[2]

Among the first topics addressed, there was the protocol of interdisciplinarity, having to compare skills in environmental design, botany, geology, geomorphology, history of the city, hydraulic engineering, civil design.[3] Valuable is the collaboration with colleagues from the College of Environmental Design of the University of California at Berkeley (USA), an institution that has formed significant operational personalities even within national and local public administrations.[4] The research in fact intends to provide original technical protocols for the towns of southern Italy; through the participatory practices, it nurtures the awareness of the growing danger of certain conditions and the possibility of mitigating among administrators, designers and users, fighting the environmental injustice of unequal training (Figs. 16.2 and 16.3) (De Falco et al. 2019).

16.3 Reflections from Interdisciplinary Analysis

By preparing the overlay of thematic maps, the difficulty of obtaining detailed data emerges in areas different from large cities, which are more frequently studied. Satellite views are often on a scale of insufficient interest, just as the local meteorological and environmental survey effort does not cover historical series. In addition, due to the characteristics of close-scale work, LIDAR returns are often imprecise and tran-scalarity is a dizzying balancing act. With peaks of urban density, total lack of planning, rules and criteria, in the case study there aren't clear and immediate keys to interpretation. Nevertheless, in Italy there is a myriad of disconnected and unvalued cultural heritage, offering new codes of interpretation also for the environmental

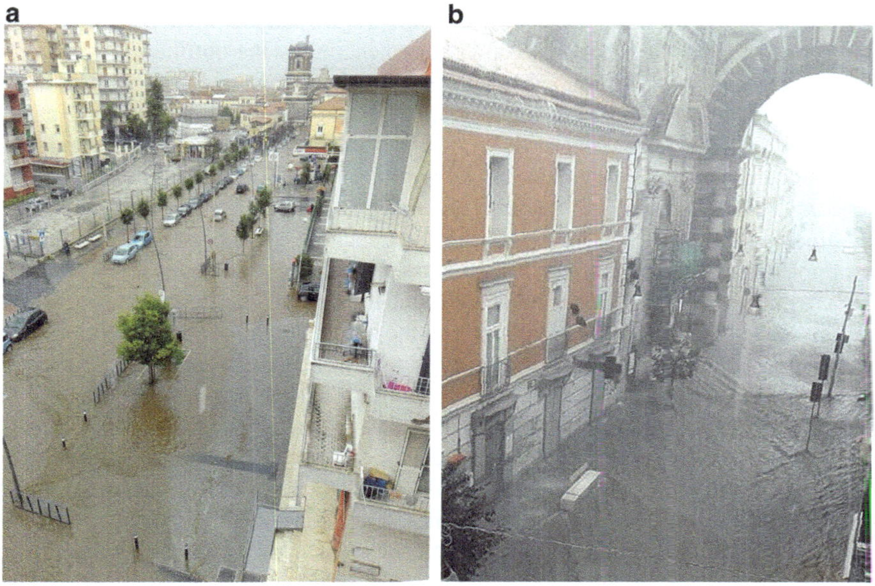

Fig. 16.2 Flooding at Porta Napoli, Aversa (CE, Italy). a November 2017; b September 2019

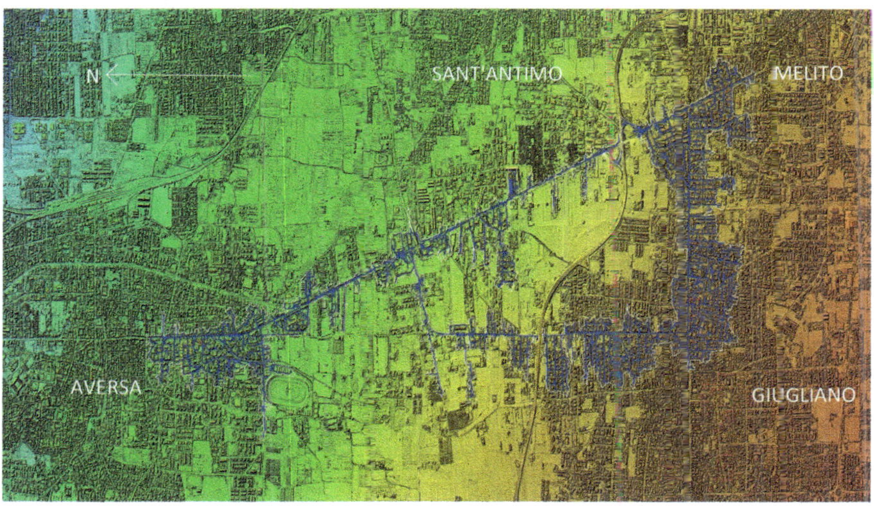

Fig. 16.3 PRIN 2015 "Adaptive design and technological innovations for the resilient regeneration of urban districts under climate change". SS7bis Mediterranean Green Street: section Aversa—Melito (southern Italy) with drainage area, by University of Studies of Campania Luigi Vanvitelli team (2019)

project network. Moreover, the study of iconography and historical chronicles helps to reconstruct the recurrence of flood phenomena, allowing outlining the evolution of the area and providing valuable information on the course of events. The examination of historical and recent cartography allows also to reconstruct the pattern of the hydrographical network generated by the intense urbanization, identifying directions of the water runoff in the geographical grid. Consequently, the drainage area is delimited as well as the sub-zone of the public areas. The infiltration map, original research aimed at the technician and the public administration following the example of the city of Seattle (USA), identifies the areas suitable for infiltration. The process combines the distance from the foundations of the buildings, the knowledge of the space between the ground floor and the aquifer with the reconstruction of the stratigraphy, in order to show the infiltration times of the water. The information is reported on different layers while an algorithm calculates a score to classify the territory into infiltration classes and identify the areas where it is possible to recharge the aquifer. The constant exchange of information between the working scales therefore runs through the whole process. The microclimatic considerations highlight the increase of intense and concentrated rainfall alternating with long dry periods, with the tendency of the phenomenon to worsen and the need to guarantee resistant plants or constant maintenance. Therefore, the study on botanical aspects considers the analysis of contrasts as a determining aspect, also with regard to phyto-purification and complex considerations on the selection of the suitable plants (Fig. 16.4).

Moreover, in the frequent cases of accelerated land consumption with spontaneous and deregulated building, the environmental functionalization through green streets is an excellent opportunity redesign of open spaces to induce a necessary semantic qualification as well as shared and sustainable lifestyles. The surplus value of the

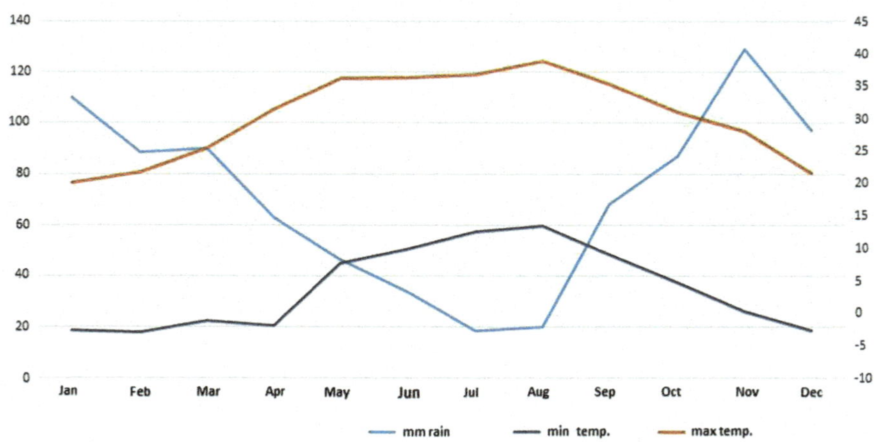

Fig. 16.4 PRIN 2015 "Adaptive design and technological innovations for the resilient regeneration of urban districts under climate change". SS7bis Mediterranean Green Street: comparison between annual rainfall trends and minimum and maximum temperatures in the case study area (Data processed by C. Donadio), by University of Studies of Campania Luigi Vanvitelli team (2019)

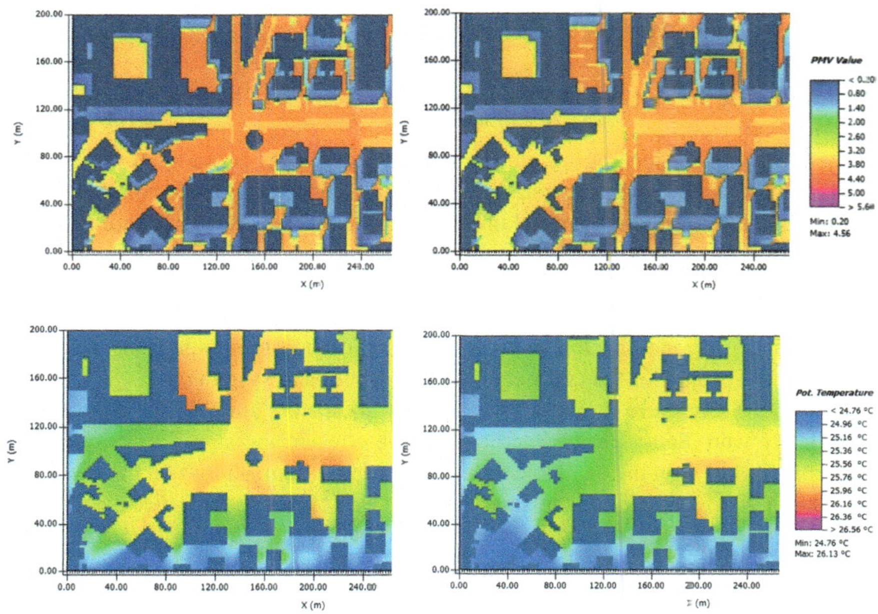

Fig. 16.5 PRIN 2015 "Adaptive design and technological innovations for the resilient regeneration of urban districts under climate change". SS7bis Mediterranean Green Street: results of fluid dynamics modelling (ENVI-met software). Comparisons between actual status and project in sample area 1 at h. 9:00 am. Above: PMV value maps; Below: maps of potential temperature values, by University of Studies of Campania Luigi Vanvitelli team (2019)

project is the redefinition of public places through configurations integrating use, functional and inclusive aspects, also stimulating the citizens' care of the equipment. This feeling can only happen through a deep awareness of the need and benefits offered. The role of local government is then to become a hinge between the scales of environmental design. Public officials in technical areas are now entitled to updated training on environmental design criteria, based on interdisciplinarity, interscalarity, use of indicators, strategic interrelation of objectives that are also performative. These roles of bureaucrats must start providing territories basic data, systematized through advanced computer tools, allowing a vision of the places that is both complex and timely, to be able to act even in distant phases and so each step is functional to the general program. By providing analyses carried out at the basin and sub-basin scale, hydraulic estimations, data collection and processing, the urban landscape can be defined with the participation of users. On the other hand, it is up to us researchers to experiment and prepare advanced protocols for effective processes, as well as for detail design, refining the existing relationship between the vast dimension and particulars, between the tools of GIS, BIM and all the modelling still tentative today. Among the objectives, there is the structuring of operational protocols for site-specific projects, with a bottom-up approach, to stimulate processes of social awareness.

16.4 Design Strategies and Indicators

By identifying the values of the territory and the chances of highlighting, it is necessary to reconnect the notable places, applying to the resulting network the ecological principle of self-regeneration, aiming at a responsiveness coming from the inter-scalar relationship. We like to call "telescopic" this process of environmental design, borrowing this definition from Belanger (2006). The reason is we continuously move from the territorial framework, to study the perimeters and characteristics of the hydrographical basin and local micro-basins, to the scale of the minute detail of the interstitial spaces available and differently connoted, and then return to build a replicable system of general rules.[5] The result is a path of continuous changes of focus, reaching the project of the components of technological systems.

In order to measure the achievable results, indicators suitable for the described case study were first identified and calculated, considering significant examples of literature.[6] Based on the available data, some were parameterized on the whole area and some were specific for the sample areas with closer scale projects. In the first case the percentage of water collected by precipitation of 30 min with a return time of 20 years, the increase in soil permeability index and the availability of safe cycling routes were considered. In the second case R.I.E. (Building Impact Reduction),[7] B.A.F. (Biotope Area Factor),[8] presence of CO_2, PMV, potential, surface and mean radiant temperature, relative humidity were added, calculated with ENVI-met fluid dynamics modelling software,[9] in addition to the values of energy consumption reduction, rainwater management, air quality, aesthetic values, CO_2, parameters calculated in dollars with I-Tree Street.[10]

The environmental assessments showed that the link between stormwater management and the climatic contribution of green street facilities requires higher amounts of green areas and tall trees than those to meet hydraulic retention and purification needs (see Fig. 16.5). The impact on microclimate changes is effective on extensions of tall trees for urban forestation actions, although this should not be confused with contributions related to the liveability of urban open spaces for the correction of CO_2 or PMV rates. By comparing these values between the present conditions and the project, it is possible to evaluate the effectiveness of the GIS model. This allows to perimeter leaking areas, to size green stormwater infrastructures (GSI), offering also localisation criteria and vulnerability indexes of critical zones, because they are both more subject to flooding and remarkable places. The model also proves to be a valuable implementation tool to define the priorities of interventions, according to administrative financial resources (Fig. 16.6).

Research through design also allowed to build an abacus of localization conditions, corresponding to types of optional equipment, as well as different scenarios in case of flooding or more frequent aridity. The locations match with remarkable places such as schools, universities, town halls, places of worship, markets, habitual meeting places, spaces of inclusion; the integrability is developed both relating to the constituent elements and to the context.

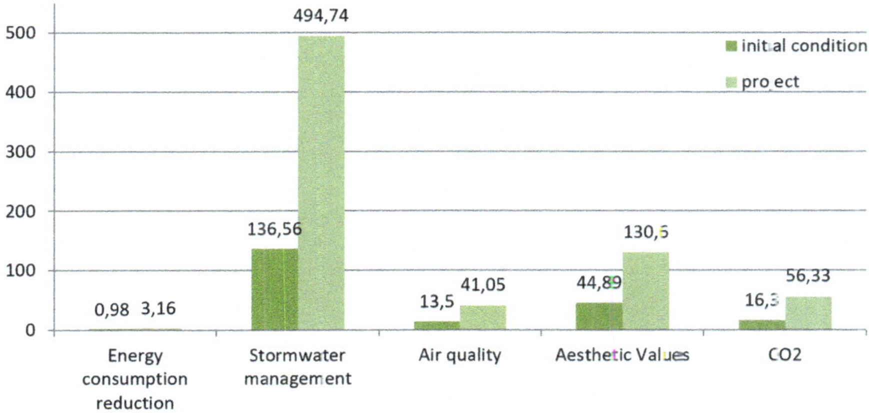

Fig. 16.6 PRIN 2015 "Adaptive design and technological innovations for the resilient regeneration of urban districts under climate change" SS7bis Mediterranean Green Street: benefits of trees on public land on an annual basis ($/tree); comparisons between actual status and project in pilot area 1 (processing with I-Tree Street software), by University of Studies of Campania Luigi Vanvitelli team (2019)

16.5 Concluding Remarks

The application of environmental research/project led to highlight some main points of reflection: first of all, the central role of local public governments as drivers of integrated projects. Further reflection confirms the indifferability of the interdisciplinary vision of the project, which integrates scientific, social, numerical and humanistic results, together with the new alliance tested in situ of devices full of uses, meanings, values, affections, acceptance, and projections. In addition, the test has shown very different local conditions requiring simplified protocols, which do not need resources comparable to those in large cities. Moreover, a quantitative survey concerns the confirmation of the insufficient contribution to air purification and the lowering of temperatures only by the trees along the roads.

Finally, yet importantly, it is the focusing of the meaning value of the network of interconnected green stormwater infrastructures (GSI), useful for the management of rainwater but potential catalysts of services for citizens. To this end, one of the inspirations within this research was the action work of the Sardinian artist Maria Lai (1919–2013), "Bonding to the mountain", the first experience of relational art at international level. When in 1981 the mayor of Ulassai committed a monument to

Fig. 16.7 Maria Lai (1981), "Legarsi alla montagna", draft

the war victims to Lai, the artist engaged in a great collective labour of community building. The idea was to bind the inhabitants for a day to their mountain, its landslides and its power, to which the village remains clinging in a symbiotic relationship between man and landscape. On September 8, 1981, citizens passed a blue ribbon from house to house, visually weaving the network of social relations of the village, tightening in knots to bind family and friends, celebrating a moment of communion and surprising themselves by the relational results produced.

In the same way, a network of multi-performance devices may remind us to belong to an urban catchment area as a system of social flows. This illustrates and refers to a complex, integrated, cultured, mutual and social vision of interconnected phenomena every day affecting us all with our places on this changing planet (Figs. 16.7 and 16.8).

Fig. 16.8 PRIN 2015 "Adaptive design and technological innovations for the resilient regeneration of urban districts under climate change". SS7bis Mediterranean Green Street: network between GSI (green stormwater infrastructures) and main public places, by University of Studies of Campania Luigi Vanvitelli team (2019)

Notes

1. Within the national research project PRIN 2015 "Adaptive design and technological innovations for the resilient regeneration of urban districts under climate change" (P. I. M. Losasso), the described research is carried out by a group of the Engineering Department of the University of Campania "Luigi Vanvitelli", coordinated by Renata Valente. The national research project includes other teams by the Federico II University of Naples, University La Sapienza of Rome, University of Florence, Mediterranean University of Reggio Calabria, Polytechnic of Milan.

2. The perimeter of the study area of 2.25 km^2 coincides with the rainwater drainage area on the stretch of road 7bis between the municipality of Melito (NA) and the Arch of the Annunziata of Aversa (CE), where sub-basins and outflow lines were identified using LIDAR data and GIS model.

3. The team includes R. Valente (Unit head), S. Cozzolino, C. De Falco, A. Di Nardo, M. Di Natale, C. Donadio (University Federico II), F. La Rocca, M. Pernetti, D. Ruberti, S. Strumia, M. Vigliotti, P. Ferrara, E. Cappelli, with the external contribution of L. A. Mozingo from the University of California at Berkeley (USA). P. Ferrara edited the elaborations on Envi-met software, E. Cappelli those on I-Tree Street software.

4. Example above all when "In the late 1970s Van der Ryn moved his field of action from design studios to public service when Governor Jerry Brown appointed him

State Architect of California, a position that allowed him to develop innovative standards for energy-efficient office buildings. Over the years he founded an interdisciplinary institute for ecological research, published articles and books, and is today a professor emeritus at Berkeley." (Castillo 2015) or even more recently the graduates who have exported good practices on green streets to local governments (Valente and Mozingo 2015).

5. A special focus of detail has been applied to the south courtyard of the Engineering Department at the Real Casa dell'Annunziata; this open space destined for rearrangement, used daily by hundreds of people, can become a living lab where people can experiment sustainable management practices.

6. Examples are the Greened Acre, used in Philadelphia (PA, USA), a run-off area using GSI (green stormwater infrastructure) devices, calculated as GA = IC (impervious coverage) × Wd (water height of at least 1″ = 2.5 cm up to 4 cm). Other types of indicators that can be imported come from the experience in Seattle (WA, USA), where the space needed to manage a leaking surface or a gallon of water (3.78 lt) and then the cost to manage a gallon of water, considering the results over two years.

7. The R.I.E. (Reduction of the Building Impact) is an environmental quality index that serves to certify the quality of construction work with respect to the permeability of the soil and greenery, promulgated by the municipality of Bolzano (Italy) and also adopted by the municipalities of Bologna and Reggio Emilia (Italy). (http://www.comune.bolzano.it/urb_context02.jsp%3FID_LINK%3D512%26page%3D10%26area%3D74%26id_context%3D4663).

8. Since 1994 the city of Berlin (GE) uses a regulation requiring a proportion of the area to be left as green space: the Biotope Area Factor (BAF) or BFF (Biotop Flächenfaktor). It expresses the ratio of the ecologically effective surface area to the total land area (https://www.berlin.de/senuvk/umwelt/landschaftsplanung/bff/index_en.shtml).

9. https://www.envi-met.com/.

10. https://www.itreetools.org/tools/i-tree-streets. Since 2006, i-Tree has been a cooperative effort between the USDA Forest Service, Davey Tree Expert Company, The Arbor Day Foundation, Society of Municipal Arborists, International Society of Arboriculture, Casey Trees, and SUNY College of Environmental Science and Forestry.

References

Bateson G (1972) Steps to an ecology of mind. University of Chicago Press, Chicago, USA

Belanger P (2006) Synthetic surfaces. In: Waldheim C (ed) The landscape urbanism reader. Princeton Architectural Press, New York, pp 239–266

Bocchi G, Ceruti M (1985) La sfida della complessità. Feltrinelli, Milano

Carson R (1962) Silent spring. Houghton Mifflin, Boston, USA
Castillo G (2015) Hippie modernism. Places J. https://doi.org/10.22269/151026. Accessed 28 Jan 2020
De Falco C, Ferrara P, Valente R (2019) Storia e Progetto Sostenibile per la Riqualificazione di un Percorso Stradale in Campania (Italia) (History and sustainable design for the requalification of a road route in Campania (Italy)). In: Garda, Mele C, Piantanida P (eds) Colloqui.AT.e Ingegno e costruzione nell'epoca della complessità, Torino, Sept 2019. Edizioni Politecnico di Torino, p 1160
European Commission, Directorate-General for Research and Innovation Climate Action, Environment, Resource Efficiency and Raw Materials (2015) Towards an EU Research and Innovation policy agenda for Nature-Based Solutions & Re-Naturing Cities Final Report of the Horizon 2020 Expert Group on 'Nature-based solutions and re-naturing cities'. European Union
Leopold A (1968) Sand county almanac and sketches here and there. Oxford University Press, Oxford
Lovelock J (1979) Gaia. A New lock at life on earth. OUP, Oxford
Morin E (1987) Il metodo. Ordine, disordine, organizzazione. Feltrinelli, Milano
Prigogine I, Stengers I (1981) La nuova alleanza Metamorfosi della scienza. Einaudi, Torino
Riano Q (2011) Landscape optimism: an interview with Chris Reed. Places J. https://doi.org/10.22269/110920. Accessed 28 Jan 2020
Sandpoint City Council, Sandpoint Urban Renewal Agency (SURA) (2012) Downtown streets plan and design guide. City of Sandpoint, p 44
Valente R (2017) Water sensitive urban open spaces: comparing North American best management practices. UPLanD J Urban Plan Landsc Environ Des 2(3):285–297
Valente R, Cozzolino S, Ferrara P (2019) Enforceability and benefits of Mediterranean green streets. SMC Sustain Mediterr Constr 10:47–53
Valente R, Mozingo LA (2015) Learning from California: environmental design tools to recover degraded built landscapes. Techne, n.10/, pp 258–267. ISSN online: 2239-0243. http://dx.doi.org/10.13128/Techne-17523

Chapter 17
Radiative Exchanges in the Dense City

Elena Garcia-Nevado and Benoit Beckers

Abstract A central question for urban sustainability is how to build cities that minimize, at the same time, the use of land and energy consumption. Future cities should be as dense as possible to meet the first requirement. However, the energy performance of dense urban areas is still under debate due to the lack of sound arguments. Since the origins of urbanism in the XIXth century, urban planners have incorporated energy considerations into their designs in a qualitative way. Nowadays, the discipline of Urban Physics is a well-established field able to provide them also with quantitative energy information to aid decision-making. This chapter explores how the current energy simulation capabilities, particularly the Finite Elements Methods, can contribute to a systemic study of energy exchanges in dense urban environments. We focus on radiative energy exchanges, which are especially challenging in dense urban morphologies. Finally, we discuss how the feedback between urban thermography and simulations can improve our understanding of the thermal behaviour of real urban environments.

17.1 Introduction

By the time we wrote this chapter, 7700 billion of human beings were inhabiting the Earth. According to the most recent projections, this population will continue growing until stabilizing around 11,000 billion people by the end of the century (United Nations 2019). The humankind faces the unprecedented challenge of building homes for 3.3 billion of new inhabitants in the next 80 years. The issue becomes even more overwhelming if the ultimate goal is providing adequate living conditions for all of them while reducing the adverse impacts of human activities on the environment.

E. Garcia-Nevado (✉) · B. Beckers
Urban Physics Joint Laboratory, Université de Pau et des Pays de L'Adour (E2S-UPPA), Anglet 64600, France
e-mail: e.garcia-nevado@univ-pau.fr

B. Beckers
e-mail: benoit.beckers@univ-pau.fr

G. Chiesa (ed.), *Bioclimatic Approaches in Urban and Building Design*, PoliTO Springer Series, https://doi.org/10.1007/978-3-030-59328-5_17

Urban areas, already the home of 55% of the world population, are expected to absorb virtually all the future population growth (United Nations 2018). Since the end of the nineteenth century, the extent of the built-up area grew faster than the urban population, resulting in a persistent decline in urban densities that lasts until today. By the end of the twentieth century, the annual growth rate of urban land was, on average, twice that of the urban population (Angel et al. 2011). If current trends in population density continue, the urban land cover by 2030 will nearly triple the urban land area circa 2000 (Seto et al. 2012). As cities have historically developed close to the most fertile lands, this urban expansion will lead to a direct loss of the prime croplands, and an even higher indirect loss of natural lands elsewhere due to the cropland displacement (van Vliet 2019). Urban expansion is hence an important cause of natural landscape destruction, not only in the areas nearby cities but also in faraway regions.

Cities are also one the main drivers of climate change, being responsible for 67–76% of global final energy consumption and 71–76% of CO_2 emissions related to fossil fuels (Seto et al. 2014). The growing urban extents and urban populations will increase the urban share in total energy use and greenhouse gas emissions. However, the future of urban energy use not only depends on the scale of urbanization, but also on the spatial patterns of urban development (Güneralp et al. 2017). Urban morphology affects, directly and indirectly, the two main sectors of urban energy use: transport and buildings (Steemers 2003). A substantial part of the energy consumed in buildings—almost half in developed countries (Pérez-Lombard et al. 2008)—is used for maintaining indoor thermal comfort. Globally, the energy use for heating and cooling is expected to increase in the next decades. Between 2010 and 2050, this increase could vary between 7 and 40% depending on the future urban densities, particularly of the rapidly growing Asian cities (Güneralp et al. 2017). These projections highlight the link between urban form and thermal urban environment as a key topic for both humans' well-being and energy sustainability.

The environmental emergency we are living in urges us to reflect on how to build our cities minimizing at the same time the use of land and urban energy consumption. Future cities should be as dense as possible to fulfill the first requisite, which would imply the containment of urban sprawl, now a global phenomenon (Sect. 17.2). Urban planners agree on the fact that higher urban densities will result in lower energy use for transportation. However, the building energy performance in dense urban morphologies is still under debate due to the lack of sound arguments. Since the origins of urbanism in the nineteenth century, urban planners have incorporated energy considerations into their designs in a qualitative way (Sect. 17.3). Nowadays, the discipline of Urban Physics is a well-established and active field that should provide them also with quantitative energy information to guide their decision-making (Moonen et al. 2012; Beckers 2016). This chapter explores how the current energy simulation capabilities, particularly the Finite Elements Methods, can contribute to a systemic study of energy exchanges in dense urban environments. The focus is on radiative exchanges, as the energy flux that poses the greatest challenges in occluded urban geometries (Sect. 17.4). Finally, we discuss how the feedback between urban thermography and simulations can improve our understanding of the thermal response of actual urban environments (Sect. 17.5).

17.2 The Shape of Cities

17.2.1 Urban Sprawl, a Global Phenomenon

'Urban sprawl' is a term used to define low-densities, scattered urban developments without systematic large-scale land-use planning (Bruegmann 2005). Generally, urban sprawl is viewed as a dysfunctional form of urban development with a large number of negative social, economic and environmental impacts (Wilson and Chakraborty 2013; Ewing 2008): high car dependency, low cost-effective infrastructures, high land use, landscape fragmentation, excess travel and congestion, poor urban live, high energy use for transportation and buildings… Despite the broad agreement about its drawbacks, urban sprawl is nowadays a global phenomenon:

- This way of understanding urban life, dominant in the Anglo-Saxon culture, has caught the collective imagination, even taking root in areas with a traditional compact urban heritage. Urban dispersion has grown rapidly in Europe since the 1970s regardless of the city size and its pre-existing morphology, as evidenced by the fact that urbanized areas increase at a faster rate than the population, stable or in slight growth (European Environmental Agency 2016). Recent studies about the Mediterranean region reported rising urban sprawl trends in the last 30 years in Rome, Barcelona, and Athens, cities where the compact model had been the norm until recent times (Tombolini et al. 2015).
- Traditionally, climate considerations used to shape the city somehow (e.g., urban morphology in extremely cold or hot regions was more compact than in tempered areas (Golany 1996). These factors seem not to be dominant anymore, and cities around the world exhibit similar urban development patterns (Lopez-Ordóñez et al. 2020). Satellite observations confirm that the urban sprawl phenomena have intensified worldwide, regardless of climatic aspects. To illustrate this trend, Fig. 17.1 presents the change in urban density between 1990 and 2015 for a representative sample of 200 cities around the world (Angel et al. 2016), located in the five major Köppen's climatic zones. On average, urban density declines in all the climates at a similar rate.
- Urban sprawl phenomena do not distinguish between rich and poor, being a rising trend in both developed and developing countries. Figure 17.2 represents the density change in the 200-city sample in (Angel et al. 2016) sorted by the Human Developed Index (UNDP 2018) of the country they belong to. This figure evidences that urban sprawl progresses similarly in all urban zones disregarding their level of human development. The extensive research addressing urban sprawl drawbacks during the last six decades has mainly focused on the developed ones

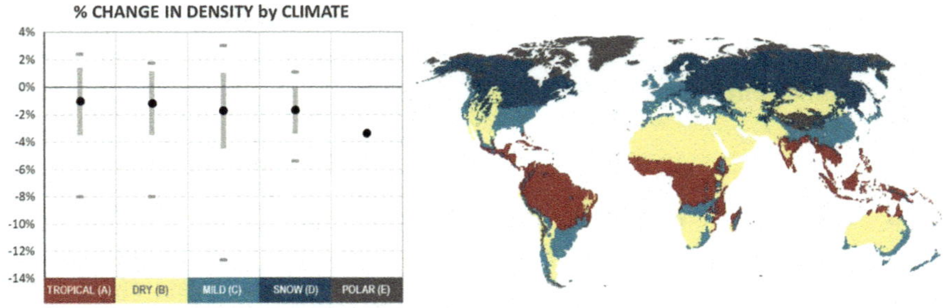

Fig. 17.1 Change in urban extent density by major Köppen's climate types (average: black dot; SD: grey bar; extreme values: grey hyphen). *Source* The authors coupling data from Angel et al. (2016) and Chen and Chen (2013)

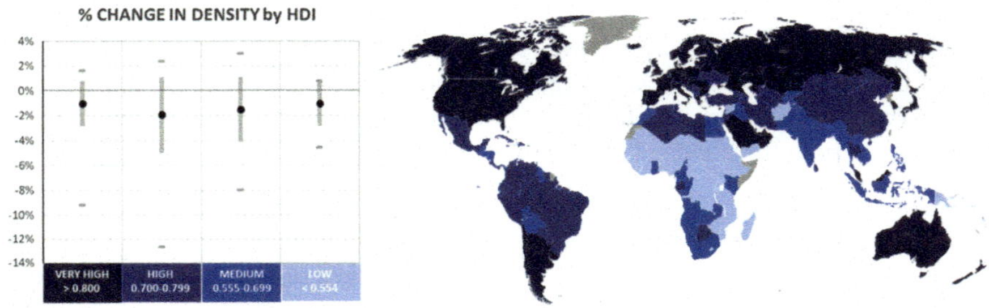

Fig. 17.2 Change in urban extent density by Human Development Index (average: black dot; SD: grey bar; extreme values: grey hyphen). *Source* The authors coupling data from Angel et al. (2016) and UNDP (2018)

(Adaku 2014). Findings in this context are not necessarily transferable to the developing countries, where most of the urban growth is happening (Nagendra et al. 2018).

17.2.2 What Is the Alternative?

Policymakers and urban planners seem to agree about the need to control urban sprawl. Since the 90 s, numerous international institutions have upheld the compact city idea as the most viable alternative, including the EU (European Commission 1990), the OECD (2012), and UN-Habitat (2017). The intent of global policymakers to control urban expansion contrasts with the generalized and sustained increase of urban sprawl trends around the world described in the previous section. This paradox evidences the lack of effective policies to fight at a small scale against the urban problems observed at a large scale (Angel et al. 2011).

Traditionally, researchers attribute this failure to the lack of a clear definition of the 'compact city' concept (Batty 2009). Generally, 'compact city' is described by a more or less extensive list of its characteristics, often as opposed to those of the 'dispersed city' (Neuman 2005). Thus, a city is compact if it has high density, mixed uses, well-defined urban boundaries, and is oriented to public transport. Conversely, a sprawled city presents low density, spatial segregation of uses, discontinuous occupation of the territory, and private transport predominance. Since cities may perform unevenly for the individual features associated with each model, this analysis approach may lead to conflicting conclusions in urban studies.

An alternative approach to avoid this controversy is the use of numerical indicators to quantify the degree of compactness of urban tissues. In this regard, one of the most used indicators—frequently the only one—is the urban density. This way, at least theoretically, when asked about how to contain the urban sprawl, a simpler answer arises: increasing the city density, whether with high-density new developments or by re-densifying the existing city (Burton 2002).

Unfortunately, defining urban density is also problematic. Strictly speaking, density is a point measure defined as the number of units of a certain entity (numerator) normalized by some measure of the area they occupy (denominator) (Batty 2009). This definition is complex because:

- Numerous types of density can be defined depending on how the entity of analysis and the reference area are accounted for (Batty 2009; Boyko and Cooper 2011; Churchman 1999). Based on the main entities analyzed, there are three generic categories of densities: building, population and housing. However, dozens of specific density indicators are used in practice [up to 23 indicators were listed in Alexander (1993)]. Additional issues arise concerning net and gross density values (Batty 2009).
- The scale of the reference area affects the values of density obtained, and hence, the conclusions drawn from the analysis. The same city may be considered as a high, medium or low-density urban environment depending on the size of the reference area. Figure 17.3 illustrates this situation for the French city of Bayonne.

Morphological indicators, such as density, reflect the third dimension of cities only to a limited extent (Batty 2009). This limitation is evidenced by the fact that very different urban layouts present the same values of density (Pont and Haupt

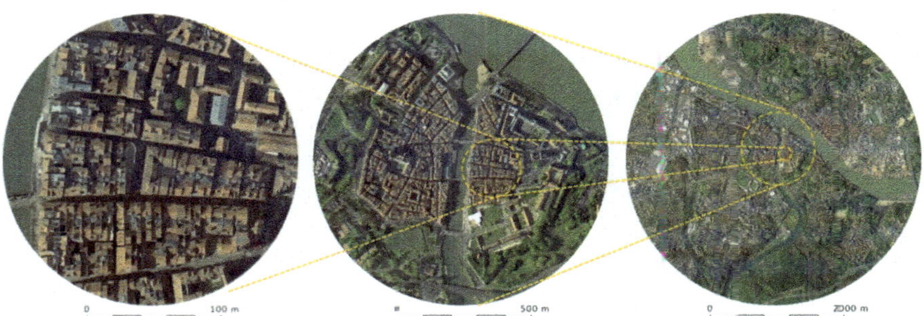

Fig. 17.3 Aerial view of the city of Bayonne in France at different scales (García-Nevado 2019)

2010). Numerical indicators are useful to characterize urban areas in some aspects. However, they cannot provide practical guidance for urban designers because one of their main tasks is, precisely, to decide on the spatial arrangement of buildings and open areas. The spatial layout of urban areas decisively affects their final performance in several aspects, including the energy-related ones (Cheng et al. 2006), affecting both building energy use and urban comfort.

Studies on the link between the city form and its energy behavior have relied on physical or numerical urban models with different levels of abstraction. A recurrent approach in the field consists of using typological urban models. The most popular ones are the urban canyon (Oke 1981) and the array of extruded volumes (Kawai and Kanda 2010), representing a simplified street and a regular grid plan, respectively. In the last three decades, this approach has allowed researchers to draw general conclusions about the energy exchanges at the urban scale while dealing with the lack of geometrical information about real environments and the technical limitations of their time (Arnfield 1990; Harman 2004). Analyses based on simple geometries, though, are unable to provide accurate quantification of energy fluxes, and consequently, to generate specific recommendations for urban and architectural design.

Since the 2000s, the advances in computational and modeling capabilities have made it possible for urban energy studies to use more and more realistic urban models. This made their results more reliable and hence more useful to guide specific interventions (e.g., energy refurbishment operations based on performance guarantee contracts). Studies on the PV potential of urban areas are a clear example in this regard. These studies need a high adherence of the urban model to the real-world features, since the actual PV potential is strongly affected by the local shading effects and the actual area available for PV panel installation (Walch et al. 2019).

Studies based on numerical indicators and typological models have both improved our theoretical knowledge about the link between urban form and energy performance. However, these approaches were too far from the physical reality of cities to be able to guide designers in practice. The former failed in representing the tridimensional nature of cities, and the latter, in reproducing their intrinsic complexity and heterogeneity. On these grounds, the use of realistic models seems to be a reasonable alternative to support designers in the energy-related aspects of their projects.

17.3 Energy Considerations in Urban Design Since the Industrial City

17.3.1 Nineteenth Century

Though cities exist since antiquity, urbanism, as we understand today, is a much more recent invention. The origins of this discipline date back to the nineteenth century, coinciding with the emergence of a set of urban proposals in response to the problems of the industrial city, ranging from city-scale utopias to more local and pragmatic solutions, such as innovative building typologies. Today, looking back at that period of urban history may be instructive, since their challenges are somehow similar to those we face now: the rapid expansion of urban areas, intense demographic growth and low environmental quality of cities.

Cities back then were crowded, polluted, poorly illuminated, and badly ventilated places. Based on their own experience, urbanists from that time knew that this situation could substantially improve if urban geometry complied with certain geometrical restrictions. The two main ones were respecting a minimum street aspect ratio (W/H) and not exceeding a maximum building depth. The first condition allows for increasing the solar radiation received and wind flow, hence leading to an improvement in daylight and air quality. The second one ensures cross ventilation in summer, favored by the pressure difference between the shaded and—now more—sunlit façades.

Following these simple principles, those urbanists could conceive urban plans for three cites which are today a reference of livable high density, vibrant urban life, and urban energy efficiency: Haussmann's plan for Paris, Commissioner's plan for Manhattan, and Cerda's plan for the Eixample of Barcelona. Several features make of Cerda's project a pioneer urban plan. One is its scientific approach, evident in his books (Cerdà 1867). Before conceiving the Eixample, Cerda had conducted a topographic study of Barcelona outskirts, and a sociological study on the living conditions in the old town. This gave him a solid knowledge of the geophysical constraints, the connectivity needs, and the desirable urban densities to meet in the project. Another pioneering feature is the attention paid to the energy-related issues. Several design decisions in Cerda's project denote a conscious control of solar energy at an urban scale: an aspect ratio of streets higher than W/H = 1, the size of the block courts, and the 45° North orientation of the urban grid to balance the sun distribution among urban façades.

To a great extent, the environmental challenge of the nineteenth century in terms of energy can be reduced to a shortwave problem. Consequently, minimal insights on the solar paths and shadow geometry were enough to guide the main energy-related decisions of the urban project. Urbanists from that time succeed because they had sufficient physical knowledge and tools to model the urban energy problem they were facing.

17.3.2 Twentieth Century

The twentieth century was a revolutionary period for architecture and urbanism. Several technical advances developed during the first decades radically changed the way of designing and constructing both buildings and cities. At a building scale, the spread of pillar and beam structures made of structural steel and reinforced concrete brought more freedom to the design of the building skin, since exterior walls were no longer load-bearing. Window size could then grow and, by the 60s, glass curtain walls were a widespread construction system (Garcia-Nevado et al. 2017). This brought more light to the interiors, but also overheating problems. However, since 1902, achieving indoor thermal comfort by artificial means was no longer a technical problem thanks to modern air conditioning systems. At a city scale, the invention of cars and elevators erased the horizontal and vertical limits for the city, opening the door to the development of high-rise building districts and endless single-family suburbs. These two urban proposals gave rise to the two trending city models of that century: the vertical Modernist city, conceptualized in the Ville Radieuse by Le Corbusier, and the horizontal Sprawl city, conceptualized in the Broadacre City by Frank Lloyd Wright. Post-war urban developments around the world were a testing ground for both city models. Soon, critical voices raised against both of them (Jacobs 1961; Real Estate Research Corporation 1974), sparking off an intense debate about the most convenient urban form.

In the second part of the century, environmental aspects became a central part of the urban debate. Olgyay, a pioneer in the fields of bioclimatic architecture and urbanism, argued for the need for designs adapted to the local climate and the use of natural energies to achieve indoor comfort in 1963 (Olgyay 1963). Since the 1950s, the rise in number and rigor of the experimental observations allowed researchers to demonstrate that cities influence climate at several scales. The most evident effect of this interaction at the urban scale is the Urban Heat Island, a phenomenon mainly driven by the increased radiative 'trapping' and heat storage taking place in urban areas (Oke 1978). By the end of the century, climatologists were able to confirm that cities also affect the climate at a global scale, playing a key role in climate change, which is leading to an increase in the average temperatures and extreme heat events worldwide (Revi et al. 2014). This warming is further exacerbated in urban areas by the UHI phenomenon and the feedback with the increasing anthropogenic heat emissions from air conditioning (Takane et al. 2019).

The environmental challenge of urban areas in the twentieth century was, conceptually, a temperature problem, and by extension, a longwave problem. Therefore, to solve it or, at least, mitigate it through design, it was necessary to be able to calculate all the urban energy fluxes respecting the actual configuration of the city, something beyond the technical possibilities of that time. By the end of the century, urbanists no longer had the tools required to deal with the level complexity of the cities of their time. They lacked operative instruments to inform decision-making regarding the thermal behavior of their projects. Consequently, the energy debate on the urban forms remained unsolved in the absence of reliable quantitative arguments.

If the nineteenth century represents the birth of urbanism, the twentieth century could be considered as the moment of its death. The architect and urbanist R. Koolhaas wondered about the paradox of the disappearance of urbanism, as a profession, at the time when urbanization was triumphing everywhere (Koolhaas 1995). His opinion was that this situation was the result of "a losing battle with the issue of quantity".

17.3.3 The Challenge of This Century

The challenge for the urbanists of this century is to deal with the systemic problem posed by urban environments. In terms of energy, this means having quantitative information about the impact of their decisions on outdoor environments, but also inside buildings. This requires being able to simulate the ensemble of urban energy processes accurately and producing compelling results for urban stakeholders. To do so, we need urban models able to capture the complexity of cities and calculations methods specifically adapted to the urban context.

Using Finite Element Methods (FEM) to simulate the urban energy exchanges is a promising approach towards this direction. The Finite Element Method (FEM) was originally developed by engineers in 1956 to solve mechanics problems in complex structures (Turner et al. 1956). Over time, applications of this numerical method multiplied beyond the structural analysis, and nowadays include mass transport, fluid flow, electromagnetic problems, and heat transfer problems. Today, the use of FEM for the resolution of thermal problems at the urban scale is possible using the current hardware capabilities.

Architects and urban designers need detailed information about the energy behavior of the building skin, as the interface between indoor and outdoor environments. This means essentially solving the heat conduction problem through the solid in-between both environments, a classical problem in the FEM field. Conduction fluxes through the building envelope are the result of the radiative and convective energy exchanges taking place over the building skin. The radiative part can be straightforwardly treated in FEM using Radiosity since both techniques share the same prerequisite: the need of a meshed geometry. The use of 'extended form-factors' allows for integrating not only diffuse but also specular reflections in Radiosity algorithms (Sillion and Puech 1989). Given the low inter-visibility between building surfaces, sparse matric representations (Aguerre et al. 2017) and importance-driven algorithms (Aguerre et al. 2019a, b) could help to reduce computational costs.

Through appropriate design, architects and urban planners can effectively control the magnitude of radiative and conductive fluxes to improve outdoor and indoor environmental conditions. Hence, simulation tools should always ensure an accurate calculation of these two energy exchanges. Convection-related aspects, on the contrary, rarely influence urban design (except for the case of high-rise buildings). Therefore, on a first stage, estimating convection fluxes based on sensible convective heat transfer coefficients may be accurate enough for FEM simulations. On a second stage, simulations of the buoyancy-driven ventilation could help designers to modify their projects to enhance indoor and outdoor summer comfort. Coupling FEM and CFD methods would be necessary to assess the impact of cross-ventilation and wind flow on outdoor and indoor spaces (Blocken 2015).

Despite the great potential of FEM simulations for the study of urban thermal processes, it is not an operational working method in their current state. The first steps to becoming a useful tool for designers include the development of:

- A specific framework adapted to the particularities of the work at the urban scale and compatible with designers' logic.
- Case studies including urban energy-responsive designs where FEM simulations help in the decision-making (in a way not achievable with traditional tools).

17.4 Simulating Radiative Exchanges in a Dense Urban Environment

Energy simulations should be able to help urban designers in one of their main challenges: the design of dense and energy-efficient cities, both future and existing. Energy studies are especially complicated in dense urban environments, mostly due to the complex radiative interactions taking place between buildings. Knowing the actual urban geometry is key for accurate radiation calculations and for conceiving feasible urban interventions. Section 17.4.1 explores how to develop urban models suitable for FEM analysis and coherent with the architectural logic. Section 17.4.2 discusses the impact of the level of detail of urban simulations, regarding not only the model definition but also the consideration of the energy exchanges. Section 17.4.3 presents three examples of urban interventions to enhance the energy performance of dense urban environments that require a thorough assessment of radiative exchanges. These three questions will be illustrated based on two case studies located in dense historical districts: one in Bayonne (France) and the other in Cordoba (Spain). See (Garcia-Nevado 2019) for more details.

17.4.1 Considerations on the Urban Modeling for FEM

City models are basic work tools for design and energy analysis, two domains that should be strongly related, especially at the urban scale, where the quantification of phenomena is less intuitive for architects. However, models developed for one of these applications are very difficult—sometimes impossible—to reuse for the other one. Usually, the energy analysis of an urban project starts with the remake or reprocessing of the urban model (Panczak et al. 1998). Therefore, simulation results are hardly integrable into the 3D models managed by urban designers and, hence, in their decision-making process (Billen et al. 2014). This situation fosters the persistent gap between urban energy knowledge and urban planning (Mills et al. 2010). In this section, we aim to contribute to overcoming that limitation by developing a framework for the generation of urban models that can be useful for the design of an urban environment and, at the same time, for its energy assessment with FEM.

The basic prerequisite for FEM is meshing. For 3D problems such as ours, mesh elements are polyhedrons. In practice, the two most common types are tetrahedrons and hexahedrons, although there are no theoretical constraints in this regard for the

FEM implementation (Roca Navarro 2009). Here we provide some arguments based on the architecture reasoning to tilt the balance in favor of hexahedrons.

The more basic one is the recurrence of the right angle in buildings, resulting from the intersection between vertical and horizontal planes. Besides, unlike tetrahedrons, hexahedrons allow for obtaining structured meshes, more coherent with the construction logic. Additionally, the choice of structured hex-meshes has two additional advantages from the computational point of view: a more efficient information storage system (Lo 2018) and a reduction in the minimum number of elements needed to represent a certain entity (e.g., one hexahedron instead of five tetrahedrons to model a single brick). Both aspects may be important in the context of urban studies, which can easily exceed a hundred thousand of elements for a single street.

The main architectural elements (such as roofs, walls, floors or windows) often comprise several layers, with different thicknesses and material properties. The use of hexahedrons eases the analysis of their separated behaviors, being possible to control the number of elements within the thickness to study its transversal temperature profile in detail. The layered-configuration of buildings may arise two issues concerning FEM mesh: the occurrence of strongly deformed mesh elements (e.g., window glass pane of 1000 × 1000 × 6 mm) and the presence of adjoining mesh elements differing greatly in size. Within the FEM community, there are some techniques fully implemented to deal with these problems, typically, the combination of 3D solids with shell elements standing for the thinnest built entities.

An important geometrical aspect of 3D models for FEM is the compatibility of the mesh (also called congruence or conformity). In principle, meshes for FEM have to be compatible, meaning that adjacent elements must share the contact face completely (Lo 2018). Over the time, difficulties in meeting this requirement in the boundaries of certain problems (e.g. cracks) led to the development of techniques to deal with non-compatible meshes in FEM [e.g., using the gluing method, developed by Quiroz and Beckers (1995)]. Then, if the full congruency of the model is not mandatory, what level of congruency the urban model should meet? A good compromise between the architectural and FEM logics seems to be to respect the mesh conformity rule at the building level but not at the urban block level. This choice reproduces a typical dynamic of urban developments, where blocks are usually a 'collage' of different buildings. Additionally, this approach facilitates the modeling process and avoids re-meshing the model after changing or eliminating a particular building.

The final question about FEM mesh generation regards the 'discretization philosophy', that is, the rules driving the mesh of volumes. A common workflow in FEM consists of meshing a geometry previously modeled in CAD using automatic mesh generators [e.g., Gmsh by Geuzaine and Remacle (2009)]. These standard tools allow for discretizing arbitrary geometries into a high-quality mesh that can be refined according to the user's preferences, all at a reasonable time-cost (minutes). For architectural applications, it would be convenient to use ad hoc mesh generators developed upon the architectural logics to ensure the further readability of the simulation results. Unfortunately, such tools are not yet available. These tools should desirably allow for defining an initial mesh based on basic architectural principles that could be post-refined preserving its architectonic coherence if needed:

- Buildings consist of an assembly of vertical and horizontal planes (leaving aside their roofs). These planes could act as cutting planes, slicing the whole building. The resulting parallelepipeds could be beams or pillars in actual buildings, commonly associated with thermal bridges. Similarly, the mesh could incorporate additional structural elements.
- Openings for windows and doors are weak spots from the structural point of view, being frequently reinforced by elements such as lintels and sills, or even framed. This meshing proposal opts for the latter option, as the more general case. To comply with the congruency prerequisite, the vertical and horizontal planes of the frame will re-subdivide the model.

Following the above-described framework, Acuña et al. (2018) generated an urban model for of Tonneliers St, placed in the historical district of Petit Bayonne. Figure 17.4 depicts the elevation and cross-section of several of its buildings. The model succeeds in reproducing the main building features and material changes (timber-frame structure, granite arcades and windows) respecting the mesh congruency for each building. This model, manually developed using CAD tools, was the base for several thermal and daylight studies (Bugeat et al. 2019; Duport et al. 2019). In this sense, this work constitutes the proof of concept of the method proposed. However, implementing it in real urban projects is still un-realistic since the hand-by-hand modeling is too time-consuming. Procedural techniques could be a promising way to explore in this regard. The grammar rules used by these methods, usually conceived to generate surface models, could be adapted to generate volumetric urban models, intentionally discretized for FEA. Additionally, these techniques would allow for including "plausible" building interiors into the urban model, something hardly achievable by hand.

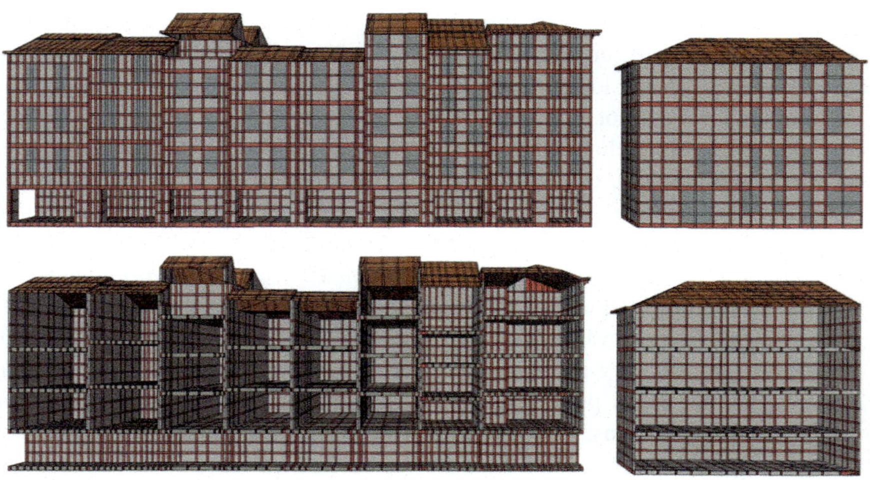

Fig. 17.4 Elevation and cross section of the urban model Tonneliers St. conceived for FEM (Acuña et al. 2018)

17.4.2 *Level of Detail for Radiative Studies*

The geometric information of an urban model can be enriched with non-spatial data, often referred to as 'semantics'. The amount and quality of geometric and semantic information included in a model define its Level of Detail (LoD) (Biljecki et al. 2015). This aspect plays a key role in the simulation of the radiative exchanges in urban environments, both in shortwave and longwave, especially in dense geometries.

Shortwave studies

Estimations about the solar potential of urban surfaces are especially sensitive to the geometrical definition of the model in dense urban environments. Differences in height between built elements strongly affect the magnitude and spatial distribution of solar irradiances over urban surfaces. The vertical irregularities of cities have a higher impact in occluded environments and for low solar elevations since buildings may easily block sunrays (i.e. in winter and/or high latitudes). Studies based on uniform-height models generally overestimate the solar potential over roofs (Beckers et al. 2010). Within urban canyons, especially if they are deep, this deviation is more complex to assess and requires a case-by-case analysis. To illustrate this aspect, we show here the solar irradiances received by urban surfaces of Tonneliers St. on a sunny spring day, obtained from simulations under several LoD scenarios (Fig. 17.5). Results show that simulations based on uniform-height models may overestimate

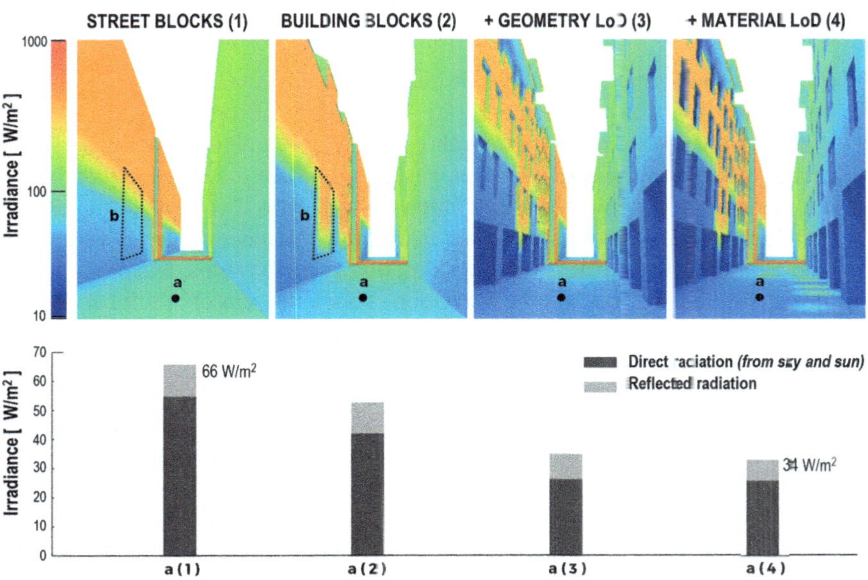

Fig. 17.5 Solar irradiance over Tonneliers St. surfaces simulated in radiance for four different LoD on April 23 at noon. From 1 to 3: an average reflection coef. for all urban surfaces. Simulations 4: individual reflection coef. for each surface

irradiances in some areas (e.g., +20% in point 'a' over the ground) and underestimate them in some others (e.g., zone 'b' at the lower part of the south façade).

The setbacks or projecting elements of façades also have a remarkable impact on street solar gains. The main effect of this 'horizontal roughness' is a reduction in the direct solar radiation impinging over urban surfaces, now in shadow. This effect is especially noticeable with high solar elevations (in summer and/or low latitudes), but still significant in the intermediate seasons. See, for instance, the extent of the shadows cast by overhangs, arcades, and window setbacks over Tonneliers St. façades on a spring day (Fig. 17.5-3). These elements also reduce the sky view factor of urban surfaces, therefore limiting the diffuse radiation they receive from the sky vault. In deep urban canyons, the diffuse contribution may represent a major fraction of the solar energy received through the year over some street surfaces (e.g., North façade or ground). For instance, the solar irradiance on the point 'a' decreases by up to 33% due, mostly, to roof overhangs (Fig. 17.5-3).

Setting the appropriate geometrical LoD of the urban model allows for obtaining accurate predictions just on the direct radiation received by outer urban surfaces from the sun and the sky. However, to obtain accurate estimations of their total solar gains, urban models should also incorporate information about the optical properties of their surfaces: reflectance (ρ), transmittance ($\mathcal{T}$) and absorptance (α, being $\alpha = 1 - \rho - \mathcal{T}$). This is a challenging task because urban environments are poorly documented, extremely heterogeneous, and hard to study through fieldwork. Especial attention should be given in compact urban environments to the characterization of façades, sensitive surfaces regarding indoor-outdoor energy exchanges. Two important aspects in this regard are:

- First, the modeling of windows, which essentially means differentiating the translucent surfaces of the model ($\mathcal{T} \neq 0$) from the opaque ones ($\mathcal{T} = 0$). Windows are key elements for the energy behavior of indoor and outdoor spaces, usually poorly represented—or even absent—in urban models. Frequently, they are modeled by setting a certain window-to-wall ratio. For most of the existing built environments, though, there are no reliable reference values of this parameter (Garcia-Nevado et al. 2017). Besides, this approach helps to approximate the glazed surface area but not their position. This latter aspect is critical in dense urban environments, because of the extreme variability of solar fluxes over the façade (Fig. 17.5-4).
- Second, the determination of the actual reflectance of urban surfaces. In highly reflective dense environments, reflections account for a major part of surface solar gains (e.g., Mediterranean whitewashed streets). Even in environments with medium reflectivity, like Tonneliers St., the reflected component is still non-negligible. In our example, inter-reflections provide up to a third of the solar radiation received at the base of the canyon (Fig. 17.5-3). In the last decades, remote sensing from satellites and aircrafts allowed for obtaining relevant information with increasing resolution and spatial coverage about the reflectance of urban surfaces, particularly the horizontal ones. However, there is a gap in research

focusing on building facades, fostered by the current lack of close-range unexpensive and replicable methods for recovery of façade reflectance (Fox et al. 2018).

Longwave studies

The net longwave (LW) radiative balance between two surfaces depends on their temperature, emissivity, relative areas and positions. In an urban context, surfaces exchange longwave radiation among them and with the sky. Despite our good understanding of underlying physics, urban longwave exchanges are usually modeled in a simplistic way. After reviewing the most common building energy and urban microclimate simulation packages, Evins et al. (2014) concluded that none of them allows a straightforward assessment of the final impact of longwave exchanges on the energy balance of exterior building surfaces. The final impact of longwave radiation on building energy exchanges is smaller than the one of shortwaves. Still, they may have significant effects on indoor conditions and, therefore, on building energy needs (Yang et al. 2012).

The accuracy in the simulation of LW exchanges in urban contexts depends on three main aspects. The first relates to the modeling of the LW exchanges between the built surfaces and the sky. Simulation tools usually calculate them based on their sky view factor and assuming an isotropic temperature distribution over the sky vault, therefore, of LW radiances. Non-isotropic LW sky models are also available in literature (Nahon et al. 2019) and their use could be important for some applications (e.g. design of radiative cooling systems). The choice of a particular LW sky model may also have significant effects in the estimation of heating and cooling annual energy demands, especially critical in dry climates (Evangelisti et al. 2019).

The second key topic regarding urban LW simulations is the estimation of the surrounding surface temperatures. The self-dependence of temperatures in an urban scene requires coupling the thermal the radiative models until convergence, an iterative process that can be very time consuming. To avoid it, standard energy simulation tools make rough estimations on the surface temperatures of the surrounding environment by setting them equal to the air temperature or taking the temperature estimated in the previous time-step without any converging iteration (Frayssinet et al. 2018). These oversimplifications lead systematically to underestimations of cooling loads and overestimations of heating loads (Bouyer et al. 2011). In this regard, Evins et al. (2014) found that improving the accuracy of longwave simulation over the exterior surfaces substantially decreased the heating demands and increased the cooling demands for the same urban configuration.

Preliminary tests seem to indicate the need for detailed longwave simulations also inside buildings. To illustrate this point, we studied the thermal behavior of the concrete building in Fig. 17.6, assuming it is located in Bayonne, France (43.5 °N). FEM simulations were run in Cast3m (Charras and Di Paola 2011) for 5 days with 1-hour time-steps. Exterior surface temperatures were simulated accounting for convection exchanges (in a simplified way, based on hc coefficients), radiation exchanges with the sky and the ground (using Radiosity algorithms for SW and LW),

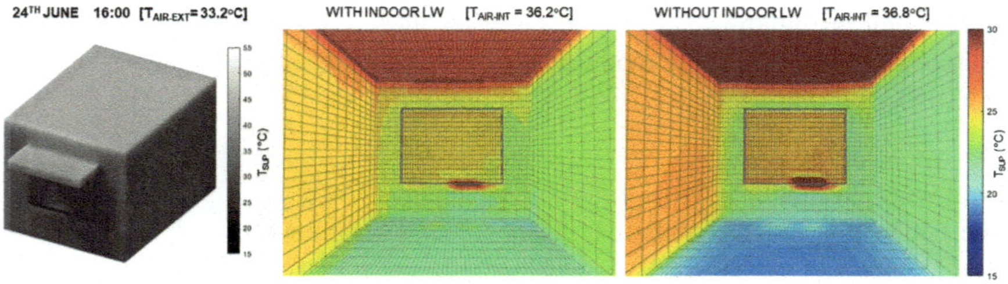

Fig. 17.6 Indoor surfaces temperatures with (left) and without (right) accounting for the indoor LW radiative exchanges

and conduction through the building envelope. As for inner surfaces, simulations were run under two scenarios. On the right, the surface-to-air energy exchange occurred just by convection, whereas on the left, exchanges by thermal radiation were also included.

Figure 17.6 depicts the simulation results on June 24 at 16:00 ST. Results show that disregarding longwave exchanges between the inner building surfaces leads to noticeable errors in the estimation of the inner surface temperatures. Differences are especially significant over the ground (up to 7 °C at that time) due to its high view factor of the ceiling, the warmest surface. Results show that longwave exchanges tend to homogenize inner surface temperatures. Disregarding indoor longwave exchanges also affect air temperatures. Differences in air temperature are smaller than for surface temperature, but non-negligible (0.6 °C at that time). Errors in the estimation of indoor surface and air temperatures affect indoor comfort and building energy use for cooling and heating. These results highlight the need for a complete radiative calculation, accounting exterior and interior LW and SW fluxes in detail.

The last major issue in LW simulation regards the emissivity properties of materials. Since the emissivity of most common construction materials is close to 1, urban surfaces are often considered as black-bodies, so that infrared reflections are not considered. Urban surfaces are also recurrently model as grey-bodies, assuming they behave as perfect diffusers with equal emissivity ($\varepsilon < 1$), and hence reflectivity, for all directions (Harman 2004). Up to date, no energy simulation tool allows for modeling the directional variation of emissivity exhibited by real materials, nor the specular component of LW reflections. These aspects hinder comparisons between simulations and urban thermographies, where grazing angles usually appear (Garcia-Nevado 2019).

Models for thermal simulations require a richer semantic LoD than solar analyses since not only the surface properties but also volume ones need to be known. The optical properties of urban surfaces both in shortwave (absorptance αSW) and longwave (emittance ε) should be defined to calculate the net radiative balance. The thermal conductivity (k, in W/m K), density (ρ, in kg/m^3) and specific heat (c, in J/kg K) of built volumes are needed inputs to account for the conduction fluxes. Increasing our—still limited—knowledge about the thermal and optical properties of urban environments is key to reduce the uncertainties regarding the results of thermal simulations.

17.4.3 Coupling Thermal and Light Simulations for a Systemic Energy Analysis

This section presents three urban interventions to test the potential of the proposed approach. These case studies echo several common proposals of urban designers, whose energy effects are hard to assess with the standard simulation tools. All the interventions require a systemic energy analysis, including thermal and light considerations, to provide specific design guidelines about their implementation in a certain urban context.

Changing the street reflectance

One of the most basic urban intervention consists of changing the visual appearance of an urban environment by painting urban façades, changing their outer cladding, or renewing the paving. Aesthetics or historical arguments often guide project decisions about the finishing of the renewed street. As this choice affects the reflectance of surfaces, designers should also be able to weigh energy-related arguments (Beckers et al. 2020).

In dense urban environments, refurbishment projects often entail repainting the opaque surfaces of streets and light-wells, usually in lighter colors. Beyond the aesthetical goals, the main aim of this operation is to improve daylighting conditions inside buildings by increasing the expected illuminance levels. The visual outcomes of the intervention are instantaneously perceivable and correspond to users' and urban designers' expectations. However, in common practice, urban stakeholders often overlook their side effects on the thermal behavior of public spaces and buildings. To discuss this latter aspect, let's take a look again at the example of Tonneliers St.

In previous work, Bugeat et al. (2019) demonstrated that painting street façades white substantially could improve the daylight conditions inside buildings. This study showed that increasing their average reflectance from 52% (actual value) to 85% (renewed value) resulted in a remarkable rise in their daylight autonomy (e.g., the DA300 raised from 18% to 45% on the first floor). Here, we explore the thermal consequences of this project decision. To this end, street surface temperatures were simulated in its current and renewed states using FEM in Cast3M [more details on the method in Aguerre et al. (2019a, b)]. Figure 17.7 shows the results obtained over the north and south façades at four different times of the day on April 23, 2017, a sunny spring day with air temperatures between 9 and 20 °C. Results evidence that the change in urban façade reflectance affects significantly the thermal response of the street. Simulations indicate that, after the renovation, both façades would be cooler all day long. Differences between the renewed and current scenario are especially important after midday and, particularly, on south facing buildings (up to −12 °C at 15:00 ST). Despite the relatively low thermal inertia of these façades, this effect

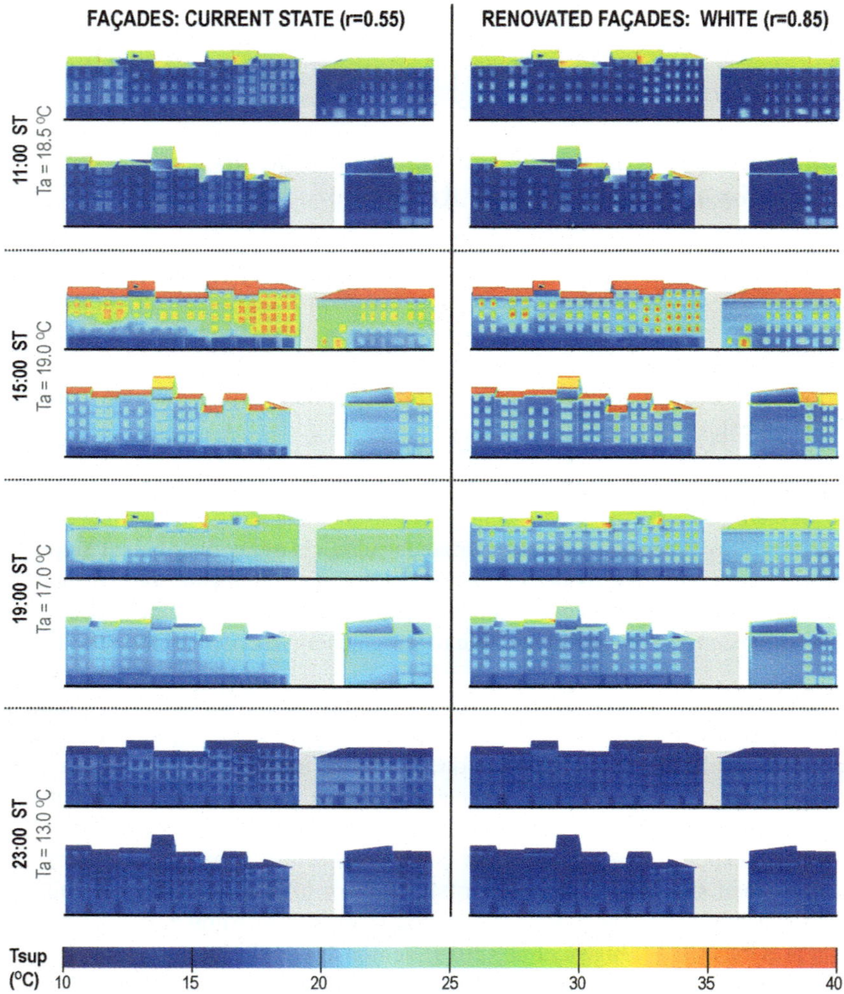

Fig. 17.7 Simulated surface temperatures over Tonneliers St. façades in the current and renovated states on April 23 at different times of the day

remains even when surfaces are not directly sunlit (up to −6 °C at 19:00 ST), even being perceptible several hours after the sunset (up to −2 °C at 23:00 ST).

These results allow for illustrating the opposing effects on building heat gains of painting façades in white. The increase in solar gains through windows overlaps to the reduction in building heat gains by conduction through the opaque surfaces. FEM simulation conducted here accounted for the combined effects of radiation, conduction and convection fluxes over the urban surfaces for a particular day. However, to guide project decisions, it is necessary predicting the global impact of this operation on the thermal conditions inside buildings over the year. Today, annual simulations in FEM start to be a realistic possibility, at a reasonable computational cost.

Strategic placing of mirrors

The strategic placement of mirrors to redirect radiation into dim-lit spaces could be an interesting case study to test the approach proposed here. This kind of intervention is mainly implemented in buildings with deep ground plans, common in dense residential areas or massive mono-functional buildings, such as hospitals or museums (Fig. 17.8). In these cases, well-designed light-wells may constitute a complementary source of daylight that can improve users' quality of life and reduce building energy consumption (Bugeat et al. 2020).

Based on measurements on scale models, Kristl and Krainer (1999) demonstrated that the best performing light-wells in deep residential buildings under overcast sky conditions were those with wide upper and narrow lower parts into which a metallic reflective wall was placed. Even in light-wells with a uniform section, only covering walls with mirrored surfaces could substantially increase the daylight autonomy in dwellings. By adding reflective heliostats, it is possible to maximize the solar radiation entering the light-well, increasing its efficiency. According to Song et al. (2018), this could be a cost-effective measure in large buildings.

Bringing more light into dim-lit spaces also means increasing their thermal loads. While the lighting contribution in these spaces is always beneficial, the thermal one may be welcomed or not depending on the time of the year. In this regard, it is important to bear in mind that reflected radiation might represent a significant energy contribution, especially when using concave-shaped reflectors (see for instance, the damages caused on cars and street furniture by the glazed high-rise buildings Vdara in Las Vegas and 20 Fenchurch Street in London). These examples evidence the

Fig. 17.8 Example of mirrored surfaces in architecture (from left to right). Reflectors in the patio of Sabadell Hospital, by Estudi PSP Arquitectura (*Credits* http://espaciosolar.blogspot.com/). Aluminium net in the metro station of Fulton Building of New York, by James Carpentier (*Credits* https://iva.velux.com/). Mirrored light-well in Museu Blau of Barcelona, by Herzog and de Meuron (*Credits* Carlos F. López Ordóñez)

need for a comprehensive assessment of mirror-based proposals, including accurate thermal-lighting simulations over realistic geometries. This is precisely one of the strengths of the proposed FEM approach.

The scope of projects including the installation of reflectors, though, rarely goes beyond an individual building or pursuits an urban goal. The heliostats installed in the mountains to redirect light to the central squares of Rjukan and Viganella—two small villages in Norway and Italy, respectively—are exceptions to the rule. To our knowledge, there are no studies at a neighborhood scale about the applicability of mirror-based interventions as an extensive strategy of an urban renovation project. This kind of analysis could be justified in historical dense urban tissues, where the possible economies of scale could make this renewal proposal more cost-effective. The urban scale could make affordable the use of selective reflectors (able to reflect the radiation in the visible part of the spectrum while transmitting infrared wavelengths or conversely, called cold and hot mirrors, respectively). The approach proposed here would allow for realistic studies over complex urban models from the visual and thermal points of view.

Urban shading devices

The use of removable textile shading devices is an interesting way of providing shade at street level, especially for locations with hot summers but cold winters, and suffering water shortages. In Mediterranean cities, usually affected by these constraints, textile sun sails appear recurrently as a street canopy placed at different heights (Fig. 17.9).

Several research works confirmed the benefits of installing street sun sails for urban comfort (Paolini et al. 2014; Kántor et al. 2018). Canopy shading devices limit the solar radiation coming into the canyon and impinging on pedestrians. Urban surfaces, now shaded, remain cooler, therefore emitting less longwave radiation (Garcia-Nevado et al. 2020a, b). This combined reduction in shortwave and

Fig. 17.9 Urban sun sails in: Traperia St of Murcia in 1905 (Credits: Twitter RaicesMurcianas), Sierpes St. in Sevilla in 1918 (*Credits:* blogtoldoschamorrosevilla.es), Preciados St and Arenal St in Madrid in 2018 (Credits: Carlos F. López Ordóñez)

longwave fluxes leads to a decrease in the mean radiant temperature (MRT), the most influencing factor on pedestrian comfort under heat stress (Kántor and Unger 2011; Shashua-Bar et al. 2011). When installed over a sufficient extent, these shading devices may also have an impact on air temperature. This effect is less significant and more difficult to assess than the one over radiative fluxes due to advection phenomena. Measurements in Shashua-Bar et al. (2011, 2009) showed slight increases in air temperature (+1 K) within a courtyard sheltered by a dark-colored mesh in Israel. Simulations in Paolini et al. (2014) predicted slight decreases in air temperature (−1.5 K) within a shallow urban canyon in Milan covered by a light-colored mesh. These opposite results suggest that the global outcomes of this intervention are dependent on the geometry and material of both the built environment and the shading devices, aspects interacting in a non-obvious way.

Though the main reason to install street sun sails is to improve pedestrian comfort under heat stress, these devices may also have a positive impact on indoor thermal conditions of buildings, and therefore, in their cooling consumption. So far, studies quantifying the latter effect extremely scarce. Potential of textile canopy shadings to reduce cooling demands in Cordoba was investigated for different street and sun sails configurations in Garcia-Nevado (2013). Simulations in this work showed reductions in cooling demands ranging from 18 to 45% due to the street sun sails. These results only reflect the effects of sun sails on cooling consumption due to the reduction in building solar gains, while the LW effects were roughly accounted for, limited by the capabilities of the simulation tool used (DesignBuilder). Due to the limitations of the simulation tool used for this study (DesignBuilder), results only account for the sun sail effects on cooling consumption concerning the reduction of solar gains, while the impact of longwave exchanges was just roughly estimated.

The installation of sun sails alters the street radiative balance in a non-intuitive way. On one hand, sun sails reduce the amount of solar radiation impinging on buildings, and hence being absorbed by their opaque surfaces (Garcia-Nevado et al. 2020a, b) or transmitted through their windows. On the other, the presence of sun sails limits the radiative cooling towards the sky vault of the sheltered surfaces. This latter effect may reduce the sun sail cooling potential, but standard energy simulation tools do not allow for quantifying this aspect. Urban energy analysis with FEM could be a promising approach to assess the multiscale impacts of these urban shading devices, simultaneously, on indoor and outdoor environments. Some interesting questions arise regarding the optimal design of sun sails and its operation: the impact of color (reflectivity) on sun sails effectiveness, the potential of low-emissivity fabrics to improve sun sails performance, the effects of removing the solar protection during the night, etc.

17.4.4 The Needed Feedback Between Measured and Simulated Thermography

A classical scientific method for the validation of simulation results consists of comparing them to measured data. Several challenges arise when it comes to transposing this approach to the study of the thermal behavior of urban environments.

Surfaces temperatures in actual urban environments experience remarkable variations in both time and space (Oke et al. 2017). Nowadays, numerical simulations allow for obtaining information on this regard at a reasonable computational cost, making it possible to study the thermal evolution of urban scenes through time. However, the reliability of simulations results is strongly dependent on the level of realism in the input data to run the calculations. Uncertainties are especially striking in the case of urban environments and concern geometry, material, and boundary conditions.

Field measurements, on the other hand, provide real data about the thermal response of urban environments at a particular time. On-site measurements are considered as an indispensable part of Urban Physics because they are the only research method able to capture the real complexity of the problem under investigation (Moonen et al. 2012). Nevertheless, to provide useful information, measurements need to last over a sufficiently long period and cover an area representative enough, two troublesome aspects in the case of urban environments.

The feedback between measurements and numerical simulations is crucial since both approaches have complementary potentials and limitations. To make this feedback effective in practice, it is necessary to select measurement and simulation methods able to produce comparable results. A combination of FEM simulations and thermography is a promising approach towards this direction. Simulations by FEM can achieve spatial resolutions similar to the thermal images obtained with digital infrared cameras (Beckers et al. 2019). In turn, time-lapse thermography provides spatial information about surface temperatures over an urban area at spaced intervals, comparable with simulation results at different time-steps (Aguerre et al. 2019a, b).

The practical usefulness of energy analysis results will depend, however, on how they are communicated. Throughout history, innovations in representation have led to innovations in the way of conceiving architecture: central perspective, descriptive geometry, solar diagram, visual rendering, etc. (Beckers et al. 2019). Thermography, whether measured or rendered, should help designers to incorporate thermal-related considerations in their projects.

To maximize their impact, thermal information should be presented using representations designers are familiar with. Perspective views, commonly used by architects and urban designers, may result particularly interesting in the urban context of infrared street measurements. The shot of thermography in perspective is an effective strategy to maximize the information obtained in a single image, allowing for the simultaneous view of both facades in compact urban canyons (Beckers and Garcia-Nevado 2019). This approach, unusual in thermographic studies, provides a global vision of the thermal urban scene that eases the understanding of the energy exchanges at a street level. Nevertheless, it increases the observation angle and the distance to the target, two aspects that may hinder the interpretation of thermography results (Garcia-Nevado 2019).

Finally, it is also necessary to develop new graphical tools to help architects and urban designers to understand the thermal behavior of urban environments. An example of these tools is the 'relative-to-air thermography' (Fig 17.10). This representation allows for visualizing the relationship between air and surface temperatures, a key aspect of urban microclimate that is not apparent in traditional thermograph (Garcia-Nevado et al. 2019). By fixing the air temperature as a reference, it is possible to draw more general conclusions on the link between the design of built environments and its thermal response since it is easier to compare urban scenes under different weather conditions. Additionally, this graphical tool provides insights about the urban convective fluxes, affected in direction and magnitude by the surface-to-air temperature difference. This work evidenced that there is room for enriching

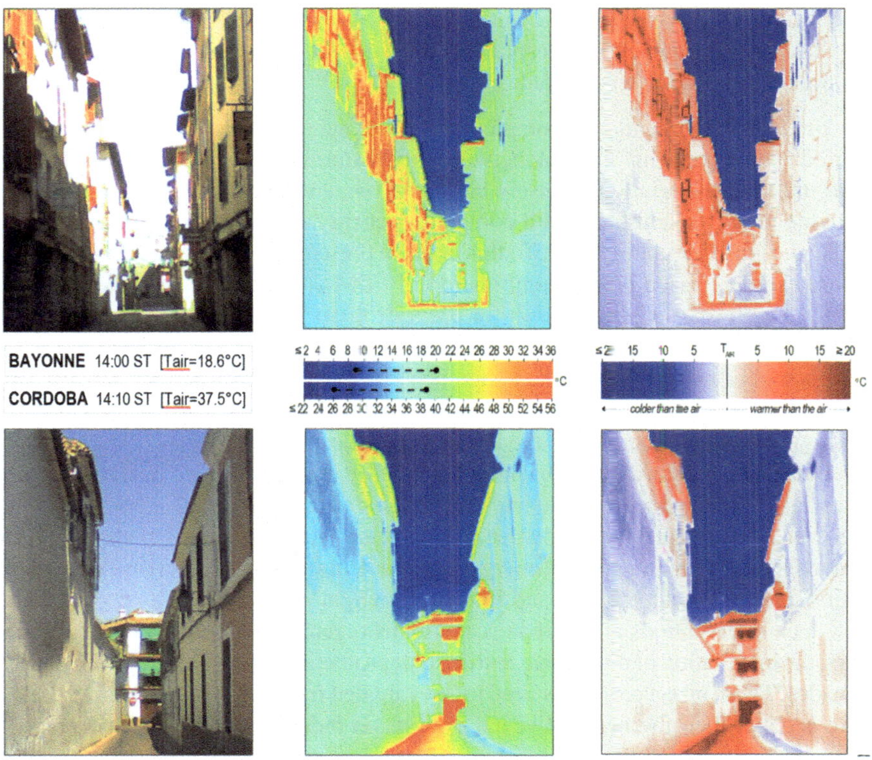

Fig. 17.10 Photography, 'absolute' thermography and 'relative-to-air' thermography in Bayonne, France (23/07/2017) and Cordoba, Spain (25/07/2017)

the insights obtained from the results of urban thermography, whether simulated or rendered, by carrying out additional graphical post-processing.

17.5 Conclusions

The future of our world is urban. The battle for sustainability is likely to be largely won or lost in cities. Urban designers and policymakers should shape cities to respond to the main challenges of contemporary society. In this sense, future cities should be:

- Dense: to host a growing population minimizing the land use while multiplying social interactions.
- Buildable with low-tech: to ensure the best environmental conditions with the minimum resource consumption and promote environmental equity (an especially important aspect in developing countries).
- Designed for a warm (or warmer) climate: to enhance urban comfort minimizing the energy consumption in response to the prospects of global warming and intense population growth low latitudes.

Urban environments are complex and heterogeneous systems that can exhibit varied energy patterns. The link between city features and its energy performance is non-intuitive, especially in densely built environments. Therefore, it is necessary to provide urban stakeholders with compelling energy information that helps them to make design-related decisions. To this end, we presented a new framework for urban energy studies based on the combination of FEM simulations and urban thermography. The former method allows for accurate quantification of energy phenomena on highly realistic urban models, while the feedback with the latter provides a validation tool of results. This mixed approach emphasizes the closeness to reality, and the graphical communication of results, two key aspects to ease the incorporation of energy-related arguments into the urban project design.

Dense cities are trouble environments in terms of energy since occluded geometries tend to impede winter sunrays to penetrate while trapping them in summer. An option to enhance their radiative behavior, and hence their global energy performance is the implementation of energy management strategies to increase winter solar gains, reduce summer loads, and improve daylight conditions. We proposed three low-tech interventions at an urban scale in this regard: the addition of mirror reflectors, the installation of removable textile solar protections, and the repainting of urban façades. These urban proposals need to be evaluated from a systemic approach, taking into account their thermal and visual impacts on indoor and outdoor environments. This requires coupled thermal-light simulations over the actual urban geometries, which are beyond the capabilities of standard energy simulation tools. The framework proposed here, based on FEM simulations and thermographic measurements, will allow for overcoming the current limitations regarding the systemic energy analysis of urban environments.

Acknowledgements This chapter gathers information of published and in-progress works by the authors and other members of the Urban Physics Joint Laboratory of UPPA (Anglet), funded by the Agglomération Côte Basque Adour and the Région Nouvelle Aquitaine. We would especially like to thank the Ph.D. students J. Acuña Paz y Miño, J. P. Aguerre, A. Bugeat and N. Duport for their contribution with some of the figures that illustrate this chapter.

References

Acuña J, Lefort V, Lawrence C, Beckers B (2018) Maquette Numérique d'une rue du vieux Bayonne pour son étude thermique par éléments finis. In: EduBIM2018: Journées de l'Enseignement de la Maquette Numérique et du BIM en France. Clermont-Ferrand, France

Adaku E (2014) Urban sprawl: a view from developing and developed countries. Afr J Geogr Reg Plan 1:193–207

Aguerre JP, Fernández E, Beckers B (2019a) Importance-driven approach for reducing urban radiative exchange computations. Build Simul 12:231–246. https://doi.org/10.1007/s12273-018-0482-4

Aguerre JP, Fernández E, Besuievsky G, Beckers B (2017) Computing urban radiation: a sparse matrix approach. Graph Models 91:1–11. https://doi.org/10.1016/j.gmod.2017.05.002

Aguerre JP, Nahon R, Garcia-Nevado E et al (2019b) A street in perspective: thermography simulated by the finite element method. Build Environ 148:225–239. https://doi.org/10.1016/j.buildenv.2018.11.007

Alexander ER (1993) Density measures: a review and analysis. J Archit Plann Res 10

Angel S, Blei AM, Parent J et al (2016) Atlas of urban expansion

Angel S, Parent J, Civco D et al (2011) Making room for a planet of cities. Policy Focus Rep., Lincoln Inst. L. Policy, pp 1–77

Arnfield AJ (1990) Street design and urban canyon solar access. Energy Build 14:117–131. https://doi.org/10.1016/0378-7788(90)90031-D

Batty M (2009) Defining density. Environ Plan B Plan Des 36:571–572. https://doi.org/10.1068/b3604ed

Beckers B (2016) Multiscale analysis as a central component of urban physics modeling. In: Ibrahimbegovic A (ed) Computational methods for solids and fluids: multiscale analysis, probability aspects and model reduction. Springer, Cham, pp 1–28

Beckers B, Aguerre JP, Besuievsky G et al (2019) Visualizing the infrared response of an urban canyon throughout a sunny day. In: Sayigh A (ed) Sustainable building for a cleaner environment. Innovative. Springer, Cham

Beckers B, Garcia-Nevado E (2019) Urban planning enriched by its representations, from perspective to thermography. In: Sayigh A (ed) Sustainable vernacular architecture. Innovative renewable energy. Springer, Cham, pp 165–180

Beckers B, Miño JAP, Lawrence C (2020) How can reflected light modify solar gains in a compact urban district? In: Sayigh A (ed) Renewable energy and sustainable buildings. Innovative renewable energy. Springer, Cham, pp 699–707

Beckers B, Rodríguez D, Antaluca E, Batoz J-L (2010) About solar energy simulation in the urban framework: the model of Compiegne. In: 3rd international congress Bauhaus SOLAR. Erfur, Germany, pp 1–5

Biljecki F, Stoter J, Ledoux H et al (2015) Applications of 3D city models: state of the art review. ISPRS Int J Geo-Information 4:2842–2889. https://doi.org/10.3390/ijgi4042842

Billen R, Cutting-Decelle A-F, Marina O et al (2014) 3D city models and urban information: current issues and perspectives—European COST action TU0801

Blocken B (2015) Computational fluid dynamics for urban physics: importance, scales, possibilities, limitations and ten tips and tricks towards accurate and reliable simulations. Build Environ 91:219–245. https://doi.org/10.1016/j.buildenv.2015.02.015
Bouyer J, Inard C, Musy M (2011) Microclimatic coupling as a solution to improve building energy simulation in an urban context. Energy Build 43:1549–1559. https://doi.org/10.1016/j.enbuild.2011.02.010
Boyko CT, Cooper R (2011) Clarifying and re-conceptualising density. Prog Plann 76:1–61. https://doi.org/10.1016/j.progress.2011.07.001
Bruegmann ROT (2005) Sprawl—a compact history. University of Chicago Press, Chicago & London
Bugeat A, Beckers B, Fernández E (2020) Improving the daylighting performance of residential light wells by reflecting and redirecting approaches. Sol Energy 207(2020):1434–1444. https://doi.org/10.1016/j.solener.2020.07.099
Bugeat A, Fernández E, Beckers B et al (2019) A multi-scale consideration of daylight in a real urban context. In: IBPSA conference, Rome
Burton E (2002) Measuring urban compactness in UK towns and cities. Environ Plan B Plan Des 29:219–250. https://doi.org/10.1068/b2713
Cerdà I (1867) Teoría General de la Urbanización, 1968th edn. Instituto de Estudios Fiscales, Madrid
Charras T, Di Paola F (2011) Utiliser Cast3M: la procedure PASAPAS. Paris
Chen D, Chen HW (2013) Using the Köppen classification to quantify climate variation and change: an example for 1901–2010. Environ Dev 6:69–79. https://doi.org/10.1016/j.envdev.2013.03.007
Cheng V, Steemers K, Montavon M, Compagnon R (2006) Urban form, density and solar potential. In: PLEA 2006—23rd conference on passive and low energy architecture. Geneva, Switzerland
Churchman A (1999) Disentangling the concept of density. J Plan Lit 13:389–411. https://doi.org/10.1177/08854129922092478
Duport N, Acuña Paz y Miño J, Beckers B (2019) Handling 3D model of a street for an urban thermal study with the finite element method. In: IBPSA—international building performance simulation association conference, Rome
European Commission (1990) Green book of urban environment. Brussels
European Environmental Agency (2016) Urban sprawl in Europe, p 140
Evangelisti L, Guattari C, Asdrubali F (2019) On the sky temperature models and their influence on buildings energy performance: a critical review. Energy Build 183:607–625. https://doi.org/10.1016/j.enbuild.2018.11.037
Evins R, Dorer V, Carmeliet J (2014) Simulating external longwave radiation exchange for buildings. Energy Build 75:472–482. https://doi.org/10.1016/j.enbuild.2014.02.030
Ewing RH (2008) Characteristics, causes, and effects of sprawl: a literature review. Urban Ecol 21:1–15
Frayssinet L, Merlier L, Kuznik F et al (2018) Modeling the heating and cooling energy demand of urban buildings at city scale. Renew Sustain Energy Rev 81:2318–2327. https://doi.org/10.1016/j.rser.2017.06.040
Fox J, Osmond P, Peters A (2018) The effect of building facades on outdoor microclimate—reflectance recovery from terrestrial multispectral images using a robust empirical line method. Climate 6:56. https://doi.org/10.3390/cli6030056
Garcia-Nevado E (2013) Toldo urbano: Posibilidades de reducción de la demanda de refrigeración. Master thesis, Universitat Politècnica de Catalunya (UPC)
Garcia-Nevado E (2019) Termografía del cañón urbano: uso de la perspectiva para una evaluación térmica global de la calle. Ph.D. thesis, Universitat Politècnica de Catalunya
Garcia-Nevado E, Beckers B, Coch H (2017) Characterization of façade fenestration for energy studies within the "Eixample" urban tissue of Barcelona. Energy Procedia 122:397–402. https://doi.org/10.1016/j.egypro.2017.07.430
Garcia-Nevado E, Coch-Roura H, Beckers B (2019) The use of thermography to highlight the relationship between air and surface temperatures in urban scenes. J Phys: Conf Ser 1343:012018. https://doi.org/10.1088/1742-6596/1343/1/012018

Garcia-Nevado E, Beckers B, Coch H (2020a) Assessing the cooling effect of urban textile shading devices through time-lapse thermography. Sustain Cities Soc 63(2020):102458. https://doi.org/10.1016/j.scs.2020.102458
Garcia-Nevado E, Bugeat A, Fernández E, Beckers B (2020b) Using textile canopy shadings to decrease street solar loads. In: PLEA 2020—35th conference on passive and low energy architecture. A Coruña, Spain.
Geuzaine C, Remacle JF (2009) Gmsh: A 3-D finite element mesh generator with built-in pre- and post-processing facilities. Int J Numer Methods Eng 79:1309–1331. https://doi.org/10.1002/nme.2579
Golany GS (1996) Urban design morphology and thermal performance. Atmos Environ 30:455–465
Güneralp B, Zhou Y, Ürge-Vorsatz D et al (2017) Global scenarios of urban density and its impacts on building energy use through 2050. Proc Natl Acad Sci U S A 114:8945–8950. https://doi.org/10.1073/pnas.1606035114
Harman IANN (2004) Radiative exchange in an urban street canyon. Bound Layer Meteorol 110:301–316
Jacobs J (1961) The Death and Life of Great American Cities. Vintage Books, New York
Kántor N, Chen L, Gál CV (2018) Human-biometeorological significance of shading in urban public spaces—summertime measurements in Pécs, Hungary. Landsc Urban Plan 170:241–255. https://doi.org/10.1016/j.landurbplan.2017.09.030
Kántor N, Unger J (2011) The most problematic variable in the course of human-biometeorological comfort assessment—the mean radiant temperature. Cent Eur J Geosci 3:90–100. https://doi.org/10.2478/s13533-011-0010-x
Kawai T, Kanda M (2010) Urban energy balance obtained from the comprehensive outdoor scale model experiment. Part I: Basic features of the surface energy balance. J Appl Meteorol Climatol 49:1341–1359. https://doi.org/10.1175/2010JAMC1992.1
Koolhaas R (1995) What ever happened to urbanism?. The Monicelli Press, New York
Kristl Ž, Krainer A (1999) Light wells in residential building as a complementary daylight source. Sol Energy 65:197–206. https://doi.org/10.1016/S0038-092X(98)00127-3
Lo DSH (2018) Finite element mesh generation. Taylor & Francis Group, Boca Raton
Lopez-Ordoñez C, Crespo I, Roset J, Coch H (2020) Urban growth vs density: The case of a low – density and hot desert climate city. In: PLEA 2020—35th conference on passive and low energy architecture. A Coruña, Spain.
Mills G, Cleugh H, Emmanuel R et al (2010) Climate information for improved planning and management of mega cities (Needs Perspective). Procedia Environ Sci 1:228–246. https://doi.org/10.1016/j.proenv.2010.09.015
Moonen P, Defraeye T, Dorer V et al (2012) Urban physics: effect of the micro-climate on comfort, health and energy demand. Front Archit Res 1:197–228. https://doi.org/10.1016/j.foar.2012.05.002
Nagendra H, Bai X, Brondizio ES, Lwasa S (2018) The urban south and the predicament of global sustainability. Nat Sustain 1:341–349. https://doi.org/10.1038/s41893-018-0101-5
Nahon R, Acuña J, Beckers B (2019) Exploring the sky longwave radiance distribution in the French Basque Country, France. In: IBPSA conference
Neuman M (2005) The compact city fallacy. J Plan Educ Res 25:11–26. https://doi.org/10.1177/0739456X04270466
OECD Green Growth Studies (2012) Compact city policies. OECD Publishing, Paris
Oke TR (1981) Canyon geometry and the nocturnal urban heat island: comparison of scale model and field observations. J Climatol 1:237–254. https://doi.org/10.1002/joc.3370010304
Oke TR (1978) Boundary layer climates. Routledge, New York
Oke TR, Mills G, Christen A, Voogt JA (2017) Urban climates. Cambridge University Press, Columbia
Olgyay V (1963) Design with climate: bioclimatic approach to architectural regionalism. Princeton University Press, New Jersey

Panczak T, Ring S, Welch M (1998) A CAD-based tool for FDM and FEM radiation and conduction modeling. SAE Tech Pap. https://doi.org/10.4271/981577

Paolini R, Mainini AG, Poli T, Vercesi L (2014) Assessment of thermal stress in a street canyon in pedestrian area with or without canopy shading. Energy Procedia 48:1570–1575. https://doi.org/10.1016/j.egypro.2014.02.177

Pérez-Lombard L, Ortiz J, Pout C (2008) A review on buildings energy consumption information. Energy Build 40:394–398. https://doi.org/10.1016/j.enbuild.2007.03.007

Pont MB, Haupt P (2010) Spacematrix: space, density and urban form

Quiroz L, Beckers P (1995) Non-conforming mesh gluing in the finite elements method. Int J Numer Methods Eng 38:2165–2184. https://doi.org/10.1002/nme.1620381303

Real Estate Research Corporation (1974) The costs of sprawl: environmental and economic costs of alternative residential development patterns at the urban fringe. Portl Reg Plan Hist

Revi A, Satterthwaite DE, Aragón-Durand F, Corfee-Morlot J, Kiunsi RBR, Pelling M, Roberts DC, Solecki W (2014) Urban areas. In: Climate change 2014: impacts, adaptation, and vulnerability. Part A: Global and sectoral aspects. Contribution of Working Group II to the Fifth Assessment Report of the Intergovernmental Panel on Climate Change. Cambridge University Press, Cambridge, UK & New York, USA, pp 535–612

Roca Navarro X (2009) Paving the path towards automatic hexahedral mesh generation. Universitat Politecnica de Catalunya (UPC)

Seto KC, Dhakal S, Bigio A, Blanco H, Delgado GC, Dewar D, Huang L, Inaba A, Kansal A, Lwasa S, McMahon JE, Müller DB, Murakami J, Nagendra H, Ramaswami A (2014) Human settlements, infrastructure, and spatial planning. In: Edenhofer O, Pichs-Madruga R, Sokona Y, Farahani E, Kadner S, Seyboth K, Adler A, Baum I, Brunner S, Eickemeier P, Kriemann B, Savolainen J, Schlömer S, von Stechow C, Zwickel T, Minx JC (ed) Climate change 2014: mitigation of climate change. Contribution of Working Group III to the Fifth Assessment Report of the Intergovernmental Panel on Climate Change. Cambridge University Press, Cambridge, UK & New York, USA, pp 923–1000

Seto KC, Güneralp B, Hutyra LR (2012) Global forecasts of urban expansion to 2030 and direct impacts on biodiversity and carbon pools. Proc Natl Acad Sci U S A 109:16083–16088. https://doi.org/10.1073/pnas.1211658109

Shashua-Bar L, Pearlmutter D, Erell E (2009) The cooling efficiency of urban landscape strategies in a hot dry climate. Landsc Urban Plan 92:179–186. https://doi.org/10.1016/j.landurbplan.2009.04.005

Shashua-Bar L, Pearlmutter D, Erell E (2011) The influence of trees and grass on outdoor thermal comfort in a hot-arid environment. Int J Climatol 31:1498–1506. https://doi.org/10.1002/joc.2177

Sillion F, Puech C (1989) A general two-pass method integrating specular and diffuse reflection. Comput Graph (ACM) 23:335–344. https://doi.org/10.1145/74333.74368

Song J, Luo G, Li L et al (2018) Application of heliostat in interior sunlight illumination for large buildings. Renew Energy 121:19–27. https://doi.org/10.1016/j.renene.2018.01.011

Steemers K (2003) Energy and the city: density, buildings and transport. Energy Build 35:3–14. https://doi.org/10.1016/S0378-7788(02)00075-0

Takane Y, Kikegawa Y, Hara M, Grimmond CSB (2019) Urban warming and future air-conditioning use in an Asian megacity: importance of positive feedback. npj Clim Atmos Sci 2:1–11. https://doi.org/10.1038/s41612-019-0096-2

Tombolini I, Zambon I, Ippolito A et al (2015) Revisiting "Southern" sprawl: urban growth, socio-spatial structure and the influence of local economic contexts. Economies 3:237–259. https://doi.org/10.3390/economies3040237

Turner MJ, Clough RW, MH C, Topp LJ (1956) Stiffness and deflection analysis of complex structures. J Aeronaut Sci 23:805–823

UN-Habitat (2017) Global expert meeting report: planning compact cities: exploring the possibilities and the limits of densification. Urban Planning—quo vadis ? Seville, Spain

UNDP (2018) Human development indices and indicators. 2018 statistical update

United Nations (2019) World population prospects 2019—highlights

United Nations (2018) World urbanization prospects 2018: highlights. New York
van Vliet J (2019) Direct and indirect loss of natural area from urban expansion. Nat Sustain 2:755–763. https://doi.org/10.1038/s41893-019-0340-0
Walch A, Mohajeri N, Scartezzini J (2019) A critical comparison of methods to estimate solar rooftop photovoltaic potential in Switzerland. J Phys Conf Ser 1343. https://doi.org/10.1088/1742-6596/1343/1/012035
Wilson B, Chakraborty A (2013) The environmental impacts of sprawl: Emergent themes from the past decade of planning research. Sustainability 5:3302–3327. https://doi.org/10.3390/su5083302
Yang X, Zhao L, Bruse M, Meng Q (2012) An integrated simulation method for building energy performance assessment in urban environments. Energy Build 54:243–251. https://doi.org/10.1016/j.enbuild.2012.07.042

Chapter 18
Urban Materials for Microclimatic Adaptation. Examples on Water Use for Cooling and Run-off Limitation

Valentina Dessì

Abstract The paper focuses on the element that has a double face in the context of climatic adaptation, the first with respect to the reduction of overheating and urban heat island, the second because it required to be managed in some moments of the year, in particular when heavy rainfalls reach the urban areas. We talk about water, in its various forms and configurations.

Keywords Climatic changes · Climatic adaptation · Thermal comfort · Water based systems · Run-off limitation

18.1 Introduction

There are two themes, faces of the same coin, which force us to make reflections and to foresee ever greater and more widespread interventions for the protection of the territories, cities and in general of the population of our Planet.

On the one hand, as firmly reported in the IPCC report (the UN climate scientific panel) in 2013, the awareness that it is extremely likely that human influence was the dominant cause of the warming observed since the mid-twentieth century. On the other hand, it is becoming increasingly clear that in the next few decades no part of the world will be spared from the effects of climate change, also because, as the IPCC special report says, even if today the structural elements that trigger global warming were modified or eliminated, the climatic inertia of the transient would cause significant changes for years to come, diversified and distributed in a completely unbalanced way in the various geographical areas.

If the issue of climate change mitigation is to be dealt at global level, the adaptation strategies must instead be defined at the local level. Strategies can be known and reinterpreted in order to be suitable for new contexts, or strategies could be proposed that in certain geographical areas are known and widespread, but in other contexts represent solutions to be still promoted and tested; that's because the impacts are

V. Dessì (✉)
Politecnico di Milano, Dept Dastu, Milan, Italy
e-mail: valentina.dessi@polimi.it

G. Chiesa (ed.), *Bioclimatic Approaches in Urban and Building Design*, PoliTO Springer Series, https://doi.org/10.1007/978-3-030-59328-5_18

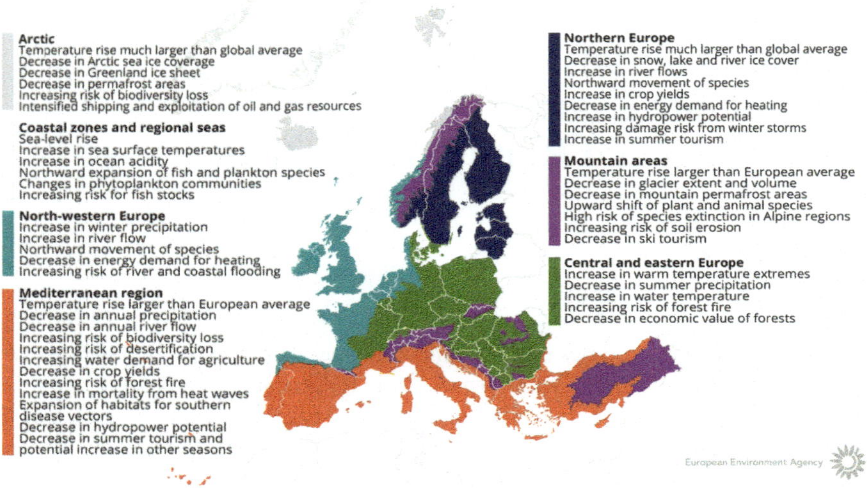

Fig. 18.1 Climate change impacts in Europe. *Source* European Environmental Agency (2017)

not the same everywhere but they change according to the climatic zone and the characteristics of the areas; Europe territory is characterized by the presence of the Arctic, coastal, mountain and Mediterranean areas with lot of climatic peculiarity.

The map in Fig. 18.1 shows a list of possible effects, some common to other areas, others different, and moreover, some typical of rural areas and others more characteristic of urban areas.

In particular, cities, as also stated by the United Nations (2004), are responsible for the consumption of at least 80% of the energy produced, and are therefore among the main culprits of climate change, but also the place where the effects occur in more evident way, especially those related to rising sea levels, the intensification of atmospheric phenomena or the gradual increase in temperatures.

However, cities can be laboratories in which the policies and practices of mitigation and adaptation to climate change are tested. Indeed, due to their ability to influence innovative processes and solutions, local governments can play a strategic role in the way climate change is addressed.

Leaving aside the structure of the institutional frame of the strategies at national and local level both as regards mitigation (Agenda 21, Covenant of mayors, PAES, etc.) and adaptation to climate change (SNACC and PNACC in Italy), it is important to observe the strategies implemented to reduce, on the one hand, the causes of climate change, and on the other, to limit the impacts of the changes, and actually make cities environmentally more resilient.

Focusing the following reasoning on the European context of the Mediterranean area there are at least two effects that are increasingly present and evident, especially in urban areas.

On the one hand it is increasingly clear that living in this age of climate change means experiencing a rise in temperatures that in cities produces an even greater impact due to the urban heat island. Especially in periods of more and more frequent heat waves of increasing intensity and duration, the risks for the normal functioning of the city and the risks for the health of the citizens increase.

The summer of 2003 will be remembered as a particularly hot period in which there were, especially in France and Italy, thousands of deaths. The following years did not go better, on the contrary … every year it is the hottest year recorded ever. However, we have equipped ourselves, adapted, both individually (for example by buying a fan or installing an air conditioner), and collectively, by trying to do it even at the wider level of the city. The latter is a longer process that involves the structure and configuration of the city, and therefore it is realistic that the outcomes are not immediately visible. If it is the city that largely generates climate change, it is now appropriate for the city to change to adapt to its effects.

The second effect is related to the precipitations; it is increasingly common that in short time large amounts of water flow onto the soil. The occurrence of these phenomena in the city has an even more evident impact compared to extra-urban areas, because the urban surfaces are largely sealed, and the water management system is not adequate anymore. Nowadays it happens that, in the event of excessive rain, the roads behave like real river beds. Also in this case a modification of the urban configuration, as far as possible in the consolidated city, is appropriate and must involve the population, the stakeholders as well as the local administrators, because the strategies can be effective at any scale of intervention.

The paper focuses on the element that has a double face in the context of climatic adaptation, the first with respect to the reduction of overheating and urban heat island, the second because it required to be managed in some moments of the year, in particular when heavy rainfalls reach the urban areas. We talk about water, in its various forms and configurations.

18.2 Water for Cooling the City

Hearing the sound of flowing water or seeing moving water from afar anticipates a feeling of freshness; however, when water is introduced into an urban configuration it is not always effective for cooling. Rivers, fountains, pools of water, water spray, intermittent sprays … are just some of the forms of water that we can find in cities and in public spaces, with different characteristic, and performances (Dessì et al. 2016). A monumental fountain has a decorative function but not an environmental role. The systems based on the use of water for cooling are based on two principles: the first is based on the possibility for an urban space for cooling the surfaces with a consequent lowering of the mean radiant temperature (MRT), the other on the water evaporation which leads to a drop in air temperature.

The choice to propose a system rather than another depends very much on the physical characteristics of the space, the functions of the space and the objective:

then we can have a water “band” on the floor that follows a path, a wall of water on a blind façade along a footpath or as a scenography in a small square; or have a portion of water spray inside a large public space. It must be said that every time water is present, it makes a space unique and recognizable, characterizing it unequivocally.

It should also be reminded that different configurations have different performances, both in terms of lowering the air temperature and thermal comfort for people, and they vary according to the volume of water and the covered surface (or rather, the ratio between surface and volume), the movement of water, and the proximity of people.

The water based systems have long been known, and were often used to cool the interior of buildings; perhaps the first experience that can be remembered on the cooling of the external space, above all because it was developed together with a field survey, is the Expo in Seville in 1992, and perhaps from that moment we can talk about strategies for passive cooling of urban spaces. At that time many strategies based on the use of water were used, both for cooling the surfaces and for reducing the air temperature, sometimes the systems were used in combination with other strategies, in any case with a visible thermal improvement especially due to the characteristics of the very hot and dry climate.

When the aim is to improve the comfort conditions of a person who is walking or relaxing for a while, in a resting activity (reading, eating, chatting, observing the flows of people, etc.), by using ponds, “ vertical water blades”, or waterways along the areas with people, it is possible to reach a temperature decrease of the flooring or the vertical elements that the person “sees”, i.e. with which there is an exchange of energy, and therefore a lowering of the MRT (which is obtained by calculating the average of all surface temperatures multiplied by the object and sky view factors). MRT is a fundamental parameter in the calculation of the various indicators of thermal comfort, such as PMV, PET, UTCI, etc. (Höppe 1999) which substantially affects the feeling of environmental well-being.

With the aim to reduce air temperature, can be proposed other systems based on the possibility of water to evaporate, i.e. to change from the liquid state to the gaseous state. This change of state requires a large amount of energy that is taken from the air in the form of sensible heat and it is converted into latent heat. The result of this adiabatic process is a drop in the air temperature and an increase in its humidity. For this reason, this cooling system is particularly effective in dry and hot climatic zones where the higher humidity and the lower temperature can be both seen as advantages.

When finally, we talk about integrated systems, we refer to the combination of different strategies that reinforce the effects of a single strategy, or enhance less effective systems. If we planned to use nebulized water under a pergola with seats or paths, this system would further cools this particular shaded outdoor “room”, as if we allowed water to flow under a porous or dense materials floor, we would help it to lower its surface temperature (Alvarez Dominguez et al. 1992, Fig. 18.2).

In any case, before planning any cooling system based on the use of water, it is advisable to check which environmental conditions exist so that it can be really effective, in particular by checking that the solar radiation values are high, while relative

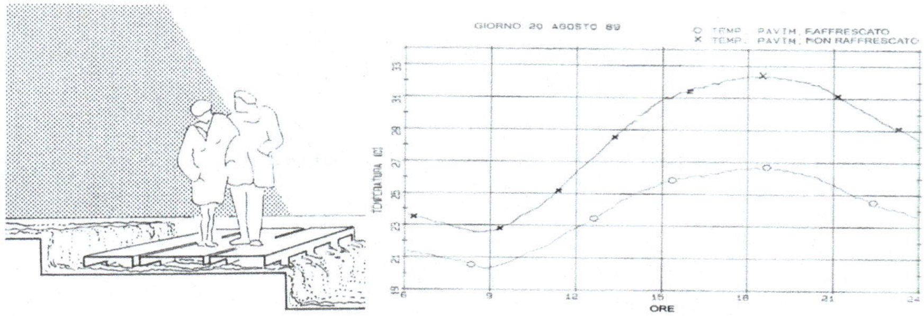

Fig. 18.2 A particular cool pavement, that combine a porous flooring material and water. The graph on the right shows the temperature difference between a conventional floor and a cool one. *Source* Alvarez Dominguez et al. (1992)

humidity values must be low. This means that often the systems have to work intermittently and designed to work when actually needed, and then be connected to monitoring systems of the climatic conditions of that particular moment.

18.2.1 Mist Water: Some Application

Among the different systems based on the use of water, this paper focuses on the nebulization, that is, the supply of thousands of very small water particles that evaporate instantly with the effect of reducing the air temperature in a short time in the place where the system is placed with an instantly evident effectiveness (Alvarez Dominguez et al. 1992). The operation of this type of system consists in "wetting" the occupants of the nebulized space without "showering" them with the drops of water not evaporated. Not evaporated drops of water could fall to the ground keeping it moist, with the risk of producing algae and fungi. Another disadvantage linked to the presence of water onto the ground is that it evaporates without reducing the air temperature but only by increasing the relative humidity values and with a probable increase in thermal discomfort. For this reason, the nozzles that distribute high-pressure water spray should be placed at an adequate height and be able to work intermittently. The water spray can come from the top with air supply systems or fixed to existing structures, or from below, involving the flooring system, or even from "free standing" structures not connected to the limits of the urban space. An example in this sense is the Bioclimatic Sphere, a large sphere with a 22 m diameter, located above a rest area created for the Expo '92 in Seville. This misting system has been realized by installing 1363 micronizers distributed among the 15 circuits (a number that allows a sophisticated regulation) with a water consumption of 10 m^3/h and a production of about 7000 kW (6 million frigories), which correspond to about 3000 small air conditioners running simultaneously (Fig. 18.3).

Fig. 18.3 Bioclimatic sphere of the Expo'92 in Seville spreading mist water from 15 circuits and 1363 micronizers. *Source* Lopez de Asian (1997)

A substantial part of the systems using nebulized water is located along paths or in rest areas, and also in this case, the Seville experience brings us many examples of areas, often shaded and further cooled with water spray, as re-proposed in Fig. 18.4.

Another interesting part is represented by the use of the strategy for resting areas, with the misting system that comes from the top or instead from the bottom, integrated below the paving. In this case the system under the flooring must often be integrated with the underground utilities, especially in those densely urbanized existing cities, where this aspect often represents a disincentive to realize them.

The "miroir d'eau" adjacent to the *Place de la Bourse* in the city of Bordeaux is now very well known. The project of revitalization of Bordeaux urban waterfront was initially aimed only at reintroducing vegetation into the space in order to attenuate the local microclimate. However, and based on the concept of addressing thermal comfort levels and reflecting surrounding facades, the "water mirror", and an incorporated on-site fog system (also based on a micro nozzle system) were installed (Fig. 18.5).

This 130-meter-long and 42-meter-wide set of granite slabs is the largest reflecting pool in the world. Conceived by the landscape architect Michel Corajoud, and developed by the water feature designers J. M. Llorca, "Le miroir d'eau" works according to a specific cycle. When the slabs are almost dry, it looks like a huge slate. A few minutes later, the fountain mechanism produces plumes of mist, creating a thick fog turning the quay in a fairyland. The whole cycle, between the delivery and the emptying of the square, lasts 26 min, and is automatically controlled.

When the fog disappears, the 3450 sm of granite is covered with 2 cm of a clear water, becoming a slightly undulating twin of *Place de la Bourse*. According to J.

Fig. 18.4 A pergola in an open space of the Politecnico di Milano, Milan, with mist water that represent an effective outdoor small cool "room"

Fig. 18.5 View from the top of the "miroir d'eau". *Source* Michel Corajoud

Fig. 18.6 View of the *Place de la Bourse* in Bordeaux from the water mirror. *Source* Michel Corajoud

Llorca, the water mirror was inspired by the regular flooding of Saint Mark's Square Venice (Fig. 18.6).

In order to avoid algae and water wastage, the water that temporarily floods the square recedes back into the slabs after a few minutes, leaving the surface dry like in any other square.

Grooves were installed in-between the granite slabs, to allow the water to be re-collected, and re-prepared for the next induced "flood". In this way, wet surfaces become part of the design of the system that increases the climatic responsiveness of the once thermally problematic public space, although in Bordeaux climate conditions are quite mild, and hot days are limited to few occasions during summer, in which the square is very popular.

18.3 Manage Excess Water Phenomena

Making a city resilient means deploying strategies and development models that guarantee the absorption of the effects of climate change by preventing the city system from going into crisis. In particular, to cope with abundant and excessive rainfall phenomena, it is necessary to provide conditions in which adequate water drainage is possible, even in the case of floods that affect the functioning of the city, and in particular its mobility, both public and private, and, of course, cycle-pedestrian (Croce 2017) . The strategies, known as sustainable drainage systems

(SuDS), a term used in the United Kingdom, known in the United States as low-impact development or Best Management Practice (BMP) and Water Sensitive Urban Design (WSUD) in Australia, are intended to increase infiltration into the soil and reduce and/or delay their entry into the sewer system.

They alleviate those problems as they were designed to store or re-use surface waters at the source. They do this by collecting surface water, even polluted water, for instance water mixed with petrol or oil from the streets, and provide natural pre-treatment of the water before it is slowly released back into the environment. SuDS may include ponds and wetlands, swales, permeable surfaces, filter strips, infiltration trenches and underground storages.

Among the systems that can delay the entry of water into the sewage system we can mention other devices, which are well integrated into the design of the public space, and therefore more compatible with an urbanized structure, namely draining floors, rain garden, gardens and water square.

More and more applications are well integrated in the different types of urban open space, from public space to private space, to school courtyards, and in most cases, the components perform different functions. In addition to those already mentioned, they include the provision of local services and the generation of an important range of habitats vital to local ecosystems. In general, it can be said that water retention structures, allow up to 90% of the infiltration in the ground, and part of this, also thanks to the presence of trees, at least 40% is disposed of by the evapotranspiration mechanisms.

Among the many virtuous cases that can be reported, it is particular important to report a process that is taking place in the United Kingdom and that concerns the work on schools and schoolyards that are prepared, especially in the parts characterized by permeable surfaces, to drain the rainwater and play an important role in case of excessive rainfall. From the environmental point of view, the importance also lies in the fact that schools, especially those related to the early grades, are very spread in the urban context; in Italy, the Ministerial Decree of 18/12/1975, on school buildings, for example, establishes that there should be a primary school within a radius of 500 m from the houses (reachable on foot in 15 min).

A grid that foresees the presence of these "equipped" spots at a short distance from each other means to guarantee in every neighborhood the presence of areas with important functions in moments of criticality that can block the life of a neighborhood, from public to private mobility, from the degradation of materials to the risk for the buildings themselves. Another element that makes this mechanism interesting is related to the involvement of the population, in this case students who learn concepts related to climate change, its effects and strategies to fight them (which can then often be repited in the backyards of their homes), as well as to the study of energy and material flows, and in general related to metabolic flows, of the context in which they live. It is worth mentioning one last aspect linked to the potential of schoolyards, because they often represent one of the few free and pervious spaces of an entire neighborhood. All the movement in the USA for the enhancement of school courtyards and outdoor learning was born precisely from the need to spend time outside for children who often do not have other opportunities (urban areas too built up and lacking urban spaces) or who prefer to spend their free time inside at home.

18.3.1 The Case Study: Melton Vale Post-16 Centre, Melton Mowbray Leicestershire, UK

The case study refers to the treatment with SuDs of an external area belonging to an English school, which is considered a best practice, and for this reason destination of visits and often published, in particular through the dissemination channels of CIRIA, association of the construction industry research and information, and on the "Adaptation inspiration book", often quoted in this paper (Capela Lourenço and Avelar 2013).

The Melton Vale Post 16 Centre (MV16) is a school in Leicestershire in the United Kingdom for students from the age of 16. The site lies on an elevated plateau off the nearest road. It was designed to be sustainable – and it was decided (according to the proposal introduced by the Landscape design team DSA) that the school had to become an instrument in teaching the students about sustainability. Currently the art, geography, biology and chemistry departments use the school environment in their teaching programmes.

The entire site, including buildings, is approximately 2.9Ha. Outside areas of the new college include grassed areas as well as, on the western side of the school, a car park at the service of both staff and students and smaller buildings (Fig. 18.7).

The building location was adjusted to enable the implementation of a solution for reducing the storm water runoff. It consists in a Sustainable Drainage system. In addition to enhancing with the large areas of new grass- and wetland to attenuate storm water, a central detention basin is used that effectively collects rainwater runoff.

A rainwater tank collects water from the roof for recycling in the building, while any excess is directed into the basin.

The car parking and access road drain the water via a swale to the main attenuation basin. Nested wetlands carry storm water to one main detention basin. Especially constructed swales lead the water to a detention basin where specific plant species can filter and clean the water. The installed SuDS also had the added advantage to clean storm water biologically in swales and ponds before being discharged into the river and public sewage system (Fig. 18.8).

The project prevents flooding even under a 100-year rainfall event with an extra allowance for climate change. The water leaves the school site via a gabion headwall. It then connects into the original (and conventionally piped underground) system of the wider school campus, which will receive much less water during an extreme rain event, due to the storage capacity at the school site.

The design process, in 2010, which involved engineer and landscape architect expert in site planning, met three main sustainable drainage principles: quantity, amenity and quality.

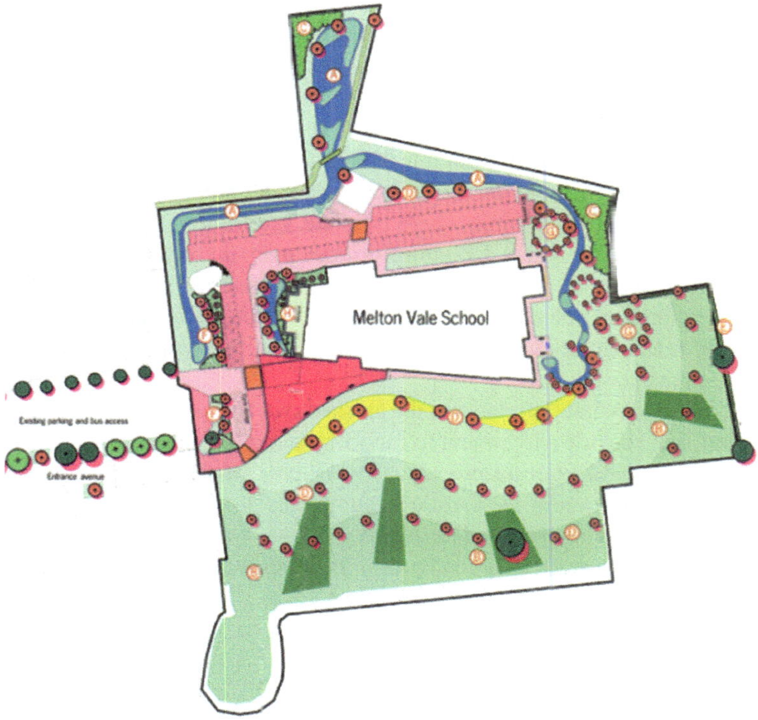

Fig. 18.7 Landscape master plan of the Melton Vale Post 16 Centre. *Source* DSA Environment & Design Ltd

Fig. 18.8 View of the swale in front of the building. *Source* DSA Environment & Design Ltd

According to the quantity principle, prior to construction, the site appeared to have generally soggy conditions, a consequence of the clay ground not presenting high soil-water infiltration potential. The site was designed to discharge runoff directly to a public sewer in a controlled manner. Actually, by using the SuDs system, the site achieved the reduction in discharge from 124 to 62 l/s.

According to the amenity principle, the school staff expressed interest early in the project's development to use the SuDS for educational purposes. It was therefore designed to be open and visible where possible and to be integrated into the site's landscape. Pedestrian access including bridges over the detention basin provides viewing platforms to allow people (including visitors to the site) close contact with the water system and biodiversity. The SuDS are thought for keeping storm water on site and releasing it back into the ground gradually, so people can enjoy it and wildlife can use it.

The quality principle, is related to the grassed swale's gentle 1:20 sloped sides and dry basins, that provide a treatment to runoff by removing sediment and debris from it through its natural filtering process. Downstream of the swales and attenuation basin, a further silt trap in the form of a sump was provided to also prevent debris from entering the public sewer. The delivery of this bio-treatment by the SuDS also meant to avoid a separate oil interceptor to trap polluted runoff entering the public sewer (Fig. 18.9).

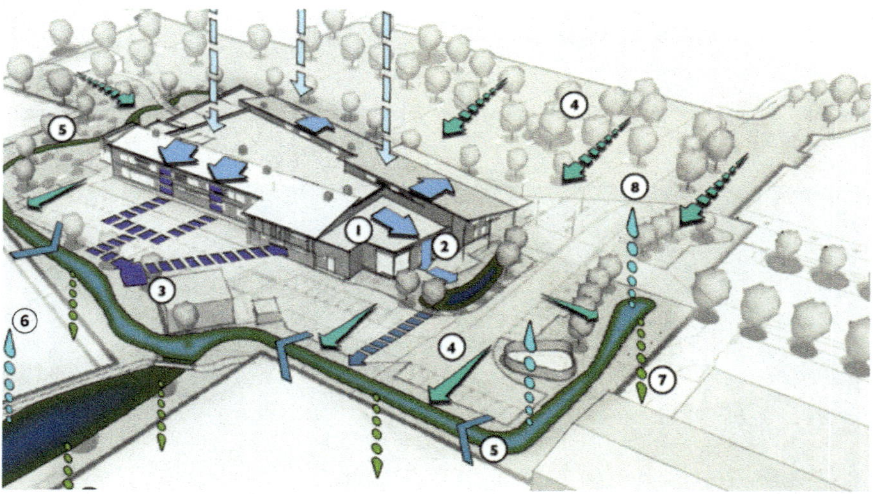

Fig. 18.9 Schematic of WSuDs for Melton Post-16 Centre that combine different strategies. 1. Rainwater is collected on the building rooftop. 2. Rooftop is conveyed via downpipes. 3. Roof water is harvested and reused as "greywater" in the building. 4. Storm water is collected from hard and soft surfaces and controlled through landscape design. 5. Swales convey collected water. 6. Water is held, using flow control, in attenuation basin. 7. Water is encouraged to infiltrate into the natural ground where it recharges groundwater. 8. Plants allow evapotranspiration of water back into the air. *Source* DSA Environment & Design Ltd

The site has been carefully monitored over that time, it is now well established and operating better than expected by the design, since swales are now fully integrated into the landscape giving the whole scheme a very natural aesthetic. The system has actually enhanced and increased the biodiversity across the site.

This project provides the expected benefits of SuDS by slowing the rate surface water run-off reaches public sewers, and in this case it reduced the rate by 50% compared to the site's previous run-off rate before installation. Besides reducing the risk of floods, SuDS are also cheaper in installation compared to conventional underground drainage systems because there is no need for pipework, connections, tanks and gullies and, importantly, expensive excavation for these hard engineering measures. In addition, swales and basins are easier and cheaper to maintain. They do not block easily and any problems (such as spillages of potential pollutants) are generally visible and can be fixed quickly.

From a contractor and client perspective, saving a very large amount of money (circa £114,000) by adopting a SuDS approach was also very beneficial and was achieved by avoiding the use of long lengths of drainage runs, the type of which is typically required in conventional drainage.

To make maximum use of the innovative techniques the curriculum of the school has been modified to allow students to use the SuDS landscape for their studies. This offers students the chance to be involved in the monitoring and evaluation of the SuDS (Fig. 18.10).

18.4 Conclusion

The two types of strategies reported in the paper refer to experiences of local realms that increase the awareness that each of us has on the opportunity to experience and realize a piece in the broader framework of climate change adaptation systems. If a few years ago talking about adaptation at an individual level represented only the possibility of buying an air conditioner for one's home, today there is the awareness that everyone, from the private citizen to the public administration, can play a role for the community to which it belongs.

As mentioned by Capela Lourenço and Avelar (2013), when we started to hear about climate change the tendency to treat them is now recognized as "Al Gore" approach which instils fear of a changing climate into people's hearts. "… knowing that a big threat is looming, but not knowing what you can do about it, leads more frequently to inaction, rather than to action. Notwithstanding the fact that initiatives like the movie 'An Inconvenient Truth' did a tremendous job putting climate change on the (political) agenda, we have learned that for encouraging people to take action, it is better to show them options, preferably as close to home as possible".

It is important to inspire and encourage people and show how their own actions can contribute to the larger goal of adaptation to climate change.

These contributions to reducing vulnerability to climate change can sometimes be small, such as a sidewalk garden as well as a nebulization system above a couple

Fig. 18.10 Students helping to collect data on the performance of the system. The art, geography, biology and chemistry departments use the school environment in their teaching programmes. *Source* https://www.susdrain.org

of benches in a square, or sometimes have a much bigger effect, such as replanting woodlands in large parts of England or connect in a blue-green infrastructure an entire town. In any case the most successful action is that one that involve people in the realization and/or in the experience of environmental resilience of the city in the moment in which the urban structure is under stress, due to summer heat waves or excessive rain fallen in a short time.

References

Alvarez Dominguez S et al (1992) Control climatico en espacios abiertos. El proyecto Expo '92. Ciemat, Sevilla, ES

Capela Lourenço T, Avelar D (2013) CIRCLE-2 adaptation inspiration book. In: Pijnappels M, Dietl P (eds) 22 implemented cases of local climate change adaptation to inspire European citizens

Croce S (2017) Strade verdi, parchi tascabili, coperture e facciate a verde per una città resiliente ai cambiamenti climatici. In: Croce S et al Città resilienti e coperture a verde. Maggioli Sant'Arcangelo di Romagna

Dessì V et al (2016) Rigenerre la città con la natura. Maggioli Ed., Sant'Arcangelo di Romagna

Höppe P (1999) The physiological equivalent temperature—a universal index for the biometeorological assessment of the thermal environment. Int J Biometeorol 43:71–75

IPCC (2013) Climate change 2013: the physical science basis. Working group contribution to the IPCC 5th assessment report of the Intergovernmental Panel on Climate Change. Changes to the underlying Scientific/Technical Assessment. Cambridge, UK
Lopez de Asian J (1997) Espacios abiertos en la Expo '92. ETSAS, Sevilla, ES

Part V
Passive Climate Control Strategies and Technologies

Chapter 19
Passive Means to Improve the Indoor Thermal Comfort

Claude-Alain Roulet

19.1 Introduction

The building is (or at least should be) designed and constructed first to bring a good indoor environment to its occupants. There could be other objectives, such as prestige, image, low cost, energy saving, real estate business, and speculation; but indoor environment quality (IEQ) should have the highest priority (Roulet et al. 2005).

The means for ensuring a good indoor environment can be roughly sorted into two categories; the passive or architectural means and the active or technological means.

Passive ways are architectural and constructive measures that naturally provide a better indoor environment quality without or with much less energy use. They are often cheap, use only local and renewable energy, and are much less susceptible to break down than active means. However, they often depend on meteorological conditions and therefore cannot always fulfil the objectives. They should be adapted to the location and therefore need creativity and additional studies in the design phase, and a design error may have dramatic consequences.

Active or technological ways control the indoor environment by mechanical actions, using energy from the market for complementing the passive ways or even for compensating low building performance. When appropriately designed, built and maintained, they are perfectly adapted to the needs. The architect does not have to take much care of them, since these are designed and applied by specialised engineers according to known technology. Flexible and relatively independent on meteorological conditions, they allow correcting architectural errors. However, the required technology is often expensive, uses much energy and may break down. Furthermore, active means require a higher maintenance input (Table 19.1).

C.-A. Roulet (✉)
Ecole Polytechnique Fédérale de Lausanne, Lausanne, Switzerland
e-mail: claude.roulet@epfl.ch

G. Chiesa (ed.), *Bioclimatic Approaches in Urban and Building Design*, PoliTO Springer Series, https://doi.org/10.1007/978-3-030-59328-5_19

Table 19.1 *Advantages* and disadvantages of passive and active ways to improve indoor environment quality

Passive ways	Active ways
Cheap	Expensive
No energy agents	Use energy agents
Don't break down	May break down
Need careful design	*Easy to design*
Control problems	*Easy to control*

Since they are cheap and energy-efficient, passive ways should be preferred, but they cannot always fulfil the comfort objectives. Therefore, the appropriate strategy is to use them as much as reasonably possible and to compensate for their insufficiencies with active systems, which will then be smaller and cheaper. This strategy often allows more freedom in choosing the type and location of active systems.

19.2 Adaptation of the Building Design to Local Conditions

The adaptation of the building design to local conditions is the base of bioclimatic architecture. The outdoor environmental characteristics (temperature, solar radiation, wind, dust, pollution, noise, etc.) change with the location of the building. Therefore, a design that is well adapted in a place may be completely unsuitable to another one: Bedouin tents, igloos, tropical huts, all well adapted to their environment, cannot be used elsewhere. This is also valid for contemporary building design: it is of course possible to compensate for environmental changes using active techniques, but this often decreases the indoor environment quality and increases the energy use. According to Chatelet et al. (1998), with all active systems swithched off, a building should be at least as comfortable than the outdoor environment.

Adaptation of the building to the environment includes the following:

- Adaptation to climate: Appropriate thermal insulation, solar protections and ventilation openings. An acceptable architecture should ensure that the building is, without any heating or cooling, at least as comfortable as the external environment. In many cases, it is easy to do better by passive means.
- Adaptation to noise: Improve acoustical insulation in noisy areas, for example by using a double skin, and installing mechanical ventilation with sound barriers.
- Adaptation to pollution: Locate air intake as far as possible from pollution sources, install mechanical ventilation with appropriate filters and ensure appropriate maintenance.

Nevertheless, it should be mentioned that clothing is the most natural first step for temperature control. A building management has justified air conditioning because full casual dress was mandatory in the company. Allowing clothing adaptation in buildings certainly improves comfort and may save much energy!

19.3 Models of Thermal Comfort

Thermal comfort criteria include the temperatures of the surrounding air and surfaces, temperature gradients, air velocity and air humidity. However, other parameters such as noise, lighting, and air quality influence the perception of the indoor climate by the occupants.

There are today two models and associated criteria for evaluating the indoor thermal comfort.

The model developed by Fanger (1982) is generally accepted and internationally standardised (ISO 2005). The main parameter is the operative (or perceived) temperature which is the temperature of an isothermal room in which the occupant has the same heat balance by radiation and convection than in the actual room, where the air temperature may present gradients and surrounding surfaces may have different temperatures. This model is based on experiments in climatic chambers, and is therefore valid for conditioned spaces (Note 19.1). It gives the optimal operative temperature for each activity and clothing.

It is now shown that it is not valid in non-conditioned spaces, i.e. naturally ventilated spaces in which active heating or cooling systems are absent or switched off.

In naturally ventilated spaces that are neither cooled nor heated, the occupants are much more tolerant as long as they can freely choose their clothing and act on the indoor climate by opening windows, closing shutters, switching fans on, etc. The so-called adaptive comfort model, developed by Brager and de Dear (1998), McCartney and Nicol (2002) and others is also now generally accepted and internationally standardised (CEN 2019). This model, based on measurements and audits all over the world, links the optimal indoor operative temperature to a weighted average of the outdoor temperature over a few past days.

Assuming that people adapt their clothing to the season, for example 0.8–1 clo (Note 19.2) in the cold season and 0.5–0.8 clo in the warm season with a progressive change in-between, and a sedentary activity (about 1.2 met (Note 19.3)), it is possible to compare the two models presented above. The upper and lower operative temperature limits to satisfy 90% of the occupants are presented on Fig. 19.1 for both models.

Note that when the space is conditioned, the acceptable temperature range is narrower than that of the adaptive model. It is obviously easier to satisfy the occupants of naturally ventilated, free-running spaces than those in a conditioned space. If the operative temperature can be kept between the adaptative model limits by passive means, there is no need to implement active means, but when such active means are installed and switched "on", the accepted comfort range is much narrower.

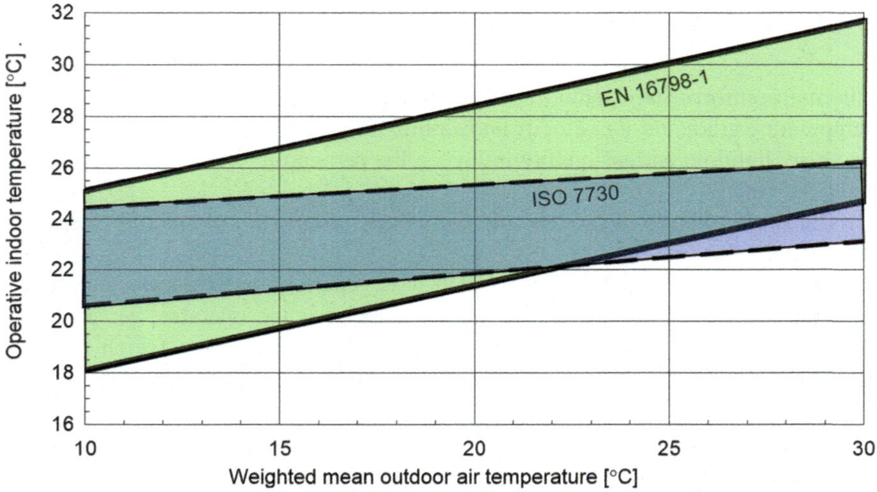

Fig. 19.1 Boundaries of an operative temperature acceptable for 90% of the occupants according to the adaptative (EN 16798) and Fanger (ISO 7730) models for 1.2 met

19.4 Passive Means to Improve the Indoor Thermal Comfort

19.4.1 Thermal Insulation

A good thermal insulation improves comfort by reducing the unpleasant effects of cold or warm surfaces. Thermal insulation of the internal space is ensured by its envelope. It reduces the heat transfer, and hence contributes to stabilise the internal temperature and reduce the energy needs for heating and cooling. To be efficient, a layer made of a specific insulation material should be part of the elements constituting the building envelope. In most cases, it is the air (or in some materials and glazing a heavier gas) immobilised in a foamy or fibrous material that ensures the thermal insulation. In glazing (and in some opaque materials too) a layer reflecting the infrared radiation emitted by warm surfaces improve the thermal insulation. Since the first oil crisis (1972) the thermal resistance of the building envelopes was multiplied by 4–10 in cold countries.

Vacuum does not conduct heat, and some fibrous or foam materials maintained under vacuum present a very good thermal insulation at low thickness, but it is very unlikely that the vacuum will be maintained for the lifetime of a building.

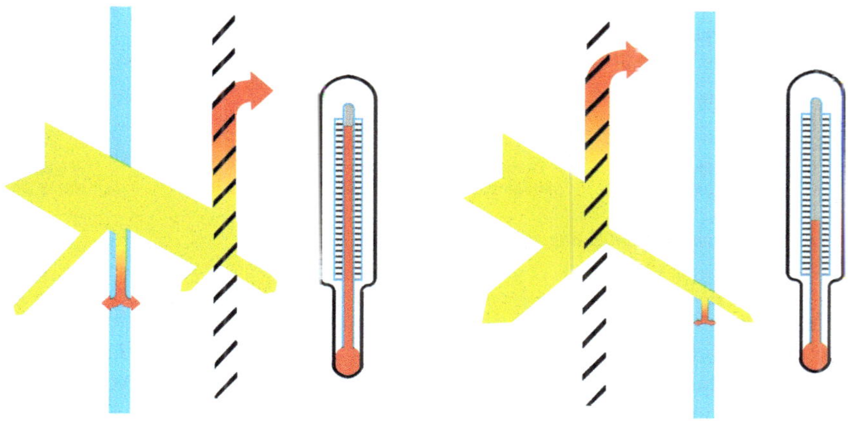

Indoor protection: the radiation absorbed by the blind or the curtain heats the inside

Outdoor protection: The radiation absorbed by the blind is released to the outdoor air.

Fig. 19.2 Solar protection should be outside to efficiently protect from the solar heat

19.4.2 Solar Protections

Solar protections avoid overheating by reducing the amount of solar energy that enters the occupied spaces. They also contribute to avoid glare.

Efficient solar protections are outside (Fig. 19.2). Solar radiation is partly reflected by the solar protection, but a large part is absorbed in it, and transformed into heat, which increases the temperature of the solar protection. If this protection is outdoors, this heat is released to outdoor air and does not enter the building. If the solar protection is inside, this heat is delivered to indoor air, thus heating the building.

External solar protection should also be wind-proof or protected behind a second skin. In this case, the space between the two skins should be well ventilated, with openings at least every second floor.

The use of sun reflective glazing decreases daylighting, hence increases artificial lighting and heat load. Indeed, the sun is a very efficient light source: it heats only 1 W/m^2 lit area for each 100 lx.

Daylight, as well as solar radiation changes with the time of the day and seasons. Therefore, solar protections shall be moveable to control the solar radiation entering the building. An automatic control with provision for individual override is preferred.

19.4.3 Solar collecting areas

For bioclimatic building design, solar collecting areas are parts of the envelope that passively collect the solar radiation to heat and light the indoor environment.

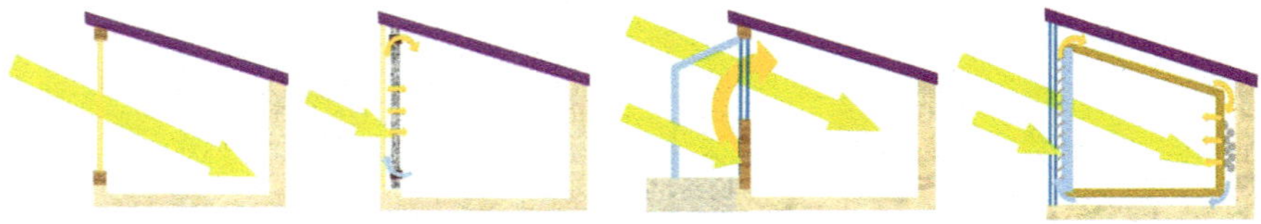

Fig. 19.3 Some solar collecting areas

These are mainly windows, but could also be sunspaces attached to—but outside the building envelope, Trombe walls, skylights, or double skins (Fig. 19.3).

These solar collecting areas should be large to contribute to passive solar heating, but also equipped with mobile solar protections to control the heat gains and avoid overheating.

19.4.4 Ventilation

Replacing the stale indoor air with outdoor air is paramount for evacuating the air contaminants produced inside, but airing at the right time can also be used to evacuate or gain heat. For this purpose, airflow rates shall be rather large (5–10 air changes an hour). Large openings, such as open windows, easily ensure such airflow rates even at no wind, by taking advantage of small density difference between inside and outside (stack effect).

For a single, large opening, the airflow rate can be roughly estimated from the Bernoulli equation and the ideal gas law, assuming a single discharge coefficient (ASHRAE 2001):

$$q_a = C_d A \sqrt{\frac{2\Delta p}{\rho}}$$

with:

A opening area [m^2]
C_d discharge coefficient ($C_d \approx 0.6$)
$\Delta \rho$ pressure difference [Pa]
ρ air density [kg/m^3]

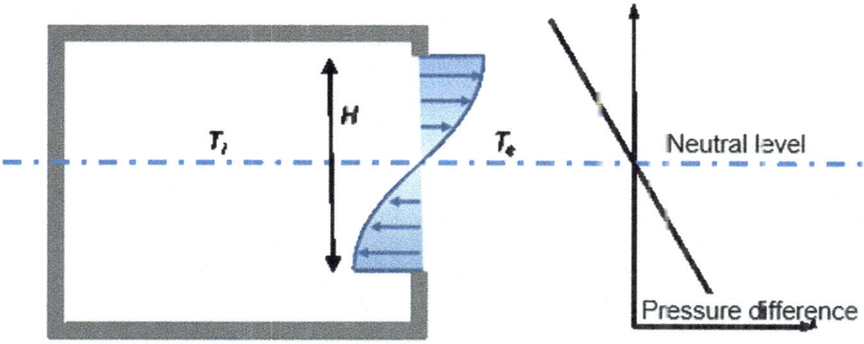

Fig. 19.4 Airflow through a single large opening, without wind, under the effect of an indoor-outdoor temperature difference

The pressure difference depends on the location and number of the openings, on the wind distribution and on the density difference between indoor and outdoor air. For some simple cases, it can be calculated.

If the ventilation takes place through a single aperture, the neutral level is approximately mid-height of that opening, so that the entering and outgoing airflow rates are equal (Fig. 19.4).

In the absence of wind, the airflow rate is driven by the stack effect only, and, assuming that the composition of the air (in particular the air moisture) is the same indoors and outdoors:

$$q_a = \frac{1}{3} H W C_d \sqrt{\frac{gH|T_i - T_e|}{T_e}} \cong 0.037 H W \sqrt{H(T_i - T_e)}$$

with:

H opening height [m]
W opening width [m]
g acceleration of Earth's gravity 9.81 [m/s^2]
T_i, T_e, internal and external air temperatures [K]

Figure 19.5 illustrates this relation. We see that the airflow rate through a window could be very large, even with small temperature differences.

When there are two openings separated by a height H, assuming that there are no additional resistance to the airflow inside, that the discharge coefficients are the same for both openings, and homogeneous air composition, we get (Van der Maas and Roulet 1991) (Fig. 19.6):

$$q_a = C_d \sqrt{\frac{A_1^2 A_2^2}{A_1^2 + A_2^2} 2gH \left(\frac{T_i}{T_e} - 1\right)}$$

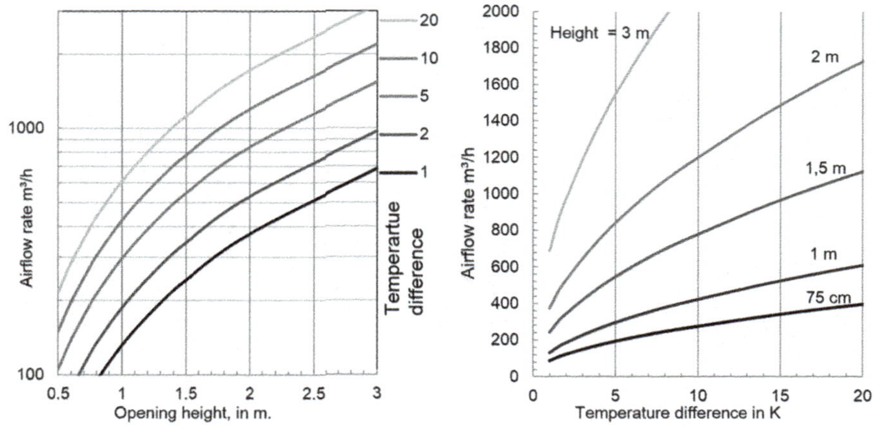

Fig. 19.5 Airflow rates through a vertical opening 1 m wide depending on its height and for several temperature differences

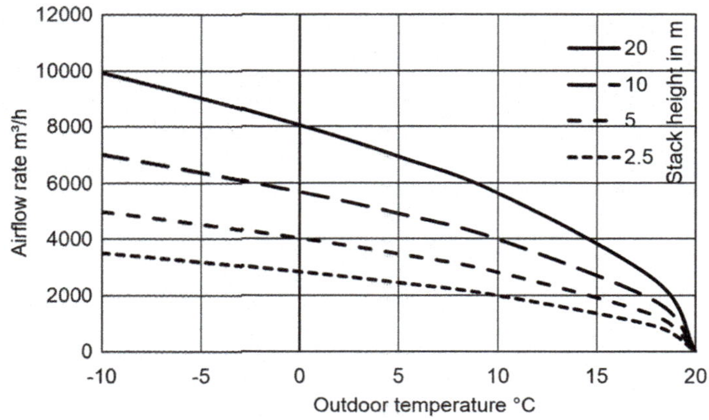

Fig. 19.6 Airflow rates resulting from the stack effect only, through two vertical, 1 m^2 openings as a function of the outdoor temperature, and for several stack heights. Indoor temperature: 20 °C

19.4.5 Thermal Inertia

The thermal inertia of a space or room is its ability to slow down and reduce changes of the indoor temperature. A building with high thermal inertia will have a naturally stable indoor temperature, despite large variations in heat gains that could result from solar gains through glazing or heavy occupancy (meeting rooms). This inertia depends on several characteristics:

1. the thermal capacity of materials in contact with the interior environment;
2. the thermal exchange between these materials and the interior atmosphere;

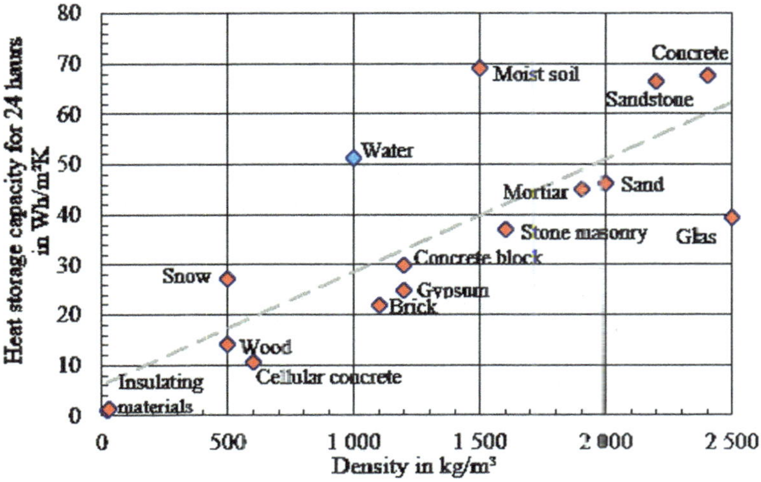

Fig. 19.7 Storage capacity of various materials based on their density

3. the thermal insulation of the building.

The daily storage capacity, i.e. the ability to accumulate heat during the day to return it at night is shown in Fig. 19.7 for various materials, assuming a large thickness.

As the specific heat of most building materials is about 1000 J/(kg K), and as thermal conductivity generally increases with the density of materials, the thermal capacity increases with the density, as shown in Fig. 19.7.

These materials can be either visible or coated with heavy, heat conducting coatings (plaster, tiling). They should not be isolated from the interior atmosphere by false ceilings, false floors, carpets, tapestry, etc. However, acoustic and aesthetic requirements should be considered. A compromise solution is to leave an important part (at least 50%) of the massive structure in direct contact with ambient air (Fig. 19.8).

19.5 Combination of Passive Means

The passive means to improve thermal comfort mentioned above, when used together in an appropriate way, become amazingly efficient. Thermal insulation, collected solar radiation controlled with an appropriate solar protection, and airing can be used to control heat flows, while thermal inertia stabilises the temperature by storing and releasing heat (Fig. 19.9).

Two widely used applications are given below.

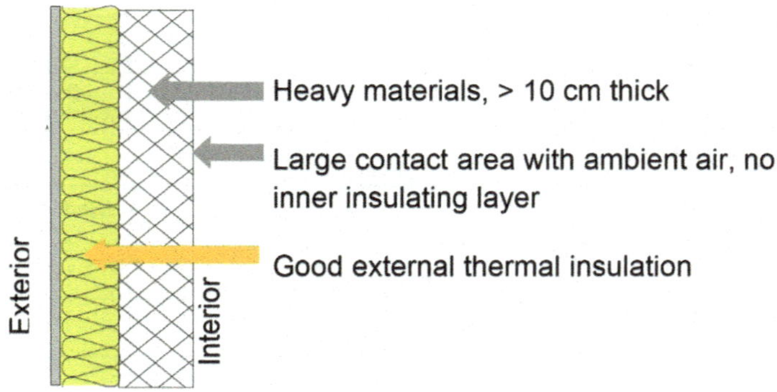

Fig. 19.8 Conditions for high thermal inertia

Fig. 19.9 Thermal inertia stabilises the temperature and solar protection, thermal insulation and ventilation can be used to control it

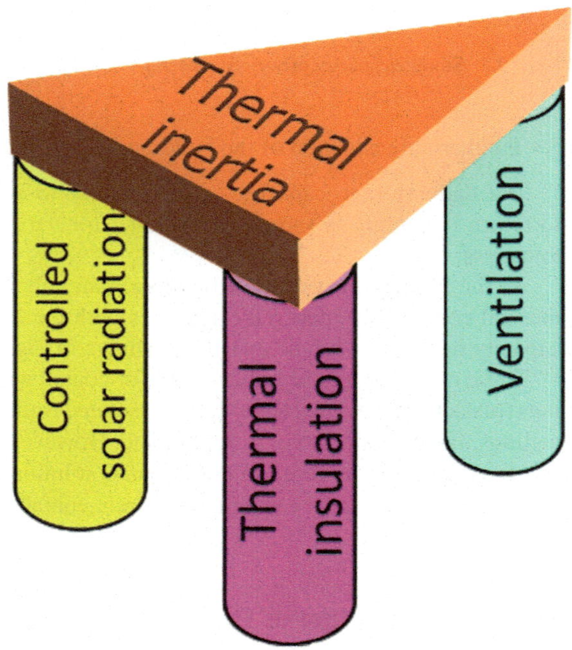

19.5.1 Passive Solar Heating

Solar energy enters freely in the building, mainly through the windows, thus contributing to heat the building. In order to take the best profit of these gains during the cold season without being overheated, the following conditions should be met (Fig. 19.10):

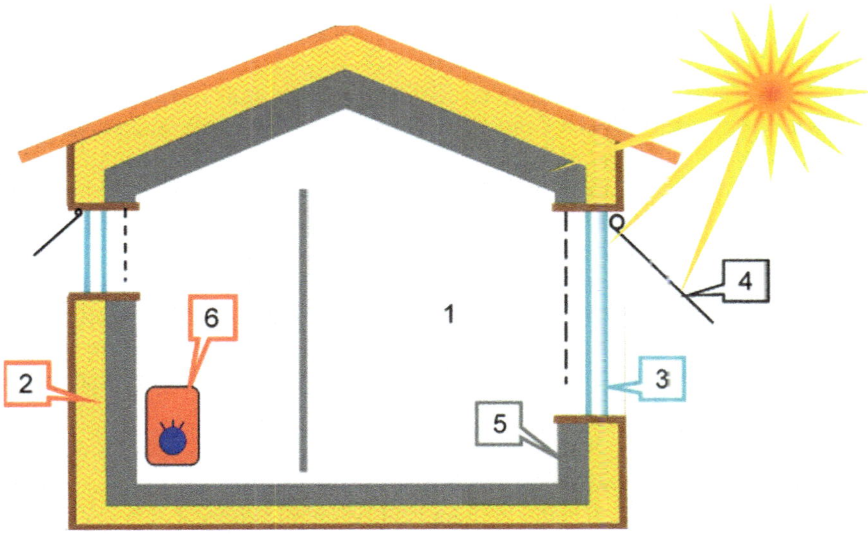

Fig. 19.10 Requirements for passive solar heating. 1. Day used spaces on sunny side; 2. Good thermal insulation; 3. Large windows on the sunny side; 4. Mobile solar protections; 5. High thermal mass; 6. Adequate heating and control systems

1. The spaces used mainly during the day should be located on the sunny side of the building.
2. The building should be well insulated, to avoid excessive losses.
3. It should have large solar collecting areas on the sunny side of the building (from south-east to south-west in Northern hemisphere). As written above, such areas are mostly windows, but can also be sunspaces or walls with transparent insulation.
4. These areas should be equipped with mobile, controllable solar protections, in order to control the gains. In winter, internal shadings are preferred, since they avoid glare but allow solar heat gains. Fixed solar protections are not ideal, since they don't allow any control: they either are not sufficient on some sunny days, or reduce daylight on overcast days. Vegetation with deciduous leaves may help by bringing a seasonal solar protection. However, it also shadows the building the whole day long, even when more daylight is necessary.
5. A large thermal inertia (heavy construction) allows to store heat, thus avoiding overheating during sunny days and maintaining a mild indoor climate during the night.
6. The heating control should cut the heating system as soon as solar gains suffice to keep a comfortable internal temperature, and switch it on as soon as necessary.

Passive solar gains can be useful to improve comfort at low cost during the heating season, but may be uncomfortable when poorly controlled, especially in mid- or hot season.

19.5.2 Passive Ventilative Cooling

Passive cooling through night-time ventilation is a comfortable, cheap, and energy-efficient way to keep the indoor environment within a comfortable temperature range during the warm season in most European climates, in particular in central Europe, north, west and higher altitudes in Mediterranean areas. In well adapted buildings, it can ensure a comfortable indoor climate in summer without artificial cooling, provided that internal heat load is not too large.

Such a strategy can be applied only in climates where the daily average outdoor temperature is within comfort limits, and where there is a significant temperature swing between night and day. This is fortunately the case in most European climates.

Principles of passive cooling are compatible with those of passive solar heating. As shown on Fig. 19.11, they are:

- avoid internal heat load by promoting daylighting and low energy appliances;
- avoid heat gains by using good thermal insulation, efficient shading systems (external and movable), and minimum ventilation rate during hot hours;
- store the remaining heat gains in the building structure, which should be heavy and in thermal contact with indoor air;
- cool the building structure with a large ventilation rate when the external temperature is lower than the internal temperature. Large ventilation rates are easily obtained by natural ventilation through windows and doors. These openings should be safe enough to remain open at night.

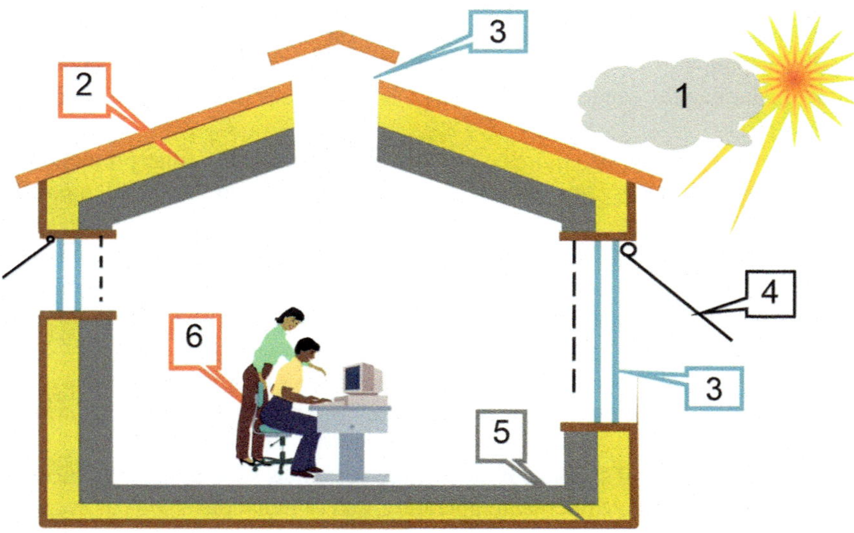

Fig. 19.11 Requirements for passive cooling. 1. Adequate climate; 2. Good thermal insulation; 3. Large openings, one as high as possible; 4. Efficient solar protections; 5. Large thermal inertia; 6. Reduced internal gains; 7. Airing

If passive cooling is not possible, mechanically driven night ventilation is an alternative that should be considered. This issue was addressed within the IEA project HybVent (Heiselberg 2002; Rajinder 2006).

19.6 Examples

Within the European research project HOPE (Bluyssen et al. 2003), 97 apartment buildings and 67 office buildings—a large part of which designed to be energy-efficient—were investigated using checklists regarding building characteristics and questionnaires addressed to the occupants regarding their perceived comfort (thermal visual, acoustic and indoor air quality) and well-being (sick building syndrome and allergies). Interpretation of the collected data showed that, on the average, occupants of low energy buildings feel better and are healthier and better satisfied with their indoor environment than occupants of other buildings (Roulet et al. 2006b).

A comparison was also performed between "ordinary" buildings and those especially designed according to passive solar architecture criteria (Roulet et al. 2006a). As expected, buildings with passive solar design present the best energy performance indices among the audited buildings, and the perceived comfort, air quality and well-being are the same or better in passive solar buildings than in the others. Indeed, among all the buildings audited within the HOPE project, most passive solar buildings are in the top group for health, comfort and energy performance.

19.7 Conclusions

Passive ways to improve the thermal comfort are thermal insulation and inertia, controlled use of solar radiation, and appropriate airing. Used in combination, these ways can be very efficient in both improving comfort and reducing the energy bill and hence also the greenhouse gases generation.

Only a design taking account of all architectural stakes—adaptation to environment and climate, space layout, structure, indoor environment quality and environmental impact—can result in a healthy, comfortable and energy efficient building. Passive architecture has the advantage that a careful design is paramount for success, and this also ensures that all stakes are considered.

19.8 Notes

Note 1 A conditioned space is a space which temperature is controlled by an active heat source or sink, such as heated spaces in winter or mechanically cooled spaces in the hot season.

Note 2 1 clo is the common winter dress in offices, which thermal resistance is 0.155m^{2}K/W.

Note 3 1 met correspond to the metabolic activity when sitting relaxed, or 58 W/m^2 skin area.

References

ASHRAE (2001) Handbook -Fundamentals. ASHRAE, Atlanta

Bluyssen PM, Cox C, Boschi N, et al (2003) European project HOPE (Health Optimisation Protocol for Energy-Efficient Buildings), pp 76–81

Brager GS, de Dear RJ (1998) Thermal adaptation in the built environment: a literature review. Energy Build 17:83–96

CEN (2019) EN 16798-1 Energy performance of buildings—Part 1: indoor environmental input parameters for design and assessment of energy performance of buildings addressing indoor air quality, thermal environment, lighting and acoustics. Brüssel

Chatelet A, Fernandez P, Lavigne P (1998) Architecture climatique. Edisud, Aix-en-Provence

Fanger PO (1982) Thermal comfort. R. E Krieger, Florida, USA

Heiselberg P (ed) (2002) Principles of hybrid ventilation. IEA ECBCS HybVent project

ISO (2005) ISO-7730: Moderate thermal environments—Determination of the PMV and PPD indices and specifications of the conditions for thermal comfort

McCartney KJ, Nicol JF (2002) Developing an adaptive control algorithm for Europe. Energy Build 34:623–635

Rajinder J (2006) Control strategies for hybrid ventilation in new and retrofitted office and education buildings (HYBVENT)

Roulet C-A, Cox C, Bluyssen P (2006a) Comfort and well-being in passive solar buildings—results from an European audit. 23rd PLEA conference, Geneva

Roulet C-A, Cox Ch, de Fernandes OE, Müller B (2005) Health, comfort, and energy performance in buildings—guidelines to achieve them all. EPFL, LESO, Lausanne

Roulet C-A, Johner N, Foradini F et al (2006b) Perceived health and comfort in relation with energy use and building characteristics. Build Res Inf 34;467–474

Van der Maas J, Roulet C-A (1991) Night time ventilation by stack effect. ASHRAE Trans 97(1):516–524

Chapter 20
Geo-Climatic Indicators to Define Local Potential of Low-Energy Technologies Including Climate Changes

Giacomo Chiesa

Abstract This chapter deals with a geo-climatic approach to low-energy technologies in climate change scenarios. Different key performance indicators (KPI) which are able to predict the geo-climatic potential distribution of bioclimatic solutions in reducing expected energy needs while guaranteeing space comfort in buildings (e.g. wind-driven, controlled natural ventilation for passive cooling of spaces) are described. This approach is further expanded to define the resilience of the above mentioned techniques in absorbing climate change impact on cooling and heating needs. The main objective of the chapter is to describe a methodological approach and introduce the aims of the geo-climatic vision. Three bioclimatic technologies based on low-energy/passive cooling (direct evaporative cooling, earth-to-air heat exchangers, wind-driven ventilation) and one low-energy/passive heating (earth-to-air heat exchangers), are taken as references. The calculation results of the local potential of these technologies based on related KPIs are described for the entire Italian peninsula, assuming both current typical meteorological years and two future climate change scenarios.

Keywords Geo-climatic applicability · Bioclimatic technologies · Passive heating · Passive cooling · Resilience to climate changes · Climatic design

20.1 Introduction

20.1.1 Bioclimatic Climate Design Tools

From its very beginnings, architectural design has been based on the instinct of human beings to respond to their needs for protection, with special regard to weather and environmental phenomena. Buildings, including envelopes and systems, satisfy this need and guarantee comfort conditions in confined spaces with respect to site conditions. The bioclimatic approach to building design correlates building

G. Chiesa (✉)
Department of Architecture and Design, Politecnico Di Torino, Turin, Italy
e-mail: giacomo.chiesa@polito.it

G. Chiesa (ed.), *Bioclimatic Approaches in Urban and Building Design*, PoliTO Springer Series, https://doi.org/10.1007/978-3-030-59328-5_20

development—aimed at fulfilling human needs—to local climate conditions, with special attention paid to comfort thus minimising the need for additional energy and maximising the usage of natural resources. Hence, it is no coincidence that quite a few basic reference texts and papers for this strategy include the words "climate and "design"/"architecture" in their titles. In these texts, bioclimatic design is regarded as a bioregionalist approach to building design (Aronin 1953; Olgyay 1963; Givoni 1969; Watson and Labs 1983). From an empirical point of view, this concept has been around since ancient times, as has been demonstrated in paleo-climatological studies, but also in proto-theoretical sources—see for example the *De Architectura* by Vitruvius Pollio (15 B.C)—even if it is only in the last century that a full scientific approach to this design methodology has been developed. Bioclimatics in architecture mainly refers to the adoption of tools and methods to maximise the potential of passive and low energy solutions in a specific location and related climate. For example, the well-known method of climate interpretation in housing design suggested by the Olgyay brothers is based on the following steps (Olgyay 1963): (i) climate data acquisition; (ii) data evaluation; (iii) adoption of calculation methodologies (e.g. shading dynamics, radiation access/protection, wind penetration/protection, envelope materials—U-value—and others); (iv) findings of methodological considerations; (v) architectural examples and practical considerations that define recommendations; and (vi) synthetic applications which outline expected improvements with respect to traditional solutions, where traditional solutions are those in which building shapes and envelopes are independent by local climatic conditions, considering that comfort conditions will be guarantee thanks to the sole mechanical equipment.

Focusing on thermal comfort, a well-recognized design methodology refers to the adoption of bioclimatic charts that define the local potential effect of low-energy technologies in reducing thermal discomfort by adopting natural cooling and heating solutions. Among the most widely used bioclimatic charts are the Olgyay chart, e.g. (Olgyay 1963), and the Givoni-Milne chart, e.g. Givoni (1969) and Milne and Givoni (1979). The former is based on a grid-diagram composed of dry bulb temperature values (y-axis) plotted as a function of relative humidity values (x-axis). A comfort zone is adopted to potentially analyse bioclimatic needs for specific weather conditions—e.g. wind needs for cooling—and the effect of different technologies—e.g. evaporative cooling effect by adding moisture in high temperatures. On the other hand, the Givoni-Milne approach is based on well-known psychrometric charts—e.g. the Carrier chart—which allows us to easily connect environmental conditions with psychrometric laws. Also in this case, specific comfort boundaries and expanded comfort limits according to the expected impacts of bioclimatic techniques are defined in the chart.

Both methods allow us to analyse a specific site and to define for that location—according to typical climatic and microclimatic conditions—methodological considerations and design recommendations, in line with the bioclimatic methodology mentioned above. Following this, additional analyses are needed in order to include, in the evaluation and design process, specific building design choices and boundary aspects—e.g. surrounding building shapes. At the early-design phase—e.g.

in building programming—different approaches were developed to correlate meta-design choices with site and climate analyses to optimise performances, as described in several site design approaches. For example, as regards bioclimatic requirements, during the building programming phase, the site microclimate matrix (Brown and Dekay 2001) can be adapted to define an approach to localise environmental/spatial units following a performance-driven methodology (Chiesa and Grosso 2017). A basic definition of the performance-driven design approach connected to environmental meta-design is reported, among the others, in Refs. Cavaglià et al. (1975) and Chiesa (2020), while a theoretical definition of the whole approach is reported in Ref. Ciribini (1968). In consideration of further design phases, additional bioclimatic methodological approaches and tools are defined to support the design and optimisation of specific building components—e.g. shading mask protractors to establish the dimensions for shading systems in the given site, see for example Olgyay and Olgyay (1957), Koenigsberger et al. (1975) and Grosso (1986). As a general consideration, it is possible to refer to bioclimatic chart methodologies as tools to analyse local climate before the definition of a specific building design spatial configuration by suggesting a priority list of locally correlated bioclimatic technological choices to be further investigated and dimensioned with additional tools. Nevertheless, being compatible with comfort analyses, bioclimatic charts may also be used to optimise design choices based on comfort and climate studies by analysing the outputs of advanced simulations, such as dynamic energy and by verifying the functionality and efficacy of chosen shapes and technology configurations.

20.1.2 A Change of Scale

In addition to the described approach, which was developed to analyse a specific site by developing specific building design actions, especially in recent years, a growing interest in developing tools for wider contexts, in line with the growing interest of the sustainable design field on smaller geographical scales is evident (e.g. at urban or regional levels). This interest was, for example, focused on during the City Futures conference held at Made Expo in 2009 on architecture and sustainability (Matteoli and Pagani 2010). This change in scale is also reflected in the bioclimatic field. Several efforts are in fact evident in recent few year to define KPIs (Key Performance Indicators) and methods that are able to calculate the climate potential of the different low-energy technologies, adopting a geographical scale. This interest is driven by the need to define regulations and suggest a holistic approach to green design by going beyond the building scale and in order to show and study the impact of climate and climate change and correlate them with design choices. Early-approaches focused on the development of applicability maps for manuals and guidelines based on the definition of boundaries in which a specific technology may be adopted in consideration of an ON/OFF analysis—see for example the applicability maps for direct evaporative cooling systems reported in Bom et al. (1999). Other approaches include the development of applicability classifications, by defining specific KPIs that are able to analyse the impact of a low-energy technology on local climates to transform discomfort hours to comfort hours or to reduce the intensity of energy needs for space cooling or heating. For example Salmeron et al. (2012) defines a method to analyse the local climatic potential of evaporative cooling according to average seasonal environmental conditions by developing a priority mapping approach for the adoption of this technology in Europe thanks to a classification ranging in the domain from "very low" to "very high" suitability. The same approach was applied to other territories including the U.S (Aparicio-Ruiz et al. 2018). In a similar way, other researchers focused on hourly-defined KPIs to map the applicability of different techniques. For example, the Climatic Cooling Potential index (Artmann et al. 2007; Kolokotroni and Heiselberg 2015) to define the night cooling potential of ventilative cooling by combining local environmental air temperatures and internal ones. The latter are defined by adopting a harmonic oscillation around a comfort threshold to include the non-simultaneous nature of sensible heat storage/dissipation in buildings when climate studies are driven without dynamically simulating a specific building. This approach can also be adopted to define the potential of fan-driven ventilative cooling, by fixing the values of air exchanges for hours (ACH), which work at night, such as the approach for mechanical night ventilation supported by the UNI/TS 11,300. In addition, it can also be modified for natural-driven ventilation. Main approaches and indicators which are able to define the applicability of a technology in a given climate are generally correlated by the adaptation of long-term evaluation KPIs to analyse thermal discomfort. These indexes may be classified according to Carlucci and Pagliano (2012) in percentage indices, e.g. the POR (percentage outside the range) approach—ISO 7730 and EN 15,251—cumulative indices, such as all Degree-hours criteria as adopted in this paper, risk indices, such as the overheating risk indicator of Nicol et al. (2009), and averaging indices, such as the averaged PPD (predicted percentage dissatisfied)—ISO 7730. These indexes were developed to analyse specific design choices supporting a building design action, or during operational phases to analyse/optimise the behaviour of a specific building. Nevertheless, they can also be adapted to study the climatic applicability of different low-energy technologies in accordance with local climate.

20.1.3 Chapter Objectives

Low-energy technologies, such as passive cooling solutions, are characterised by local-specific applicability which requires an attentive analysis from early-design phases to check their potential in a specific design site. In the past, climate has been conceived as a time-fixed variable in bioregionalism studies. In fact, climate is a typical characterisation of a place, "averaging" weather characteristics in the long term. However, in more recent years, it has become clear that climate is a variable rather than a fixed condition, and is expected to vary more and more because of the impact of climate change, thus influencing building performance during its operational lifespan. In this uncertain scenario, climate potential studies, can help us to analyse at an early stage these changes to define their impact on building indicators in order to support the development of new resilient design approaches. Furthermore, these climatic approaches can be adopted to define a geographical representation of the bioclimatic potential of technologies that may be used to develop incentive and retrofitting policies. This specific research focus may be defined as geo-climatic, connecting climatic approaches to building design and performance indicators with their geographical representation.

This chapter focuses on these specific issues and aims at demonstrating the soundness of these geo-climatic approaches in analysing low-energy applicability and its resilience to climate change. In order to pursue this objective, some authors recently developed KPIs concerning passive cooling technologies will be presented and applied to the Italian territory thanks to the development of a series of calculation tools scripted in Python. A quick comparison of current climate conditions—represented through typical meteorological years—and two future IPCC scenarios will also be presented. Given that future conditions are based on predictive models, more than one scenario needs to be used, since these data are not the real ones, but need to be treated including their statistical variations—see for example the recent paper (Zajch et al. 2020) on the climate applicability of buried pipes in North America. Nevertheless, in future studies advanced comparisons will be presented thanks to wide—ranging research into multi-climate models. For this paper, two models are adopted to show the mentioned incertitude, even if in further studies a large model database will be assumed. Furthermore, the chapter includes a detailed series of new maps that describe the climatic potential for all Italian Municipalities—represented by their statistical centroids—in consideration of wind-driven ventilative cooling, direct evaporative cooling and low-energy heating and cooling potential of earth-to-air heat exchangers.

20.2 The Geo-Climatic Proposed Methodology

In order to calculate geo-climatic bioclimatic requirements and the climate potential of different low-energy bioclimatic technologies, two main classes of indicators are adopted in this chapter: (i.) KPIs correlating climate with energy needs and/or comfort hours without including the effect of specific cooling/heating technologies—see Sect. 20.2.1; (ii.) KPIs including the cooling/heating climatic-effect of a specific technology in reducing energy needs and increasing comfort hours—see Sect. 20.2.2.

In order to give support to this approach, a specific application in Italy is described—see Sect. 20.2.3. A selection of geo-climatic indicators is introduced and applied to different locations and correlated typical meteorological years (TMY), which represent local climate conditions. Here below the chosen sample KPIs and climate data are described.

20.2.1 Climate Energy Needs—Bioclimatic Needs

The first type of indicator includes the calculation of degree-day/degree-hour indices. These indicators are linearly correlated with building energy needs, given that it is possible to roughly estimate the energy needs of a building by multiplying degree-days/hours by a building correlated factor (Santamuris and Asimakopoulos 1996), as in the energy signature characterisation. Nevertheless, thanks to the adoption of hourly defined typical meteorological years (TMY), the hourly defined version of these indexes is adopted here, with special regard to Heating Degree Hours (HDH) and Cooling Degree Hours (CDH), defined respectively according to the following expressions (20.1) and (20.2)—considering only positive values:

$$\mathrm{HDH} = \sum (\vartheta_{\mathrm{base}} - \vartheta_{\mathrm{env}})^{+} \tag{20.1}$$

$$\mathrm{CDH} = \sum (\vartheta_{\mathrm{env}} - \vartheta_{\mathrm{base}})^{+} \tag{20.2}$$

where ϑ_{env} and $\vartheta_{\mathrm{base}}$ are respectively the environmental air temperature and the base temperature, which refers to the building specific comfort balance between indoor and outdoor conditions. For this specific climate-related analysis, a base temperature of 20 °C is assumed for the HDH index (in line with Italian climatic zone calculation) and of 24 °C for CDH index. Other calculation procedures define the daily version of these indexes, in line with ISO 15927-6:2007 and/or referring to the Eurostat approach. Calculations are performed by elaborating the assumed *.epw database thanks to self-developed python coding and by data elaboration and representation in QGIS.

20.2.2 Bioclimatic Technology Climate Potential

The second type of indicator includes the calculation of the effect of a specific low-energy technology acting on typical weather conditions. In particular, for geo-climatic analyses one can adopt the residual degree-day and degree-hour indicators, in which CDD/CDH and HDD/HDH expressions are modified by including the cooling and heating effect of a low-energy technology, such as ventilative cooling, i.e. by air-exchange, or by direct evaporative cooling, or by earth-to-air heat exchangers, or passive heating, i.e. sunspaces. Also in this case, these indicators are climate-related, but the environmental conditions are mediated by the effect of a specific technology which is able to treat for example the air temperature—see the definition included in Chiesa and Grosso (2015) and Chiesa (2019a). Nevertheless, other approaches may be adopted, including the calculation of DD/DH average seasonal indexes, or even the climatic heating/cooling (dissipative) power of a technology. This chapter will focus in particular on ventilative cooling solutions, including the effect of passive downdraught evaporative cooling systems (PDEC) and earth-to-air heat exchangers (EAHE) in treating external air before space ventilation by adopting related residual CDH indexes (CDH_{res}). Furthermore, for EAHE, the impact on the winter season of pre-treating an airflow before entering HVAC systems is also analysed by calculating the residual HDH index (HDH_{res}) correlated to the reduction in the energy intensity needs.

The developed expression for PDEC is reported here below—positive values only:

$$CDH_{res\text{ - }PDEC} = \sum((\vartheta_{env} - 0.6 \cdot WBD) - \vartheta_{base})^{+} \tag{20.3}$$

where WBD represents the wet bulb depression and 0.6 a reference effectiveness of typical PDEC systems in covering the WBD. As demonstrated in previous analyses, the PDEC effectiveness may reach 80–90% for shower tower configurations, though in those cases a mixture of treated and untreated air may be needed to reduce the relative humidity of the treated airflow. A precautionary lower climatic effectiveness is hence adopted for this analysis.

As regards the EAHE technology, a virtual control system is defined by choosing for each hour the best condition produced by the environmental DH index and the EAHE residual DH index, simulating an ON/OFF control of the system. The residual DH indexes are calculated for EAHE as follows—positive values only:

$$HDH_{res-EAHE} = \sum(\vartheta_{base} - (\vartheta_{env} - 0.52 \cdot (\vartheta_{env} - \vartheta_{soil-2.5m})))^{+} \tag{20.4}$$

$$CDH_{res-EAHE} = \sum((\vartheta_{env} - 0.52 \cdot (\vartheta_{env} - \vartheta_{soil-2.5m})) - \vartheta_{base})^{+} \tag{20.5}$$

where $\vartheta_{soil-2.5}$ is the soil temperature at the depth of 2.5 m for each considered calculation hour. The adopted EAHE effectiveness of 0.52 is in line with previous analyses and refers to a sample EAHE configuration considering a humid clay soil ($\alpha = 6.13\text{E-}07\ m^2/s$), an installation depth of 2.5 m, a tube diameter of 0.25 m, a tube length of 75 m and an hourly airflow of 500 m^3/h. Furthermore, for calculating the soil temperature at the given depth for all yearly hours, the equation described by Labs (1981) and Cook (1989) was adopted. The soil surface average yearly temperature, its semi-amplitude, and the phase shift constant—parameters at the base of the Labs' equation—are calculated for all locations using the CalcSoilSurfTemp tool, an auxiliary program of EnergyPlus (U.S. Department of Energy 2018). For this paper, a heavy and damp soil and a bare moist surface are assumed to perform these calculations.

Furthermore, the effect of natural ventilation is also considered by calculating the hourly average cooling dissipative potential of CNV (controlled natural ventilation) considering only the CNV activation hours [kWh]. This calculation is based on the wind-driven ventilative cooling potential, assuming local wind velocity, but fixing a domain of difference in pressure between openings. The calculation methodology is defined by the author in a previous publication Chiesa (2019b) and is here applied to the assumed Italian climate database for a reference cooling period ranging from

June to August. Dissipation cooling potential refers only to those hours in which there is a climatic ventilative cooling potential ($\vartheta_{\text{env}} \leq \vartheta_{\text{comfort,vent}}$).

$$Q_{\text{send}} = \sum\left(3600 \cdot \left(0.6 \cdot v \cdot A \cdot \sqrt{\Delta C_p}\right) \cdot 1.2 \cdot 0.28 \cdot (\vartheta_{\text{env}} - \vartheta_{\text{com}})\right) \quad (20.6)$$

where v is the wind velocity in each calculation hour at the opening height, A is the average net area of opposite openings, ΔC_p is the difference in openings' coefficients of pressure, ϑ_{com} is the assumed comfort threshold, while 1.2 and 0.28 represent respectively the air density [kg/m^3] and the air specific heat capacity [W/(kg K)]. In this specific analysis the comfort temperature ϑ_c is assumed as 28°C, in line with early-design suggestions (Grosso 2017), in order to roughly assume a free-running comfort threshold which is nearer to potential comfort boundaries correlated to adaptive models—see Nicol et al. (2012). For the calculation of these values, reference openings are assumed at 5 m in height from the ground considering a medium/low-density urban area wind profile exponent and boundary layer height. The opening discharge coefficient is assumed equal to 0.6, as suggested for early-design stages (Grosso 2017). Calculations are based on two opposite net opening areas of 1 m^2, in order to allow a simple adaptation of results to different opening areas during design development applications. Three differences in pressure between airflow openings are assumed referring to fixed values, i.e. {0.1; 0.3; 0.5} which can be adapted by designers to different geometries and boundary conditions.

20.2.3 Adopted Climate Data and Scenarios

Both classes of indicators allow us not only to analyse the bioclimatic needs of a location given current TMYs, but also to study the expected impact of climate change (measured in the recent past or expected for future years) by calculating the absolute and/or relative differences between KPIs. For example, percentage variations in CDH and HDH may be calculated easily by comparing current and future TMYs to define the impact of one or more specific future climate scenarios on energy needs and/or on a specific bioclimatic technology potential (Zajch et al. 2020). In this chapter, the effect of climate change is analysed by defining the variations in frequency distribution of the mentioned KPIs for the adopted list of locations with reference to two future climate series (A1B—balanced emphasis on all energy source scenarios for an integrated world—and A2—more divided world scenarios) in the year 2050. Climatic data are produced for all Italian Municipalities (ISTAT 2018) adopting the METEONORM software—see technical specifications (Meteotest 2017)—for a total of 7978 *epw files for each TMY series (current, reference years 1991–2010 for irradiation and 2000–09 for temperature, and the 2 future scenarios). It is important to remember that the number of Italian Municipalities can change due to administrative decisions, e.g. merging. Those climate databases are hence processed adopting the mentioned methodology and results are described in the f Sect. 20.3.

20.3 Results and Discussion—Italian Peninsula Study Case

20.3.1 Geo-Climatic Heating and Cooling Needs

The Italian Peninsula is analysed by assuming its Municipality statistical centroids in order to map the geo-climatic impact of low-energy cooling technologies in accordance with the described methodology. In addition, the local geo-climatic indicators for space heating and cooling (DH) are also defined. Focusing on the latter, Fig. 20.1 shows the distribution of HDH_{20} and CDH_{24} indices for current adopted TMY database. These maps clearly illustrate that heating needs are higher in the northern and alpine areas and in municipalities that are nearer the Apennines in central and southern Italian areas, while Apulia, Sicily and Sardinia have some of the lowest HDH values. From a cooling point of view, the map on the right shows how Alpine spaces and some of the mountain locations in south-central Italy are characterised by low cooling needs, while Apulia, southern Sardinia, several Sicilian areas and some specific central Italian locations have high cooling needs. Furthermore, the Po valley, especially southern Lombardy and Emilia-Romagna are characterised by both medium-high heating and cooling needs. In these maps, it is clear that both heating and cooling geo-climatic indicators are not homogenously distributed in the Italian peninsula and therefore require specific local approaches and attentive mapping to define proper policies and regulations.

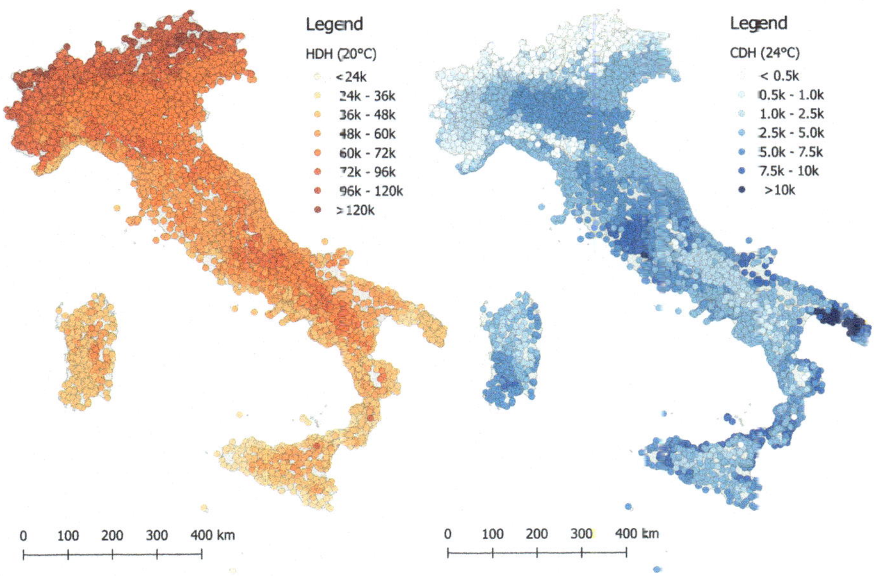

Fig. 20.1 The geo-climatic HDH_{20} (left) and CDH_{24} (right) distribution for the 7978 Italian municipalities—calculation based on TMY (1991–2010 radiation; 2000–09 temperature) elaborations

20.3.2 Geo-Climatic Potential of Low-Energy Technologies

Focusing on the technology-correlated geo-climatic indicators which are able to describe the local potential of specific low-energy solutions, Fig. 20.2 (left) analyses the potential impact of PDEC technologies in reducing the local CDH_{24} index by calculating the CDH_{res}. The map shows how a direct evaporative cooling solution may considerably reduce the air temperature below the set environmental threshold with a consequent reduction in sensible cooling energy needs. Nevertheless, dehumidification may be needed after this treatment although the adoption of a low PDEC efficiency (0.6) in this analysis limits the amount of hours in which relative humidity breaches comfort limits. This figure illustrates a very high potentiality in those locations in which lower original CDH were calculated, but also a very high potentiality in reducing the original requirements to very low CDH_{res} classes. Nevertheless, in Apulia, some parts of Sardinia and Sicily this technology may have limitations in the number of applicability hours since local wet bulb depression is not sufficiently wide enough because relative humidity is too high—see also the analysis of a limited number of locations in Mediterranean climates reported in Chiesa et al. (2019).

Figure 20.3 shows the distribution of residual HDH_{res} (left) and CDH_{res} (right) with reference to the virtual air-treatment of an EAHE system. By comparing these maps with the ones in Fig. 20.1 it is possible to analyse the geo-climatic potential of EAHE. This analysis underlines that this technology is very efficient in reducing virtual climatic cooling needs in almost all locations, even if small residual needs may be envisaged especially in some of the hottest coastal areas in Sicily and in

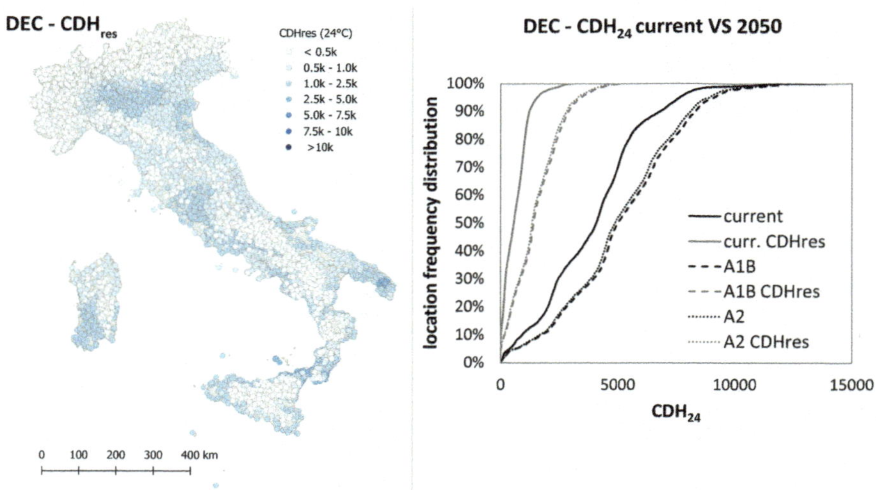

Fig. 20.2 (left) Geo-climatic applicability of PDEC systems mapped thanks to the residual $CDH_{res\text{-}24}$ indicator to be compared with the map on the right in Fig. 20.1 (right) the frequency distribution of locations according to local CDH_{24} and $CDH_{res\text{-}24}$ (PDEC) for current and future considered scenarios

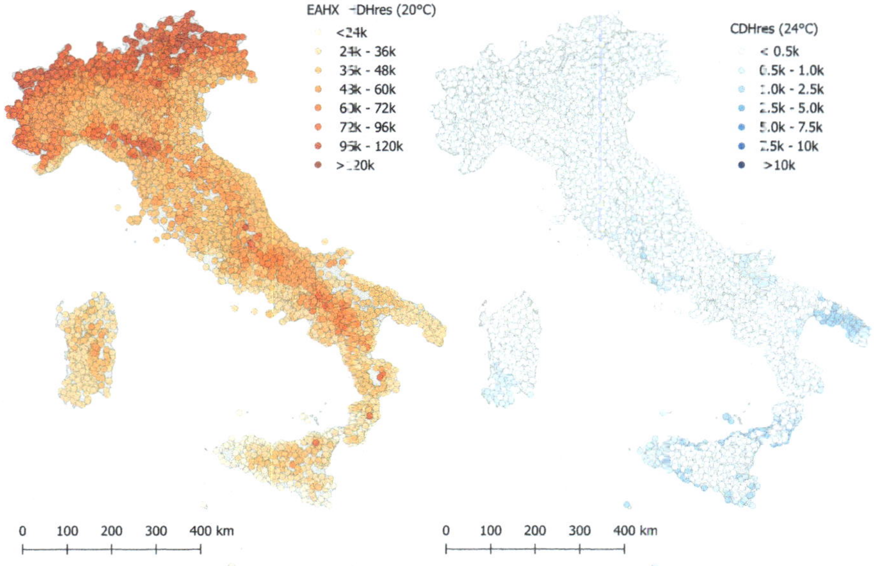

Fig. 20.3 The geo-climatic potential of EAHE considering $HDH_{res\text{-}20}$ (left) and $CDH_{res\text{-}24}$ (right) for the 7978 Italian Municipalities—calculation based on current TMY (1991–2010 radiation; 2000–09 temperature)

central Apulia. In all cases more than 72% of climatic CDH_{24} values can be covered, although this analysis does not include the calculation of relative humidity boundary conditions. Furthermore, as regards heating, this technique may reduce the heating climatic demand as a pre-heater system rather than fully fulfil the need. These results are in line with other studies which focused on Italian locations, as in Ascione et al. (2011). It should be noted that this reduction can be considerable and cover an average of about 23.9% of the local HDH_{20} index, with a maximum value of 44.4% and a minimum one of 10.4%.

Finally, Fig. 20.4 illustrates the potential heat dissipation due to wind-driven air exchanges for different values of fixed differences in pressure between openings. Reported values are hourly averages based on ventilative cooling activation hours. These maps show a very high potential in thermal dissipation due to wind-ventilative cooling, especially when a correct design of related technical elements allows us to reach high differences in pressure, at least for prevailing wind directions. High potentials are underlined in Sardinia, Sicily, Apulia, southern Campania, Basilicata, and in central Italian Regions. In addition, with medium-high differences in pressure, the Po valley also shows good potential for ventilative cooling dissipation.

These figures not only show the cooling potential of wind-driven ventilation, DEC and EAHE, but also the fact that low-energy technologies are characterised by

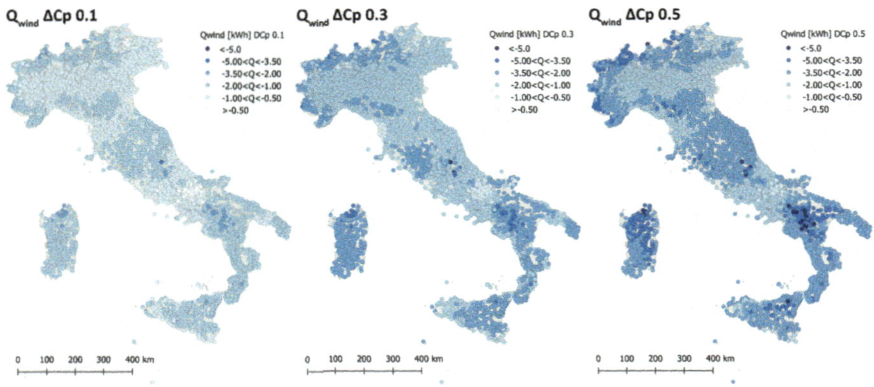

Fig. 20.4 The geo-climatic potential of wind-driven ventilation considering the expected heat gain dissipation averaged for activation hours for the 7978 Italian Municipalities—calculation based on current TMY (1991–2010 radiation; 2000–09 temperature). Various differences in pressure between openings are defined: 0.1 (left); 0.3 (centre); 0.5 (right)

an irregular geo-climatic distribution which requires attentive mapping analyses to positively orient design choices at the early-design phase towards the most suitable local technological choices.

20.3.3 The Impact of Future Climate Scenarios

In addition to current TMY analyses, the geo-climatic potential behaviour in future climate scenarios is analysed for the same low-energy technologies and in consideration of the climate-related heating and cooling demand (DHs). Figure 20.5 shows the calculated HDH_{20} for the two selected climate change scenarios for year 2050. Both scenarios follow a similar trend with a small reduction in general heating-correlated need for the whole Peninsula, even if for a few mountain locations a countertrend is evident (west of Piedmont). By analysing the frequency distribution of local HDH_{20}—see Fig. 20.7 (left)—changes between the continuous black line (current TMY) and the dotted black lines for 2050, data describe the dimension of this slight reduction. These lines confirm the generalised shift toward lower HDH_{20} values for the whole Italian territory. On the other hand, Fig. 20.6 illustrates the CDH_{24} variations for A1B 2050 and A2 2050 scenarios. The maps show an increase in this geo-climatic indicator that is correlated to space cooling needs, even if this change is not always homogenously distributed throughout the whole Peninsula. During the summer season a small difference between the two future climate scenarios can be seen, with a major effect on the increase of cooling needs for the A1B scenario. An increase in cooling geo-climatic needs is underlined for the whole Po valley, with special regard to Emilia-Romagna while the medium mountain sites are also interesting. Central Italy is characterised by non-homogenous results, with

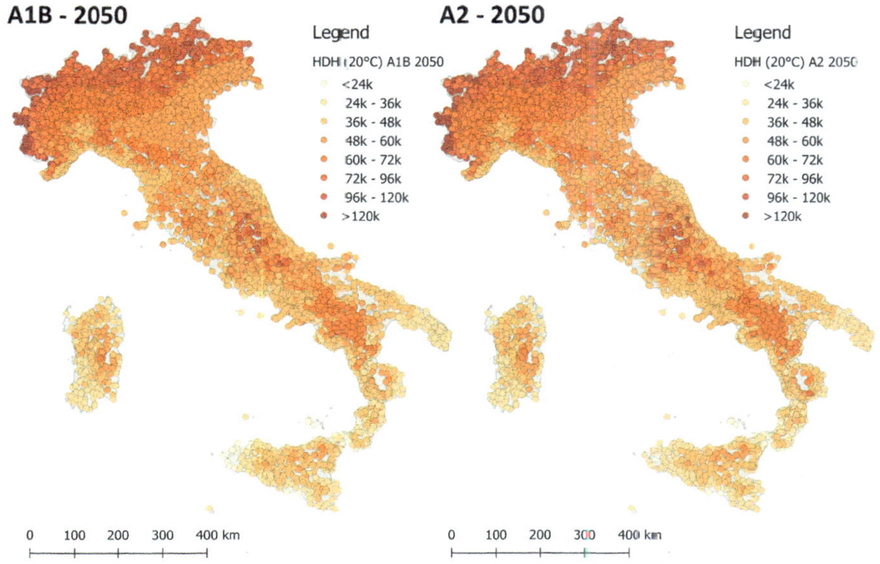

Fig. 20.5 The geo-climatic HDH_{20} distribution for the 7978 Italian Municipalities—calculation based on TMY elaborations for 2050 considering the A1B IPCC cases (left) and the A2 ones (right)

an evident increase in the Marche Region. Furthermore, Campania and the Gargano area in Apulia also show a considerable increase in CDH_{24}, as do the coastal locations in Sicily and Sardinia (north-east and south-west). In Fig. 20.7, the graph on the right shows the changes in frequency distribution of cooling degree hours. Differences between the continuous black line (current TMYs) and the dashed black ones which represent future scenarios report the cooling need increase described above. For example, at present about 70% of locations are characterised by a CDH_{24} below 5000, while in 2050 only 52% of locations are below 5000 CDH_{24} for the A2 scenario and only 50% for the A1B one.

With reference to the effect of PDEC systems, Fig. 20.2 (right) shows an increase in the residual CDH_{res} index for both scenarios. This reduction in PDEC potential is partially ascribable to the growth in CDH_{24} values even if the rise in geo-climatic cooling needs also translates the treated air after a virtual PDEC treatment to higher temperatures that in several cases do not reach comfort conditions. Furthermore, in humid locations a low WBD does enable us to reach PDEC effectivity. For these reasons, a reduction in the percentage of PDEC covering potential of climatic CDH is evident.

Focusing on EAHE systems, Fig. 20.7 reports the frequency distribution of locations in consideration of, on the left, the local HDH_{20} and HDH_{res} indexes and, on the right, the local CDH_{24} and CDH_{res} indexes. The heating-referred graph shows that climate changes do not considerably affect the pre-heating potential of EAHE systems given that the difference between the continuous and dashed lines are only

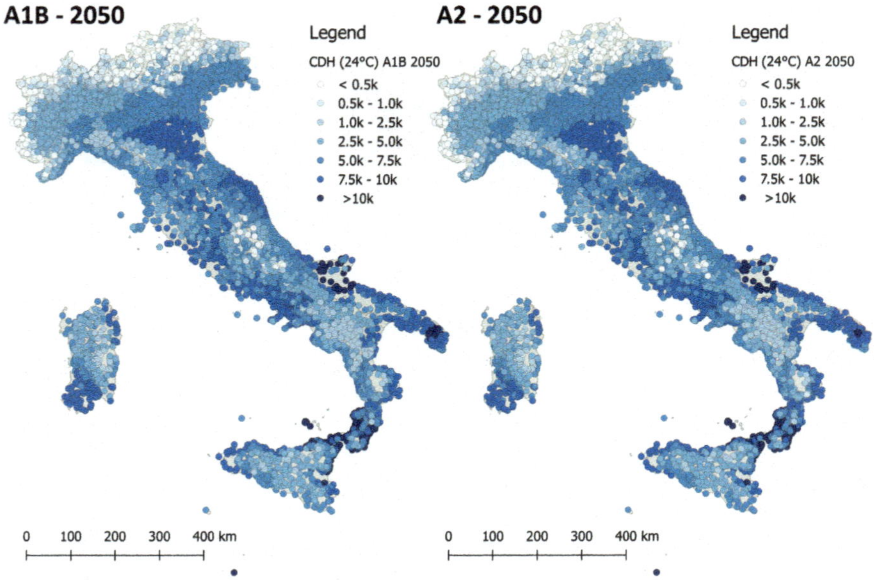

Fig. 20.6 The geo-climatic CDH_{24} distribution for the 7978 Italian Municipalities—calculation based on TMYs elaborations for 2050 considering the A1B IPCC cases (left) and the A2 ones (right)

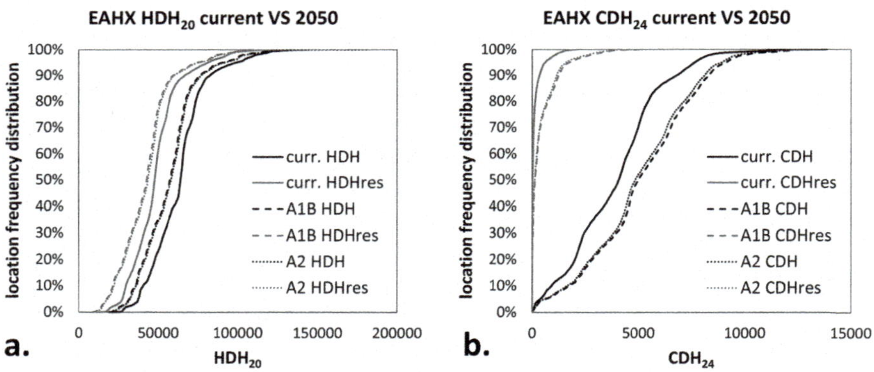

Fig. 20.7 Frequency distribution of locations according to local HDH_{20} and $HDH_{res\text{-}20}$ CDH_{24} and $CDH_{res\text{-}24}$ (PDEC) for current and future considered scenarios

slightly reduced in current and future databases. Focusing on the cooling analysis, the increase in ambient temperatures which is in the climate-related cooling need indicator (CDH_{24}) also affects the frequency distribution of EAHE residual CDH_{res}. A slight reduction in the covering percentage of climate degree hours is envisaged, even if in all cases it only impacts slightly on the high potential of this technique. However, for the hottest locations, EAHE may not be sufficient to cover on its own

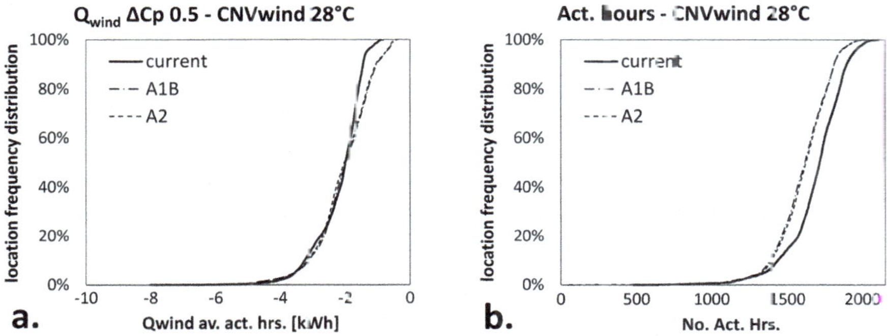

Fig. 20.8 Frequency distribution of locations according to local Q_{wind} values averaged on the local number of activation hours (left) and frequency distribution of number of activation hours of wind-driven ventilative cooling (right) for current and future climate scenarios

all cooling needs in the light of future climate trends, even if this preliminary analysis shows that EAHE technology has a higher resilience than PDEC.

Considering the variations due to climate changes in the expected wind-driven heat dissipation, Fig. 20.8 (right) shows a reduction in activation hours—an average loss of about 87 h per location in the A1B scenario and about 80 h for the A2 scenario—even if this outcome is not homogenously distributed. Actually, in some locations an increase in the number of activation hours is evident. Roughly speaking, these changes are correlated to the local increase in ambient temperature and show a reduction in hours with ambient temperatures below the natural ventilation assumed threshold for the hottest locations. A similar argument is reported in Fig. 20.8 (left) which shows the frequency distribution of heat-dissipation average hourly potential. Also in this case a slight average reduction in the cooling potential of ventilative cooling is evident, even if these changes are not homogenous and in some locations a countertrend is reported.

20.4 Conclusions and Further Developments

This chapter introduced one specific sub-research field in bioclimatic architectural design which is connected to the development of geo-climatic KPIs and analyses which are able to define at geographical scales the climatic applicability of different low-energy technologies based on hourly-defined weather and climatic databases. Each technology is studied in relation to one or more typical design configuration(s) that support early-design choices and the definition of policies/regulations to valorise the diffusion of passive/hybrid systems. Actually, the correlation between climate and cooling/heating system potential is very local specific and needs to be investigated from the building programming phase in order to define the correct toolbox of required technological elements. The geo-climatic mapping approach is of great help to designers in making choices. Furthermore, this approach has proven to be effective in showing the impact of climate change on expected geo-climatic potential thus helping architects to analyse at an early stage the resilience of the chosen set of low-energy bioclimatic technologies. Additional studies are needed and are under development with special regard to the adoption of a large set of climate databases focusing on the hybridisation between geo-climatic technological and climatological studies in order to include reanalysis databases and a large set of climate change scenarios. Further analysis is needed to include urban impact on local climate with a focus on the effect of urban heat island on climatic potential. Further analysis on different passive technologies is also required.

Focusing on the chosen sample territorial analysis, this chapter underlines that direct evaporative cooling and earth-to-air heat exchangers are low-energy solutions which are able to increase the potential of ventilative cooling in almost the entire Italian territory, even if its potential may vary locally such as has been illustrated in the maps. Furthermore, EAHE is also able to preheat an airflow, to support the HVAC system during the winter season by reducing the difference in temperature between the environment and the comfort threshold. As regards wind-driven ventilation for summer cooling, medium-to-high potential levels are evident for almost the entire territory when a sufficient difference in pressure is assured. In the climate change scenarios A1B and A2, Italy is subjected to a general reduction shift in HDH_{20}, while in the summer months an increase in climatic cooling needs comes to the fore, even if its distribution is less homogenous. In a similar way, the heating potential of EAHE is maintained at almost optimum levels with negligible reductions. On the other hand, as far as cooling is concerned, a general reduction in PDEC percentage covering of original CDH_{24} is evident, while this technique is still a valid cooling option in most locations. As regards the EAHE cooling potential, a slight reduction is noticeable in some locations, but this technology shows a high level of resilience to climate change. Wind-driven ventilative cooling is also slightly affected by climate change and shows a slight reduction in its potentiality in some locations, but a countertrend in others. In general, this technology proves to be highly resilient though a slight decrease in number of activation hours is reported.

Additional studies are under development to increase the coverage all of Europe and other areas, such as North America. Further research is under development that combines the variations in the geo-climatic potential of EAHE in several climate change scenarios and the effects of the same climate variations on sample building energy needs by adopting dynamic energy simulations.

References

Aparicio-Ruiz P, Schiano-Phan R, Salmerón-Lissén JM (2018) Climatic applicability of downdraught evaporative cooling in the United States of America. Build Environ 136:162–176

Aronin JE (1953) Climate and architecture. von Nostrand Reinhold, New York

Artmann N, Manz H, Heiselberg P (2007) Climatic potential for passive cooling of buildings by night-time ventilation in Europe. Appl Energy 84(2):187–201
Ascione F, Bellia L, Minichiello F (2011) Earth-to-air heat exchangers for Italian climates. Renew Energy 36:2177–2188
Bom GJ, Foster R, Dijkstra E, Tummers M (1999) Evaporative air-conditioning, applications for environmentally friendly cooling, World Bank Technical Paper No. 421, The World Bank, Washington
Brown GZ, Dekay M (2001) Sun, wind & light. John Wiley & Sons, New York
Carlucci S, Pagliano L (2012) A review of indices for the long-term evaluation of the general thermal comfort conditions in buildings. Energy Build 53:194–205
Cavaglià G, Ceragioli G, Foti M, Maggi PN, Matteoli L, Ossola F (1975) Industrializzazione per programmi. RDB, Piacenza
Chiesa G (2019a) Calculating the geo-climatic potential of different low-energy cooling techniques. Build Simul 12:157–168
Chiesa G (2019b) Climatic potential maps of ventilative cooling techniques in Italian climates including resilience to climate changes. In: IOC Conference Series: Materials Science and Engineering, vol 609. IOP Publishing, pp 032039
Chiesa G (2020) Technological paradigms and digital eras Data-driven visions for building design. Springer, Cham
Chiesa G, Grosso M (2015) Geo-climatic applicability of natural ventilative cooling in the mediterranean area. Energy Build 107:376–391
Chiesa G, Grosso M (2017) An environmental and technological approach to architectural programming for school facilities. In: Sayigh A (ed) Mediterranean green buildings & renewable energy. Springer, Cham, pp 701–716
Chiesa G, Huberman N, Pearlmutter D (2019) Geo-climatic potential of direct evaporative cooling in the mediterranean region: a comparison of key performance indicators. Build Environ 151:318–337
Ciribini G (1968) Brevi note di metodologia della progettazione architettonica. Edizioni Quaderni di Studio, Torino
Cook J (ed) (1989) Passive cooling. MIT Press, USA, Cambridge
Givoni B (1969) Man, climate and architecture. Elsevier, Amsterdam
Grosso M (1986) Dinamica delle ombre, 2nd edn. Celid, Torino
Grosso M (ed) (2017) Il raffrescamento passivo degli edifici in zone a clima temperato, Maggioli, Sant'Arcangelo di Romagna
Koenigsberger OH, Ingersoll TG, Mayhew A, Szokolau SV (1975) Manual of tropical housing and building. climatic design. OrientBlackswan Private Limited, Hyderabad
Kolokotroni M, Heiselberg P (2015) Ventilative cooling. State-of-the-art review, IEA EBC ANNEX 62, Aalborg University, Aalborg
Labs K (1981) Regional analysis of ground and above-ground climate, ORNL/Sub-81/40451/1, Oak Ridge National Laboratory
Matteoli L, Pagani R (eds) (2010) City futures—architecture design technology for the future of the cities. Hoepli, Milano
Meteotest (2017) Meteonorm handbook part I. Meteotest, Bern
Milne M, Givoni B (1979) Architectural design based on climate. In: Watson D (ed) Energy conservation through building design. Architectural Record Book, New York
Nicol F, Hacker J, Spires B, Davies H (2009) Suggestion for new approach to overheating diagnostics. Build Res Inf 37(4):348–357
Nicol F, Humphreys M, Roaf S (2012) Adaptive thermal comfort Principles and Practice. Earthscan, London
Olgyay V (1963) Design with climate: bioclimatic approach to architectural regionalism. Princeton University Press, Princeton
Olgyay A, Olgyay V (1957) Solar control and shading devices. Princeton University Press, Princeton

Salmeron JM, Sánchez FJ, Sánchez J, Álvarez S, Molina LJ, Salmeron R (2012) Climatic applicability of downdraught cooling in Europe. Archit Sci Rev 55:259–272

Santamuris M, Asimakopoulos D (eds) (1996) Passive cooling of buildings. James and James, London

U.S. Department of Energy (2018) Energy Plus version 8.9.0 Documentation. Auxiliary Programs, Cap 12, pp 176–180

Vitruvius Pollio M (15 B.C. about) De Architettura (italian edition (2002) Architettura, Biblioteca Universale Rizzoli, Milano)

Watson D, Labs K (1983) Climatic design; energy-efficient building principles and practices. McGraw-Hill, New York

Zajch A, Gough W, Chiesa G (2020) Earth-air heat exchanger potential under future climate change scenarios in nine North American cities. In: Littlewood J, Howlett RJ, Capozzoli A, Jain LC (eds) Sustainability in energy and buildings. Springer, Cham, pp 109–119

Chapter 21
Greening the Building Envelope

Katia Perini

Abstract Four out of five European citizens live in urban areas and their quality of live is influenced by the environmental conditions. Greening the building envelope provides important ecosystem services, improves microclimatic conditions and reduces the energy demand for air conditioning. The chapter presents research results aimed at identifying benefits and challenges of vertical greening systems. In order to evaluate if alternative low-cost and low-maintenance vertical greening systems can be developed, the case of mosses is presented.

Keywords Environmental quality · Urban areas · Green envelopes · Ecosystem services · Economic and environmental assessment

21.1 Introduction

Over 50% of the global population currently lives in urban areas; this proportion, according to the United Nations, is expected to increase to 68% by 2050 (United Nations and Division 2018). Cities are particularly exposed to climate change and environmental problems due to the impact of anthropic activities. Flooding, heat and drought, in particular, are hazards which are increasingly characterising the urban areas (Commission of the European Communities 2005; Field et al. 2014). Climate change and anthropogenic pressures have altered the functions of ecological systems Nelson et al. (2013) and have consequently modified the flow of ecosystem services in terms of their scale, timing and location (Evans et al. 2014). In urban environments, additionally, the negative effects of climate change are amplified by settlement features (impervious surface, buildings, transport infrastructure, socio-economic activities) (Perini and Sabbion 2017).

Greening systems can improve environmental conditions in cities, and thus quality of life of citizens by providing multiple ecosystem services (Dover 2015; Milliken 2018). Nature based strategies, such as urban forestry and greening systems for the

K. Perini (✉)
Dipartimento Architettura e Design, Università degli Studi di Genova—Scuola Politecnica, Genoa, Italy
e-mail: katia.perini@unige.it

G. Chiesa (ed.), *Bioclimatic Approaches in Urban and Building Design*, PoliTO Springer Series, https://doi.org/10.1007/978-3-030-59328-5_21

building envelope, can be implemented to curb a wider urban heat island phenomenon having regional scale impact on energy demand, air quality and public health (Perini and Magliocco 2014; Perini and Roccotiello 2018; Taha 1997). Green infrastructure (GI) can contribute to mitigate the negative effects of climate-related hazards, including storm surges, extreme precipitation, and floods (Perini and Sabbion 2017). Planning, developing, and maintaining GI can integrate urban development, nature conservation, and public health promotion (Benedict et al. 2012; Dover 2015; Milliken 2018).

The chapter presents an overview on the latest research results aimed at quantifying the ecosystem services provided and the most relevant environmental and economic costs and benefits. Research results show the multifunctionality of greening systems and performances, in terms of microclimate regulation (outdoor comfort), air quality improvement, providing at the same time cultural and social services for resilient cities. In addition, the case study of mosses is presented, as an alternative low-cost and low-maintenance greening system.

21.2 Vertical Greening Systems Main Characteristics

Vertical greening systems include many different ways to vegetate building envelopes, with different configurations, layers, materials, plant species, growing methods, installation costs, maintenance requirements, etc. (Table 21.1; Figs. 21.1 and 21.2). According to the literature (Fernández-Cañero et al. 2018; Köhler 2008; Perini et al. 2012), vertical greening systems can be classified considering the growing method as living wall systems (Fig. 21.1) and green façade (Fig. 21.2). In the first case, the growing method of living wall systems is hydroponic, i.e. since plants grow without or with a very small amount of organic substrate, an automated irrigation system provides water and nutrients. In the second case, climbing plants or shrubs root in the ground in front of a building or in planter boxes filled with substrate, covering directly or indirectly (i.e. when plants grow on a supporting mesh) the surface. For living wall systems, different materials can be used for the (thin) supporting structure. In the case of cloth system, felt or other textile material are connected to a waterproofing layer working as support (e.g. PVC panel) and create small pockets to host plants. Modular panels are made of plastic planter boxes filled with organic substrate, foam panels, etc. (Fernández-Cañero et al. 2018). As according to Table 21.2, which summarizes the main characteristics of vertical greening systems, living wall systems allow integrating a wide range of plant species (including also vegetables) to cover rapidly a façade but require periodic/constant maintenance. Green façades require less maintenance but, depending on the plant species used and on the surface to cover, the growing speed can be a limiting factor. As an example, an Hedera helix will require more than 10 years to cover a 7 storey building.

Table 21.1 Main characteristics and performances of vertical greening systems (Fernández-Cañero et al. 2018; Perini et al. 2020)

	Green façades	Indirect green façades	Hanging indirect green façades	Living wall cloth systems	Living wall modular panel systems
Scheme	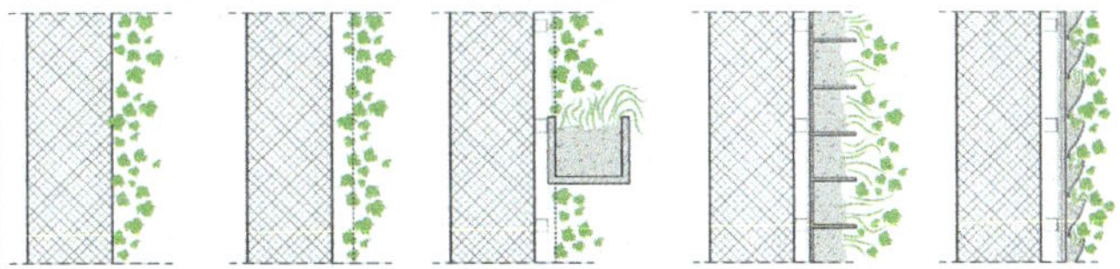				
Layers/materials/supporting structure	Creepers planted in the ground and cling directly to the wall	Light support structures for creepers of stainless steel or similar (cables, nets, trellis)	Planter boxes at different heights connected by means of light supporting structures for creepers	Container elements (galvanized steel, polyethylene, or recycled plastic) with organic substrate	Textile or nonwoven felt with pockets. Hydroponics
Performances	Aesthetic, passive energy-savings, biodiversity improvement			Aesthetic, passive energy-savings, marketing, biodiversity improvement, vegetable production	
Growing method	Organic substrate			Hydroponic cultures with/without organic substrate	
Vegetation	climbing and hanging plants		Climbing, hanging plants and shrubs	Epiphytic, lithophytic and bromeliads, ferns, succulent, herbaceous, small shrubs, climbers and even vegetables	
Growing speed	Slow	Medium-slow	Medium-fast	Fast	
Irrigation	Periodically depending on plants and climate			Computerized irrigation (1–5 l/m2/day)	
Maintenance requirements	Low		Low-medium	Medium-high	

Fig. 21.1 Living wall system designed by Patrick Blanc, The Driver, London, UK

Fig. 21.2 Direct green façade in Milan, Italy

Table 21.2 Ranges of installation and maintenance costs for the main vertical greening systems (Perini et al. 2020; Perini and Rosasco 2016; Rosasco 2018)

	Green façades	Indirect green façades	Hanging indirect green façades	Living wall—cloth systems	Living wall—modular panel systems
Installation costs (€/m^2)	22–39	127–270	190–365[a]	210–590[a]	
Maintenance costs (€/m^2/year)	2–5	2–5	5–7.5	40–100	

[a]Depending on system conception, material and other factors

21.3 Ecosystem Services Provided by (Vertical) Greening Systems

Ecosystem services are the benefits provided by nature to society and economy, they include biodiversity, food, fibre, water resources, carbon sequestration, and recreation (Millennium Ecosystem Assessment (Program) 2005). The living conditions of cities depend on such services, since a healthy natural environment provides services as clean air and protection against floods (TEEB 2011).

Ecosystem services can be divided into four categories (TEEB 2011): provisioning services, regulating services, habitat (or supporting services), and cultural services. Provisioning services are the material or energy outputs from ecosystems, e.g. food, raw materials, fresh water, medicinal resources. Regulating services are particularly relevant in urban contexts, as they include the regulation of air quality, noise, climate, carbon sequestration and storage, moderation of extreme events, waste-water treatment, erosion, prevention and maintenance of soil fertility, pollination, and biological control. The Millennium Ecosystem Assessment defined cultural ecosystem services as "the nonmaterial benefits people obtain from ecosystems through spiritual enrichment, cognitive development, reflection, recreation, and aesthetic experiences", e.g.: aesthetic, spiritual, physical and psychological benefits, recreation and tourism. Finally, habitat services regard the living spaces for plants and animals to enhance the biological diversity of species and the variety of genes within populations (Millennium Ecosystem Assessment (Program) 2005; Milliken 2018; Perini and Sabbion 2017).

In the challenge of improving cities conditions, green areas as well as green envelopes play a key role, due to the ecosystem services they can provide. Indeed, as summarized in Fig. 21.3, the ecosystem services provided by green infrastructure, among which are green envelopes, regard several fields. Green envelopes can insulate, shelter and shade, contributing to energy management (Coma et al. 2017, 2016). By directly collecting, storing and evaporating rain water (through sustainable urban drainage practices), greening systems can improve water management (Ballard et al. 2007). Evaporation and shade are among the most relevant mechanisms of plants which allow regulating urban climate (Pérez et al. 2018). Green

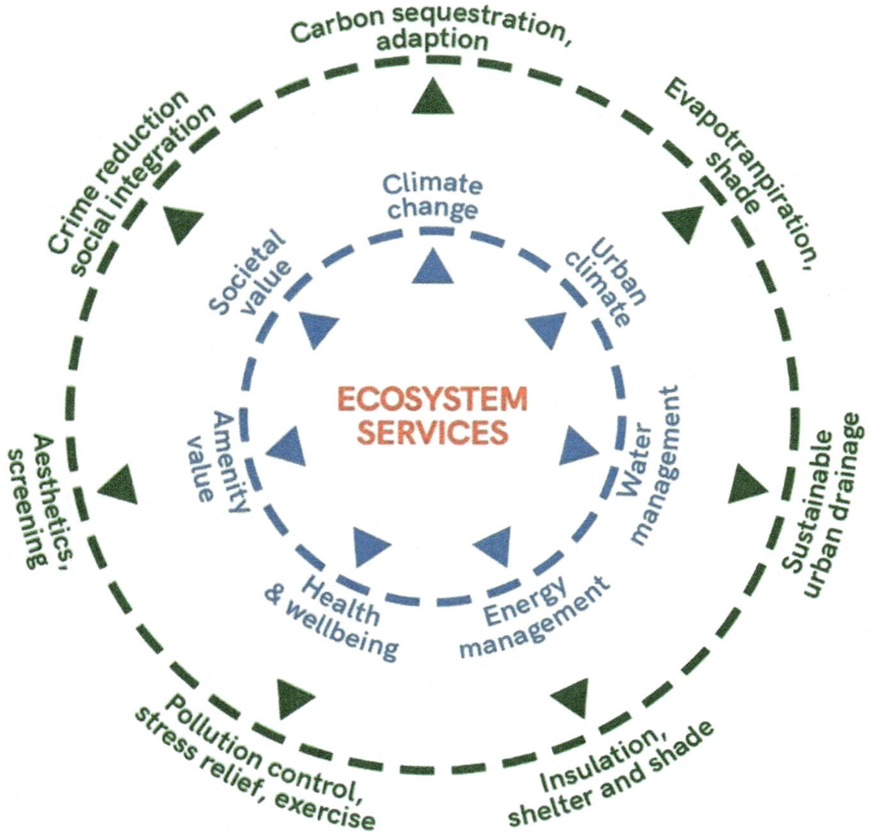

Fig. 21.3 Ecosystem services which can be delivered by green infrastructure in urban areas (Dover 2018)

infrastructures contribute to mitigation and adaptation to climate change (European Environmental Agency 2015; Perini and Sabbion 2017). The integration of vegetation in cities can improve health and wellbeing, by controlling pollutants, stress relief and exercise (the latter is more connected to the integration of green infrastructure such as greenways) (Fjeld et al. 1998; Perini and Sabbion 2017; Ulrich 1984). Green infrastructures have also relevant amenity and societal values. Urban vegetation plays an important role for human psychological well-being. In the Eighties, Wilson (1984) defined the term 'biophilia' as « the urge to affiliate with other forms of life » , stressing that human are instinctively bonded to other living systems. Green envelopes are now recognised as means to increase the property value of buildings, due to their aesthetic value (Rosasco and Perini 2018).

The effects of vegetation on wellbeing and human health are also related to the improvement of environmental conditions of cities. The urban heat island phenomenon can be mitigated also by means of vegetated surfaces, which intercept solar radiation and thus decrease the warming of hard surfaces (Petralli et al. 2006; Susca et al. 2011). Worth mentioning that, in order to obtain a relevant effect, a wide surface of the same area/district have to be greened (Rosenzweig et al. 2006).

The effects of a vertical greening systems can be connected to urban climate regulation and to energy management. The benefits provided by greenery, in terms of passive energy saving in building and indoor comfort, are well known and reported (Besir and Cuce 2018). These benefits depend on basically four mechanisms: the shade effect, the cooling effect, the insulation effect, and finally the wind barrier effect (Pérez and Perini 2018). The shade effect is mainly related to the solar radiation interception provided by plants; the cooling effect takes depends on the water evapotranspiration process from the substrates (evaporation) and plants (transpiration); the insulation capacity is related to the characteristics of the layers/materials used (as described in the paragraph 1.1). The wind barrier effect refers to the capacity of a green layer to modify the direct wind effect on the building (Coma et al. 2017; Pérez et al. 2018).

Urban pollutants (NO_x, CO, SO_2, VOC, etc.) cause relevant damages to human health. Phytoremediation is shown to have a potential to mitigate such environmental pollutants in the air; in addition, plants are able to decrease the concentration of airborne particulates, gaseous pollutant such as VOC, NO_x, CO_2, CO and Ozone (Abhijith et al. 2017; Janhäll 2015). Computational Fluid Dynamic (CFD) simulations show that the effects of vegetation on fine dust particles deposition and dispersion and uptake of gaseous pollutants (CO_2, NO_2, SO_2) depend on several parameters, including the site geometry (e.g. urban canyon or open road), meteorology and air quality, varying across space and time (Janhäll 2015). The specific characteristics of plants play a key role: for example, the PM10-PM2.5 collecting capacity of plant species significantly varies depending on leaves shape and surface (micromorphology) (Perini and Roccotiello 2018).

21.4 Economic and Environmental Impacts

Vertical greening systems can improve the environmental quality of dense urban areas, providing a wide range of ecosystem services, as described above. The types of vertical greening systems, green façades and living wall systems, entail different environmental and economic costs, which are not always balanced by the benefits provided (Ottelé et al. 2011; Perini 2018; Perini and Rosasco 2013).

Life cycle assessment methodology—i.e., according to the ISO14040:2006 the compilation and evaluation of the inputs (e.g. raw materials, water, energy), outputs (e.g. emissions in the air, waste) and the potential environmental impacts of a product system throughout its life cycle—can be used to compare environmental impacts, deriving from construction, installation, maintenance, and disposal, of a vertical greening system and the environmental benefits provided.

According to the studies developed in the field, the most influencing aspects to be considered to evaluate the environmental impact of vertical greening systems are the materials and components used (data collected by means of an inventory analysis), and related dimensions and weight, which depend on the greening system type. For example, supporting system made of steel mesh, wood trellis, plastic mesh (indirect green façade) or felt, HDPE planter boxes, geotextile, etc., for a living wall system. Studies show that the material stage contributes in a range of 43–97% to all the environmental impact categories (Pan and Chu 2015). The selection of recycled materials can reduce the environmental burden of a system.

In addition to the materials used, also the maintenance needs can play an important role. The water demand of living wall systems (irrigation, around 3 L/m2/day) influences the total environmental burden especially in terms of the impact categories eutrophication potential and freshwater aquatic ecotoxicity (Pan and Chu 2015). Plant species used for vertical greening systems require pruning, from twice to several times per year. In addition, if the durability of the materials used as supporting structure is not adequate, panels have to be replaced or fixed, which implies raw materials extraction, transport, production, construction, etc., with a consequent increase of the total environmental burden produced by a system (Ottelé et al. 2011).

The economic costs of vertical greening systems vary depending some factors like: location (urban or peripheral area, local climate, etc.), type of plants and system (direct, indirect, living wall). The most influencing and variable costs are related to installation and maintenance (Table 21.2): besides the costs of the system (panel, materials, plants, etc.), also the maintenance of plants and vertical systems play an important role in the global cost of the system within lifespan.

In order to evaluate the environmental and economic impact of a system, the benefits which can be provided can be taken into account. In the LCA conducted to evaluate the environmental impacts of vertical greening systems, the benefits calculated are mainly related to the potential energy reduction through summer cooling (Ottelé et al. 2011; Pan and Chu 2015), as also the air cleaning potential (Feng and Hewage 2014). In the case of a cost benefit analysis, also the increase of building plaster facade durability, which reduces the frequency of intervention thanks to a protective action given by the leaves and other layers involved, and the real estate value increase can be play an important role (Rosasco 2018). The latter is quantified by several authors, e.g. Hunt (2008) states that the premium of a house value can vary from 3.9 to 15.0%.

Studies demonstrate that a vertical green layer can reduce the energy demand for air-conditioning, depending on the climate (Alexandri and Jones 2008), on the greening system, façade orientation, etc., with energy saving rates for the living wall and the indirect green façade of respectively 58.9 and 33.8%, in comparison to a reference system (Pérez et al. 2018). According to Yoshimi and Altan (2011) an average daily electricity saving during summer in Hong Kong of 16% was recorded.

The results of an LCA conducted by Ottelé et al. (2011) on four different vertical greening system for a life span assumed of 50 years, show that direct green façades have a very small influence on the total environmental burden of an envelope, since they are made of vegetation only. In the case of living wall systems, the benefits (cooling potential calculated for Mediterranean area) can balance the environmental burden but, in the case of a cloth system (Table 21.1), the environmental burden can be too high. These results are confirmed by also by Feng and Hewage (2014).

Pan and Chu (2015) show that the environmental burden (in particular in regard to abiotic depletion of fossil fuels) of a vertical greening system installed in Hong Kong, can be paid back in 20 years thanks to a 16% energy savings for air conditioning during summer.

Cost benefit analysis can be used to evaluate if installation and maintenance costs within the lifespan are balanced with economic private benefits such as energy savings, lifespan increase of building plaster, and higher real estate value.

Rosasco and Perini (2018) show, in a sensitivity analysis of a VGS installed in a south façade of a building office located in Genoa, that the economic sustainability of vertical greening system is strongly influenced by the installation and maintenance costs and energy savings. In this field, public administration can play a key role by providing tax incentives, to promote the implementation of these systems at urban scale.

Worth mentioning that considering both the economic and the environmental assessment of greening systems, social and other environmental benefits (such as carbon reduction, better urban hydrology, biodiversity and habitat creation, and reduction of UHI) should be taken into account, although these benefits are less relevant in economic terms and are not always quantifiable, they are very important because they improve the quality of urban environments and alleviate climate change (Perini 2018; Rosasco 2018).

21.5 Case Study: Mosses as Innovative Biomaterials

Vertical greening systems provide ecosystem services in urban areas but, economic (initial and maintenance costs) and environmental costs may limit their diffusion, as described above. Direct green façades, based on climbing plants without supporting structure, require a reduced initial investment and environmental impact, but a plant species could take several years to cover a building envelope (Bellomo 2003).

With the aim to identify low-cost, low-maintenance and fast growing alternative green envelope for large-scale application, the possible use of mosses was evaluated by an inter-disciplinary study between plant biology and architecture (Perini et al. 2020). Indeed, mosses have low requirements in terms of growing substrates, low amount of water and nutrients needed, and high desiccation tolerance. These characteristics could potentially allow a wide diffusion on existing buildings with low budget availability, such as residential buildings in low income suburbs and neighbourhoods, industrial areas, etc.

The research Perini et al. (2020) evaluated the ability of mosses to grow on building materials and low-cost materials, instead of using specifically designed supports. These are: cement plaster, lima plaster, terracotta, slate, quartzite and capillarity matting. In order to compare the growing ability on the mentioned materials,

Fig. 21.4 Tests in the growth chamber, at the University of Genoa (DISTAV), photo by Paola Castellari

tests were performed under controlled environmental conditions (growing chamber, Fig. 21.4); the Barbula unguiculata specie was selected by researchers working in the field of botany (Fig. 21.5), due to its resistance to light radiation and dehydration.

The results obtained so far show that the use of mosses-based vertical greening, with a single layer and low-cost materials, could allow overcoming the limits deriving from the installation costs and maintenance needs of many greening systems available on the market (Perini and Rosasco 2013).

The growing ability of moss depends mainly on the physical characteristics of the materials; indeed, the most performing material in terms of covering rate is capillary matting, followed by cement plaster > lime plaster > terracotta brick > slate > quartzite. Figure 21.6 shows the growth of moss on some of these material. The second most relevant parameter is the water retention capacity of the (supporting) materials and its ability to homogeneously distribute water to mosses. Therefore, the design of an irrigation system is important for moss growth: low and constant water supply provides faster moss biomass development and reduces water waste. Further research will be oriented to the study of these two man aspects: water distribution and adhesion of moss mixture. In addition, research is needed to specifically quantify the ecosystem services which could be provided by moss envelopes in urban areas (Perini et al. 2020).

Fig. 21.5 Barbula unguiculata, the moss species selected for the study. Photo by Paola Castellari

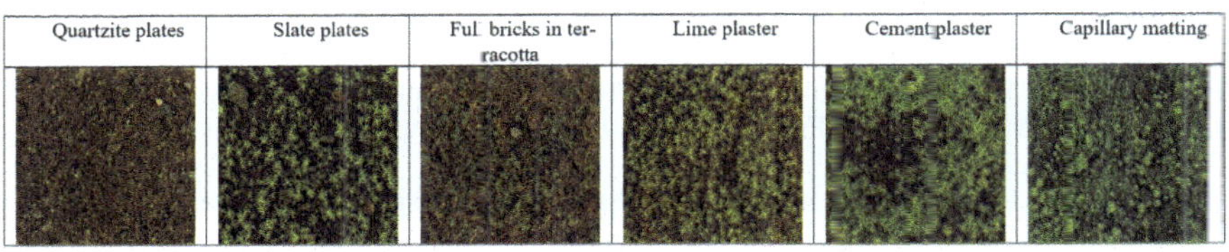

Fig. 21.6 Mosses growth on the materials analysed

21.6 Summary

Green envelopes can improve the (environmental) quality of dense urban areas by providing a wide range of ecosystem services. These regard several fields: energy management, water management, urban climate, climate change (mitigation and adaptation), societal and amenity value and health and wellbeing, as described in the chapter. However, vertical greening systems can entail high economic and environmental burden, which is not always balanced by the benefits. Life cycle assessment methodology can be effectively used to demonstrate if a vertical greening system can be considered environmentally sustainable or not. Cost benefit analysis allow evaluating the pay-back period of a system to verify the economic sustainability. Energy saving for air conditioning is the most influencing benefit to consider in order to balance the environmental burden produced by a vertical greening systems. Economic investments can be supported by tax incentives and balanced also by real estate increase and longevity of façade.

The case study presented on the use of moss as green envelope show that low-cost and low-maintenance alternatives can be explored. Using such vegetation, with low requirements in terms of growing substrates, water and nutrients needed, and high desiccation tolerance, could allow a wide diffusion of green envelopes on existing buildings with low budget availability, such as residential buildings in low income suburbs and neighbourhoods, industrial areas, etc.

Acknowlegements The research was supported by PRA 2019 funding (Progetti di Ricerca di Ateneo 2019), provided by the University of Genoa–DAD.

References

Abhijith KV, Kumar P, Gallagher J, McNabola A, Baldauf R, Pilla F, Bro-derick B, Di Sabatino S, Pulvirenti B (2017) Air pollution abatement performances of green infrastructure in open road and built-up street canyon environments—a review. Atmos Environ 162:71–86. https://doi.org/10.1016/j.atmosenv.2017.05.014

Alexandri E, Jones P (2008) Temperature decreases in an urban canyon due to green walls and green roofs in diverse climates. Build Environ 43:480–493. https://doi.org/10.1016/j.buildenv.2006.10.055

Ballard BW, Kellagher R, Martin P, Jefferies C, Bray R, Shaffer P (2007). Site handbook for the construction of SUDS. CIRIA

Bellomo A (2003) Pareti verdi: linee guida alla progettazione/ Antonella Bellomo. Esselibri, Napoli

Benedict MA, McMahon, ET, Fund, MATC (2012) Green infrastructure: linking landscapes and communities. Island Press

Besir AB, Cuce E (2018) Green roofs and facades: a comprehensive review. Renew Sustain Energy Rev 82:915–939. https://doi.org/10.1016/j.rser.2017.09.106

Coma J, Pérez G, de Gracia A, Burés S, Urrestarazu M, Cabeza LF (2017) Vertical greenery systems for energy savings in buildings: a comparative study between green walls and green facades. Build Environ 111:228–237. https://doi.org/10.1016/j.buildenv.2016.11.014

Coma J, Pérez G, Solé C, Castell A, Cabeza LF (2016) Thermal assessment of extensive green roofs as passive tool for energy savings in buildings. Renew Energy 85:1106–1115. https://doi.org/10.1016/j.renene.2015.07.074

Commission of the European Communities (2005) EUR-Lex—52005DC0718—EN

Dover JW (2018) Chapter 1.1—Introduction to urban sustainability issues: Urban Ecosystem. In: Pérez G, Perini K (eds) Nature based strategies for urban and building sustainability. Butterworth-Heinemann (Elsevier), Oxford, United Kingdom, pp 3–15. https://doi.org/10.1016/B978-0-12-812150-4.00001-X

Dover JW (2015) Green infrastructure: incorporating plants and enhancing biodiversity in buildings and urban environments, 1st edn. ed. Routledge, London; New York
European Environmental Agency (2015) Exploring nature-based solutions. The role of green infrastructure in mitigating the impacts of weather and climate change-related natural hazards
Evans CD, Bonn A, Holden J, Reed MS, Evans MG, Worrall F, Couwen-berg J, Parnell M (2014) Relationships between anthropogenic pressures and ecosystem functions in UK blanket bogs: linking process understanding to ecosystem service valuation. Ecosys Serv 9:5–19. https://doi.org/10.1016/j.ecoser.2014.06.013
Feng H, Hewage K (2014) Lifecycle assessment of living walls: air purification and energy performance. J Clean Prod 69:91–99. https://doi.org/10.1016/j.jclepro.2014.01.041
Fernández-Cañero R, Pérez Urrestarazu L, Perini K (2018) Chapter 2.1—vertical greening systems: classifications, plant species, substrates. In: Pérez G, Perini K (eds) Nature based strategies for urban and building sustainability. Butterworth-Heinemann (Elsevier), Oxford, United Kingdom, pp 45–54. https://doi.org/10.1016/B978-0-12-812150-4.00004-5
Field CB, Barros VR, Mastrandrea MD, Mach KJ, Abdrabo M-K, Adger N, Anokhin YA, Anisimov OA, Arent DJ, Barnett J, others (2014) climate change 2014: summary for policymakers. climate change 2014: impacts, adaptation, and vulnerability. Part a: global and sectoral aspects. Contribution of working group II to the fifth assessment report of the intergovernmental panel on climate change, pp 1–32
Fjeld T, Veiersted B, Sandvik L, Riise G, Levy F (1998) The effect of indoor foliage plants on health and discomfort symptoms among office workers. Indoor Built Environ 7:204–209. https://doi.org/10.1177/1420326X9800700404
Hunt D (2008) POW/MIAs & John McCain (4). Nation 287
Janhäll S (2015) Review on urban vegetation and particle air pollution—deposition and dispersion. Atmos Environ 105:130–137. https://doi.org/10.1016/j.atmosenv.2015.01.052
Köhler M (2008) Green facades—a view back and some visions. Urban ecosys 11:423–436. https://doi.org/10.1007/s11252-008-0063-x
Millennium Ecosystem Assessment (Program) (Ed) (2005) Ecosystems and human well-being: synthesis. Island Press, Washington, DC
Milliken S (2018) Chapter 1.2—ecosystem services in urban environments. In: Pérez G, Perini K, (eds) Nature based strategies for urban and building sustainability. Butterworth-Heinemann (Elsevier), Oxford, United Kingdom, pp 17–27. https://doi.org/10.1016/B978-0-12-812150-4.00002-1
Nelson EJ, Kareiva P, Ruckelshaus M, Arkema K, Geller G, Girvetz E, Goodrich D, Matzek V, Pinsky M, Reid W, Saunders M, Semmens D, Tallis H (2013) Climate change's impact on key ecosystem services and the human well-being they support in the US. Front Ecol Environ 11:483–893. https://doi.org/10.1890/120312
Ottelé M, Perini K, Fraaij ALA, Haas EM, Raiteri R (2011) Comparative life cycle analysis for green façades and living wall systems. Energy Build 43:3419–3429. https://doi.org/10.1016/j.enbuild.2011.09.010
Pan L, Chu LM (2015) Energy saving potential and life cycle environmental impacts of a vertical greenery system in Hong Kong: a case study. Build Environ. Https://doi.org/10.1016/j.buildenv.2015.06.033
Pérez G, Coma J, Cabeza LF (2018) Vertical greening systems to enhance the thermal performance of buildings and outdoor comfort In: Pérez G, Perini K (eds) Nature based strategies for urban and building sustainability. Butterworth-Heinemann, Oxford, United Kingdom, pp 99–108. https://doi.org/10.1016/B978-0-12-812150-4.00009-4
Perini K (2018) Life cycle assessment of vertical greening systems. In: Pérez G, Perini K (eds) Nature based strategies for urban and building sustainability. Butterworth-Heinemann (Elsevier), Oxford, United Kingdom, pp 333–340. https://doi.org/10.1016/B978-0-12-812150-4.00030-6
Perini K, Castellari P, Giachetta A, Turcato C, Roccotiello E (2020) Experiencing innovative biomaterials for buildings: potentialities of mosses. Build Environ 172:106708. https://doi.org/10.1016/j.buildenv.2020.106708

Perini K, Magliocco A (2014) Effects of vegetation, urban density, building height, and atmospheric conditions on local temperatures and thermal comfort. Urban Forest Urban Green. https://doi.org/10.1016/j.ufug.2014.03.003

Perini K, Ottelé M, Haas EM, Raiteri R (2012) Vertical greening systems, a process tree for green façades and living walls. Urban Ecosyst 16:265–277. https://doi.org/10.1007/s11252-012-0262-3

Perini K, Roccotiello E (2018) Chapter 3.4—Vertical greening systems for pollutants reduction. In: Pérez G, Perini K (eds) Nature based strategies for urban and building sustainability. Butterworth-Heinemann (Elsevier), Oxford, United Kingdom, pp 131–140. https://doi.org/10.1016/B978-0-12-812150-4.00012-4

Perini K, Rosasco P (2016) Is greening the building envelope economically sustainable? an analysis to evaluate the advantages of economy of scope of vertical greening systems and green roofs. Urban Forest Urban Green. https://doi.org/10.1016/j.ufug.2016.08.002

Perini K, Rosasco P (2013) Cost–benefit analysis for green façades and living wall systems. Build Environ 70:110–121. https://doi.org/10.1016/j.buildenv.2013.08.012

Perini K, Sabbion P (2017) Urban sustainability and river restoration: green and blue infrastructure. Wiley, Construction Sustainability

Petralli M, Prokopp A, Morabito M, Bartolini G, Torrigiani T, Orlandini S (2006) Ruolo delle aree verdi nella mitigazione dell'isola di calore urbana: uno studio nella città di Firenze. Rivista Italiana di Agrometeorologia 1:51–58

Rosasco P (2018) Chapter 4.4—Economic benefits and costs of vertical greening systems. In: Pérez G, Perini K (eds) Nature based strategies for urban and building sustainability. Butterworth-Heinemann (Elsevier), Oxford, United Kingdom, pp 291–306. https://doi.org/10.1016/B978-0-12-812150-4.00027-6

Rosasco P, Perini K (2018) Evaluating the economic sustainability of a vertical greening system: a cost-benefit analysis of a pilot project in mediterranean area. Build Environ 142:524–533. https://doi.org/10.1016/j.buildenv.2018.06.017

Rosenzweig C, Solecki WD, Slosberg RB (2006) Mitigating New York City's heat island with urban forestry, living roofs, and light surfaces. New York City regional heat island initiative. New York State Energy Research and Development Authority (NYSERDA). New York City

Susca T, Gaffin SR, Dell'Osso GR (2011) Positive effects of vegetation: urban heat island and green roofs. Environ Pollut 159:2119–2126. https://doi.org/10.1016/j.envpol.2011.03.007

Taha H (1997) Urban climates and heat islands: albedo, evapotranspiration, and anthropogenic heat. Energy Build 25:99–103. https://doi.org/10.1016/S0378-7788(96)00999-1

TEEB (2011) TEEB Manual for cities: ecosystem services in urban management. In: UNEP and the European Union (ed) The economics of ecosystems and biodiversity. Manual for Cities: Ecosystem Services in Urban Management

Ulrich R (1984) View through a window may influence recovery. Science 224:224–225

United Nations, DESA Population Division (2018) World urbanization prospects: the 2018 revision

Wilson EO (1984) Biophilia. Harvard University Press, Cambridge, Mass

Yoshimi J, Altan H (2011) Thermal simulations on the effects of vegetated walls on indoor building environments. In: Proceedings of Building Simulation, pp 1438–1443

Chapter 22
Shading Architectures—Bioclimatic Approach to "Well Tempered" Civic Spaces

Giovanni Scudo

Abstract In relation to the growing impact of urban heat island, thermal well-being in outdoor spaces becomes a relatively new environmental problem architectural designers have to take in charge. Bioclimatic approach to architecture can give a substantial contribution to achieve liveable conditions in outdoor spaces as well as to maintain multi sensorial comfort conditions, sustain natural rhythms, and stimulate human senses. Adaptive approach to thermal comfort in inner spaces—recently adopted by UE—is opening the road to more sensitive comfort approaches also for outdoor spaces. Green structures contribute to mitigate urban microclimates through a multi scale "climate control strategies": shading, air temperature reduction, setting off and canalising cold breezes. Architects have now a responsibility to develop bioclimatic-based design to promote mutual support among natural and social communities of living beings as a tool for co-existence and progress. A toolbox can supply architects a basic knowledge on green systems morphology and metabolism in order to develop sensibility and bioclimatic design capability by shaping accordingly the green/built environment since the preliminary design phases. Basic knowledge of such a toolbox are: bioclimatic urban and building design tools; morphology, metabolism and sociology of green systems; building physics; principles of urban climatology; sensorial experiences, and evaluation of comfort conditions in outdoor spaces.

Keywords Shading architecture · Bioclimatic approach · "well-tempered" outdoor spaces

---

G. Scudo (✉)
Politecnico di Milano, Milan, Italy
e-mail: gianni.scudo@gmail.com

G. Chiesa (ed.), *Bioclimatic Approaches in Urban and Building Design*. PoliTO Springer Series, https://doi.org/10.1007/978-3-030-59328-5_22

22.1 Introduction: Environmental Crisis and Ecological Regeneration

Trans-scalar design, realisation and maintenance of built contexts had in the past very low sensibility and evaluation of their environmental fallouts. Global urbanisation has exerted enormous impacts on ecosystem matrixes (bio diversity, ground, water, air, agriculture, energy), causing the current crisis, which has brought to high risk the environmental balance of many ecosystems and of a large part of mankind in the planet.

Nowadays, architecture has to take in charge complex environmental problems, which were not considered in its modern and contemporary programmes, developed within the "without limits growth" paradigm (Scudo 2009). Nevertheless, some very important pioneer contributions have been done since the half of the XIX century such as:

- the British utopians—More, Swift, Morris[1]—and the anarchist movements (Kropotkin 1973), (Bookchin 1974) to be reread in the context of Italian contemporary environmental debates (Bottero 2005);
- the Anglo Saxon Bioregionalism contribution—Munford, Geddes, Berg and others—to be reread within the framework of the 'Italian territorial' school (Magnaghi 2010, 2011; Fanfani and Saragoza 2011)

In the context of the current global climate crisis mitigations and adaptations policies, promoted by international Institutions (UN COP 21, Paris 2015 9), European Union Environmental Energy Directives, National and Local Government Policies, environmental citizen associations, aim mainly to improve materials and energy efficiency to reduce environmental pressure. Only recently, efficiency begin to be structurally associated to 'sufficiency'[2] and 'inclusive wellbeing'.[3]

22.2 Adaptation and Inclusive Thermal Comfort in Outdoor Spaces

Inclusive comfort processes, based also on a 'sufficiency' concept, affect all environmental regeneration aspects of human settlements, with special attention to outdoor spaces (streets, squares, courts, parks) which are the new, old rooted places to meet and aggregate (Consonni 2000).

Multi-cultural aggregation policies, design and management of urban landscapes contribute to the wellbeing in outdoor spaces through various strategies: soft mobility, presence of water bodies, urban furniture's, and green structures (Cooper Moore 1984), which play a very important role, largely undervalued so far.

22.3 Shading Architecture

Bioclimatic performance is a characteristic of site-specific architecture projects. It can be evaluated using the concept of 'shading architecture' following nature and life rhythms (day/night; winter/summer) within ecosystem coevolution (Magnaghi 2010), as the opposite to 'zero shading' architecture, which follows global growth commercial rhythms: night/day, summer/winter according to market demand.

Shading architecture is a structural aspect of the bioclimatic approach, which "takes shape" from site topography, anisotropic field of local microclimatic forces (solar and thermal radiation, air and radiant temperature, wind, relative humidity) ground and vegetation, along with historical tradition and demand of inclusive housing in comfortable environment with high level of vitality (defined as a performance indication of liveability).

22.4 Vitality, Consonance and Thermal Comfort

Lynch defines vitality through three levels of performance: sustenance, security, consonance (Lynch 1990).

The first two are predictable (even if urban security in general—including health—is still an open problem, at least in the Italian context). The consonance is linked to its field of actions: maintain multi sensorial comfort conditions, sustain natural rhythms, stimulate senses.

In shading architecture, consonance performance concerns environmental comfort in a trans-scalar continuum from a room to outdoor spaces.

The adaptive approach to thermal comfort in indoor spaces—recently adopted by UE—is opening the road to more sensitive comfort approaches also for outdoor spaces.

The adaptive model, which proposes a positive correlation of summer indoor comfort temperature in a given day to the average outdoor temperature in the previous weeks, allows for designing buildings with lower energy use when compared to a restrictive interpretation and application of statistic models, such as Fanger's. (Pagliano and Zangheri 2010). Furthermore, thermal (and environmental) comfort in outdoor spaces is much more influenced by behavioural and psychological aspects.

In outdoor spaces, human behaviour is more variable and random than in indoor spaces, and expectations to change comfort conditions can play an important role. As an example, if I am walking in a sunny street in midday summer and I see a green structure non far away, I feel better: my thermal balance did not change, but the view of a green structure anticipates the cool sensation I will have in a short time. Since our thermal sensors are not distance receptor, we have to rely on other senses (in this case, seeing) to give us comfort clues in advance.

22.5 Green Structures as Bio-Climatically Sustainable Matrixes in Outdoor Spaces

To be environmentally efficient, 'sufficient' and 'inclusive', green structures must involve all spatial scales, from "room" to townscape, and be organised as matrixes of different types: linear green networks for "sweet" mobility (cyclist and pedestrians lanes), connecting extended green outdoor spaces (squares, parks, courts) to multi-functional agricultural and forestry landscapes (regional and national parks).

Particularly, agricultural landscapes must have a new life as "common goods", facilitating public access without disturbing their activities and economy.

To reach this ambitious aim, it is important to implement various strategies such as: "political measures to set up green matrixes; curbing the tendency to escape from congested cities to apparently more rural hinterland; providing multi-functional green structures as integrated and inclusive character of any urban community".[4]

In this renewed continuity and symbiosis of green structures with the urban environment, a new aspect needs to be evaluated besides the functional performance: the multi sensorial perception, which stimulates "environmental pleasure" as a sophisticate ecological value (Pallasma 2007).

"There is an extra delight in the delicious comfort of a balmy spring day as I walk beneath a row of trees and sense the alternating warmth and coolness of sun and shade".[5]

Green structures contribute to mitigate urban microclimates and heat island through a multi scale "climatic service" from interior spaces, to buildings, neighbourhood, city, landscape. Efficiency of air temperature mitigation can be in the order of 3–6 °C at the urban landscape scale (great urban parks, e.g., Tier Gartner in Berlin, or regional agricultural parks, e.g. South Milano Park), and in the order of 1–3 °C at local scale (local and" pocket" parks, street, squares, courts, green walls and roofs).

Green urban structures deeply modify local microclimates (or "Local Climatic Zones" defined by Oke) and give a substantial contribution to reduce thermal stress in open spaces through the following strategies:

- shading: foliage and canopy shade solar radiation, producing different intensity shaded zones as a function of form and density of foliage;

At mid-latitudes (35–50° lat. North), it is always convenient the use of deciduous trees which have genetically a seasonal bioclimatic behaviour, shading in summer and allowing partial solar radiation transmission through skeleton in winter (in the range of 40–80%).[6]

Deciduous shrubs, shading opaque and transparent building facades give an important bioclimatic contribution to radiation and heat control.

- air temperature reduction is produced by the effect of vegetation shading and by the effect of evaporation and transpiration from leaves, branches and ground. Evaporation process cools the air near the foliage by the mechanism of heat absorption in the phase change water/vapour process.

A big tree (i.e., a 25/30-m-high beech with a foliage surface of around 7000 sm.) evaporate about 400 water litres per day, resulting in a cooling effect similar to that of 18 small domestic air conditioning. Cooled air effect is close to canopy and is dispersed by air movements. Therefore, evaporative cooling effect by trees is function of trees concentration, mainly if they are planted at two levels of height (oasis effect).

- cool green surfaces; leaves temperature is usually equal, or a little less than, the air temperature; hence, the temperature of foliage exposed to solar radiation is generally much lower than that of irradiated mineralized surfaces (asphalt, grey stone, bricks). Vegetated green surfaces (green walls and pavements, trees) constitute, therefore, a relatively "cool sink" (about 1–2 °C less than air temperature).
- Breezes generated by the effect of temperature difference between vegetated cool areas and mineralized hot zones can be set off in winter and canalise in summer with respect to buildings, neighbourhood, and urban layout. Convective breezes can be induced in green street and places, in neighbourhood parks, in urban greenbelt.[7]

In addition to the impact on microclimate and comfort, green structures have several other important environmental functions:

Reduction of air particulate and gas pollution; shelter belt (protective plantation screens); protection against radioactivity; promotion of urban ventilation and protection by winds, noise reduction, oxygen release, and the above mentioned "consonance" linked to physical and mental health.[8]

22.6 Basic Knowledge and "Tool Box" to Preliminary Design of Green Structures Integrated to the Built Environment

The large contribution of green systems to microclimatic control and thermal comfort in outdoor spaces, in the context of the global climate crisis, calls for architects' responsibility to develop ecologically based design. This will promote "mutual support among natural communities of living beings as tool for co-existence and progress"[9] and equip themselves with a "tool box" to design integrated green systems as a contribution to the ecological regeneration of built environment.

The toolbox has to supply architects with the basic knowledge on green systems morphology and metabolism and to develop sensibility and capability to give bioclimatic shapes to the new green/built environment metabolism in the preliminary design phase.

The "tool box" has to support designers with:

A. Knowledge and practice of the bioclimatic approach to 'civic architecture urban networks' (Los 2013, 2019) and of bioclimatic design tools (Givoni 1969, Olgyay 1981; Serra 1997; Bottero et al. 1984; Szokolay 2008; Rogora 2012; Grosso

2017), which help to represent site climatic forces to "give bioclimatic shape" to multiscale "shading architecture", with particular attention to the mutual interactions vegetation/built systems.

B. Basic knowledge on morphology, metabolism and sociology of green systems. Architects must approach the "mutual support" among human and "natural" communities through detailed overlap mapping of built and green requirements underling potential future performances of new green hybrid systems in a symbiosis perspective. In particular, the toolbox has to supply:

   a. basic knowledge on plants dimensional and metabolic characters (shape, height, canopy diameter, growing time, leaves metabolism, dimension and LAI—Leaf Area Index-, skeleton, deciduous period, solar radiation transmission in different seasons).[10]
   b. basic knowledge of climatic/environmental conditions which stimulate the correct growing of vegetal community integrated into architecture (water, cold, wind, ground, atmosphere) (Navés Vinas 1992; Chiusoli 1985).

C. Basic knowledge of building physics as can be found in handbooks targeted to architects (Butera 1995, 2012)
D. Basic principles on urban climatology such as the ones included in the above-mentioned handbooks. To extend urban microclimates knowledge, the fundamental contribution on Urban Climates (Oke et al. 2017) can be read.
E. Knowledge, sensorial experiences, and evaluation of comfort conditions (or reduction of the thermal stress) in outdoor spaces (Hoppe 1999; Dessì 2007a, b). Direct experience on outside comfort conditions is very important, because we have a consolidated experience to feel and evaluate comfort in indoor space, but we lost the experience in outside spaces. Therefore, it is important to promote and practice "bioclimatic walks", experiencing different Urban Local Climates thermal sensations accompanied by some simple measures (air, surface temperatures, wind, RH). Thermal comfort evaluation in urban spaces is relatively complex, because we have to calculate the energy balance person/environment (evaporative, thermal and radiant fluxes,) which change place by place when the person moves, modifying continuously the view factor. Many programmes are available, at different complexity level[11], together with bioclimatic indicators[12] used in different fields of activities from health monitoring to urban design.

22.7 Green Outdoor Places Design Through Mutual Support of Vegetal and Human Communities

Outdoor spaces with their heterogeneity, living environments, attraction capability are the connective system of human societies and constitute those cultural, vital and liveable milieus where mutuality with vegetal communities play a fundamental role.

These milieus have been strongly compromised by the development of new global urban "cyber virtual" spaces. The origin of this compromise is a very different conception of the relations man/nature, built environment/natural cycles.

On one side, an environmental local knowledge/conception interprets man/nature relations as synergic and co-evolutive and requires, for outdoor spaces, mutuality and consonance, phased on natural cycles and biological rythms.

On the other side, a globalizing anthropocentric conception interprets this relation on the basis of a functional performance: "it is winter or summer according to the desire to take a bath in a fake see shore or skiing in a simulated snow covered mountain".[13] In this case, biological rhythms are completely alterated and, therefore, the symbiotic and emotional relations with green structures completely marginalized and used as a simple "décor".

Bioclimatic gradients (hot, warm, tepid; shadow, half shadow, sun…), which contribute to the sensitive richness of local places, tend to disappear and leave space to artificial dimensions, which degrade environmental experiences and produce emotional typologies and architectures to be replicated and sold everywhere.

Architects have to develop a new feeling toward the "Plant Nation" and provide themselves with a toolbox to design neo–symbiotic processes between green systems and the built environment.

The toolbox methodological contributions help also to stimulate design environmental imagination, integrating architecture conventional characters (Utilitas, Firmitas, Venustas) with emerging environmental characters in order to relieve the dominant visual perception constraints and enter in a multisensory design culture, able to design liveable and consonant outdoor spaces, ensemble of intangible and beneficial fluxes, which are part of our local cultural values canalised by built environment shapes and materials in our collective behaviour.

Notes

1. My idea of architecture is a broad concept because it embraces human life whole environment; (…) it represents the whole of modifications and elaborations operated on the earth surface with the view of human needs (…) Each one of us is committed to watch and preserve earth landscape right order, everyone with the portion up to him to avoid to pass on our sons a treasure which is less of that passed on by our fathers. W. Morris, Prospectus of Architecture in Civilization, 1881 p. 3, 4; quoted in Bottero (2005).
2. Sufficiency is a condition in which services basic needs (house, energy, water, food etc.) are satisfied in a socially fair and inclusive way respecting planet ecological limits (Pagliano and Erba 2019).
3. UNEP, Inclusive Green Economy, 2016.
4. Munford (1964).
5. Heshong (1979), p. 19.
6. Native vegetal associations have a natural shading pattern that is completely phased with cooling demand of building and settlement of the site (Olgyay 1981, Chap. 7).

7. Urban forestation has a great tradition. The "green belts" were often a must in the XIX and XX Centuries Urban planning (paradigmatic examples: London and Frankfurt).
8. A short review of green structure environmental services ca be found in the classical handbook Bernatsky 1978.
9. Mancuso 2019, articolo 8 p. 125.
10. Leaf as bioclimatic masterpiece in vegetal system climatic adaptation process is well synthetized in Tucci 2008 pp. 65–91.
11. The most diffused are: Solene, Radtherm, Raymen, Townscope, COMFA+, Envimet. For a synthetic description see Dessi V.
12. Humidex, PET—Physiological Equivalent Temperature. For a short description of bioclimatic indicators, see WMO/WHO 2015.
13. Barbara (2000) a. op. cit. p. 307.

References

Barbara A (2000) Storia dell'architettura attraverso i sensi. Bruno Mondadori, Milano
Bernatsky A (1978) Tree ecology and preservation. Elsevier Scientific Publishing Company, Amsterdam
Bookchin M (1974) I limiti della città. Feltrinelli, Milano
Bookchin M (1984) l'ecologia della, libertà edn. Antistato, Cesena
Bottero M (2005) Progetto Ambiente. Libreria Clup, Milano
Bottero M, Rossi GC, Scudo G, Silvestrini G (eds) (1984) Architettura Solare. Clup, Milano
Butera FM (1995) Architettura ed ambiente, Manuale per il controllo della qualità termica, luminosa ed acustica degli edifici, Etaslibri, Milano
Butera FM (eds) (2012) Confort, energia ed ambiente. In: Almanacco dell'Architetto, da un'idea di Renzo Piano, Costruire l'Architettura, Proctor, Bologna
Chiusoli A (1985) Elementi di Paesaggio. Edagricole, Bologna
Consonni G (2000) Dalla radura alla rete: ragione moderna ed ospitalità nel mondo. Milano
Cooper Marcus C (1984) Peoples places: design guidelines for urban open space. Van Nostrand Reinholds, new York
Dessì V (2007a) Strumenti e metodi per la valutazione del confort termico negli spazi urbani. Il Progetto Sostenibile 16:37–45
Dessì V (2007b) Progettare il comfort urbano. Sistemi Editoriali, Napoli
Fanfani D, Saragoza C (2011), Il bioregionalismo nelle esperienze italiane ed europee, Il Progetto Sostenibile 29:22–29
Givoni B (1969) Man climate and architecture. Elsevier Amsterdam
Grosso M (2017) Il raffrescamento passivo degli edifici. Maggioli editori, S. Arcangelo
Hawkes D (2008) The environmental imagination: technics and poetics of the architectural environment. Routledge, London
Heshong L (1979) Thermal delight in architecture. MIT Press, Cambridge
Hoppe P (1999) The physiological equivalent temperature—a universal index for the biometeorological assessment of the thermal environment. Int J Biometeorol 43:71–75
Kropotkin P (1973) La società aperta. Antistato, Cesena
Lorenzini G (1983) Le piante e l'inquinamento dell'aria. Edagricole, Bologna
Los S (2013) Geografia dell'Architettura. Progettazione Bioclimatica e disegno architettonico, Il Poligrafo, Padova

Los S (2019) Città e paesaggi come sistemi simbolici, Artena Anarchist Press
Magnaghi A (2010) Il progetto locale. Verso la coscienza di luogo, Bollati Boringhieri, Torino
Magnaghi A (2011) Il progetto locale: coscienza di luogo e autosostenibilità, il Progetto Sostenibile 29:22–29
Mancuso M (2019) La nazione delle piante—Un nuovo patto per la terra, G.us. Laterza & Figli—la Repubblica, Bari-Roma
Mumford L (1960) Landscape and Townscape, article in "Landscape" 1960. Reprinted in: The Highway and the city, Secker e Warburg, London 1964
Navés Vinas F (1992) El àrbor en la giardinerìa y paisajismo. Ediciones Omega, Barcellona
Oke TR, Mills G, Christen A, Voogt JA (2017) Urban Climates. Cambridge Univ. Press, Cambridge
Olgyay V (1981) Progettare con il clima—un approccio bioclimatico al regionalismo architettonico. F Muzzio, Padova (ed originale 1962)
Pagliano L, Erba S (2019) Energy sufficiency in (strongly intertwined) building and city design—examples for temperate and mediterranean climates. In: ECEEE summer study proceedings 2019, 10
Pagliano L, Zangheri P (2010) Comfort models and cooling of buildings in the Mediterranean zone. Adv Build Energy Res 4(1):167–200. https://doi.org/10.3763/aber.2009.0406
Pallasmaa J (2007) Gli occhi della pelle: l'architettura ed i sensi. Jaka book, Milano
Poli D (2013) Agricoltura paesaggistica, visioni, metodi, esperienze. Firenze University Press, Firenze
Rogora A (2012) Progettazione bioclimatica per l'architettura mediterranea. Wolters Kluver Italia, Milanofiori, Assago (MI)
Scudo G (1998) La vegetazione domestica e il controllo del microclima. L'architettura naturale 1:25–35
Scudo G (2009) Crescita verso sviluppo. In: Clementi M, Dessì V, Lavagna M (eds) La Rivoluzione Sostenibile: Territorio, città, architettura. Maggioli S, Arcangelo
Scudo G (eds) (2012) Ambiente esterno. In: Almanacco dell'Architetto, da un'idea di Renzo Piano, Proctor Edizioni, Bologna
Scudo G, Ochoa de la Torre JM, Spazi verdi Urbani (2003) La vegetazione come elemento di progetto per il comfort ambientale negli spazi urbani. Sistemi Editoriali, Napoli
Serra Florensa R, Coch Roura (1997) L'energia nel progetto di architettura.CittàStudiEdizioni—UTET, Torino
Silvestrini G (1998) Interazioni energetiche tra verde ed ambiente costruito. L'architettura naturale, 1:20–25
Szokolay S (2008) Introduction to architectural science: the basis of sustainable design. Architectural press, Oxford
Tanizaki J (1982) Libro d'ombra. Bompiani, Milano
Tucci F (2008) Teconologia e natura. Gli insegnamenti del mondo naturale per il progetto dell'architettura bioclimatica. Alinea editrice, Firenze
WMO (2015) World Meteorological Organization/WHO—World Health Organization (2015) Heatwaves and Health Guidance on Warming-System Development

Part VI
Technological Representation and Digital Innovation

Chapter 23
Energy and Digital Scenarios. Is the European Buildings Future Smart?

Chiara Tonelli

Abstract Two apparently distinct scenarios have an effect on the architecture, engineering and construction industry: the energy and the digital transitions. Technology evolves at a never before seen pace, transforming the ways we visualise and understand the world, and providing new tools for designing and building architecture. This paper explores the impact renewable energy and digital technologies are having on the construction industry and the new possibilities they offer.

Keywords Energy transition · Digital changes · NZEB · Sustainability · Smart city

23.1 Introduction

Contemporary society is facing with two apparently distinct scenarios, that have an effect on the architecture, engineering and construction industry (AEC): the energy and the digital transitions.

The former deals with the shift to a wide electrical use in transportation, home functioning, and energy production by renewable sources, involving strategies for low energy consumes. The latter engages people's lifestyles and way to work, buildings production and construction methods.

Today, technology evolves at a never-before seen pace, transforming the ways we visualise and understand the world, and providing new tools for designing and building architecture.

In this evolution, the majority of our buildings, which ought to mirror an evolving society, are seemingly not responding to the transformations affecting our daily lives, remaining almost unchanged since the Second World War.

Which effect should digital have on the energy efficiency of our homes and cities?

How the latest innovations in smart technology can be applied in the construction industry?

Is citizenship prepared to exploit these digital opportunities?

C. Tonelli (✉)
Department of Architecture, Roma Tre University, Rome, Italy
e-mail: chiara.tonelli@uniroma3.it

G. Chiesa (ed.), *Bioclimatic Approaches in Urban and Building Design*, PoliTO Springer Series, https://doi.org/10.1007/978-3-030-59328-5_23

This paper explores the impact renewable energy and digital technologies are having on the construction industry and the new possibilities they offer.

23.2 The Energy Transition

The energy transition is part of the wider transition to sustainable economies through the use of renewable energies, the adoption of energy saving and sustainable development techniques.

Thanks to the greenhouse effect human life could exist on the earth planet. Unfortunately, human activities' pollution is modifying this effect, bringing the global temperature rising.

Buildings use a huge amount of energy during their operation. According to the European Commission Statistic Institute (Eurostat 2020), buildings account for 39.6% of energy consumption in the European Union, more than any other sector, including transportation (30.5%) and industry (25.8%). The residential buildings account for 26.1% of the total. The vast majority of the energy used in buildings is due to heating and cooling systems (85%). Moreover, the construction sector in Europe accounts for more than 40% of the total carbon emissions (Ecofys Document 2010). Among the 265 million Europe's buildings, 240 million (90%) are housing. Thus, our cities, and above all our dwellings, are the main responsible of the climate change.

In this precise moment, a global emergency caused by the Covid-19 virus, in a matter of a few months, transformed the entire world. The pandemic spread of the virus has convinced many governments to introduce exceptional measures to contain the disease, leading to the temporary closure of many businesses and extensive restrictions on travel.

An important consequence of this emergency is the impact on Air Quality, that improved dramatically in 2020 April. Data collected by the European Space Agency (ESA) reveal the decline of air pollution, specifically nitrogen dioxide concentrations (NO_2), over Italy (ESA 2020).

A Harvard study has found that long-term exposure to air pollution may significantly increase the risk factor for many predicted to die from Covid-19 (Harvard University 2020). Efforts to reduce pollution post-Covid-19 emergency is thus essential to decrease vulnerability in the population, also to the virus.

While satellites found decreases in one air pollutant, that does not mean the air is free of all pollution (Committee for the Coordination of Statistical Activities - CCSA 2020), being housing over occupied during this period.

If the actual tendency remains unchanged, the prevision for the near future is critical: in the retail sector, for example, the electricity requested has doubled in the period between 1980 and 2000, and it is expected to increase up to 50% by 2050 (Pilkington et al. 2011). European Union tried to enhance buildings performance and limiting their energy use through the Energy Performance of Buildings Directive (EPBD) and the related recast Directive, aiming at the drastic reduction of buildings greenhouse gas emissions of 80% by 2050, through a step-by-step definition of minimum requirements that will lead to the Nearly Zero Energy Buildings (NZEB) wide spread (Building Performance Institute Europe 2015).

Nevertheless, improving energy performance and reducing their environmental impacts in new buildings can be achieved by a simple two steps approach: reducing the energy demand and exporting energy optimally (Andaloro et al. 2010; Voss et al. 2011). Reduce the energy consumption of building is achievable using simple measures such as thermal insulation material for the building envelope (Voss et al. 2012) and designing properly the building in terms of orientation and ratio between opaque and transparent surfaces (Butera 2013), according to the local climate and microclimate (Tonelli and Grimaldo 2014).

It is clear that energy efficiency alone is not enough (Erhorn and Klutting 2012). On the other hand, it is also necessary to consider that buildings must provide a comfortable and healthy indoor environment to users (Sesana and Salvalai 2013; ISIAQ-CIB 2004).

Also because of this Covid-19 global alert, more than ever the awareness of how important is to live in comfortable and healthy environments has broaden. It has been estimated that European people spend about 90% of their time in both public and private indoor environments, such as homes, workplaces, schools, transportation vehicles, gyms (World Health Organization 2013).

So, the Indoor Air Quality has a great impact on health and quality of life. For many people, the health risks from exposure to indoor air pollution may be greater than those related to the outdoor one.

Anyhow, energy efficiency, environmental impacts and thermal comfort usually influence each other in an opposite way but should all have encompassed in sustainability visions (Bluyssen 2010; Yang and Wang 2013). Thanks to technological advancements it is nowadays possible to measure and control all these parameters: our dwellings are being transformed by home automation technologies and these approaches are predicted to become more prevalent in the future.

23.3 From Research to Practise

The challenge to reach a widespread NZEB standard involved many paths, from certifications, to legislations, to competitions or tests and prototypes. Among all this trials, the most spectacular and effective practice should be considered the Solar Decathlon Competition (Tonelli and Bellingeri 2016).

It is an international competition, established by the United States Department of Energy in 2002 in Washington DC. Since then, the competition moved to Madrid, Los Angeles, Beijing, Versailles, Colombia, Morocco, Dubai, Budapest, with many hundred thousand people involved, thanks also to the important communication and social awareness activity that teams, and organizations are required to develop.

Each edition involves 20 multidisciplinary teams selected internationally and led by universities. Teams are required to design, engineer, build and run zero energy residential prototypes, powered by the sun, assembled in 10 days in a solar village, a small smart city, based on the Internet of Things (IoT). A full set of sensors is installed in every competing housing prototype and sensors are all connected to a centralized data-logger, transmitting real-time performance data (comfort and energy balance) over the web. Every team is required to simulate the tasks of daily home use, according to a competition calendar that establishes what is to be accomplished daily, leaving it up to the teams to decide how and when to combine these activities, such as laundry, showers and cooking.

Hence, user behaviour becomes one of the parameters influencing the result of the performance, and therefore of the competition ranking.

Actually, the final result of the competition is a combination of measured performance contests (energy balance, comfort conditions, house functioning) and juried contests (architecture, energy efficiency, sustainability, engineering, affordability, communication and social awareness, innovation).

The competition originally targeted single-family homes. In the 2014 edition, in Versailles, France, the competition moved towards a more urban and local approach, asking teams to develop and build modules to densify urban settlements in their country of origin.

The Roma TRE University team was selected three times for participating in Solar Decathlon:

- in the 2012 edition, with its prototype called "MED in Italy", fast-assembly lightweight energy efficient housing tailored to the Mediterranean climate, overall awarded 3rd;
- again in 2014, with its prototype called "RhOME", namely "a home for Rome", overall awarded 1st;
- and for the 2020 edition, in Dubai UAE, that has been suspended for the Coronavirus pandemic, with a proposal for a housing prototype called "MOON".

Since two of the main points of the competition rules ask to emphasize ease and speed of assembly and disassembly, this competition provides for the mandatory use of advanced prefabrication. This involves Building Information Modeling (BIM) design instruments, and Computer Added Manufacturing machines.

Since sensors, IoT, smart village, digital design and production are some of the keywords of this competition rules, thus we can affirm that this competition is based on digital exploitation in order to attain energy efficiency and therefore we can move describing the second scenario taking place today.

23.4 Digital Transition

The digital changes are affecting not only our daily lives, but also all the economies, including the construction one. In this sector, in the last few years, we have assisted at big changes in the way to conceive the design project, to build it, and to measure, control and run buildings (Tonelli and Bellingeri 2016).

The ordinary way to envision an architectural project had been strongly modified by the BIM workflow. The BIM model, with the components defined in 3D, offers the possibility to foresee problems, so as to improve the design and reduce the time for their solution during production.

Building performances, such as phase shift and damping of the thermal wave, lighting, luminance, day light factor and glare are easy to be simulated in a reliable output, as guide to assess what has been conceived. It is not too wide diffuse the employ of such instruments, but it is anyway possible to experiment innovative approaches matching different softwires, to test strategies and evaluate results in order to enhance the design solutions.

The construction phase, although the traditional way to build still remains the most employed in the AEC industry, should address and be always more related to advanced prefabrication and assembly on the construction area.

The most cutting-edge way to fabricate building components or, moreover, entire buildings, is the Construction 3D Printing (C3DP), in development since the mid-1990s. The 3D printing has steadily gained pace since 2012 and the sub-sector of C3DP is beginning to mature. Now C3DP refers to various technologies, with the main ones being extrusion (concrete, wax, foam, polymers, carbon fiber), powder bonding (polymer bond, reactive bond, sintering), and additive welding. Advantages of these tools can be identified as faster construction, lower labour costs, increased complexity, accuracy, and less waste produced.

A number of different approaches have been demonstrated up to now, which include on-site and off-site fabrication of buildings and construction components, using industrial robots, gantry systems and tethered autonomous vehicles. Demonstrations of construction 3D printing technologies to date have included fabrication of housing, construction components (cladding and structural panels), bridges and civil infrastructure, follies, and sculptures.

On 2016 Chattanooga startup Branch Technology launched an architectural competition challenging participants to use the company's patented cellular fabrication technology to design a 3D printed single-family home that "rethinks traditional aesthetics, ergonomics, construction, building systems, and structure from the bottom up" (Scott 2016). First place went to WATG Chicago, with "Curve Appeal," a swooping structure that resembles some sort of natural rock formation. With this project they have indicated the direction in which architectural theory is turning, taken a break from creating the next tallest tower and quietly revolutionized the way we think about homes.

In Dubai a new 3D printed office was recently built, printed out layer by layer over 17 days at a cost of $140.000. Dubai hopes the project will kick-start its plans to transform the sheikhdom into an incubator for emerging technologies.

I am among the sceptics that these tools can give as complex and performing results as our houses require. However, I came across Bill Gates' statement of 1981, when he supposedly uttered, in defence of the just-introduced floppy disk: «640 KB of RAM limit ought to be enough for anybody».

The digital age is set to cause more upheavals than previous technology revolutions because change is happening faster than ever before and is fundamentally altering the way we live. 4 days were required to download 10 million apps from the newly opened app store in 2008. We predict technology will continue to follow this trend. It will not be long until a new app is downloaded at such incredible rates that 50 million consumers will be reached in days. The telephone technology, in the last century, took 75 years to reach a similar number of customers (Interactive Schools 2018).

23.5 Digital Role for the Existing Building Stock Retrofit

The technological evolution of digital will allow to permit greatly easily to link design output to production machines. But ones again this concerns the construction of new buildings, as above the way to design a NZEB. Nevertheless, is digital able to supply the existing real estate retrofit and refurbishment?

It is clear, today more than in the previous ages, that the main urgency does not consist on how to build future homes (as we wrote above it is well known how to solve dramatically this issue), but on how to retrofit the old and energivorous existing real estate. In Europe, 180 million of the existing residential buildings have been built after 1945 (77%). In Italy 12 million buildings exist. 7,4 million has been built in between 1950 and 1991 (the year of the 1st law that deals with energy efficiency) (ISTAT 2011). So that, more than 60% of the constructions were built before limitations of energy uses and carbon emissions, with corresponding high-energy consumption.

In the post Second World War period the Italian economic boom and the technological advances allowed to split the building envelope from the role to create indoor comfort conditions. Comfort was devoted to HVAC, forgetting all the elements that characterized the Italian Architecture, since ever: *genius loci*, or in English the relation between shape, orientation and building techniques with local climate and materials. The upsetting result consists of: imitation of archetypes beyond the Alps, prevalence of glazed facades, housing models far from our culture, and big amount of energy use that with the rising costs of fossil fuels has become increasingly unsustainable and unaffordable.

It is important to underline that to produce 1 °C of cold air requires 3 times more energy than to produce 1 °C of warm air. This means that to avoid discomfort from summer conditions is demanding 3 times more energy than in winter ones. This concerns seriously Italian and south European latitudes. Many techniques have been developed since the first Reverse Engineering/Retrofit method thought by a TUM research, based on large scale, timber panels for substantial improvement of the energy efficiency of a renovated building (Heikkinen et al. 2010). It was a systemized modernization process from survey, planning, production off-site to assembly on-site as a consistent structure along a digital based workflow.

Although this or other similar technologies have already been tested in several European countries, they are not feasible to be used in the Italian context, characterized by specific geometric features of the building stock, by cold and hot climate requirements and by seismic constraints, which are not relevant for most of European countries. The Ri.Fa.Re. Project (Malacarne et al. 2016) presents an innovative timber-based prefabricated panel for the energy refurbishment of the existing Italian buildings façades that does not aim only at developing a product but also at adopting a new process, based on parametric design and lean construction principles. The main outcome is a concrete kit for energy refurbishment suitable for the Italian context, including a technological prefabricated but customizable product and a more efficient process to decrease time and cost wastes during the whole process.

23.6 Less Automation, More Information

A perfectly designed and engineered house will likely not work properly if its inhabitants do not know how to use and live in it. It has been studied that in some cases, there are NZEB in which users behave and consume in the exact same way as they would in non-efficient homes (Keesee 2005), as if they do not recognize the direct relationship between their behaviour and the performance of the house (Tonelli and Converso 2014).

The active role of users has proven its effectiveness in several successful experiences. One of the first documented is the Empire State Building energy retrofit, in which a number of measures to increase the passive and active behaviour of the building itself (i.e., new windows, more efficient systems) saved a yearly amount of $4.393.796 and user awareness reduced 8% of consumption costs, $386.709 (Rocky Mountain Institute 2012).

In residential use, this number would evidently increase (Janda 2011), but its exact determination is difficult, because the ratio that is attributable to user behaviour can vary according to context strategies.

The impact of human awareness on reducing energy consumption is particularly evident with cars. Currently, cars are equipped with increasingly information systems that tell the driver in real time the impact of his driving behaviour. The simple evidence of immediate consumption and the projection of the remaining distance have proven to stimulate drivers to change their behaviour and adopt a less energy-consuming driving style. Some authors (Kurani 2007; Inbar et al. 2011) also identified in particular cases a "game-effect" (Deterding et al. 2011) whereby social rewarding is used in strategies to promote eco-driving. This relationship between different players in a competitive environment, based on shared monitoring, is already present in the Solar Decathlon competition. Thus, each competitor can see in real time the performance of others and observe the differences on how his house is performing.

Despite architecture is still about designing static environments, devices and sensors increasingly embedded into buildings provide dynamic information on its behaviour: comfort conditions, energy balance, and also user's interaction. Sensors are not new in buildings. Lights and gates embed sensors since decades. Even our smartphones have many sensors. However, smart homes are not very intelligent simply because they are pre-wired accordingly to algorithms and decisions made by the systems designers, rather than the dwellers. In contrast to such homes, which are not able to adapt structurally over time, what we designed to participate in Solar

Decathlon, was a 3D model, accessible on web, that allow to navigate in real time and back in the time, named "DWELL!" (play on words that, extended, could sound like "do well to dwell well!").

Following the philosophy of "more information less automation", we created a building/technological systems synergy that allows to obtain a better balance between comfort and consumption, having an active participation of the users through a data visualization guiding the inhabitant to a correct practise in the house. It is also an interactive point-and-click model of the house where the user can visualize what's going on in real-time through simple animations (changing colours, point clouds, opening/closing windows and doors, etc.) and popup windows (Converso et al. 2013).

Likewise, we believe that true improvement begins with improving people, not just the places they live. It is possible to design homes with high levels of efficiency, but to get an effective energy saving is necessary that dwellers can be aware of how the house works and the extent of consumption related to their behaviour in order to learn how to use the house in the best way to get the comfort most suitable for themselves with the minimum energy consumption.

The same happens at an urban level, where advances in technologies ranging from sensors to broadband, artificial intelligence and big data are making smart cities a reality. Nevertheless, the smart city of the future will have little hope for sustained prosperity if its people are not themselves smart (Tonelli 2015). It also takes thoughtful citizens to correctly interpret data produced by smart devices. Indeed, the success of cities depends greatly on the ability of individual citizens to intelligently engage with each other and their environment (Tonar and Talton 2019). ICT and IoT are already transforming cities around the world into hyperconnected networks and it is thanks to these emerging technologies that during the Covid-19 pandemic our lives remain efficient and productive: we assisted to the conversion of our homes in offices, schools, restaurants, and leisure environments in the time of few days. Covid-19 virus restriction measures created a more innovative, educated, talented, resilient and empowered citizenry: smart working entered in our homes and fast accelerated the process of digitalization.

In 2019 Roma Tre University has been selected again to participate in Solar Decathlon, in the Middle East 2020 edition (SDME). The Covid-19 emergency stopped the competition and it is still unknown if or when it will be possible to start working again. For SDME we proposed MOON, a housing prototype designed for a harsh, adverse, hot and very humid climate, like the one in Dubai, which we looked at as a new world we need to explore and understand. We liked to think of ourselves as astronauts, leaving for a new planet, arriving where we need to conceive new neighbourhoods, built with the most advanced technologies at a sustainable and human scale. The MOON project will convey space technologies into a human habitat, considering it a challenge that of building in such a harsh climate (a same approach was foreseen by NASA in the 3D Printed Habitat Challenge for Mars (NASA 2020). The concept will combine innovation with tradition, with a strong outlook onto the future. The house will be shaped as a closed volume, protected from the sunshine and the dust by a double skin, yet open towards an inner courtyard, shaded and climatized by vegetation able to live through humidity.

The urban concept enhances local resources, optimizing passive and active energy strategies, densifying and encouraging electrical and shared mobility, as tools for reducing harmful emissions and activating sustainability. MOON for SDME is the dream project, a dream lunar world that can finally connect with the Earth. A new world: technologically smart, able to offer a sustainable opportunity to the future city of Dubai. Pursuing again the claim "Less automation, more information" (Tonelli 2013) who guided the home control of RhOME prototype, we were supposed to design a system able to give awareness to its users on the ecological footprint of their behaviour. The device will be a dashboard based on Machine Learning technology, grounds on five "cycles": Energy; Environmental features (air quality); Water consumption; Waste production; Food consumption. It would been designed for measuring and monitoring in real time the ecological footprint of the house on the basis of the environmental data and the behaviour of people living in the house, predicting future scenarios and effects.

23.7 Conclusions

We are at a turning point dictated by many new needs. We must look for new possible paths and meet change without fear of it. These are in synthesis the different ways that the architect profession must try to follow in order to meet the real needs of today: to review operational and organizational methods, to redefine project requirements, also recovering a tradition from the past, and lastly to keep in mind the demographic and social reality, in constant and violent modification. In parallel, knowledge increases. Life is facilitated by the digital revolution, a very fast and upsetting revolution, which, if on the one hand helps, on the other ousted the less qualified from the world of work.

A Google algorithm has narrowly beaten the world's best player in the ancient Chinese board game of Go, reaffirming the arrival of what its developers say is a ground-breaking new form of artificial intelligence. This application have learned by millions of played games, thanks to the Machine Learning technology, but to win have created a new strategy, not belonging to what studied (Barricco 2018).

We ought to consider creativity as the only true quality that can make a difference between man and machine. But, perhaps, creativity, the soul of humanity, of some in humanity, is about to be reached by machines. This drastically reduces the supremacy of the human being, because it was the real difference between an evolved society and one still linked to manual work. In the Chinese board game Go, inventing a strategy not present among those learned, represents in a nutshell a creative act. This, if it become a machine potential, will affect principally the architects' job.

The Solar Decathlon competition give us a lesson on how to face this digital revolution in the AEC industry, putting people, students in particular, at the centre of the challenge. It is an opportunity to fully tackle a new kind of design process, using an integrated and multidisciplinary design team, made by researchers, students and technicians. The digital tools could be considered as shared culture enhancing exchange and interaction between different disciplines and people. The intention is to challenge the existing distinction between design conception and design development, working towards a process based on form resolution. The process will therefore be based on a cyclical structure, rather than a linear sequence. The ultimate research goal of the interaction and sharing system overcomes the current distinction between communication and design development environments, by integrating as much live design data as possible into the management system. These are the premises that allow for the proposal of a model for customized interventions based on advanced prefabrication, implying the combination of digitally controlled models with CNC manufactured components, starting from a shared digital model.

In the process the fabrication manufacturing model is constantly interfaced with energy analysis platforms, which take into account all the involved parameters, with models made to teach the team on the numerical impact of the measures adopted for passive strategies, with both stationary and dynamical simulation.

But in our view, all this does not come to an end with construction. A bigger part of energy savings is possible during life cycle management which is an emergent and increasingly important domain in digital systems for the building sector.

The approach to home automation has gone beyond domotics, in order to put people at the core of the system. Awareness is the main objective of a control structure that generates a continuous stream of data. In the Solar Village of each edition of the competition, a local server is bundled to a network of wireless sensors linked to all kinds of appliances (HVAC, lighting, kitchen appliances, monitoring comfort conditions, movement and position of doors and windows, security webcams). This will not just assure remote control, but also transfer data to a remote server, online, where a home portal provides an integrated view of the house network, where the link between systems normally considered as separate is crucial.

The web space is based on social networking technologies. If it integrates into the smart city, it will allow a comparison between different users and data sharing across multi-family buildings and neighbourhoods or different areas of the city or the country itself.

This approach belongs to gamification, an emerging field of research, to encourage energy conservation behaviour in house occupants.

An innovative serious game should be developed to promote reduced energy consumption and carbon emissions by changing social tenants' energy efficiency behaviour.

Nevertheless, although smart infrastructure helped to facilitate value creation, we consider people still the protagonists in the urban drama.

References

Andaloro APF, Salomone R, Ioppolo G, Andaloro L (2010) Energy certification of buildings: a comparative analysis of progress towards implementation in European countries, Energy Policy 3(10):5840–5866. Building Performance Institute Europe, 2015. Europe's Buildings under the Microscope. http://bpie.eu/wp-content/uploads/2015/10/HR_EU_B_under_microscope_study.pdf

Barricco A (2018) The Game, Einaudi

Bluyssen PM (2010) Towards new methods and ways to create healthy and comfortable buildings. Build Environ 45:808–818

Butera F (2013) Zero-energy buildings: the challenges. Adv Build Energy Res 7(1)

Converso S, Tonelli C, Bellingeri G (2013) More documentation, less automation. A strategy of user involvement in domotics at Solar Decathlon 2012. In: Architectural education and the reality of the ideal: environmental design for innovation in the post-crisis world, vol 61. EAAE Transactions on architectural education, pp 717–726

Committee for the Coordination of Statistical Activities-CCSA (2020) How COVID-19 is changing the world: a statistical perspective. https://unstats.un.org/unsd/ccsa/documents/covid19-report-ccsa.pdf

Deterding S, Dixon D, Khaled R, Nacke L (2011) From game design elements to gamefulness: defining "gamification". In: MindTrek'11 Proceedings of the 15th International Academic MindTrek Conference: envisioning future media environments, ACM, New York, 9–15

Ecofys Document (2010) http://www.ecofys.nl/com/news/pressreleases2010

Erhorn H, Klutting H (2012) The path towards 2020: nearly zero energy buildings, REHVA J 12–15

ESA (2020) Coronavirus: nitrogen dioxide emissions drop over Italy. https://www.esa.int/ESA_Multimedia/Videos/2020/03/Coronavirus_nitrogen_dioxide_emissions_drop_over_Italy

Eurostat (2020) http://ec.europa.eu/eurostat. Accessed 27 Sept 2020

Harvard University (2020) COVID-19 PM2.5. A national study on long-term exposure to air pollution and COVID-19 mortality in the United States. https://projects.iq.harvard.edu/covid-pm

Heikkinen P, Kaufmann H, Winter S, Larsen K (2010) TES EnergyFaçade–prefabricated timber-based building system for improving the energy efficiency of the building envelope. mediatum.ub.tum.de

Inbar O, Tractinsky N, Tsimhoni O, Seder T (2011) Driving the scoreboard: motivating eco-driving through in-car gaming. In: CHI 2011 workshop gamification: using game design elements in non-game contexts

Interactive Schools (2018) https://blog.interactiveschools.com/blog/50-million-users-how-long-does-it-take-tech-to-reach-this-milestone

ISIAQ-CIB (2004) Performance criteria of buildings for health and comfort CIB Publication 292, CIB task group TG 42, Rotterdam. The Netherland

ISTAT (2011) L'Italia in 150 anni. Sommario di statistiche storiche 1861-2010, Roma

Janda KB (2011) Buildings don't use energy: people do. Architec Sci Rev 54:15–22

Keesee M (2005) Setting a new standard: the zero-energy home experience in California. In: Proceedings of Solar World Congress, Orlando, FL, International Solar Energy Society

Kurani K (2007) Impact of in-car instruments on driver behavior. In Presentation at the eco-drive workshop, Paris

Malacarne G, Pasetti Monizza G, Ratajczaka J, Krausea D, Benedetti C, Matt DT (2016) Prefabricated Timber Façade for the energy refurbishment of the Italian building stock: the Ri.Fa.Re, Project. Energy Procedia 96:788–799

NASA (2020) https://www.nasa.gov/directorates/spacetech/centennial_challenges/3DPHab/index.html

Pilkington B, Roach R, Perkins J (2011) Relative benefits of technology and occupant behaviour in moving towards a more energy efficient sustainable housing paradigm. Energy Policy 39(9):4962–4970

Rocky Mountain Institute (2012) Empire state building retrofit surpasses energy savings expectations. http://blog.rmi.org/ blog empire state retrofit surpasses energy savings expectations#savings

Sesana M, Salvalai G (2013) Overview on life cycle methodologies and economic feasibility for nZEBs. Build Environ 67:211–216

Scott C (2016) Branch technology announces winners of freeform home design challenge. Published on April 25, 2016, by 3D design/3D printing. https://3dprint.com/131259/branch-technology-design/
Tonar R, Talton E (2019) Smart cities are built by smart people, not smart things, forbes. https://www.forbes.com/sites/ellistalton/2019/07/09/smart-cities-are-built-by-smart-people-not-smart-things/#c84da7c75d62
Tonelli C (2013) An overall sustainable approach to shape the citizenship of tomorrow: the solar decathlon competition. In: Architectural education and the reality of the ideal: environmental design for innovation in the post-crisis world, vol 61. Thessaloniki, pp 505–516
Tonelli C (2015) Abitare Domani. Sfide e opportunità per la Smart City. In: City memory people. Book of Acts of the 9th Congress "Virtual city and territory". RomaTrE-Press
Tonelli C, Bellingeri G (2016) Strategie per l'alta efficienza energetica in clima Mediterraneo, ISBN:978-88-96386-52-1
Tonelli C, Converso S (2014) Digital mirror: a method to shape smart citizenship. Energy Build 83:173–180
Tonelli C, Grimaudo M (2014) Timber buildings and thermal inertia: open scientific problems for summer behaviour in Mediterranean climate. Energy and Build 83:89–95. https://doi.org/10.1016/j.enbuild.2013.12.063
Voss K, Musall E, Lichtmeß M (2011) From low energy to net zero energy buildings: status and perspectives. J Green Build 6:46–57
Voss K, Sartori I, Lollin R (2012) Nearly-zero, net zero and plus energy buildings. How definitions and regulations affect the solutions. REHVA J 6(49):23–27
World Health Organization (2013) Combined or multiple exposure to health stressors in indoor built environments. http://www.euro.who.int/__data/assets/pdf_file/0020/248600/Combined-or-multiple-exposure-to-health-stressors-in-indoor-built-environments.pdf
Yang R, Wang L (2013) Multi-zone building energy management using intelligent control and optimization. Sustain Cities Soc 6:16–21

Chapter 24
The Green Design Approach Digital Innovation Facility: BIM and New Industrial Processes

Maria Antonietta Esposito and Filippo Bosi

Abstract The chapter relates bioclimatic with the development of digital design in the field of Design Science, recalling some salient points of the origins and development of the concepts of bioclimatic from its origin in the search for alternative energy sources for space missions over the 1960s. From the studies in this field have resulted in new calculation methods and energy and characteristics physical environment assessment. Energy research developed in parallel with distributed computational capabilities and the software that allowed them to automate not only the calculations but also parametric simulations. A chronological summary of this evolution is illustrated with references to the main modeling and energy simulation software. Some experimental projects developed over the last 20 years are presented, demonstrating the state of the art of Green Design with evidence relating to the actual performance in use conditions. In advanced areas such as the development of tertiary buildings and terminals, the three areas of sustainability are combined to obtain more sustainable projects not only on an environmental basis but also on economic and social ones. The phenomena are driven by market and management needs; it is evidenced by the increase in standards and life cycle certifications aimed not only at products but also at the overall management of the project and activities over the life cycle. With the evolution of CAD towards BIM platforms since the second half of the 1950s, the possibility of creating interoperable platforms useful for various purposes has opened up: in a first phase, it was possible to import data processed by specialized software and subsequently to integrate them into the field of modeling. With the perspectives opened by platforms on the web, a new way of designing and producing is started, fully compatible with the digital environment. Introducing these types of innovation that enable the full digitalization of processes appears to be the most plausible operating scenario for Green Design as it allows the connection and implementation of lean and efficient management of the project-production chain in all phases of the cycle of life and in perspective to create new project areas aimed at the digital construction of buildings directly from the digital model (Digital Twin).

M. A. Esposito (✉) · F. Bosi
Department of Architecture, University of Florence, Florence, Italy
e-mail: mariaantonietta.esposito@unifi.it

Toscana Aeroporti Engineering, Florence, Italy

G. Chiesa (ed.), *Bioclimatic Approaches in Urban and Building Design*, PoliTO Springer Series, https://doi.org/10.1007/978-3-030-59328-5_24

24.1 State of Art

24.1.1 Theories and Methods Evolution

In the 2nd post-war period, atom, physical and engineering scientists put to good use their competencies integrated them into the Los Alamos laboratory, obtaining results not only in the thermonuclear research program but also in the development of supercomputers and pioneering research in Physics, energy and materials science:

> The main focus is on the history of the world.
>
> Laboratories - other key theme for the years 1942-1956 include supercomputing advancements, fundamental biomedical and health physics research, high explosives research and development, reactor research and development, pioneering physics research, and the development of the field of high-speed photography

(McGehee and Garcia 1999; de Dear et al. 2013; McGehee et al. 2015).

The line of research was oriented above all for space missions, but it also represented the first time in which renewable energies are considered a resource for the project.

In the '60s following the address given by President Kennedy: "*First, I believe that this nation should commit itself to achieve the goal, before this decade is out, of landing on the moon and returning him safely to the Earth.*" (President John F. Kennedy 1961) it takes shape the technological and industrial scenario that leads to the conquest of the moon in competition with Russia. The challenge starts with the first tests on the new vectors to launch human beings into space. Astronaut Neil Armstrong, who set foot on the lunar surface on 16 July 1969, defined the event "*a small step of a man a great step for humanity*". The goal achieved would have been unimaginable only a few years earlier.

The design of the lunar module—and later of all means and space satellites—were based on the use of solar energy, because it was the energy resource available in space. Reaping energy from other sources traditionally used for vehicles on earth was excluded due to the impossibility of transporting other fuel—in addition to that required by the rocket to reach orbit—and make the necessary corrections during the mission.

Solar energy and that derived from the atom enter space research and the industrial sphere first for the realization of projects for aerospace missions and then, with the spread that aerospace innovations, in all industrial sectors.

However, the aspect that determined the scientific revolution of which we see the effects today was the scientific and propulsive interest given by the algorithmic thought, as well as the disturbing character of what now appears to be an undisputed domain of digital computing (de Dear et al. 2013; Zellini 2018).

For example, the first tools to evaluate physical-environmental parameters were the charts that were soon automated thanks to the development of calculation automation tools (in the 1950s) and personal computing (in the 1970s). Computers that have enabled us to greatly enhance our capacity for modeling phenomena, performance simulation and design of artificial environments. Table 24.1 presents a timeline of

Table 24.1 Energy research milestones and main events

Time frame	Research team or project	Features
1935	Hoover Dam, Boulder, CO	The largest hydroelectric power plant is built within the Hoover Dam. The gross power of the plant amassed to 2.4 Mw after an uprating project concluded in 1993, with a maximum capacity of 2080 megawatts. (Wikipedia, accessed 02/11/2019)
1942	Enrico Fermi/Fermi Lab	First Nuclear fission reactor
1945–1956	Los Alamos Laboratory, USA	Weapons research and development of supercomputing, material science, Energetic research etc.
1961–1972	NASA—National agency for science and aerospace, USA	Moon space program or Apollo program. Achieved its goal of crewed lunar landing. The Apollo program spurred advances in multiple technology fields beyond rocketry and human spaceflight, such as energy use optimization research, avionics, telecommunications, and computers including use of the solar radiation to produce energy (NASA 2004)
1963	Bioclimatic map of olgyay	Design methodology which considers factors of human physiology, climatology and building physics. Its target is achieving comfort in buildings through efficient use of available resources. Maps heat, humidity and potential comfort achievement strategies for designers. Visualizes climate data and puts them in relation to comfort conditions. (Olgyay 1963)
1966	ASHRAE	First edition of the standard 55

(continued)

Table 24.1 (continued)

Time frame	Research team or project	Features
1969–1976	Givoni's psychrometric chart	Used to plot the temperature/humidity points over the whole year. Illustrates aspects of thermal comfort conditions and widely used by engineers and other professionals in the HVAC field (Milne et al. 2009)
1973	FEP—Finalized energy plan	Based on the studies by Los and Pulitzer, proposes the solar approach that will produce a vast research program. It is focused on research for alternative energy sources. Gave positive results and started spreading environmental awareness
1974	Joseph lyndmayer	Development of the first silicon photovoltaic cell for harnessing solar power, enhancing efficiency
1977	Action environment, L. Matteoli and Politecnico di Torino	The studies introduce environmental performance-based design at the onset of the project
1977	Evolution Architecture, S. Los and N. Pulitzer	Considers alternative technologies and industrial production for housing systems also to address comfort requirements
1979	FACT Forum Architecture Communication Territoire	European Conference at Lusanne Polytechnique on early quantitative studies on the efficiency of passive solar systems of greenhouses equipped with Trombe walls
1979	Experimental studies relating to the internal environmental comfort of Givoni-Milne	Definition of the psychrometric chart for the evaluation of the physical-environmental parameters that influence well-being. It's now possible to quantitatively evaluate significant improvements in comfort based on measurements of internal temperature, air velocity and relative humidity. Requirements to achieve measurable quality levels with verification calculations during design and with instrumental tests in place and in conditions of use are defined (VV. AA. 1989). Performance specifications are drawn thanks to the refinements made to the Givoni-Milne chart.

(continued)

Table 24.1 (continued)

Time frame	Research team or project	Features
1980	New Hampshire, USA	The first wind farm is built at Crotched Mountain, consisting of 20 30 kW wind turbines. Now dismissed it started the Wind Energy harvesting process locally and throughout the whole United States
1981–1990	Studies led by Grosso, Peretti, Gangemi and project design concepts made by Esposito-Veninata	Theories for the holistic analysis of buildings and transfer of performance-based design and requirements research to the industrial components sector
1981	The bioclimatic map of Givoni-Milne also	Introduces the association between geographical coordinates and hourly data. The design strategies recommended in the bioclimatic chart that is being investigated in the current work, include: passive solar heating, active solar heating, conventional heating humidification, internal gains, shading, high thermal mass with and ventilation, and air conditioning with conventional dehumidification
1981	Dagget, California	Solar One—the first large scale solar-thermal power plan—opens in California
1989–1999	NASA applied physics laboratory	The Applied Physics Laboratory began operations in 1989 as the Optical Instrumentation Laboratory, and name change occurred in 1999. For 20 years, the lab supported ground processing for the Space Shuttle Program. The laboratory invented tools and systems within its interdisciplinary team: involving fluids, heat transfer, material properties, optics, mechanics and other areas. Many problems can be solved with a combination of mathematical modeling and simple, inexpensive experiments The Applied Physics Laboratory often tackles problems using electronics and software components as part of the solution. These experiments need data acquisition and control. Instrumental development work in the lab generally requires software and avionics support. Integrated labs in development for spaceflight

(continued)

Table 24.1 (continued)

Time frame	Research team or project	Features
1996	Hydrogen future act (US Government 1996)	Hydrogen Future Act is passed to promote hydrogen power development. Specified funding levels for Hydrogen Program Specified evaluation of integrating fuel cells with photovoltaic systems for hydrogen production or with systems for hydrogen production from solid waste via gasification
1966, 1981, 1992, 1994, 2004–today	ASHRAE 55 standards	Commonly known as "Comfort Chart", establishes the ranges of indoor environmental conditions to achieve acceptable thermal comfort for occupants of buildings. First published in 1966, and since 2004 has been updated every three to six years. The most recent version of the standard was published in 2017

Source Data search contribution by F. Bosi

Energetic research milestones and main events, offering a synthetic description of the main features of the milestone (Tables 24.1 and 24.2).

In the construction sector, the scientific question relating to bioclimatic begins with the famous book by Victor Olgyay (1963) Design with Climate, which was the first and most important of the reference texts of the last 60 years, until today a theoretical reference base for professionals, teachers and students. In the book, Olgyay investigates the impact of climate—divided into 4 main regions—on building design and explains the effects of orientation, airflow, geographical location, on materials and built forms. His analysis is derived from principles and concepts of various disciplines such as Biology, Engineering, Meteorology and Physics. The analysis of Olgyay bases the bases of the bioclimatic theory demonstrating how a scientific and analytical approach to the climate-based project can lead to better results not only from an aesthetic point of view but also from environmental functioning. The first bioclimatic map of Olgyay (1963) allows evaluating heat, humidity and potential comfort strategies (Fig. 24.1).

In the USA, the trend of architectural design with the climate initiated by aerospace research on solar energy had a great development. Thanks to the connection with the themes promoted by the Hippy societal protest movement of the 1970s that has had proposed new models of life and alternative human settlement, the theme was spread in multiple experiments. Radical architects—such as the Italian Paolo Soleri who builds Arcosanti as an ideal city project combining architecture and nature in Arizona—have developed experimental projects on an urban scale inspired by integration into the environment, population density and less land consumption, elimination of cars and promotion of collective transport. The social aspect is already at that time an important part of the sustainability of projects and products. Today it is fully included as a pillar of sustainability (Murphy 2012), along with the environmental and financial one in the European vision for the problem. Regarding the energy aspects for buildings, the hypothesis nZeB—Near Zero-energy Buildings (Hermelink et al. 2013) was finally formulated, which classifies the project objectives on the objective of neutrality or building energy producer.

In Switzerland, at the Lausanne Polytechnic, the FACT '79 scientific conference presented the first experimental studies with the efficiency data of passive solar systems equipped with greenhouses integrating Trombe walls (FACT 1979).

In Italy Sergio Los and Natasha Pulitzer—returning from the United States in full energy crisis due to the OPEC decision to reduce oil extraction as retaliation to the Middle East crisis—successfully propose the solar approach producing a vast applied research program as national Finalized Energy Plan (1973). In fact, throughout Western Europe, the energy crisis also led to new awareness and strategies to the search for new sources of supply, including alternative ones, which also

Table 24.2 Building energy modeling software timeline, presents the major and most diffused innovations in the field

Time frame	Research	Features
1963	BRIS	Developed by the Royal Institute of Technology in Stockholm. The tool calculates temperature variations in a room by using a system of heat balance equations solved by an iterative method
The 1960s	Work by T. Kusuda	Focus on load calculations and energy analysis Simulation of heat and mass transfer, sided by airflow studies
1961–1990	TMY2 Weather data US National Renewable Energy Laboratory (NREL)	1961–1990 data are derived from the 1961–1990 National Solar Radiation Data Base (NSRDB)
1970s	DoE 2 software	Developed by James J. Hirsch & Associates (JJH) in collaboration with Lawrence Berkeley National Laboratory (LBNL). Worked mostly with the United States Department of Energy (USDOE) and funded by a wide range of industry organizations (http://doe2 com/)
1975	TRNSYS and Simulation Studio	Transient System Simulation Tool, it uses a graphic interface to show variations and perform simulations of transient systems
1986	SPARK	Short for "Simulation Problem Analysis and Research Kernel". It is an object-oriented environment for graphically assembling custom models of complex building energy systems

(continued)

Table 24.2 (continued)

Time frame	Research	Features
1989	Neutral model format for energy simulation	Later developed into IDA and ICE engines, constitutes the base for modern simulation
1998	IDA Engine, ICE and ESBO	Is used to model the building, its systems, and controllers. The target is ensuring the lowest possible energy consumption and the best possible occupant comfort (De Dear & Brager, 1998)
2001	Energy + and DesignBuilder	DesignBuilder is one of the most common User Interface for Energy +. Simulates the thermal behaviour of the building envelope that was designed as a Research Institute of Renewable Energy of Yazd University. Includes modules for: CFD—Computational Fluid Dynamics, Natural Lighting DesignBuilder LEED and ASHRAE 90.1 Appendix G 3D DesignBuilder modeller DesignBuilder HVAC modeling DesignBuilder optimization EnergyPlus DesignBuilder simulation DesignBuilder Economic Evaluation DesignBuilder display EMS by DesignBuilder Italian legislation with DesignBuilder UNI 11300
2005	eQuest	Building energy use analysis tool. first enacted under the Energy Policy of 2005 (EPACT), later extended to the Emergency Economic Stabilization Act of 2008 and most recently extended and modified
2008	Autodesk Ecotect	Previously available as an independent beta tool; Building performance is simulated since project onset, combining the traditional representation of the project with analytical results and environmental data in one interactive visualization. Since 2015 is integrated into Autodesk Revit

(continued)

Table 24.2 (continued)

Time frame	Research	Features
2010s	Building life cycle cost programs	Life Cycle Costing (LCC) analysis tools to identify cost-optimal building design solutions yet not widely used with confidence. They provide computational support for the analysis of capital investments in buildings. Typically, BLCC is used to evaluate alternative designs that have higher initial costs but lower operating costs over the project life than the lowest-initial-cost design. It is especially useful for evaluating the costs and benefits of energy and water conservation and renewable energy projects

Source Data search contribution by F. Bosi, TxP Research, 2019)

gave positive results and started austerity measures that introduced and began to spread greater environmental awareness. Their book *The architecture of evolution*, which illustrates the housing system between building industrialization and alternative technologies, published in the series "*Studies on building industrialization*" on the thirteenth International Exhibition of Building Industrialization (SAIE). (Los and Pulitzer 1977) it is still a best seller on Amazon (1,295,051 copies sold up to the moment of writing).

Subsequently the book by Lorenzo Matteoli *Action Environment* (1977) and the studies carried out at the Politecnico di Torino (Grosso et al. 1981–1990) and in other universities in Italy take up the theoretical line of this approach and illustrate it as a method reference that considers the resources, requirements and environmental performance at the basis of the development of the project.

In the same years, studies on urban and building scale are joined by experimental studies relating to internal environmental comfort (Milne & Givoni 1979; Guyot and Cros 1979; McIntyre 1980; Wu 1980; Queffelec and Trebbi 1981) with the definition of the psychrometric chart for the evaluation of the physical-environmental parameters that influence well-being (Crump and Harms MJ 1981; Allard and Santamouris 1998). Givoni's psychrometric chart (1969, 1976) makes it possible to quantitatively evaluate significant improvements in comfort based on measurements of internal temperature, air velocity and relative humidity, to define the contours of design strategies and the characteristics dominated by the building envelope (Tanabe et al. 1987).

At the University of Florence was based on the early on-site testing laboratory for comfort belonging to the National Scientific Laboratory List. The laboratory was entitled *Laboratorio per la Qualità Edilizia* (LQE) and it was equipped with state-of-the-art portable instruments to test buildings on site. Tests were conducted during operations following the currents national standards and data related to comfort zone

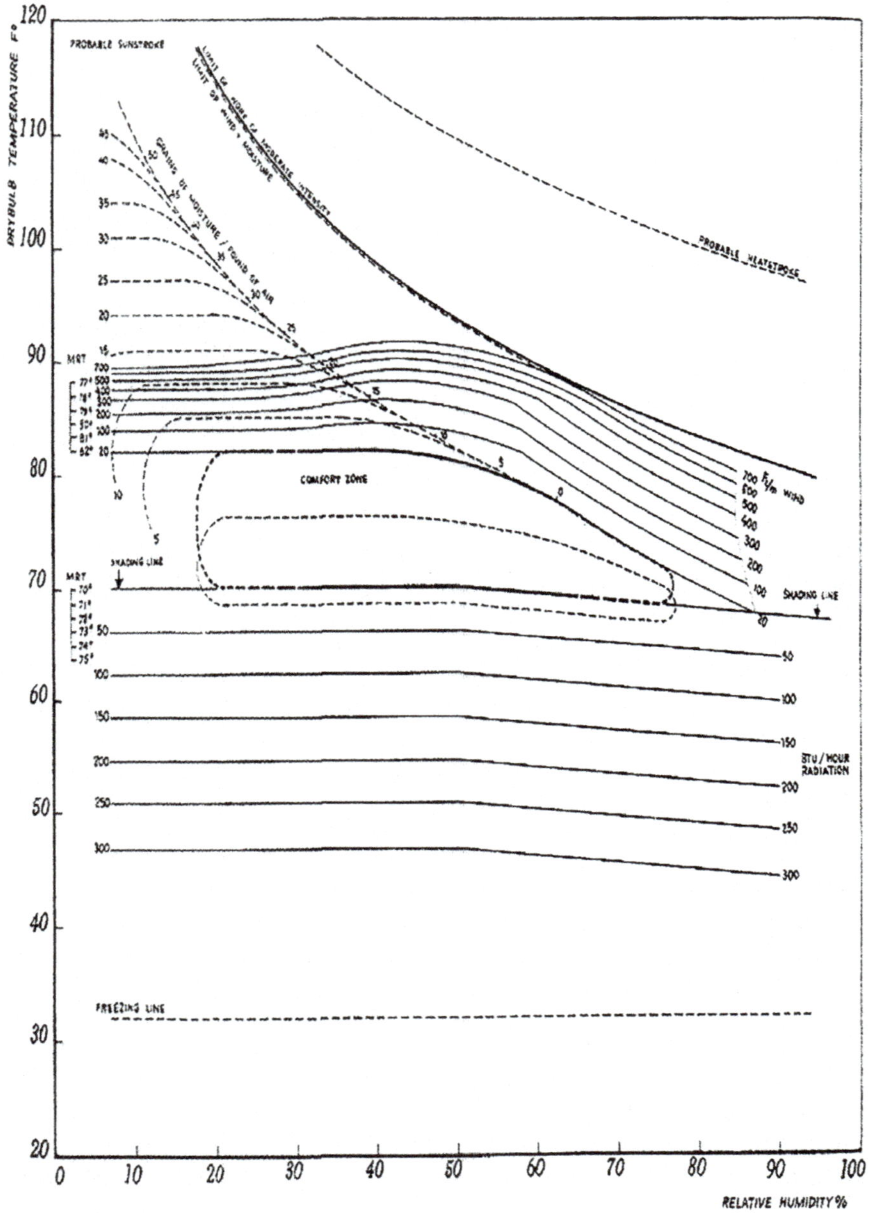

Fig. 24.1 Bioclimatic chart (Olgyay 1963). *Source* https://www.slideshare.net/tiwoze/design-with-climate-bioclimatic-approach-to-architectural-regionalism-victor-olgyay-ebook-ppt-epub-download-free

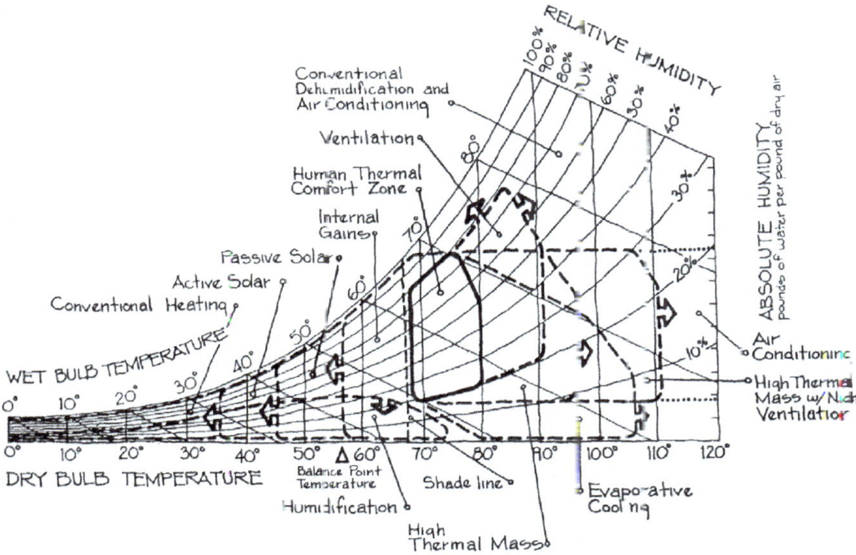

Fig. 24.2 Givoni-Milne bioclimatic chart © 1981 (1981 accessed 10/28/2019). *Source* https://www.slideshare.net/komalbhallu/bioclimatic-design-at-the-site-planning-scale

gathered were used to study better design solutions. With these studies, we began to define the requirements to achieve measurable quality levels with verification calculations during design related to use conditions (Del Nord et al. 1989). The availability of this technical basis allows to draw up performance specifications for environmental aspects and to check the results but above all to test them using instrumental verification procedures in place based on shared technical standards and with instruments that process measurements on the basis of the paper algorithms bioclimatic of Givoni-Milne (Fig. 24.2).

The 1981 bioclimatic chart by Givoni-Milne, which also combines geographical coordinates and hourly data, provides the basis for a dynamic assessment of comfort (https://www.ibpsa.us/sites/default/files/publications/SB04S3B3.pdf, accessed 10/28/2019).

The bioclimatic chart continues to be subject to refinement during the 1980s (Watson and Labs 1983) and at the beginning of the 1990s, it is widely tested and validated.

The US literature of the 90s is very rich and shows a vast scenario of applications especially in the residential sector with regard to Design Strategies for Residential Buildings using Psychrometric Chart-Processes (Eads 1979). The in-door quality of the project can be measured but the scenario of the external environment remains unclear.

In Italy the theories, guidelines and manuals for holistic analysis of buildings were also developed (Balcomb et al. 1979; Gangemi 1994, 1997, 2001) as well as early bioclimatic project concepts aimed at industrialized systems (Veninata and

Esposito 1979; Veninata et al. 1980) multiply with respect to each specific renewable resource: solar energy, wind, etc. (Grosso 1998; Esposito 2004; Grosso 2011) and by relationship between architecture and comfort (Grosso et al. 1998)

With the aid of the 1979 GM-Chart it is possible to calculate and evaluate the design strategies of the set of significant environmental physical-technical parameters of passive solar heating (Arens et al. 1980):

- active solar heating and convection heating
- humidification and dehumidification
- internal thermal gain
- shading
- thermal mass with and without ventilation (Trombe walls)
- evaporative cooling
- ventilation
- conditioning with conventional dehumidification.

The Department of Energy (DoE) of the US Government develops DoE2, a public access software to calculate the strategies listed above and simulate performance for residential buildings. With the spread of the software, the strategies are coded and widely shared internationally.

The 1981 issue and the subsequent developments of 1992 and 1994 of the American ASHRAE 55 standards (the so-called "Comfort Chart") further facilitated the diffusion of the methodology of analysis and calculation of thermo-hygrometric comfort.

At the end of the 1990s, the methodology for assessing internal physical-environmental performance in four phases was established at an international level:

1. Search for climatic data (temperature, humidity, sunshine, ventilation, etc.) and simulation of internal conditions (overlap with the comfort zone of the GM chart), a combination of internal and external conditions;
2. Procedure for evaluating results from overlapping data (DoE2 Program + TMY2 Weather data) calculation and time display;
3. Comparison of results with the GM chart;
4. Definition of design strategies based on bioclimatic data.

The latest developments in the diagrams recently gave new momentum in the United States to bioclimatic analysis applied to buildings. Among the main contributions, W. Brown reviewed the Olgyay diagram and implemented the considerations made by Givoni (Givoni 1998) on the ASHRAE diagram (Fig. 24.3).

All the developed knowledge and the great popularity of the energy approach have contributed since 1967 to generate methodologies and calculation tools for the modeling and energy simulation of buildings. The calculation tools in the thermo-mechanical and energy sector of buildings are automated and aimed at dynamic simulation. They are divided into two groups: software for the design of technical installations and software for the energy simulation of buildings that evaluate energy performance with the related seasonal or annual consumption. Energy simulation

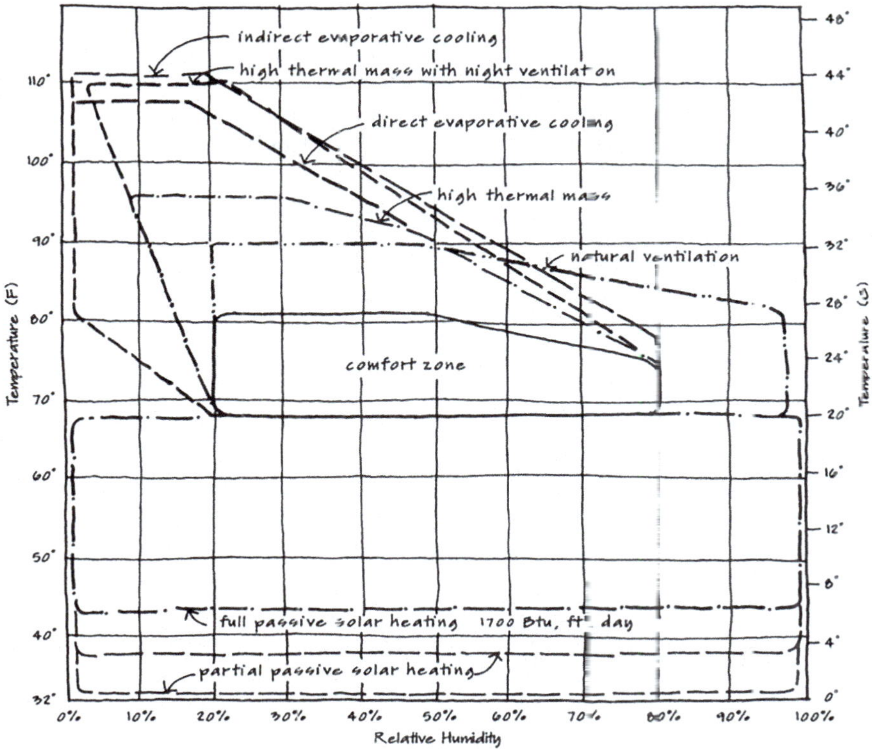

Fig. 24.3 Bioclimatic chart by Brown (2004)

software use numerous physical models for energy representation of the building so that they can be used in all phases of the building's life and for every fraction of time considered. The calculation algorithms are of two types: stationary and dynamic. The former is used to give a standard energy rating to attribute an energy performance to contain consumption. The dynamic calculations based on the resolution of the thermodynamic equations are instead used not only for the verification of the functioning of the systems but also to verify thermal conduction in transient regime, internal and external radiation to the environments with low or high wavelength, external shading, air flows to and from the external environment and between internal environments, the calculation of free thermal gains, etc. (Marinosci and Seprini 2019).

Calculations results can now be integrated into BIM (Building Information Modeling) platforms, although we must always remember that in energy modeling it is not necessary to have advanced definition levels of the architectural model, indeed the geometry of the model is drastically simplified. Furthermore, at present, it is not possible to model some morphologies such as volumes confined by curved surfaces because the simulation is based on one-dimensional heat transfer. Some elements are not relevant (e.g. a pillar) but for example, it is always necessary to define the volume to be calculated. In fact, it is mathematical, that the abstract representation of the thermophysical behaviour of the technical elements and therefore the geometric description has the only meaning of providing data (measurements) for the calculation. In the construction of the model, the climatic input data are very significant and must be appropriately chosen from qualified sources. Similarly, it must be considered that the thermal performance of building components and systems can change over time, but this aspect cannot be represented in the Building Energy Modeling (BEM).

The state of the art is in fact referred to as the BEM of the building's overall energy modeling. It is a versatile methodology based on multi-objective tools that can be used both in the design of new buildings and for the restoration of existing ones, to verify the fulfilment of standard requirements, for environmental certification, for tax relief or incentives offered by energy suppliers, as well as for real-time control of buildings. BEM methodology is also used on a large scale to analyze and define regulations or structure decisions in environmental and energy policies (US DOE 2019).

Whole-Building Energy Modeling (WBEM) is a versatile, multi-purpose tool used in new building and retrofit design, code compliance, green certification, tax qualification, utility incentives, and real-time building controllable is also used in a large-scale analysis to develop building energy efficiency codes and inform policy decisions.

However, the software landscape in this area is always evolving especially in relation to the integration with BIM and to the difference that it can support. A summary of the evolution in this field with the main software is shown in the tables (Fig. 24.4).

To complete the state of the art in this brief summary of the fundamental steps of transferring the results of advanced research to the industrial sectors in this field, it is also necessary to introduce the third pillar of the concept of sustainability concerning the economic and financial aspect. The concept is closely related to that of resources, energy ones, as the cost significantly affects the design and technical choices of any energy project.

The studies focused on the concept of cost in the life cycle as a recognized reliable method to address the economic dimension of the problem in design choices. This is a method of financial derivation with which designers are not yet very familiar. The basic concept of the LCC concerning energy and its use and conservation during the life cycle is aimed at highlighting the less efficient alternatives (scenarios) that entail higher costs than the efficient ones over the considered time frame. The method, therefore, is not based on the construction cost but adds to it the projection of energy and maintenance costs in the life cycle according to the current financial rates (Addison and Associates 1999). Therefore, the choice of the most economically advantageous solution—required by the legislation of public works in Italy—can be highlighted by the evaluation of future costs (ENPV—Expected Net Present Value) for energy and maintenance during the operation of the buildings. Life-Cycle Costing (LCC) analysis is recognized to reliably identify cost-optimal building planning solutions, yet it is not widely used with confidence (De Neufville and Scholtes 2011).

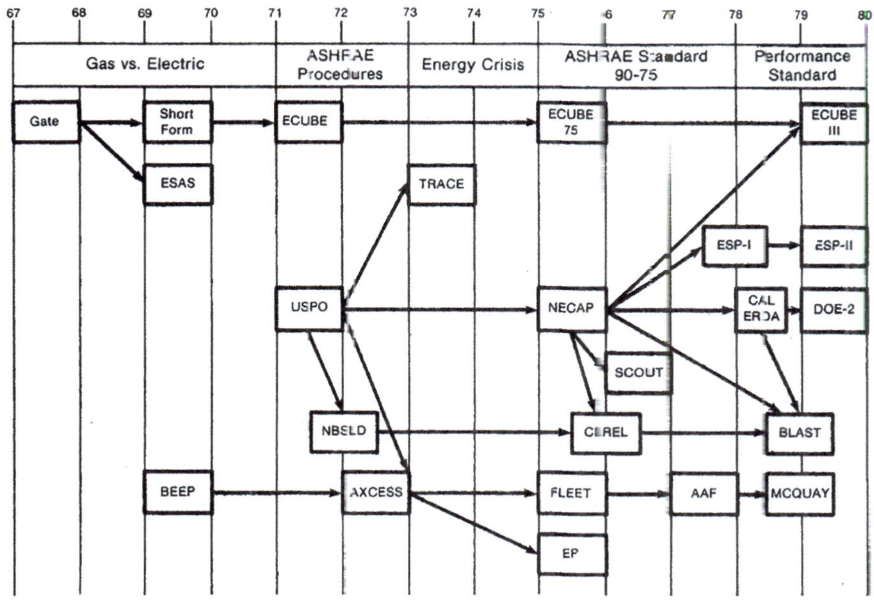

Fig. 24.4 History of energy analysis computer programs 1967–1980 (cit. Judkoff et al. 2008)

24.1.2 Technologies

In parallel with the development of theories and methods for alternative energy and the environment since the late 1950s, design automation tools fit into the production processes of the project had been developed, starting a profound transformation of operational practices.

Just 60 years ago, Dr Patrick Hanratty created the first numerical control system, which later became Computer-Aided Design (CAD). This innovation has revolutionized engineering, architecture and the entire production environment for which the importance of digital has proved to be strategic in order to push the transformation of production processes.

CAD evolution line is fairly well known (Table 24.3) but it is useful to underline the parallel chronology with respect to the evolution of the concepts, the parametric control and environmental modeling software because these technologies converge and merge at some point in radical innovation. Today this innovation has reached the complete digitalization of the design, manufacturing and project management processes.

Design Science theory addresses an important issue that arises from the digitalization of the process concerning the need for knowledge sharing within project information and use of the data in the most effective way. Software is still inadequate to support this goal. Studies will focus in fact on the ontological aspects of information management (Carrara et al. 2009, 2017).

Table 24.3 CAD to BIM transition Timeline

Timeframe	Software/Milestone	Features
1957	READY	Designed by Patrick Hanratty, it was the first CAD-based programming system on the market will be the future CAD
1960	Sketchpad	Designed by Ivan Sutherland, it was the first-ever to use a graphical user interface; users wrote with an optical pen on a XY-dot plotter display; allowed for the first time to constrain properties in a drawing, creating the concept of "objects" and "instances"
1971	ADAM	Designed by Patrick Hanratty, this interactive design, drawing and production system was written in Fortran and designed to work with virtually any computer; it was a huge success so much that it was updated to work on 16 and 32 bit computers: today 80% of CAD programs can be traced back to the roots of ADAM
1970s	CATIA	Developed by Dassault systems on a base code from Lockheed Martin, introducing the first forms of 3D drawing and representation
1980	IGES	The software transfers the 3D drawings between different CAD software; since the STEP format was released in 1994, IGES has not been updated, but it is still used by many technical offices
1982	AutoCAD	Designed by Autodesk, this was the first 2D CAD design software developed for PCs instead of mainframes or minicomputers
1987	Pro/Engineer	PTC Creo was the first conventional CAD software that took up the ideas of Sketchpad (interactive, easy to use, fast); introduced feature-based parametric solid modeling and the use of constraints; represents a huge turning point in the history of CAD because it completely eliminated any competition: the other systems had been written in Fortran while Pro/Engineer works under different UNIX and Windows dialects which make it fast and easy to use
1994	Autodesk AutoCAD R13	Autodesk software is compatible with 3D models
1994	Industry Foundation Classes	Platform-neutral, open-source data model. Intended to describe architectural, building and construction industry data
1994	STEP	IGES replaced as a new format to use when transferring 3D models from one CAD software to another; in 1994 there was the first release of the STEP format that made it an international standard for models

(continued)

Table 24.3 (continued)

Timeframe	Software/Milestone	Features
1995	SolidWorks 95	Developed by Dassault Systemes—Another software that has been successful due to its ease of use; allowed engineers to take advantage of 3D CAD technology
1995	Solid Edge	Developed by SIEMENS and developed as a Product Lifecycle Management (PLM) software; offers many features for Windows; provides solid models, assemblies and supports 2D orthogonal visualization: it is the answer to SolidWorks success
1999	Autodesk Inventor	Autodesk chooses a new direction with the aim of making the SW easier and more intuitive to use. It allows the creation of complex assemblies in record time; is still in use; has raised the bar a lot in the world of CAD
2000	Revit	Originally developed by Charles River Software, the software allows users to design a building and structure and its components in 3D, annotate the model with 2D drafting elements, and access building information from the building model's database
2001	Navisworks	Formerly Jetstream in its previous iterations is used primarily in construction industries to complement 3D design packages (such as Autodesk Revit, AutoCAD, and MicroStation). Navisworks allows users to open and combine 3D models, navigate around them in real-time and review the model
2002	Autodesk Revit	Autodesk acquires Revit and starts developing its own advanced BIM tool
2003	Generative Components	Parametric design software developed by Bentley systems; users can interact with elements either by modifying geometries or applying "rules" to the elements
2006	Digital project	BIM solution based on CATIA
2008/2010	Parametricist manifesto	Neologism developed by Patrik Schumaker, Parametricism offers a new approach to architecture on the basis of advanced computational design tools and techniques
2012	Formit	Innovative tool that blends BIM capabilities and conceptual design
2012	Autodesk 360	Operate CAD on the cloud with a remote access client-server system

(continued)

Table 24.3 (continued)

Timeframe	Software/Milestone	Features
2011	Solibri + dRofus	BIM can control complex rules and regulations in the model and carry out technical checks of various kinds. dRofus is a database management system (DBMS) capable of normalizing and integrating different databases in a common environment interfaced with BIM. The databases created by the specialized models (e.g. with the design builder modules) can be used in BIM modeling
2014	Autodesk design suite	Autodesk in addition to Revit has acquired various other specialist proprietary parametric modules, laying the foundations for a common data environment exchanged by various disciplinary models. The focus shifts to data integration capabilities in the models. CAD is enriched with non-geometric data integration functions and clash detention becomes BIM building information modeling

Source Data search by Bosi, TxP Research 2019)

The functions of CAD software become increasingly sophisticated and, with regard to the possibility of managing environmental data, in particular in the USA with the support of the Department of Energy (DoE) of the Government at the Collaborative Agent Design Research Center (CADRC) of California University (Pohl and Myers 1992) is studied the way to automate the energetic verifications (lighting and heating) of the project laying the bases of what will be the parametric future of the software for the environmental control. The development for automated control of physical parameters of the project soon exceeds the geometric aspects and is also extended to other dimensions such as the physical-environmental ones. These studies were promoted above all by the International Institute for Advanced Studies in Systems Research, Informatics and Cybernetics, published since the early years and at the 6th InterSymp Conference held in 1992, until the 2000s, in Baden-Baden in Germany.

Going back to the common thread of environmental design, when it can be defined as Green Design?

In fact, environmental design especially if focused on the energy problem, it is only part of the problem faced by Green Building approach: as discussed above it is a subset of a more complex holistic concept of sustainability. Green Design which is dealing with the whole sustainability approach to the project not only looking at the energy and environment aspects. It is part of a renewed conception of the building process structured on circularity and holistic sustainability criteria. As indicated above it refers to the most advanced vision in Europe based on three fundamental dimensions: environmental, social and financial (Fig. 24.5).

Fig. 24.5 The Three Pillars of Sustainability. *Source* EU policies graphicized by TxP Research)

Environmental aspects are in fact related not only to the internal environment but also to the external and natural environment to the different scales that the project involves.

Therefore, Green Design is not only referable to a design that is sensitive to the environment and to energy but must respond to the various demands of sustainability. It is possible to say that is that "*The single planning action that can't overlook 'knowledge' of its environmental, economic and social consequences*" (Forlani 2014), so they must be assessed from the point of view of sustainability to be considered Green.

As we have seen in the evaluation of sustainability aspects of economic and social variables enter in the form of different sets of measurable parameters belonging to the various disciplines involved in the project. Therefore, in order to associate the parametric evaluation with Green Building process in the digital environment, it must have characteristics that are suitable for managing and integrating different data (such as data content, logical and formal structure) belonging to sets that can be modeled, interpreted and/or simulated with logics and algorithms of a different nature.

From this point of view BIM is presented as an enabling platform supporting the interested parties of the project in the different phases of the development allowing a remote interaction and the use of data in real-time preventing or helping the resolution of contradictions and errors already from the early stages of the project, increasing its efficiency, estimating its accuracy, and optimizing planning (Ferrandiz et al. 2017;

Eastman et al. 2012). In fact, BIM is not just software but more like a technological-managerial approach to the project that responds to digital logic. This is a very complex concept that implies a different organization of processes oriented towards digital work environments of a collaborative nature, based on the sharing and interoperability of data based on international standards. BIM environment is based on a simplified approach and universal access to data for the creation of project information thanks to a common environment of data generated by different specialist models (Forgues et al. 2016).

To apply BIM design processes that must be industrialized. It means to adopt a significant increase in organizational requirements. The state of the art in project management is based on Lean Thinking methodologies (Lean Design + Lean Construction) and therefore on the objective of eliminating errors, waste, interferences and conflicts in the design solutions for the realization and management of the works until the end life cycle. This avoids the major cost factors derived from them (Kiviniemi 2018).

The state of the art of digital technologies also opens the perspective of digital fabrication. This industrial approach introduces and accelerates the direct transition from model to manufacturing (e.g. 3D printing and robotics) which combined with BIM will radically transform the construction process into the future (Esposito and Bosi 2018).

The combined use of modeling and digital fabrication has begun a process of document obsolescence of the project that is no longer aimed at supporting communication between humans but between the machines that generate the design and the machines that generate the product. A digital ecosystem of this type could be identified with the concept of Industry 4.0 also in Construction as formulated for other industry sectors.

The conventional codes expressed in the notations of the drawings (symbols, dimensions, etc.) as the modeling is sent directly to the production are no longer necessary, the data underlying them exist in the common environment and can be recalled by the various uses of BIM. This will mean a definitive shift from document-based processes to model-based processes, that is also the need for a radical change in the training of architects and engineers that unfortunately is often still based on the old document paradigm (Kiviniemi 2018).

From modeling to fabrication, introducing the rapidly growing digital fabrication (e.g. 3D printing and robotics) combined with BIM will radically change the construction processes in the future (Esposito and Bosi 2018). Their combined use will possibly make drawings (as currently used) obsolete as the production will go straight from digital model to production, implying a final move from document-based design to model-based fabrication.

One central element of State-of-the-Art Project Management is the Lean Design methodology and eliminating waste in the form of design errors and clashes, which often are a major factor in the cost and schedule over-runs.

This also implies a need for a radical change in the architectural and engineering education, which unfortunately is still often based on the old document-based paradigm (Kiviniemi 2018).

24.2 Evidence from Experimental Projects

To understand the relevance of the change occurring in the scientific and technological scenario and its impact on the project design it is useful to illustrate some examples referring to experiences from the last 20 years that have achieved successful projects regarding Green Design objectives. It may help to understand the correlation with the indicated phenomena, newly adopted approaches and the technological artefacts of the scenario.

The first case study concerns the Post Tower in Bonn (Germany) built-in 2003 to a design by Helmut Jahn. The building is 163 m high, has an elliptical shape oriented according to the prevailing winds and can exploit them to operate the bioclimatic ventilation of the building without the aid of mechanical systems for the whole year. The project was awarded in 2013 for the years of perfect functioning in operation in relation to the expected thermo-hygrometric performance. The lightweight construction technology consists of a steel and concrete structure for the floors and a double skin of high-performance glass for the casing. The minimalist design is not only due to aesthetic intent, but it is the result of the optimization of materials in a perspective of minimization of resources (basic concept of the Green Building). Furthermore, the choice of construction technologies and materials are based on the concept of circularity, favoring those that are 100% recyclable such as steel and glass. Consequently, the Post Tower design approach is holistic, establishing the goal of minimizing the use of energy with respect to typical consumption of office buildings from the early stages of conception. The innovations of architectural and technological solutions have been developed in various subsystems of the building organization: first of all, they concerned the building envelope with the design of the double-skin façade solutions, with shading equipment inside. Then, in the use of operable windows for individual regulation of natural ventilation and for the introduction of fresh air for offices from the double skin interspace or also through soundproofed ducts with heating and cooling function. Secondly, they concerned the plant and systems engineering, in particular for the natural cooling of the air through two underground wells at a temperature of 12 °C (Low-pressure Heat Exchange (LHX), Indoor extract AI, R convection thermal gradient fan replacement of high-pressure, heat-recovery systems) (Grosso et at. 2015) (Table 24.2) and the activation of the thermal mass of exposed concrete ceilings: furthermore the mechanical ventilation was decentralized due to heat recovery used to air-condition the indoor roof garden with the exhausted air of the offices. The building does not provide any vertical channeling for the air that circulates in the double skin and through the atrium where the exhausted air is conveyed. Heating, ventilation and cooling are individualized in every environment using digital sensors. Regarding the double-skin façade, this typological solution has had a very wide diffusion, but not always with the same results. The technical solution adopted is based on energy modeling: it allows flexibility of use through maneuverability without mechanized apparatus, being equipped with a system of shading which is adjusted according to the solar declination and which are the elimination of energy gains due to direct radiation. The cavity contains the system that extends over the entire height of the tower and works to extract air thanks to the chimney effect. The accelerated air circulation inside it from this effect combined with the wind-induced pressures/depressions on the facades reduces the temperature differential. The glass in the façade contains metal and is double-chambered, containing argon gas to obtain better insulation (Fig. 24.6).

How is all this complex operation controlled? The designer, making use of his experience as a sailor, used sensors and electronic control equipment: the digitalization of the management system is based on numerous physical and environmental

Fig. 24.6 Post Tower, Bonn (Jahn Architects 2003) (Courtesy of Jahn Architects, Chicago)

Table 24.4 Italian references for airports environmental impact assessment (from: Esposito 2010)

ICAO
Annex 14 volume I - Aerodromes design and operations
Annex 14 volume II - Heliport
Annex 16 volume II - Enviromental Protection Aircraft Noise
Annex 16 volume II - Enviromental Protection Aircraft Engine Emission

IATA
Airport Development Reference Manual

FAA
FAA AC 150/5070-6B – Airport Master Plan
FAA AC 150/5300-13 – Airport Design
FAA AC 150/5320-14 – Airport Protection Aircraft Noise
FAA AC 150/5360-13 – Planning and design Guidelines for Airport Terminal Facilities

ENAC - Regolamento per la Costruzione e l'Esercizio Degli Aeroporti
D.Lgs 17 gennaio 2005 n.13 Attuazione della Direttiva 2002/30/CE relativa all'introduzione di restrizioni operative ai fini del contenimento del rumore negli aeroporti Comunitari
D.Lgs 9 maggio n.96 Revisone della parte Aeronautica del codice della navigazione
D.Lgs 15 marzo n.51 Disposizioni correttive ed integrative al D.lgs 9/05/2005 , recante la revisione della parte Aeronautica del codice della navigazione

factors such as external and internal temperature, wind speed and pressure differentials. Each Post Tower room is equipped with a digital control unit (smart room control). The Post Tower casing is also inspired by the sail due to the use of lightweight hi-tech technologies for the exploitation of wind pressures and depressions as a conduction element for the flow of air inlet and outlet with controller support digital used by racing boats (Fig. 24.6).

The use of optimized energy has enabled savings of 30% to be achieved compared to normal office buildings in Germany (Table 24.4; Fig. 24.7).

Furthermore, the use of geothermal energy in the Rhine basin by means of heat pumps for heating and cooling uses the concrete floors for distribution by means of coils (Fig. 24.8).

The use of natural elements to create a semi-confined garden in the attic has been made possible at that latitude thanks to the reuse of the heat from the exhaust air extracted from the offices and thanks to a wooden floor that also dampens the effect of albedo caused by solar radiation. The landscape effect increases the comfort condition of the occupants, a social aspect that is not negligible in a public building.

In conclusion, it can be said that in this case, the holistic approach digitized to Green Design, associated with the environmental engineering calculations previously carried out in the design phase prefigured the range of technical solutions that have been developed on a technical-scientific and not just a formal basis.

The second case study concerns a project of an airport terminal, still designed by Jahn Architects; the international terminal of Bangkok in Thailand: the Suvarnabhumi Airport. The 2005 design was opened to the public very fast in 2006, it is in continuous development thanks to the ingenious typological construction system

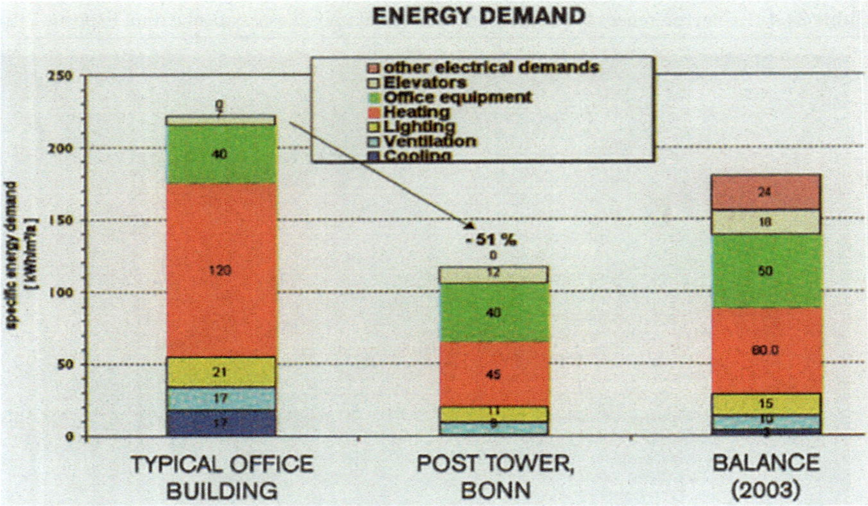

Fig. 24.7 Energy saving of approximately 30% at Post Tower, Bonn over conventional office buildings (Jahn Architects 2003) (Courtesy Jahn Architects, Chicago)

characterized by maximum flexibility and low construction and management costs, consistently with the main objective of financial sustainability posed by the airport management client. The approach is totally innovative from an architectural, engineering and environmental standpoint and has been aimed at maximum integration between the disciplines by implementing a collaborative development scheme to prevent design changes and errors. In addition, the designer has focused on the innovation of technologies, systems and materials. The climate of the area is hot and humid with temperatures ranging from 25 to 30 °C and characterized by high rates of very high relative humidity which posed a difficult problem of energy consumption for air conditioning (Table 24.5; Fig. 24.9).

Consequently, the terminal casing was designed with transparent parts characterized by 95% radiation reduction to avoid excessive thermal gain and consequent overheating of the air volumes inside. The opaque parts are made with EFTE (Etilene Tetra-Fluoro Etilene) polymeric layering of stretched sheets on 40 m high portals, made of reticular tubular steel. The relative lightness of the casing has allowed the structural sections to be reduced, saving raw material and ultimately the built-in energy of the structure from the point of view of the life cycle assessment (Fig. 24.10).

The terminal has a linear type that develops like a grid and flows into a central access area (main concourse). Among the arms (piers) that house the boarding points (gates) at an altitude lower than the ground floor, there are gardens that exploit the presence of water in the ground originally a reclaimed swamp.

The challenge of sustainability, both in terms of engineering and environmental solutions, consisting of the swampy features of the land and the economic challenge

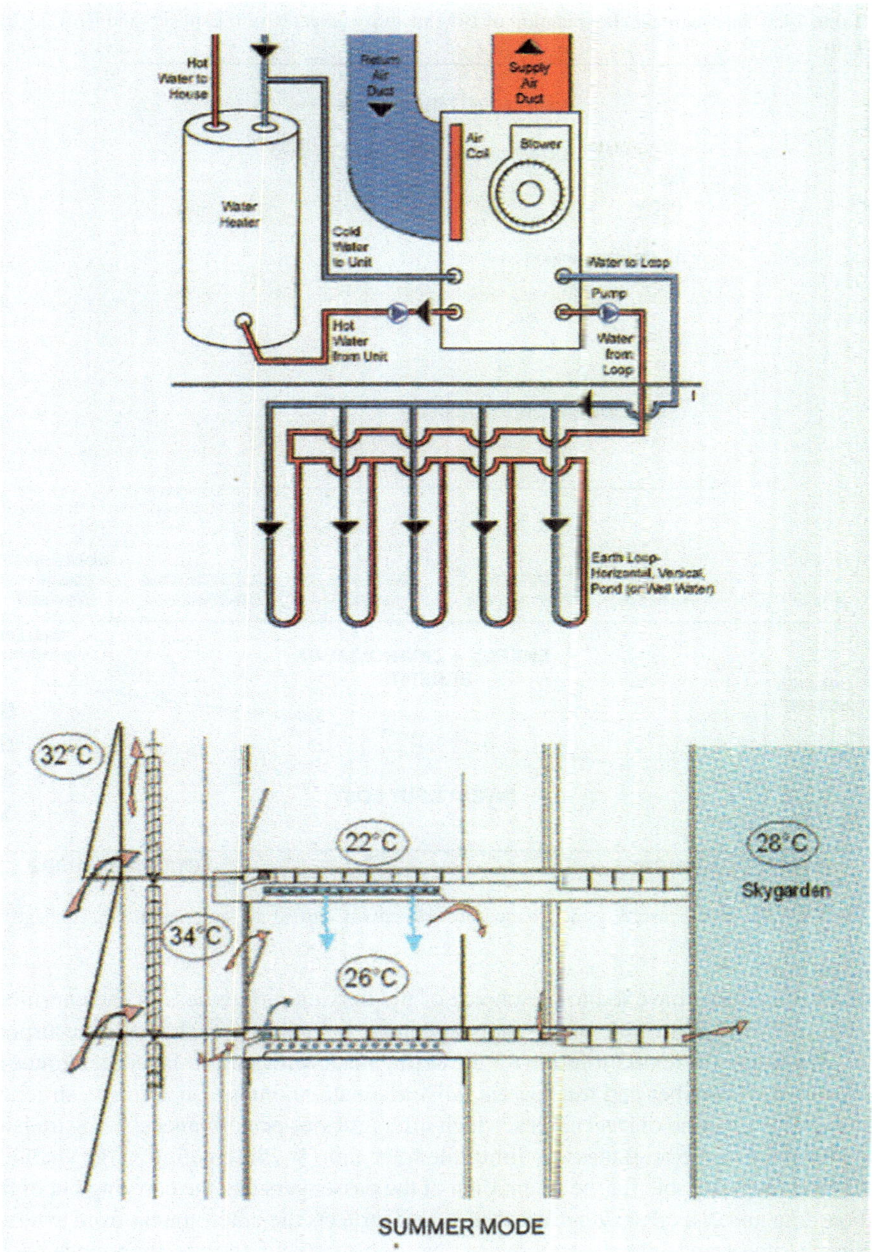

Fig. 24.8 Geothermal heat energy form Rhine river, Post Tower, Bonn (Jahn Architects 2003) (Courtesy Jahn Architects, Chicago)

Table 24.5 International classification of BIM maturity levels (from: Esposito and Bosi 2018; p 130)

APPALTARE DIGITALE

LOD italiani e internazionali a confronto

FASI	Esigenziale Piani	Fattibilità	Definitivo	Esecutivo	Contabilità	Collaudo	Restauro
ITA (UNI 11337 Parte 4)	A	B	C	D	E	F	G
UK (PAS 1192-2)	1	2	3	4	5	6	
USA (AIA 2008/2013: BIM Protocol Exhibit)	100	200	300	350	400	500	

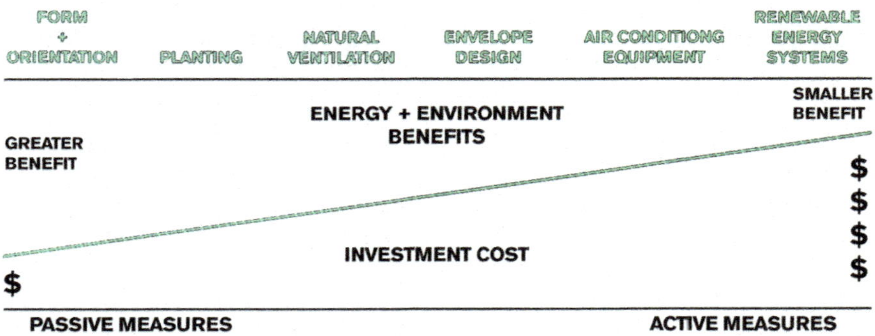

Fig. 24.9 Passive measures vs active measures benefits (Courtesy of Jahn Architects, Chicago)

to contain costs, have led to the choice of the maximum lightness of the enormous structure and architecture of the landscape in the holistic approach inside the airport.

Regarding the textile solution for the casing made with a triple-layer membrane of Teflon™, glass fiber and transparent polycarbonate, mounted on a tensile structure consisting of a grid of steel cables, which offers various performances: being translucent it diffuses the brightness but limits the solar gain (1–2%) creating a high visibility environment suitable for the orientation of the passengers and the carrying out of the boarding and disembarking processes. It also protects the environment from external aircraft noise and internal sound reverberation of sound sources (announcements, etc.) due to its soundwave absorbing properties (Figs. 24.11 and 24.12).

The most interesting Green Building feature is an innovative approach to physical-environmental engineering for indoor air conditioning, aimed at obtaining the greatest thermo-hygrometric comfort opposed to the lowest possible energy consumption and

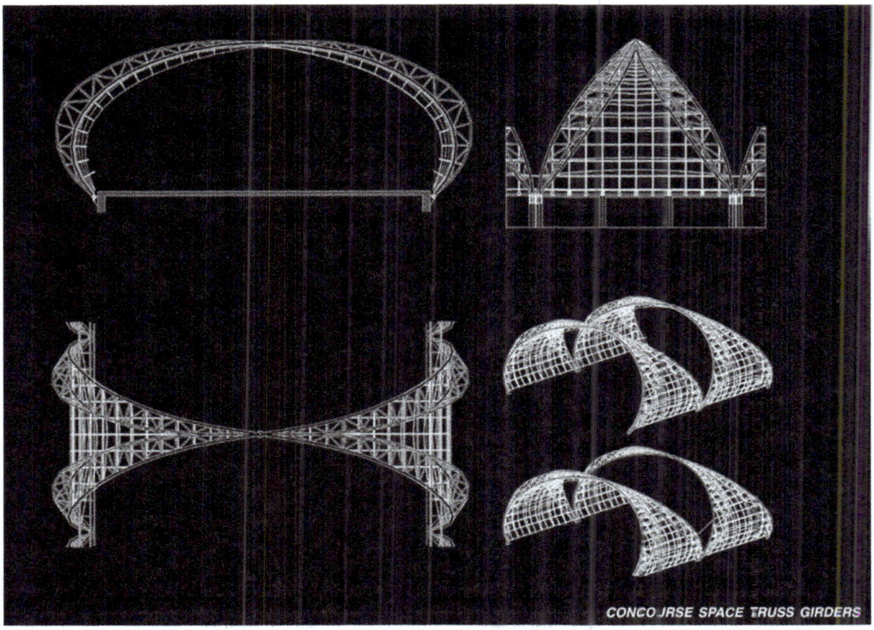

Fig. 24.10 Truss Girders simulation for structural efficiency (Jahn Architects 2006) (Courtesy of Jahn Architects, Chicago)

operation costs. As mentioned above the location in this climatic zone would have involved consumption and therefore very high management costs with conventional systems. The design team made use of computational fluid dynamics simulation tools, arriving at a never-before-seen solution: the horizontal partition of the air layers at different temperatures. In fact, during the simulations, it was noticed how the need for conditioning only concerned the lower layers of the 40 m-high air volume. Therefore, it was decided to divide the volume into a conditioned area and a non-conditioned zone by reducing the volumes to be treated by 50%. The lower area where the passengers circulate and the embarkation and disembarkation processes take place is the one that must meet the comfort criteria. The lower area of the air volume is served by two systems: one on the floor and a ventilation system with a low cooling speed distributed with delivery and return ducts within the required height range of about 2.5 m (Figs. 24.13 and 24.14).

Since hot air tends to rise, a natural stratification of the air layers is achieved it keeps the fresh air in the lower range where it is needed. This innovation has become typical of newly conceived airports (Fig. 24.15).

In conclusion also, in this case, the Green Design was managed with a holistic approach to the disciplines involved, with the aid of digital simulation modeling tools and finally with experimental tests on the 1:1 model of the solution. The results of the project show a 35% savings in energy consumption for air conditioning compared to

Fig. 24.11 Suvarnabhumi Airport textile envelope site work (Jahn Architects 2006) (Courtesy Jahn Architects, Chicago)

a conventional system, which corresponds to an annual saving of 65,000 tons of CO_2. Indeed, these are the large numbers that arise in airport projects. For such a reason the Green Building issue has received increasing attention from the Civil Aviation industry which has produced sector-wide policies at an international level, project strategies and airport management, project and technological innovations and finally certification protocols for all the significant aspects of the problem. An example is the birth of studies and environmental impact assessment that is due to the Civil Aviation sector in which some of the main causes of disturbance such as noise and pollution have been studied since the late 1960s (Convention of Buenos Aires 1968; ICAO—International Civil Aviation Organization, Annex 16). The most advanced holistic good practices regarding Green Building are still observable today in civil aviation and airports (Table 24.4).

Another example with reference to the Green approach in the airport field involves not only the infrastructure project but also the managerial aspects (Transportation Research Board 2012a and 2012b). Thanks to a series of sector directives, the scenario of European Civil Aviation is certainly the most advanced not only with respect to other industrial sectors but also confronted to other continents. Most European airports have joined the ACA—Airport Carbon Accreditation certification promoted by ACI Europe (Airport Companies International). European civil aviation is the most technologically advanced, the most oriented to the environmental, social and

Fig. 24.12 Suvarnabhumi Airport textile envelope (Jahn Architects 2006) (Courtesy Jahn Architects, Chicago)

economic sustainability of the globe. In fact, the European regulation package of the sector is based on ICAO addresses—SAP (State Action Plans) on the reduction of CO_2 emissions:

- ETS Emissions Trading System
- SES Single European Sky
- SESAR Single European Sky ATM Research
- Clean Sky

Each measure provides for the achievement of measurable targets for 2030 through European and national policies. Among the objectives pursued is the reduction of the environmental impact by 10% (environmental sustainability, the cutting of traffic management costs by 50% (economic sustainability), the increase in both flight and ground safety (social sustainability). By 2050, the goal of European civil aviation is complete de-carbonization, that is zero impact, which for some aerospace companies will probably be anticipated by 2030. In fact, to date, 44 airports have reached level 4 of Neutrality and over 102 the other 3 levels out of the 4 levels envisaged by the ACA protocol promoted by ACI—Europe (Airports Company International) signed by 200 airports worldwide.

Fig. 24.13 Suvarnabhumi Airport radiant floor directly removes heat radiation striking the floor; air displacement systems with controlled air stream supplies cool air to the floor low velocity (Jahn Architects 2006) (Courtesy Jahn Architects, Chicago)

Today, 44 European airports are operated by carbon-neutral models and another 103 are accredited at the other levels of Airport Carbon Accreditation. The decade of how they will work on their promise to become Netzero for their carbon emissions by 2050 (or even 2030 in some cases)—now committed formally by ACI EUROPE and undersigned by close to 200 individual airports.

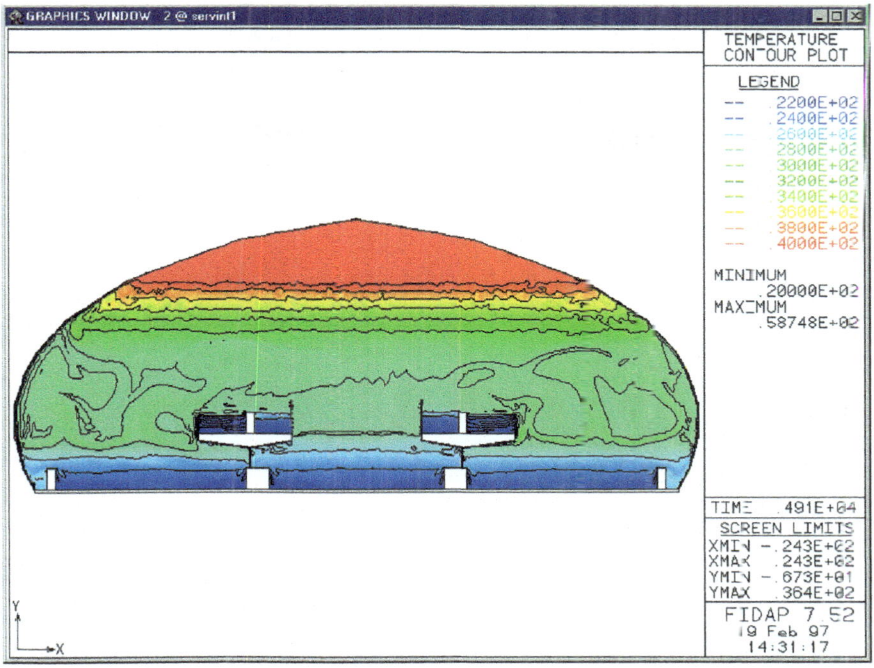

Fig. 24.14 Simulation of air stratification of natural ventilation at Suvarnabhumi Airport (Jahn Architects 2006) (Courtesy Jahn Architects, Chicago)

These measures also have an impact at the international level, making the system more competitive thanks to the environmental rules imposed on the single European sky. A totally integrated system (SESAR) has been created and is operational between the member states with centralized air traffic control (ATM) at Eurocontrol, with routes optimized with sophisticated algorithms to reduce consumption and therefore CO_2, take off and more gradual landings and with classes of aircraft gradually less and less polluting. These policies have multiplied industrial investments to make airplanes increasingly silent and with increasingly lower emissions of the main pollutants such as nitrogen oxide (NO), an oxidizing substance which also has paramagnetic characteristics, it reacts with combustible and reducing materials and contact with air is transformed into nitrogen dioxide (NO_2); unburned hydrocarbons (HC), carbon monoxide (CO) and carbon dioxide (CO_2).

In the USA has been also defined the Envision® protocol introducing a systemic approach for understanding and interpreting the complex reality of the infrastructure design in the relationship between technical and human agents aimed to sustainability evaluation of infrastructure projects within the global trends. The infrastructure project is subject to the resource constraints of multiple public administration bodies and decisional levels. Such constraints create problems and delays at different design steps when not properly managed. Ratings systems and tools intended for buildings

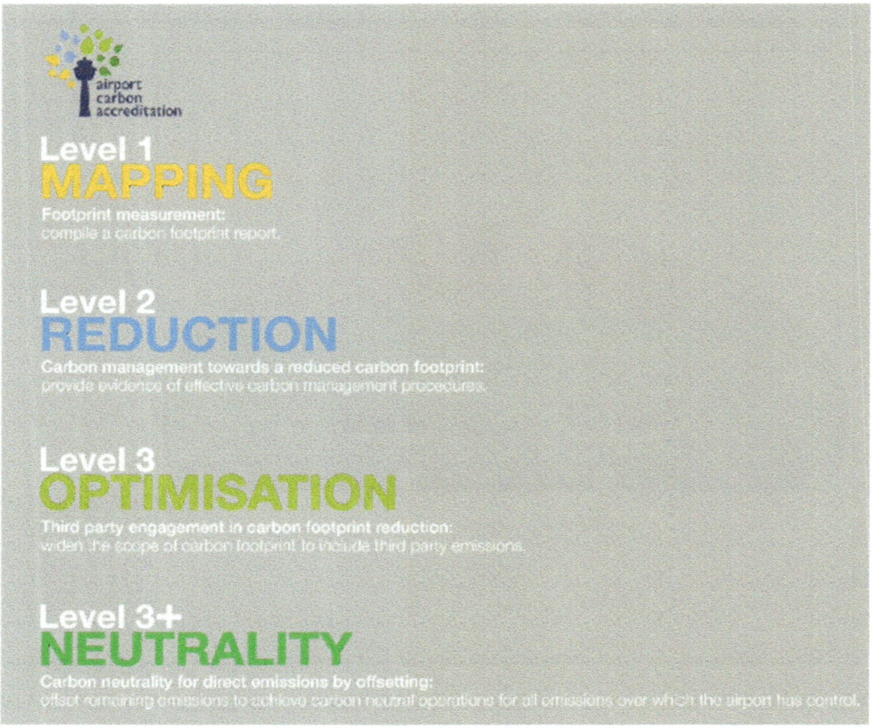

Fig. 24.15 ACA airport carbon accreditation protocol levels (ACI Europe)

are not designed for this context and cannot adequately assess the extensive external benefits and impacts infrastructure has on a community. The basic requirement to be provided is the efficiency and long-term sustainability of the communities the infrastructure serves. Envision is a framework that provides the guidance needed to support this systemic approach in the planning, design and delivery of sustainable and resilient infrastructure. The protocol works as decision-making guide based on a set of performance and flexible measures. The protocol allows higher performance through more sustainable choices in infrastructure development. The framework provides a flexible system of criteria and performance objectives to aid decision-makers and help project teams identify sustainable approaches during planning, design, and construction that will carry forward throughout the project's operations and maintenance and end-of-life phases. It provides sustainability metrics for all types and sizes of infrastructure to help users assess and measure the extent to which their project contributes to conditions of sustainability across the holistic full range of social, economic, and environmental indicators.

In such a highly regulated, integrated and advanced context, there is also the contribution to the ground of the built infrastructure. Airport projects are traditionally the result of interdisciplinary teams and therefore the use of organizational methods and digital technologies for project management has been more stringent, also due to the need imposed by design times that are always very short and meet narrow financial logics. The main problems to be faced in the project development are on the one hand the modeling of the terminals, which are building bodies formed by complex systems that must be integrated. Therefore, performance simulation to comply with international standards as regards spatial-functional aspects, of safety and technological-environmental are required. The specialists in the various engineering sectors have developed and used various specific software for decades, while the architects who have been called since the '70s to develop solutions increasingly suitable not only to make boarding and landing processes more efficient terminal but also to improve the passenger experience. Such an objective is pursued trying to integrate the various functions in the spaces where they are to be carried out according to precise rules. The rules can be controlled automatically from CAD to BIM thanks to the possibility of directly using specialized engineering data in the modeling of architectural solutions (Fig. 24.16).

The substantial innovation that occurred with the transition to the digital process enabled by BIM was such that the common data environment can be used even after the construction of the works for the management during the operations of the terminal and for the continuous maintenance and transformations that define service cycles of its useful life (Fig. 24.17).

This aspect contributes substantially to the verification of the economic sustainability of the work.

24.2.1 Modeling and Fabrication

To have a digital model that integrates the various aspects to keep under control in the design solutions allows to also extend controls to the verification of constructability and to the manufacture of the product. BIM can produce the data for the Product Information Document (PID) that is the documentation on all information relating to a product extended to its environmental performance throughout the entire life cycle and disposal if it is developed and implemented at an adequate level of specification of the model. The concept of Level Of Development (LOD) is articulated in coded levels that identify the maturity of the model from the point of view of the completeness of the data and the breadth of the domain of the information that it can generate in relation to defined uses. Therefore, the development of the model is to be understood as a concept relating to the use and not absolute. This means that there is a greater investment in the initial phase of the project in order to have the data necessary

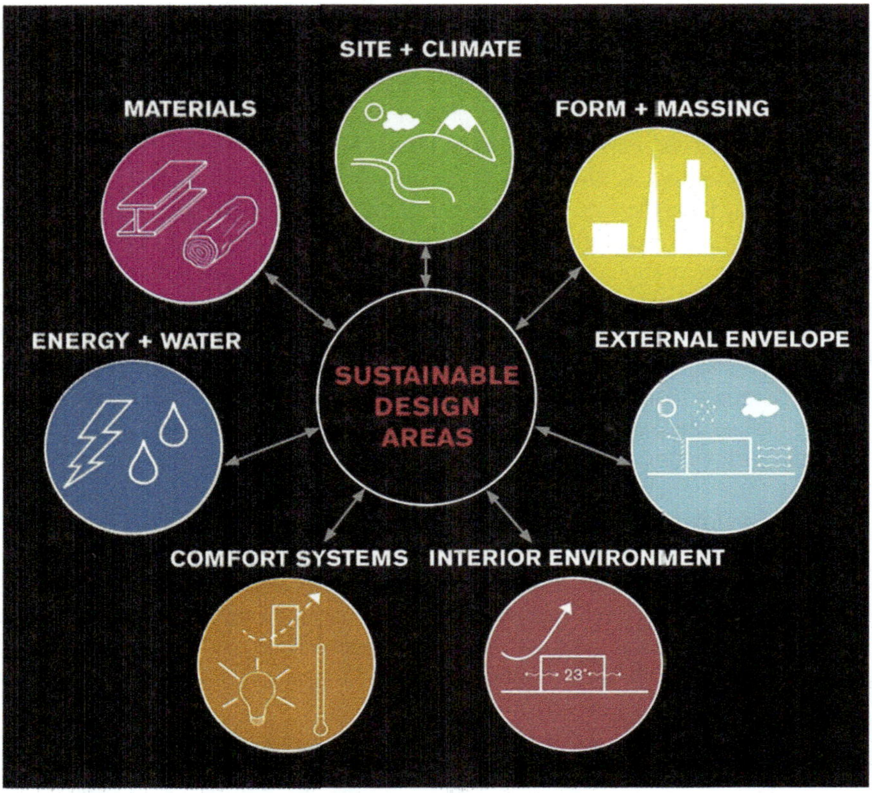

Fig. 24.16 Sustainable design concept map (Courtesy of Jahn Architects, Chicago)

to structure the modeling and the simulation of performance. If this is not the case, digitization will turn out to be a boomerang that can amplify the extent of the damage of any error. This aspect can be calculated in terms of the project's financial risk. The problem is clearly described by the Regnier, Davis and Mc Leanny curves showing the design phases confronted with cost in definitively pointing out the need to anticipate financial investments to create more information on early design stage (Fig. 24.18).

Italy has defined a specific technical regulation on BIM with the UNI 10337: 1-10 series covering all the aspects of the processes and skills necessary for the reorganization of processes in the digital environment (Esposito and Bosi 2018) (Table 24.5).

The digitalization of the process provided that an adequate level of maturity has been developed prefigures the possibility of direct manufacture of the product. In many industrial sectors this step is already implemented regardless of the complexity of the product. The greater investment made in the design development phase—as highlighted in the MacLeamy curve (MacLeamy 2004)—with the complete IPD development opens this possibility drastically reducing the variables

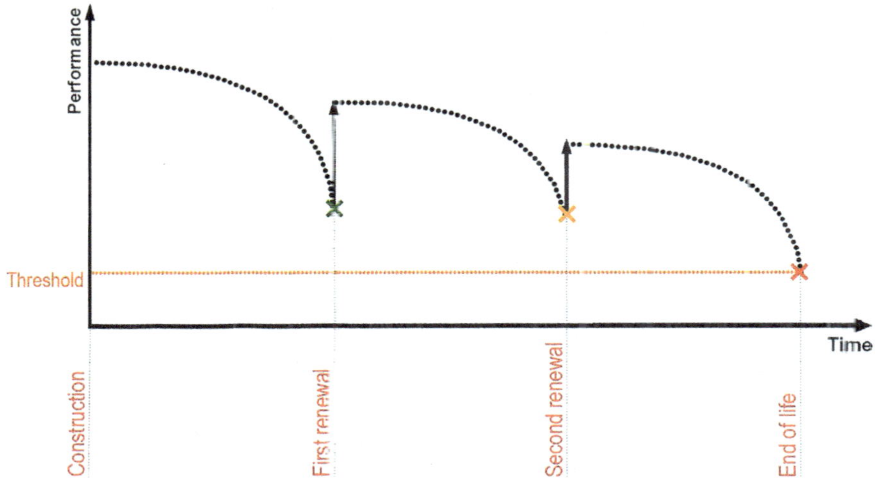

Fig. 24.17 Building life cycle time terminal/performance chart (TxP Research graphic processing from ACRP Report 068, 2012)

and errors and making possible an industrial product planning. This planning does not merely concern construction technologies—which become a defined variable of the system—but above all the different processes redefined and coordinated in terms of quality, time and costs.

Digital fabrication is, therefore, one of the possible uses of BIM that must be defined before the start of the digitalization process in order to field the resources the necessary skills (Esposito and Bosi 2018, pp 47–49). In fact, the adoption of digital platforms such as BIM in construction processes is a function of the purposes of digitization that can relate to the most diverse aspects, from the Maintenance Plan to the management during use—both as regards spatial-functional aspects and of flows both in relation to security and emergency plans. Similarly, it is possible to use BIM to track changes, reconfigurations of the terminal or for final design in the technological-constructive field or even to predict digital fabrication. In fact, in addition to three-dimensional control and design and three-dimensional coordination, BIM can support management uses designed to support the documentation of design responsibility in various fields, from civil engineering analysis, to energy analysis, structural analysis, fire safety, to evaluation of sustainability according to the most diverse certification protocols, both holistic (LEED, BREEAM, Itaca, ENVISION, Building System; ARCA, etc.) rather than with specific certification protocols such as the mandatory energy or acoustic performance. All potential BIM uses can be combined to support the RUP (Responsabile Unico del Procedimento). The procedure manager can use it for the preventive checks as well as for the final validation of the project as required by the Code of Public Works in Italy. With reference to the current and most widespread uses of BIM, such as for example the planning of activities, site analysis, cost analysis, time schedule and operational planning of the

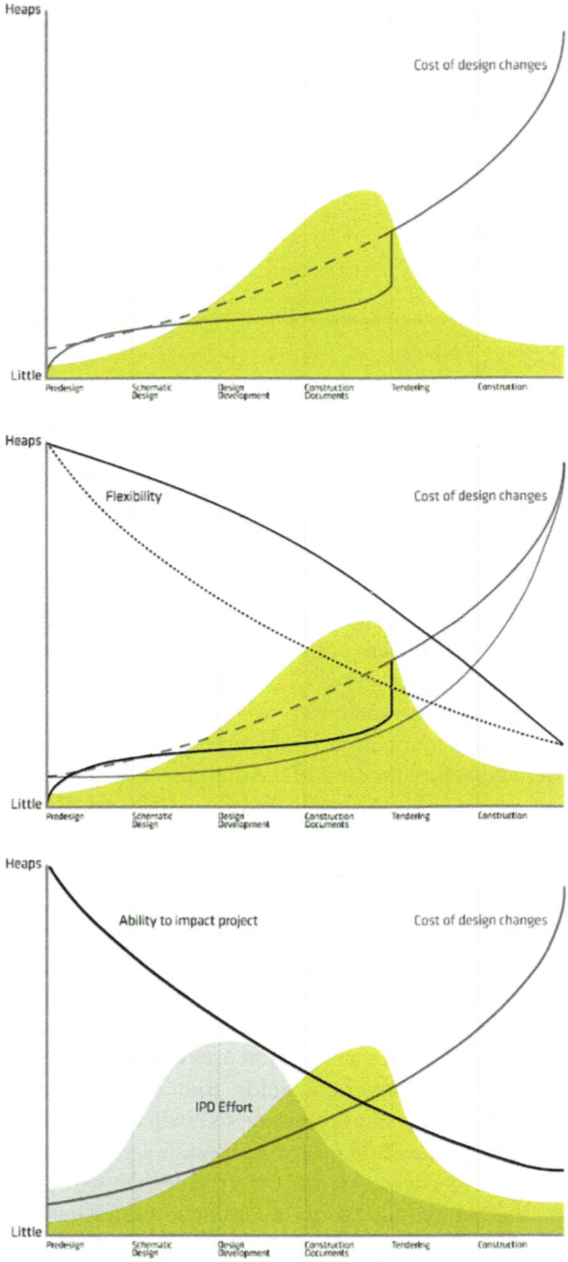

Fig. 24.18 BIM maturity curves by Regnier (2011); Davis (2010) and MacLeamy (2004) (from: Esposito and Bosi 2018; p 50)

site. It is finally possible to adopt lean and error-free techniques (Lean processing). In the technological and organizational scenario described, the possibility of holistic management of the project required by the Green Building vision and necessary to face the theme of sustainability as indicated in the state of the art and in the European vision is more concretely defined.

Green Design can be defined as a point of balance between the objectives of sustainability in the various areas and can be identified through one or more uses of the combined BIM. This definition takes place in the early stages of performance planning of the intervention and implemented during different levels of development necessary for the various disciplinary aspects. In fact, it is important to note that not all aspects require advanced levels of development to be properly addressed. For example, energy modeling has the sole purpose of defining the orientation and the shape of the building can be limited to essential levels and addressed simply by using the correct climate data. If, on the other hand, we want to tackle the most advanced energy modeling with a bioclimatic perspective and with the aim of eliminating most of the mechanical plant as in the example of Bohn's Post Tower (Jahn 2003) we will have to foresee a greater articulation of the development of modeling by introducing dynamic aspects and other uses of BIM for which a greater number of data made available in the shared environment (Common Data Environment, CDE) by the designers of the various disciplines are needed.

Finally, if the use of BIM is digital manufacturing, data will have to be available on the models of systems and components once defined in the drawings and fabrication sheets (shop drawings) (Fig. 24.19).

However, if Green Design really is the goal, it implies a mature technological-constructive ecosystem, created through investments by producers to provide data

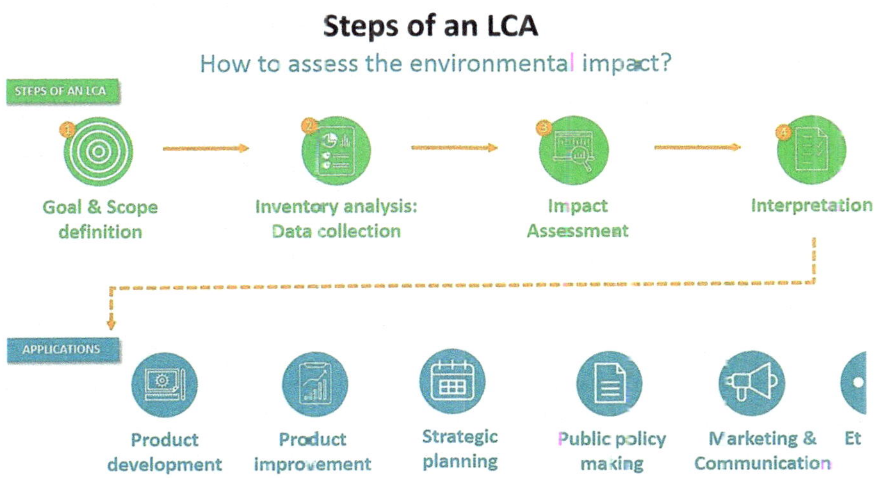

Fig. 24.19 Steps of LCA. *Source* Liberal Dictionary, accessed on 20/10/2019; http://www.liberaldictionary.com/wp-content/uploads/2019/02/l-c-a-6002.jpg

according to the most advanced European standards in this direction, such as the LCA—Life Cycle Assessment based on the ISO 14000 series (EU 2003). In accordance with the ISO 14040 and ISO 14044 standards, the Life Cycle Assessment is divided into 4 distinct phases:

(1) definition of the objective and field of application
(2) inventory analysis
(3) assessment of the life cycle impacts
(4) interpretation of the life cycle (Fig. 24.14).

The possibility of evaluating the product in environmental terms is based on inventory data that are for example also by energy costs in each geographical area (ENEA 2010). In Construction, unlike other industrial sectors, manufacturers consider the supply of product data as a secondary aspect of investment in innovation. The complexity of the LCA assessments implies instead an accuracy of the data and their interoperability in view of their use in the BIM environment. For this reason, international standards have been developed for the representation of product data that are the equivalent of the technical specifications of the products. Standards are an important part of Product Data Technology (PDT) and are used as a specification to manage the flow of information in the life cycle of a product. The standards also provide a tool to describe product data throughout their life cycle and are independent of any hardware/software system. They are useful both for exchanging files between different computer systems and for implementing and sharing product databases and for storing data. In ISO (International Standard Organization) the Technical Committee ISO TC184 "Systems of automation and industrial integration" was established for the computerized representation of product and process data.

The data standards of the entire life cycle of the product available today are:

- ISO 10303 Representation and exchange of product data (known as STEP)
- ISO 13584 Libraries of parts—information model for dictionaries of terms that ISO 10303 can refer to (known as PLIB)
- ISO 15926 Integration of the process data lifecycle including the oil and gas—a reference data library that defines the terminology for products that are used in process plants (known as RDL)
- IFC (Industry Foundation Classes), used in the civil and construction sector, used for the creation of the Common Data Environment (CDE) in BIM.

Most of these standards were developed before the Web imposed the requirement of transmissibility of data as a basis for information. In this sense the possibility of exchanging data remotely is critical for the quality of processes and digital modes. In fact, these standards are now being updated to take advantage of Web technologies, such as RDF and OWL (which we will see later). The standards are written in a specific language called EXPRESS, which was designed to construct information models as abstract but unique representations of real-world artefacts. EXPRESS models are operable from a computer and can be interfaced with existing engineering systems, but they are not useful for creating and managing information that can be used directly by mankind. The data so structured and transmissible are placed on the Web. To operate on the Web a Uniform Resource Identifier (URI) of the object is associated with them. The identifier is unique and allows you to track down any object that has this label that also works as an address in the Web. So, the URI has three functions: describing and locating and accessing. To locate you use the domain that is a name given to the page. The page may or may not be accessible because it is its URI begins with the HTTP://protocol. If the page is not accessible the Web provides a default message: HTTP error 404, which means: you have not found anything. If the URI is accessible and contains descriptions of objects, these are enclosed in files in standard formats (pdf; .html, .gif or .wmv. Etc.) i.e. they are de-referenced. The de-referencing does not allow one to obtain the object by itself if it is a physical object but provides its description or digital model.

The connection between objects potentially at the base of the World Wide Web is the most recent frontier opened by the so-called Semantic Web that dynamically creates intelligent links in real-time thanks to Disruptive Technologies present in the digital ecosystem. An example of that is Distributed Ledger (DL) quests and technologies like the so-called Bloch Chain that allows the creation of an architecture of unchangeable and shared information in a chronological register that can grow to infinity recording all the variations and the author of the same. The applications of the DL have multiplied in all sectors even if the most known is linked to the creation of cryptocurrencies that can be exchanged through transactions between the holders of enabled tokens as nodes of the system to which they also provide IT resources such as CPU, band, memory. This is a technology that must be fully shared and synergistic between the participants of the process to which it applies. Transaction security is the basis of trust in its operation. Technology pushes the virtual potential of processes by allowing data digitization, decentralization, disintermediation, transfer traceability, transparency/verifiability, the immutability of the register and programmability of transfers. DL technologies offering an alternative in terms of security, reliability, transparency and costs to databases and registers managed centrally by recognized and regulated authorities (public administrations, banks, insurance companies, payment intermediaries, etc.) (Wikipedia, accessed 10/02/2019) have disruptive features for the processes and sectors that adopt them and lead to a profound reorganization of the information systems related to them. In fact, the information is considered indisputable for the participants in the process to which a Blockchain is applied (Banca d'Italia 2016) (Fig. 24.20).

In practice, many of the transactions that currently require processes and bureaucratic verification and validation systems such as the authorization and acceptance processes of the project by the public administration could be completely transformed and streamlined, above all thanks to new collaborative approaches that enable the introduction of these technologies and a more and more advanced automation (Carrara et al. 2009; Fioravanti et al. 2017; PRIN 2017).

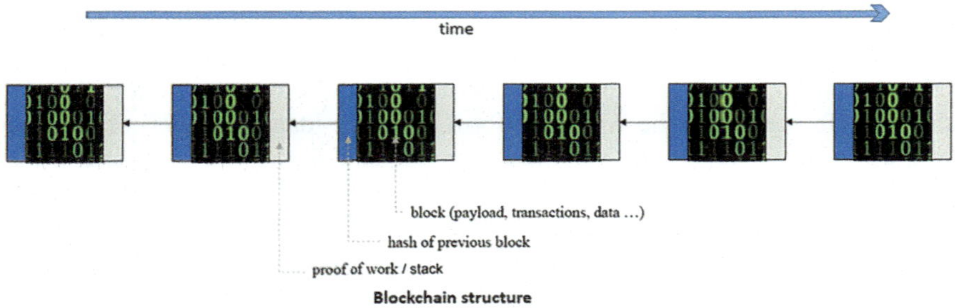

Fig. 24.20 Blockchain structure (PRIN 2017)

The possibility of creating the digital model for manufacturing on which to perform performance simulations, project verifications, construction management (Giretti et al. 2011) opens concretely. This possibility responds to the concept of the Digital Twin of the final product (Ciribini et al. 2016). Web inventor Tim Berners-Lee, an English physicist, states that "Cool URIs don't change", which explains that the impact of web architecture on product data is fundamental. Each product can be identified by a URI that could be stamped on the lower part of the product (so that it can be traced on sight or better that it could be kept on an RFID tag (radiofrequency mark) so that it can be traced by the search engine. It is the Internet of things that can communicate on the web by composing more complex and articulated meanings based on requests specified in terms of digital and/or object requirements. (https://www.w3.org/Provider/Style/URI (accessed 10/02/2019). It is the start of a new way to design and produce fully compatible with the WebIM digital environment (Fig. 24.21).

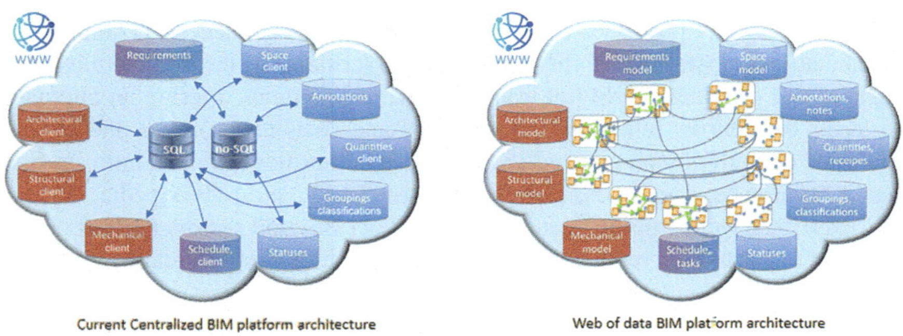

Fig. 24.21 Semantic web in construction WebIM (PRIN 2017)

24.3 Conclusions

The origins of bioclimatic are found in the search for alternative energy sources for space missions, from studies in this field new methods of calculation and evaluation of energy and physical characteristics of the environment have risen. Energy research has developed in parallel with the distributed computational possibilities and the software that has made it possible to automate not only calculations but also parametric simulations.

With the evolution of CAD towards BIM platforms, the possibility of creating interoperable platforms has opened: in an early stage, it was possible to import data processed by specialized software and subsequently to integrate them into modeling. With the perspectives opened by platforms on the web, a new way of designing and producing is started, fully compatible with the digital environment. Introducing these types of innovation that enable the full digitalization of processes appears to be the most plausible operating scenario for Green Design as it allows the connection and implementation of lean and efficient management of the project-production chain in all phases of the cycle of life and in perspective to create new project areas aimed at the digital construction of buildings directly from the model (Digital Twin).

Appendix

Building Energy Modeling Tools, Energy Research, Events and CAD-to-BIM comparative timeline (Figs. 24.22 and 24.23).

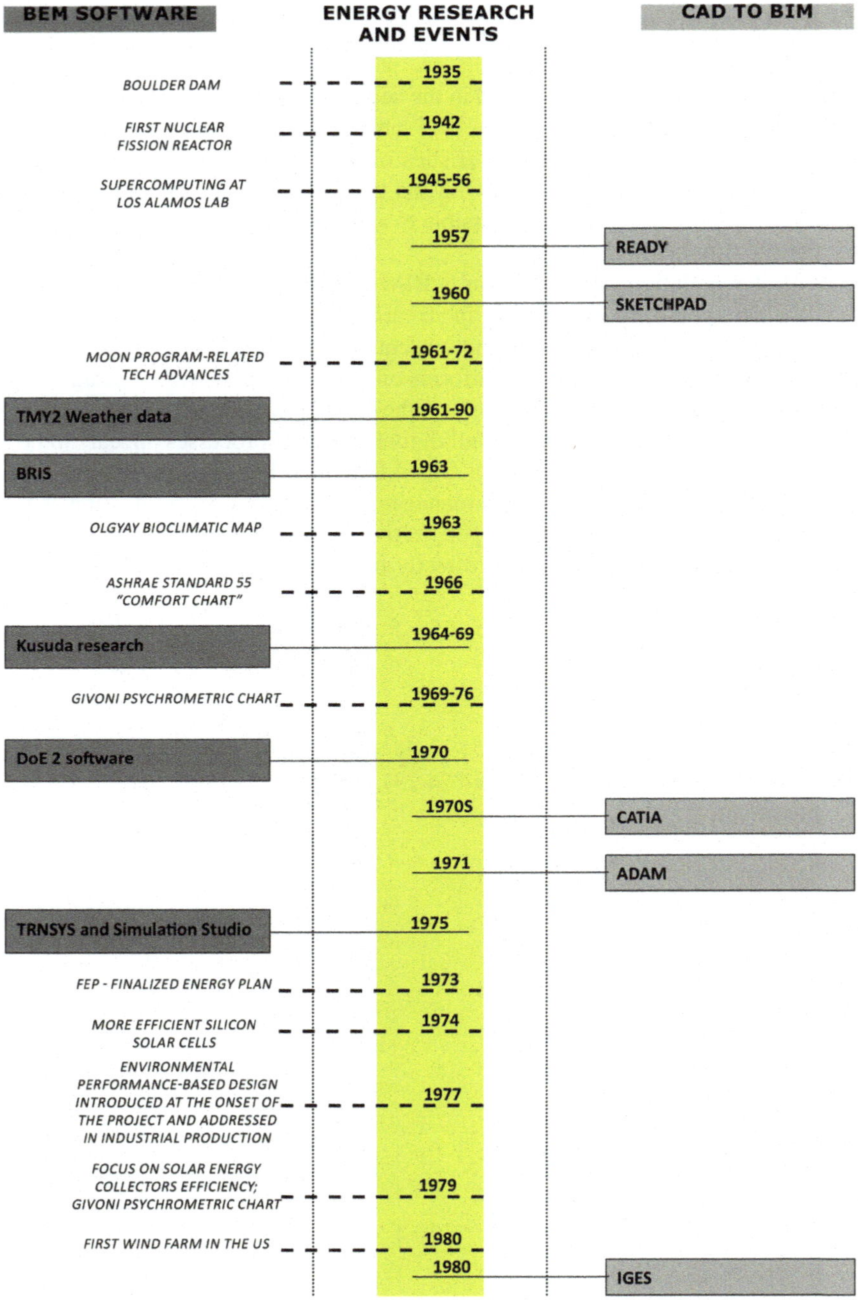

Fig. 24.22 Building Energy Modeling Tools, Energy Research, Events and CAD-to-BIM comparative timeline: 1936–1980. *Source* elaboration of F. Bosi; TxP Research 2019)

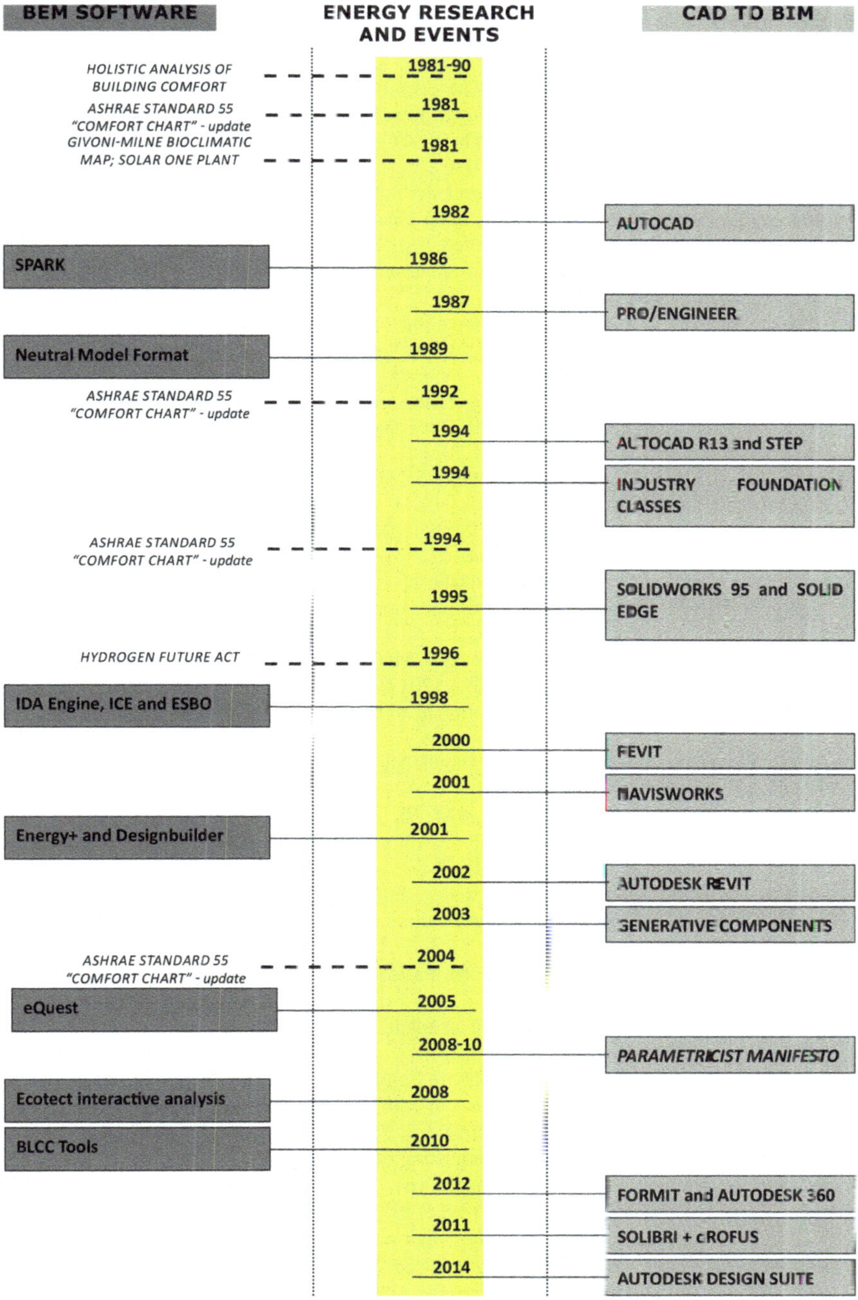

Fig. 24.23 Building Energy Modeling Tools, Energy Research, Events and CAD-to-BIM comparative timeline: 1980–2014 (Design by F. Bosi, TxP Research, 2019)

References

Addison M (1999) User-friendly life cycle costing: the BLCC procedure in an Easy-to-use spreadsheet. Available via semantic scholar. https://www.semanticscholar.org/paper/%22User-Friendly%22-Life-Cycle-Costing%3A-The-BLCC-in-an-Addison/c0932c6bb25948f0e613d1bf3bcf4b608807ceb0. Accessed 12 Oct 2019

Allard F, Santamouris M (eds) (1998) Natural ventilation in buildings—a design handbook. James & James, London, ISBN: 9781873936726, pp 195–254

American Society of Heating, Refrigerating and Air-Conditioning Engineers (1981) ANSI/ASHRAE Standard 55: thermal environmental conditions for Human occupancy. ASHRAE, Atlanta, GA

American Society of Heating, Refrigerating and Air-Conditioning Engineers (1992) ANSI/ASHRAE Standard 55: Thermal Environmental Conditions for Human Occupancy. ASHRAE, Atlanta, GA

American Society of Heating, Refrigerating and Air-Conditioning Engineers (1994) ANSI/ASHRAE Standard 55: Thermal Environmental Conditions for Human Occupancy. ASHRAE, Atlanta, GA

Arens E, Gonzalez R, Berglund L, McNall PE and Zeren L (1980) New bioclimatic chart for passive solar design. United States

Balcomb JD et al (1979) New energies, technological means and architectural expressions. In: FACT 79 Conference's Seminar, Losanne, 27 Jan (1979)

Banca d'Italia (2016) La tecnologia blockchain nuove prospettive per i mercati finanziari. Roma 21 giugno. https://www.bancaditalia.it/pubblicazioni/altri-atti-convegni/2016-tecnologia-blockchain/index.html. Accessed 10/02/2019

Brown W, De Kay M (2004) Sun wind and light. McGraw Hill, New York

Carrara G, Fioravanti A, Loffreda G, Trento A (2017) Knowledge collaboration design. In: Theory techniques and applications for collaboration in architecture, vol. 1. Gangemi Editore spa, Rome, pp 55–90

Carrara G, Fioravanti A, Nanni U (2009) Knowledge-based collaborative architectural design. Int J of DS & Tech. 1630–7267

Ciribini A, Mastrolembo-Ventura S, Paneroni M (2016). Implementation of an interoperable process to optimise design and construction phases of a residential building: A BIM Pilot Project. Automation in Construction, pp 0926–5805

Crump RW, Harms MJ (eds) (1981) The design connection. Energy and technology in architecture, Preston Thomas Memorial Series in Architecture. Essays delivered at a Symposium at Cornell University, Nov 1978

de Dear R, Akimoto T (2013) Progress in thermal comfort research over the last 20 years. Indoor Air 23:442–461

De Dear RJ, Brager GS (1998) Developing an adaptive model of thermal comfort and preference. ASHRAE Trans 104:145–167

De Neufville R, Scholtes S (2011) Flexibility in Engineering Design, MIT Press, Cambridge MA

Del Nord et al. (1989) Controllare la qualità in edilizia. Quaderni di ricerca del Dipartimento di Processi e Metodi della Produzione Edilizia, n 3, Polistampa, Firenze

Eads WG (1979) Testing, balancing, and adjusting of environmental systems, sheet metal and air conditioning contractors national association, Inc., Vienna VA

Eastman C, Teicholz P, Sacks R, Liston K (2012) BIM handbook: a guide to building information, modelling for owners, managers, designers, engineers and contractors, 2nd edn. Wiley, New York

ENEA (2010) DEPUIS—Design of environmentally friendly products using information standards environmental testing laboratory, Arizona State University

Esposito MA (2004) Progettare con gli scenari ambientali. In: Sala M (ed) I percorsi della progettazione per la sostenibilità ambientale. Alinea Editrice, Firenze, pp 256–258

Esposito MA, Bosi F (2018) Tecnologie del progetto di architettura. Dida Press, Rimodellazione di progetto e fabbricazione, p 190

EU Communication on Integrated Product Policy (COM (2003) 302)
EU Environment https://ec.europa.eu/environment/ipp/lca.htm. Accessed 10 Jan 2019
FACT forum architecture communication territoire (1979) Palais de Beaulieu, Losanna 26–28 Gennaio
Ferrandiz J, Banawi A, Peña E (2017) Evaluating the benefits of introducing "BIM" based on Revit in construction courses, without changing the course schedule. Springer GmbH, Germany
Fioravanti A, Novembri G, Rossini F (2017) Improving proactive collaborative design through the integration of BIM and agent-based simulations. In: Novembri G (ed) (2017) Shock! sharing of computable knowledge, vol 1, pp 103–108, Bruxelles and Rome:eCAADe and DICEA, Sapienza—Università di Roma, Rome, 20–22 Sep 2017
Forgues D, Poirier E, Staub-French S (2016) The Impact of BIM on Collaboration: An Action-Cognition Perspective
Forlani MC, Mastrolonato L (2014) Edilizia sostenibile e risorse locali , Techne, p 194
Gangemi V (1994) L'ambiente risanato. La bioarchitettura per la qualità della vita. Ed. Clean, Napoli
Gangemi V (1997) Ricerche ed esperienze nell'ambito della bioarchitettura. In: Muzzillo F (ed) Tecnologie di recupero ambientale. BE. MA, Milano
Gangemi V (2001) Emergenza ambiente. Clean, Napoli
Giretti A, Carbonari A, Vaccarini M, Robuffo F, Naticchia B (2011). Interoperable approach in support of semi-automated construction management. In: Proceedings (Part 1) of ISARC 2011. Seoul: Secretariat PEOPLE-X Inc., vol 2011, pp 266–271
Givoni B (1969) Man, climate and architecture. Applied Science Pubblisher, London
Givoni B (1976) Man, climate and architecture, 2nd edn. Applied Science Publishers Ltd, London
Givoni B (1998) Climate considerations in building and Urban design. Van Nostrand Reinhold. the USA
Grosso M (1998) Design Guidelines and technical solutions for natural ventilation. In: Allard F, Grosso M, Fracastoro M, Vincenzo G, Simonetti M, Chiesa G (2015) A hybrid passive cooling wall system: concept and laboratory testing results. In: Energy Procedia 0:0(2015), pp 1–6
Grosso M (2011) Il raffrescamento passivo degli edifici in zone a clima temperato, Maggioli, Sant'Arcangelo di Romagna (RN)
Grosso M, Chiesa G, Nigra M (2015) Architectural and environmental compositional aspect for technological innovation in the built environment. In: Heritage and Technology. Mind Knowledge Experience, pp 1572–1581
Grosso M, Scudo G, Dessì V (1998) Architettura e comfort. Ambiente Costruito 1998:3
Grosso M, Peretti G (1990) Flussi d'aria e ventilazione naturale degli edifici, UNCSAAL, Milano
Guyot A, Cros A (1979) Reconcilier L'Architecture avec la forme, la matiere et l'energie. FACT 79 Conference, Losanna 28 gennaio
Hermelink A, Schimschar S, Boermans T, Pagliano L, Zangheri P, Armani R, Voss K, Musall E (2013) Towards nearly zero-energy buildings. Definition of common principles under the EPBD, Final report
Jahn H (2014) +Energy. Essay presented at: world architecture conference series: Jahn Architects Airport Architecture, 28 Mar 2014. Università degli Studi di Firenze
Jahn Architects (2003) Project for the Deutsche Post Tower, Bonn, Germany
Jahn Architects (2006) Project for Suvarnabhumi Airport, Bangkok, Thailand
Judkoff R, Wortman B, O'Doherty B, Burch J (2008) A methodology for validating building energy analysis simulations. NREL/TP-550-42059 National Renewable Energy Laboratory, Golden, CO. https://www.nrel.gov/docs/fy08osti/42059.pdf. Accessed 10 Jan 2019
Kennedy JF (1961) Speech before a joint session of Congress on May 25, 1961
Kinviniemi A (2018) Foreword. In: Esposito MA, Bosi F 2018) Tecnologie del progetto di architettura. Rimodellazione di progetto e fabbricazione, Dida Press, pp 9–11
Los S, Pulitzer N (1977) L'architettura dell'evoluzione. Ed. Luigi Parma, Bologna
Marinosci C, Seprini G (2019) Software di simulazione energetica dinamica per gli edifici, In: Ingenio

Matteoli L (1977) Azione Ambiente. Edizioni Libreria Cortina, Torino
McGehee ED, Garcia K (1999) Historic Building assessment for the Department of Energy Conveyance and Transfer Project, Volume 1. Los Alamos National Laboratory, Historic Building Report, No. 1778
McGehee ED, Garcia K (2015) An assessment of cold war-era buildings at technical area 46: phase II Los Alamos National Laboratory. Report Environmental Programs, LA-UR-15-20296. Issued: 2015-01-27 (rev.1)
McIntyre DA (1980) Indoor Climate. Applied Science Publishers, London, UK
MacLeamy P (2004) BIM Maturity curve, reworked by Davis D (2011). https://www.danieldavis.com/macleamy/, Accessed 20 Sep 2019
Milne M et al. (2009) Climate consultant 4.0 develops design guidelines for each unique climate. In: Proceedings of the American Solar Energy Society, Buffalo New York, 2009
Milne M and Givoni B (1979) Architectural design based on climate. In Donald Watson (ed) (1979) Energy conservation through building design, McGraw-Hill, New York
Murphy K (2012) The social pillar of sustainable development: a literature review and framework for policy analysis. Sustain Sci Practice Policy 8(1):15–29. https://doi.org/10.1080/15487733.2012.11908081
Olgyay V (1963) Design and Climate, Princeton University Press
Peretti G (1981) Energia Progetto, Celid, Torino
Peretti G, Pagani R, Grosso M (1981) Energia/Involucro, Celid, Torino
Pohl J, Myers A (1992) Requirements for computer based design. In: Advaces in computer aided design, 6th international conference on systems research, informatics and cybernetics, Baden-Baden Publishing, USA
PRIN (2017) A distributed digital collaboration framework for small and medium-sized engineering and construction enterprises. On-going 3 years research
Queffelec C, Trebbi JC, CSTB Centre Scientifique et Technique du Batiment (1981) Pronciples et formes de l'habitat bioclimatique. In: Researche & Architecture, no 46
Regnier C (2011) BIM Maturity curve, reworked by Davis D (2011) https://www.danieldavis.com/macleamy/, Accessed 20 Sep 2019
Tanabe S, Kimura K, Hara T (1987) Thermal comfort requirements during the summer season in Japan. ASHRAE Trans 93:564–577. Tempe, AZ
Trasportation Research Board (2012a) Case examples. ACRP—Airport Cooperative Research Program, Syntesis 33. Airport Climate Adaptation and Resilience, Washigton D.C., pp 7–17
Trasportation Research Board (2012b) Common motivations. In: ACRP—Airport cooperative research program. Report 068—Guidebook for Evaluating Terminal Renewal Versus Replacement Options, Washigton D.C., pp 4.1–4.6
United States department of energy, office of energy efficiency and renewable energy (2019) BEM building energy modeling software. https://www.energy.gov/eere/buildings/building-energy-modeling. Accessed 10 Jan 2019
USA department of energy, Sandia Laboratories, University of California Los Alamos Scientific Laboratory (1978) Passive Solar Buildings: A Compilation of data and Results. contract AT (29-1)-789
Veninata V, Esposito MA (1979) Bioclimatica residenziale: ipotesi di sfruttamento di un sistema energetico passivo nella progettazione di un insediamento residenziale industrializzato (in clima caldo secco). Dissertation Università degli Studi di Firenze, Florence
Veninata V, Esposito MA, Mosetti A (1980) Bioclimatica residenziale: ipotesi di sfruttamento di un sistema energetico passivo per edifici residenziali industrializzato in clima caldo secco. Project for "Il sole e l'Habitat" competition, INARCH, Rome
Watson D, Labs K (1983) Climate design: energy efficient building principles and practices. McGraw-Hill, New York, p 37
Wu H (1980) The potential use and application of oscillating fans in extending the summer comfort envelope, Research Rep
Zellini P. (2018) la dittatura del calcolo, Adelphi

US Government (1996) Hydrogen future—act of 1996. 42 USC 12401 note. 42 USC 7238 note. Public Law 104–271. 104th Congress

National Aeronautics and Space Administration (NASA) Lyndon B. Johnson Space Center (2004) Benefits from Apollo: Giant Leaps in Technology https://www.nasa.gov/sites/default/files/80660main_ApolloFS.pdf. Accessed 20 Sep 2019

Chapter 25
Simplified Methods and Tools for the Representation of Energy in the Project of Architecture

Alessandro Rogora

25.1 Introduction

Designing is the attempt to see beyond. It is an attempt to take under control the quality of a space and to make it appropriate for humans. It should be, but several times moving around famous buildings we have the perception that in many cases the visual quality was the main element considered by architects in the design, but—at the same time—the missing control of "other" qualities in the building was clear and the result was an unbalanced feeling of nice and unpleasant that is difficult to decode and understand.

As humans we are made of meat and blood and the perception we have from our senses cannot be easily separated into visual and acoustic, tactile and thermal. As humans we have a global synesthetic perception of the space and we express a synthetic evaluation of our deep feeling.

The Architect is defined as the responsible of construction. The world ἀρχιτέκτων (arkhitekton), comes from two Greek words arkhi (head, leader) that underline the value of the leadership and the responsibility of the designer while the world tékton is referred to his/her technical abilities.

It is quite clear that nowadays architects need technicians and consultants to help them in their work to give help on different topics and to give contributions in specific fields; nevertheless the main problem is still the communication between the different technical operators. The new BIM software try to help the communication using a common platform for data interchange giving the opportunity to have a single file that all the technicians can share. This technical solution is effective during the executive phase, but it is not usable during the conceptual phase and the genesis of the design proposal is still in charge of a single professional who has the ability to define and control the design concept.

A. Rogora (✉)
Politecnico di Milano, DAStU, Milan, Italy
e-mail: alessandro.rogora@polimi.it

G. Chiesa (ed.), *Bioclimatic Approaches in Urban and Building Design*, PoliTO Springer Series, https://doi.org/10.1007/978-3-030-59328-5_25

Fig. 25.1 Three-dimensional spiral as represented by architects (staircase at Musei vaticani in Rome) and engineers (plot of a 3d spiral)

As known the most effective decisions in terms of energy behaviour are the one taken in the initial design phase and the attempt to correct these decisions later will produce a bad design forced to work better through high tech orthopaedic prostheses (note 1). It means that to take good decision during the conception of building design is a critical issue and that our effort should be to prepare adequate tools usable in this phase. On the contrary, nowadays most of the researches and tools on energy saving in architecture are focused on advanced analysis and simulation, optimization of single components and systems, while very few (almost none at all) are focused to help designers during the building design conception (note 2) (Fig. 25.1).

The problem is probably related to the difficulties to find out an appropriate code to link the visual design approach of architects to the numerical one typical of the technical engineering sciences. Both talk of the same problem, but the way to represent energy and the information reported are strongly different and there is a sort of incommunicability between the users. This difficult communication is related to the way each of the disciplines represents the world. In his PhD thesis titled "Fìsica aplicada a l'ensenyament de l'Arquitectura" (note 3) prof. Jaume Roset clearly highlighted the problem asking the question: "Are architects stupid?". It was a provocative question due to the difficulty to communicate with students and professionals he met during his professional career as teacher of Building physics at the ETSAB-UPC at Barcelona. Prof. Roset jumped directly to the key problem saying that building physics and architecture use different codes to represent the world and that there is a missing link and a lack of communication between the two disciplines. The attempt to use the scientific code to communicate with (or to teach) architects produce a poor result and give the students no real improvement in their academic training. In our universities it is easy to find examples of professors teaching Building Physics (it happens also with other subjects, of course) who are disappointed of the poor results of their teaching. The general proposal is to increase the amount of credits for the students to give them extra information and more exercises, but the

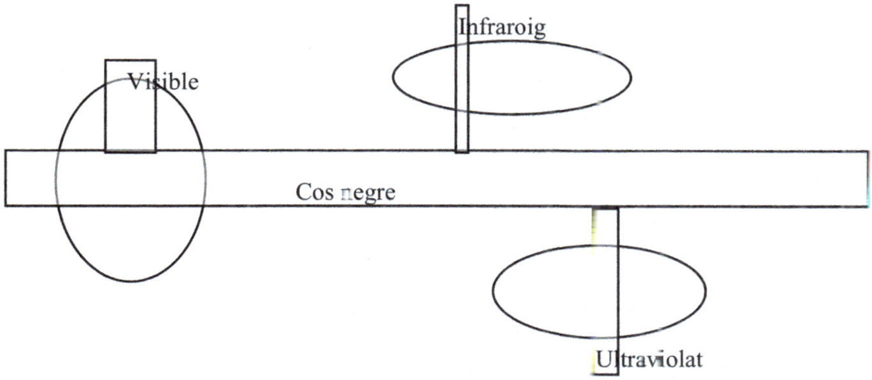

Fig. 25.2 Building physics approach and architectural newer fit properly; as professors and researchers we should work to give the students a better information on the topic adapting our teaching to their needs

result generally does not change. The idea of modifying the teaching method is seldom considered, even if it is probably the main cause of the failure. In the thesis by prof. Roset the approaches of building physics and architecture are respectively represented by a rectangle and an ellipse. It means that even if we try to properly overlap the two shapes, we will never get a perfect coincidence. In general, the overlap is only partial (e.g. a vertical rectangle coupled with a horizontal ellipse) and it is the teacher that has to make the best to fit the students' necessities.

How is it possible to link the analogic approach of the designers to the digital one of the engineers and consultants? This is an interesting issue that several researchers tried to solve in the last 40 years (and even before, to tell the truth) or, at least, since when the computers were available as a support for architectural design (Fig. 25.2).

25.2 Software for Analysis and Simulation and Tools for Design

In the last years extraordinary software for building simulations made possible to walk through buildings before their construction and advanced numerical analysis verified the mechanical resistance of a structure to simulated earthquakes … but what about the help these software can give to an architect during the design phase? How is it possible to help the designer during the conception of the building? And then, is this phase a sort of black box where no aid is possible if not the personal experience of the designer that must be verified soon after the design is made without any possible help? If we accept that the only available option is to verify a building design only once it has been detailed, the risk of time consuming in trial and error approach forces us to use and adopt over simplified and "secure" design solutions to avoid nuisance, but it would mean clipping the wings of architecture.

Several options have been proposed to overcome the problem: software for early analysis of the energy behaviour of buildings that can be used in the very first stage of the design, simplified tools for the representation of some invisible forces (e.g. solar energy, wind, radiation, etc.) and technics for counter intuitive approach in building design.

As usual all the three options have positive and negative aspects that is interesting to highlight. The software for early analysis of building performances had a golden age in the 90s, soon after the period of the simplified calculation tools of the 80s (Fig. 25.3).

In the 80s no personal computers were available and the architects were looking for simplified tools (nomogram or manual calculations) to adequately estimate the performance of advanced passive systems and components. It was the time in which the Los Alamos Lab, as well as the Wisconsin University produced simplified manual methods to estimate the monthly energy saving adopting some passive building components Bottero et al. (1983). Few years later the France Method 5000 made possible a simplified analysis of a solar building using a mix of datasheet and nomogram. Other experiences like the LT method (designed for a student competition

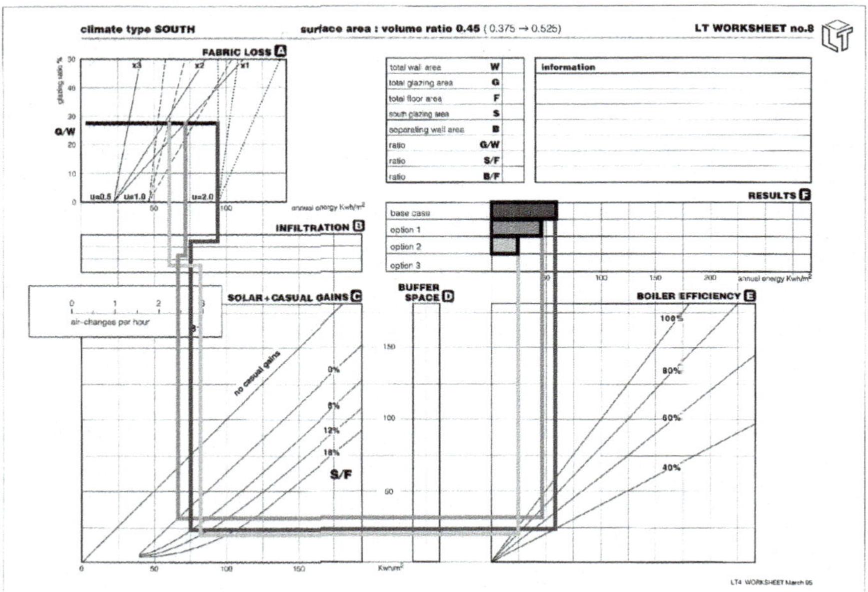

Fig. 25.3 The LT method was designed to help the students in a design competition. It is based on graphic input and nomograms and it is helpful to compare different design proposals Baker and Steemers (1996)

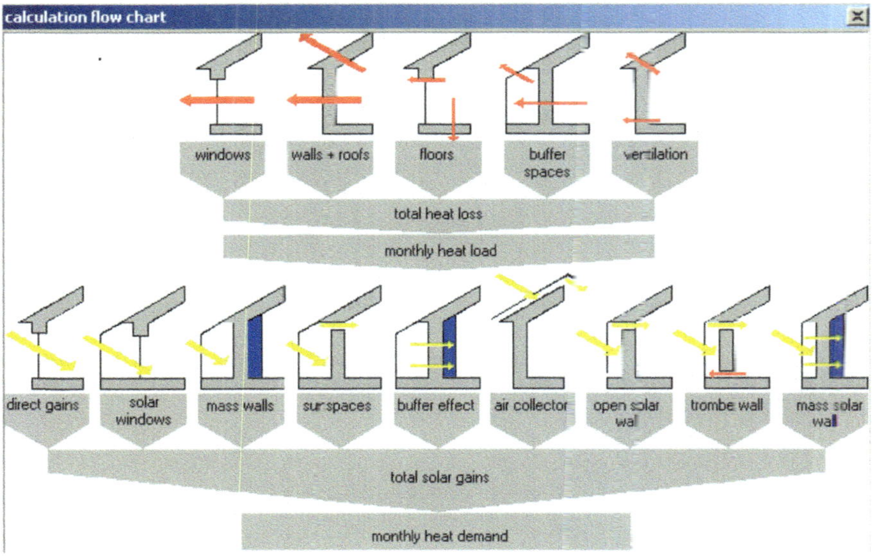

Fig. 25.4 Screenshot of an input window for the electronic version of the simplified code Method 5000

in Europe) showed the potential use of nomograms to roughly estimate the energy saving and the payback time of different design alternatives (Fig. 25.4).

Soon later, the diffusion of personal computer made possible to perform accurate evaluations of complex buildings using thermal nets and finite difference analysis calculations. The problem was not any more related to the power for calculation but to the user interface. The software for advanced analysis required experienced users and these people were the same "few" architects or energy consultants able to run such software, to decode the results and to make sense of the numerical values. On the other hand the hypothesis was to give all the architects the possibility to use software for energy analysis in their professional life.

The problem was that architects use an approach to building design that is far from the one used by engineers in software design and the advanced codes had no real interaction with the design process. It is not a mere question of user interface; for sure the original DOS system was much different and difficult to use than the Macintosh or Windows ones, but the problem was much deeper and more critical. Architects use analogies instead of numbers, relative values instead of absolute ones and design is much closer to the analogic then to the digital approach.

One interesting proposal—far from the representation of energy—was proposed by P. Galle and L. B. Kovács in The Logic of Worms: A Study in Architectural Knowledge Representation (note 4) and later in other articles titled The logic of Walking and the logic of Plaza published on the same magazine. In these researches Galle and Kovács showed the genesis of walking patterns in an empty space due to the social behaviours. The proposed tool is a proper design tool in which there is

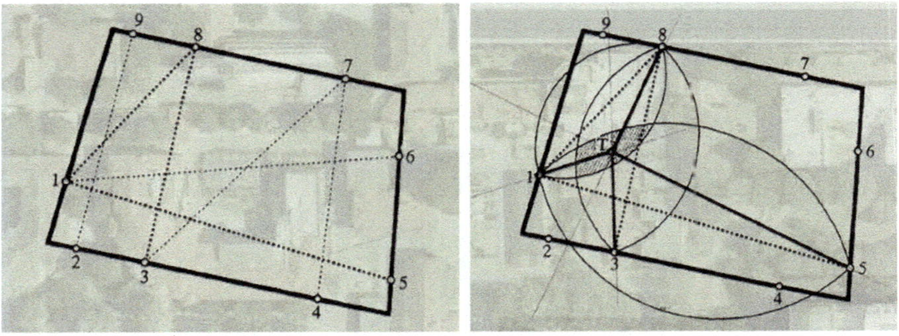

Fig. 25.5 Transformation V => Y of the walking paths (*Credits* Galle and Kovács (1992). The logic of walking: representing design knowledge on pedestrian traffic nets. Environment and Planning B: Planning and Design)

a set of rules, but saving a level of "conditional freedom" in the decision process (Fig. 25.5).

25.3 Software for Preliminary Analysis and Graphic Tools

An incredible advancement was made by the research group of Rafael Serra and Helena Coch at the ETSAB—UPC with the development of software CLA (note 5), acronym of the Spanish words Clima Lugar y Arquitectura (Climate, Location and Architecture) but in Catalan the word "cla" means "clear". The software CLA used a revolutionary approach based on the so called dimensional analysis: the idea was to describe the building in a qualitative way using an architectural like language. The software asked the designer to choose between 5 values of each parameter and the synthesis was a complex, dynamic energy analysis. The compactness of a building could be defined as Very low, Low, Average, High and Very high, like many other values that have an architectural value and a physical translation. The description of the building was made using a set of questions with a reference, graphic image for each of the five categories and at the end the software used the answers to perform a sophisticated numerical analysis and to produce reliable results even if the building was still under development in the mind of the designer. The idea to couple a qualitative description of the building (form and technology) with a sophisticated calculation engine and with an interesting representation of the results was absolutely great (Fig. 25.6).

Other similar software were produced few years after the CLA was released (between the others the Hungarian 6B and the Italian DESA), but after a few years the interest slowly diminished.

The rapid development of computer graphic and CAD software made these tools no longer attractive, not only because they were developed on old fashion operative

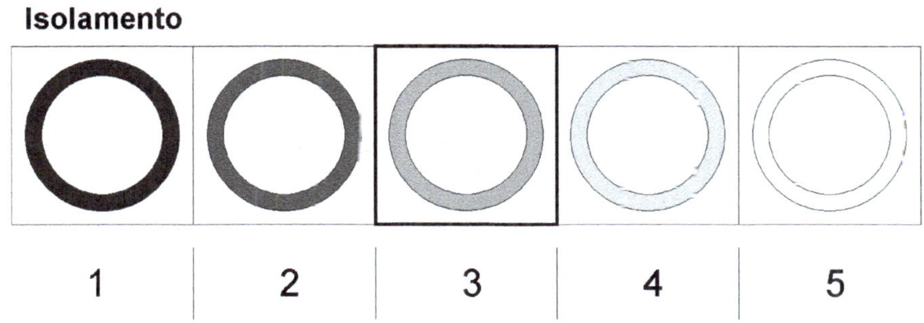

Fig. 25.6 Data input regarding the insulation level—software CLA (Clima, Lugar y Arquitectura)

system (CLA only runs under DOS), but also because the interest was oriented to the incredible new options offered by 3D software and photorealistic rendering that dazzled the designers giving them the feeling of a new promise land for building design Serra (1989), Serra and Coch (1997).

Energy was once again confined into a corner, because architects had new available tools to feed their visual approach to building design. The next step of the CAD software was to incorporate energy simulation using added plug-in, reinforcing the idea of independency between building design and energy functioning (that was performed soon after the design was ready). If we exclude some examples of enlightened professionals energy—as well as sustainability—had substantially no role in the conception of buildings. At the same time some teachers of Building design still try to maintain the visual approach as the centre of the design. The hypothesis of a necessary "aesthetic sustainability" that should be of key importance for Sustainability (energy or environment) represents an attempt to save the leadership of the visual approach. The missing contribution of the Building design is not to have indagated the aesthetic consequences of sustainability (that is something strongly different).

Another and totally different approach was used in the attempt to represent the energy qualities in the environment. The general idea is that these tools can be used to give to architects some pieces of information about energies and other invisible forces that are spatially orientated and variable through the time. The sun charts coupled with the shadow masks make visible the sun position at a given latitude on a sheet of paper and the effect of surrounding elements in term of shading. Some researches worked on the improvement of these tools in order to add local information. Advanced sun charts containing information on the amount of locally available solar radiation, coupled with values of air temperatures were developed and used. Another interesting graphic tool is the so called bioclimatic Map; it represents the attempt to make visible some invisible (energy) forces in the design area. The idea was to use graphic signs to represent wind direction, speed and frequency using colours to describe the period of the blowing wind (red for summer and blue for winter), as well as the sunrise and sunset in different period of the year and the elements that reduce the solar collecting, the amount of radiation falling on differently oriented surfaces and all the possible

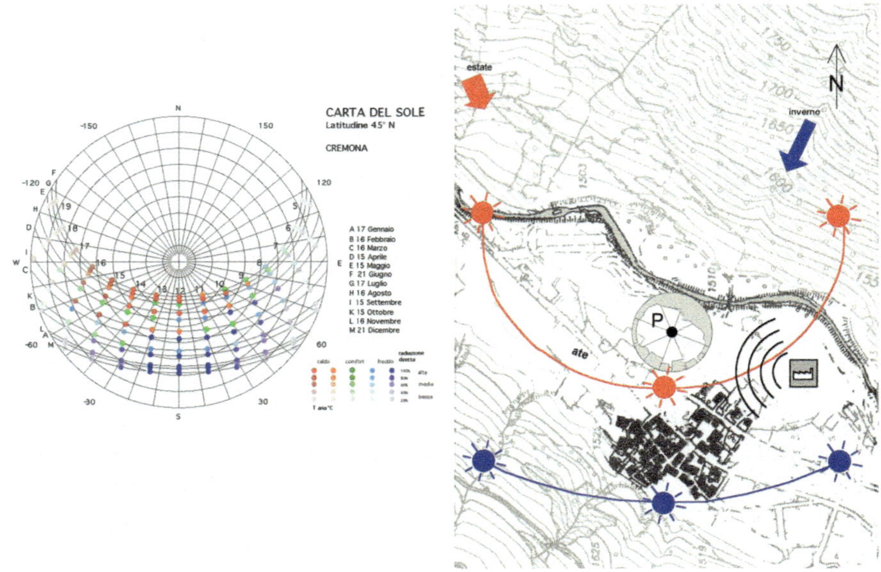

Fig. 25.7 Improved solar chart and bioclimatic map for a given location

pieces of information that can affect design choices using a graphic representation on the same CAD layer where the building is represented (note 6).

Unfortunately these tools are not computer based, but require a boring, manual "procedure" to produce and to use the charts. This is a great limit for the diffusion of such tools, but on the other hand it is also a sort of insurance, because effortlessly activities seem to affect scarcely design decisions (Fig. 25.7).

25.4 The Meaningless of kWh and the Effectiveness of the m^2 in the Design Process

For an architect there are numbers and numbers. Numbers that have a value and numbers that have no one. A nice example are the kWh/y that describe the energy to heat a building in one year. The number could be 20,000 kWh/y or 40,000 kWh/y for a villa, but only a few architects could appreciate the difference. On the other hand the surface of a flat could be of 65 m^2 or of 90 m^2, but this difference is absolutely clear even to a young professional. KWh/y are visible only on the energy label, m^2 are experienced daily. The difference of an apartment with the surface of a few meters square more or less can be substantial, while the difference of some thousands kWh/y is not significant (in general it does not mean anything to the designer).

Several attempts have been made to make physically perceivable the impact of a building on the planet as well as to make clear how every single design choice

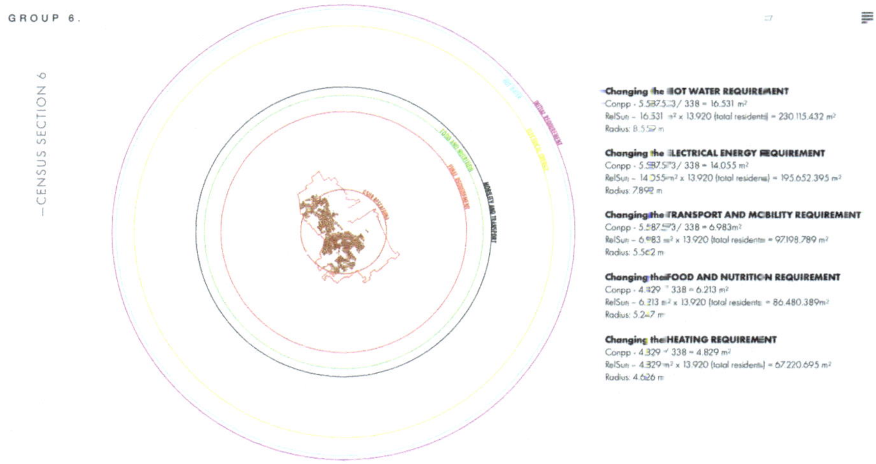

Fig. 25.8 Land impact to support human behaviour

affects this impact. One example at urban scale can be found in the Imitation Game, a research currently carried on at the DASTU Department at Politecnico di Milano University. The idea is to represent the impact of people living in a small town in terms of m^2 necessary to support their lifestyle. Six categories of consumptions have been defined: Reside, Nutrition, Moving, Dressing, Communication, Wellbeing, plus a seventh to take into consideration the services (infrastructures supporting our society and way of living). For each category the amount of energy (separated into electrical, thermal and into kms for the category moving) was first calculated and then converted into m^2 of usable land necessary to produce the required energy in a renewable way (wood to produce thermal energy, sunflower to produce biofuel for mobility and electricity). The representation of the impact, and the effect of each choice in term of land "footprint" was used to persuade students to look for appropriate design choices and to modify their personal behaviour (note 7) (Fig. 25.8) Rogora et al. (2019).

25.5 Generative Rules and Counter Intuitive Approach to Building Design

The last proposal is related to the definition of a counter intuitive approach in building design. The profession of the architect is generally considered to be both technical and artistic, but the latter cannot be if the former is not solved. Sometimes, on the contrary, the design is the result of personal intuition (successful or not) and it is justified in term of art even if the usability or the functioning is poor and not satisfactory at all. If we consider that achieving comfort is essential in building design (if we partially exclude very few representative buildings in which communication and symbolism are the main issues), it is necessary to adopt a counter intuitive approach to avoid mistakes and misunderstandings.

The counter intuitive approach was tested for architecture lighting and has the objective to highlight the main elements to consider in the design of buildings. It allows to define a sort of lighting DNA code of the building which precedes the definition of the building form. Once the lighting genetic code has been defined each choice is no longer completely free or, in any case, it constrains the subsequent choices. It does not mean that the architectural design is limited to simple shoe boxes in which to perform specific activities, but that designer is forced to declare the environmental objectives of the project before defining the form and once declared such goals cannot be disregarded. Rogora (1997)

The counterintuitive approach represents an interesting and largely unexplored area of research from which we expect a significant advancement of knowledge and new tools to aid building design (Fig. 25.9).

Notes

Note 1

R. Serra ed H. Coch, L'energia nel progetto di architettura (titolo originale Arquitectura y energìa natural), edizione italiana a cura di A. Rogora, G. Scudo, Utet-Città Studi editrice, Torino, 1997.

Note 2

Rogora A., In La qualità ambientale degli edifici, a cura di S. Piardi e D. Faconti, cap. V, pagg. 485–516, Maggioli editore, Rimini, 1997.

Note 3

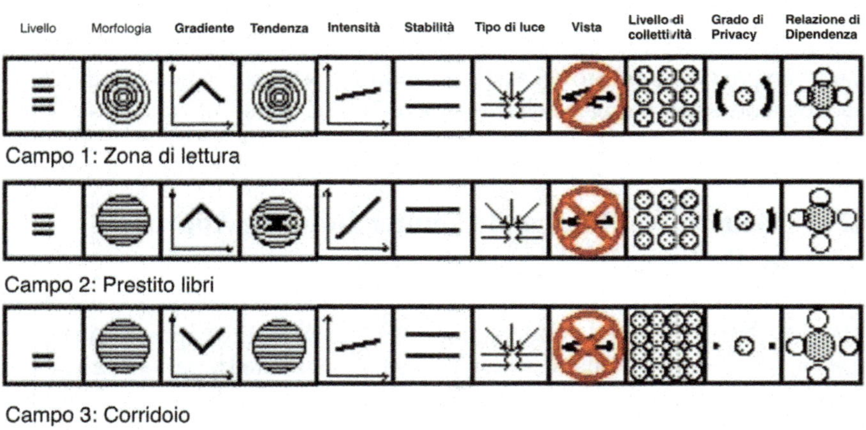

Fig. 25.9 Example of lighting conditions in the main hall of the Library at Rovaniemi (Alvar Aalto). The definition of the lighting qualities was used to define the counterintuitive approach in building design

The PhD thesis by prof. Jaume Roset Fìsica aplicada a l'ensenyament de l'Arquitectura was presented in 2006 at the ETSAB-UPC, Director of the Thesis prof. Rafael Serra.

Note 4

Galle P. and Kovács L. (1992) The Logic of Worms: A Study in Architectural Knowledge Representation, Environmental and Planning B, 19, 5–31.

Note 5

Serra R., Clima, lugar y arquitectura: manual de diseño bioclimatico, Centro De Investigaciones Energéticas (1 de julio de 1989), Madrid, España.

Note 6

Rogora A., Progettazione bioclimatica per l'architettura mediterranea, Wolters Kluver editore, Milano, 2012.

Note 7

Rogora A., Carli P., Trevisan A., The Imitation Game. The game as experience of a sustainable project, in Project Essays and Researches, Edited by Giuseppe De Giovanni and Francesca Scalisi Pro-Innovation: Process, Production, Product, Palermo University Press Palermo (Italy), 2019, pp. 123–130.

References

Baker N, Steemers K (1996) LT Method 3.0—a strategic energy-design tool for Southern Europe. Energy Build (Elserier) 23(3):251–256

Bottero M, Rossi GC, Scudo G, Silvestrini G (1983) Architettura solare—tecnologie passive ed analisi costi-benefici. Clup, Milano

Galle P, Kovács L (1992) The logic of worms: a study in architectural knowledge representation. Environ Plann B 19:5–31

Piardi S, Faconti D (1997) A cura di. La qualità ambientale degli edifici, a cura di. Maggioli editore Rimini

Rogora A (1997) Luce naturale e progetto. Maggioli editore, Santarcangelo di Romagna

Rogora A, Carli P, Trevisan A (2019) The imitation game. The game as experience of a sustainable project. In: De Giovanni G, Scalisi F (eds) Project essays and researches, pro-innovation: process, production, product. Palermo University Press, Palermo, Italy

Serra R, (1989) Clima Lugar y arquitectura: manual de diseño bioclimatico. Centro de Investigaciones Energéticas (1 de julio de 1989), Madrid, España

Serra R, Coch H (1997) L'energia nel progetto di architettura (titolo originale Arquitectura y energìa natural). Edizione italiana a cura di A. Rogora, G. Scudo, Utet-Città Studi editrice, Torino

Chapter 26
Kinetic Adaptive Façades. A Systematic Review of Technological and Adaptive Features

Rosa Romano

Abstract Kinetic envelopes can be defined as technological solutions which are able to move, open and close themselves, increasing the reactive capacity to incident solar radiation or other forms of environmental energy stresses, typical of a traditional facade system. Starting from this definition the chapter will provide a systematic review of these innovative technological solutions and their performance features, with a focus on materials, sensors and mechanical devices that is possible using to guarantee microclimatic control based on the dynamic interaction between the external and internal environments, achieving the target nZEB.

26.1 Introduction

As indicated by numerous international researches (Feng et al. 2016; Tang and Wu 2011; Fang et al. 2014; Friess and Rakhshan 2017), the architectural envelope is responsible for around 40% of the energy consumption of existing buildings, with incidence percentages that increase in the summer months due to the presence of façade systems built without using insulating materials that have adequate thermal mass and integrating appropriate shading device systems. These envelopes are therefore unable to guarantee good phase shift and thermal inertia values and tend to increase overheating inside the building.

Furthermore, the European Energy Performance of Buildings Directive 2010/31/EU (EPBD) and the Energy Efficiency Directive 2012/27/EU, in addition to indicating the need to construct nZEBs in all member states from 2018, give a precise definition of the summer and winter performances of the architectural envelope, emphasizing how the problem of energy consumption for heating and cooling can only be overcome by choosing appropriate materials and technological solutions. Besides, it is interesting to note that the nZEB objective can be achieved not only by improving the thermo-hygrometric performance of the building envelope and the energy efficiency of the heating and cooling systems, but also by integrating

R. Romano (✉)
Department of Architecture, University of Florence, Florence, Italy
e-mail: rosa.romano@unifi.it

G. Chiesa (ed.), *Bioclimatic Approaches in Urban and Building Design*, PoliTO Springer Series, https://doi.org/10.1007/978-3-030-59328-5_26

technological systems for the production of renewable energy into the building and therefore into the opaque and transparent vertical closing elements.

For these reasons, it is clear that research on innovative façade systems is becoming a priority on the international scene and its aim is to transform the architectural envelope from a static system into a dynamic one capable of actively and passively controlling the energy flows through it, while at the same time guaranteeing the ability to produce thermal and/or electrical renewable energy to reduce the overall energy consumption of the building to zero. Experimentation with new façade technologies (vertical and horizontal, opaque and transparent) has thus become one of the emerging issues in the construction sector and in research into process and product innovations aimed at energy optimization in contemporary buildings. In other words, the international debate attributes the success or failure of international energy and environmental policies to the experimentation and implementation of innovative façade systems.

Moreover, a focus on the need to reduce summer energy consumption, especially in the geographical areas of southern Europe, has led to assessments of the need to develop technological solutions capable of dynamically controlling incident solar radiation at micro and macro scale, such as: envelope layers made with innovative insulating materials (PCMs, reflective insulators, etc.) or with a high mass coefficient (wood fiber panels, cork, etc.); transparent closing systems that integrate glass with autonomous selective properties (photochromic, prismatic, holographic, etc.) or connected to the presence of electric actuators (thermochromic, electrochromic, etc.); static or dynamic shielding systems; architectural components (balconies, roofs, windows, etc.) which are able to move thanks to the presence of more or less complex handling devices, changing spatial configuration and consequently thermo-hygrometric performance, in relation to external climatic conditions. Therefore, the need for innovation in envelope system technologies is characterized by an approach that focuses on two main aspects:

1. the study and development of innovative materials that allow an intrinsic variation of the thermo-physical properties of closing systems, not necessarily related to the presence of actuation systems;
2. the design and testing of kinetic components and/or façade systems, namely those that are able to move through time and space thanks to the presence of mechanical devices, ensuring variable geometric configurations that can affect the energy performance of the building.

Referring to other works by the author on opaque and transparent materials that belong to the first category (Romano 2020) the chapter will deal only with kinetic façade systems that, due to the possibility of changing shape and location, are able to affect the overall energy performance of the building, configuring new aesthetic scenarios capable of influencing the architectural trends of the coming decades.

The aim is to identify the main characteristics of a new generation of envelope systems, which we could define as "eco-efficient, environmentally interactive and advanced bioclimatic" (Benham 1969), namely able to guarantee microclimatic control based on dynamic interaction between the external and internal environments and able to manage, by moving or changing form, the energy flows through them (heat, light, sound), thanks to the presence of manual or digital regulation systems connected to fixed or variable devices (sunshades, opening/closing of windows, air vents, etc.). The main feature of this new generation of envelope systems is dynamic anisotropy, understood as the capacity to offer differentiated energy responses depending on the exposure and orientation of the façade, local climatic conditions and user needs.

26.2 The Envelope as a Skin and Membrane

Michael Davis introduced the concept of a building's skin in 1981, defining a technological component equipped with microsystems capable of absorbing, radiating, reflecting, filtering and transferring energy and information from one space to another. The architectural envelope was described as a system capable of changing conformation in relation to external thermo-physical stresses, going from transparent to opaque, changing colour, and varying its optical and geometric properties (Davis 1981). The etymological comparison between the biological term "skin" underlined the osmotic capacity of façade systems to regulate and sometimes generate energy flows, continuously mediating the relationship between humans and the environment. In this sense, the architectural "membrane" or "skin" of the building organism transforms over time from a two-dimensional and unidirectional system to a three-dimensional and bidirectional element with a clear environmental matrix. In other words, it evolves from a simple cladding and protection element into an instrument that has a relationship and interfaces with the outside world, with metabolic and morphological functions designed to actively control the thermal-functional dynamics characteristic of advanced buildings. Consequently, scientific investigation pertaining to the architectural envelope in recent decades has been geared towards the study and definition of its "techno-organic" qualities, through the use of innovative materials and technological solutions capable of guaranteeing, through stratified aggregations, evolutionary plasticity and selective resilience to the differentiated stimuli coming from the external environment.

Furthermore, in recent years technological research has investigated new experimentation frontiers capable of reaffirming the osmotic quality of an exchange process that concerns energy flows that have passed and exchanged right through the envelope (Altomonte 2004; Ad van der Aa et al. 2011). These studies demonstrate how the vertical closure surfaces can be equipped with systems designed to ensure a dynamism that allows them to manage passing energy flows in the same way as a biological organism. From the screening system of the Arab World Institute by Jean Nouvel to the dynamic screenings of Al Bahar Towers by Aedas Architects, the new frontiers of experimentation in architecture seek to propose new models of living in which the building organism is also capable of autonomously ensuring the comfort of its users. In this sense, the evolution and dissemination of Information Technology Control systems (e.g. home automation and Building Management

Systems) to transfer the potential of systems equipped with artificial intelligence to the building scale, has ensured the regulation of space also in the absence of human users and in relation to a whole series of requirements that guarantee optimization from the functional and physical perspective of the built space.

The articulated eco-tech envelope systems, in the form of environmentally responsive walls (Ferguson et al. 2007; Meagher 2015) and interactive walls (Velikov and Thün 2013), have been designed to independently select the environmental stimuli, transforming them into reactions (thermo-hygrometric, lighting and acoustic) that are expressed in the changing ergonomics of the indoor space. This epistemological examination is made explicit in the study of adaptable and adjustable "organic" systems in the form of biological skins (Vermillion 2002; Loonen et al. 2015) and multifunctional skins (Stazi 2019), such as absorbent, radiant, reflective, filtering and transfer devices (thermal, luminous, aerial). The main typological combinations are (i) "single-layer" systems (single skin façades), (ii) multilayer systems (multiple skin façades) and (iii) systems integrated with solar control devices, such us shading devices inside or outside the façade. In all three cases (Fig. 26.1) the kinetic technological systems, over the course of the day and night, and in relation to radiation and external temperature, are able to move, rotate or deform one or more of their parts to accommodate, repel or deflect the most diverse atmospheric events in relation to indoor comfort parameters.

Therefore, it is important to underline the tendency, recorded in the last half century, to overcome the static concept of closure (vertical and horizontal) of the built space, and to define and parameterize new concepts of dynamism and kinematism, characterized by the possibilities of imagining the building as an autonomous organism equipped with artificial intelligence and able to regulate itself. The "envelope-skin" thereby becomes the key element to regulate a building's energy needs, especially in the absence of a user responsible for its continuous management.

26.3 Kinematics in Architecture

Kinematics is a distinctive element that has always featured in architectural components and envelope systems: just think of the possibility of opening and closing doors and windows to control the microclimatic conditions (air quality, temperature, lighting and noise) inside buildings.

In conjunction with the industrial revolution, the possibility of integrating engines and mechanical equipment used in the automotive and industrial sectors into the building structure paved the way for new design visions, such as those of Archigram or Jan Prouvè, who even envisaged that the building system could be transformed into a technological organism capable of moving from one place to another according to the needs of users. With the spread of electronic control systems, which have replaced and/or integrated the manual devices typical of traditional architecture, the capacity for movement has become an increasingly connotative element of complex architectural systems: the utopia of equipping the building with a cognitive system similar to

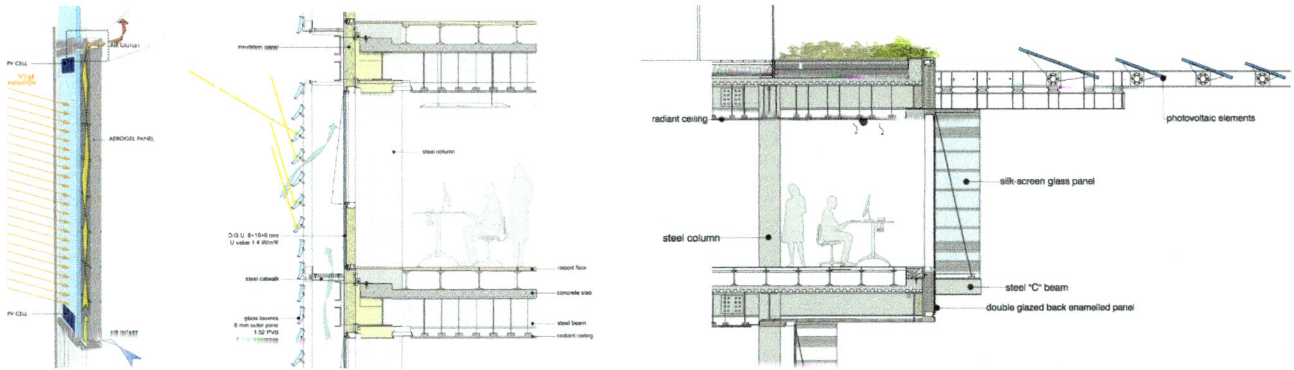

Fig. 26.1 Examples of multifunctional skins: **a** solar wall diagram of Sculpture Building and School of Art Gallery, Yale University, 2007, © Kierantimberlake Associates Llp; **b** east façade detail of SIEEB, Beijing, 2006 © MCA; **c** south façade detail of SIEEB, Beijing, 2006 © MCA

the neurological apparatus of a living organism has become real. Electronic devices and digital connection systems allow for long- and short-term changes. In both cases movement is linked to the complexity of the technological solution chosen (characterized by the presence of interconnected modular elements) and to the somewhat significant energy expenditure necessary for the variation. The envelope, and sometimes entire parts of the building, are therefore given the possibility of transforming, moving through space to assume different configurations, even in the absence of direct human control.

Furthermore, kinetic architectures can be defined as technological solutions which, thanks to the presence of sensors (which perform the "hypothalamic" function reactive to external and internal stimuli) and mechanical devices, are able to move, open and close themselves, increasing the reactive capacity to incident solar radiation or other forms of environmental energy stresses, typical of a traditional building system. Following the classification provided by Ramzy and Fayed (2011) we can define the kinetic architectural systems in relation to (Fig. 26.2):

1. the configuration of the architectural system which, taking into account the indications of the MIT Kinetic Group, can be defined as: (i) Deployable kinetic structure (e.g. tents, movable structures like caravans, mobile homes, capsule units and balloons, as well as movable towns like aircraft carriers and oversees ships/airplanes); (ii) Dynamic kinetic structure (e.g. the Stadium of Phoenix University by architect Peter Eisenman built in Glendale, Arizona, United States in 2003–2006); (iii) Embedded kinetic structures (e.g. responsive façades and roofs).
2. the control techniques integrated into the architectural system and identifiable as: (i) Internal control; (ii) Direct control; (iii) Indirect control; (iv) Responsive indirect control; (v) Ubiquitous responsive indirect control; (vi) Heuristic responsive indirect control.
3. the functional purpose of the architectural system: (i) transportable buildings (e.g. the pneumatic dome by Lomazzi, realized in Italy 1968; the m.poli Kiosk by Brut Deluxe realized in Madrid in 2006; the Tea house by Kengo Kuma and Associates built in Frankfurtin 2007); (ii) automated connection systems (lifts, platforms, bridges, etc.); (iii) ecological buildings capable of rotating to maximise, for example, passive and active solar collection (e.g. the Villa Girasole by Angelo Invernizzi built in Italy in 1935; the Heliotrope House by Rolf Disch realized in Freiburg-im-Breisgau in 1994; the Rotatable Housing Cube by Sturm und Wartkeck Architekten realized in 1996); (iv) internal partition elements that can be packaged or rotated to ensure multiple configurations of the indoor space (e.g. the Hinged Space Housing by Steven Holl realized in Fukuoka in 1991; the K-Space by 6a Architects designed in London in 2008).

Focusing only on kinetic systems able to influence the energy-environmental behaviour of the building, the following macro-groups (Fig. 26.3) can be identified, distinguished in relation to the size of the architectural element with respect to the scale of the building system:

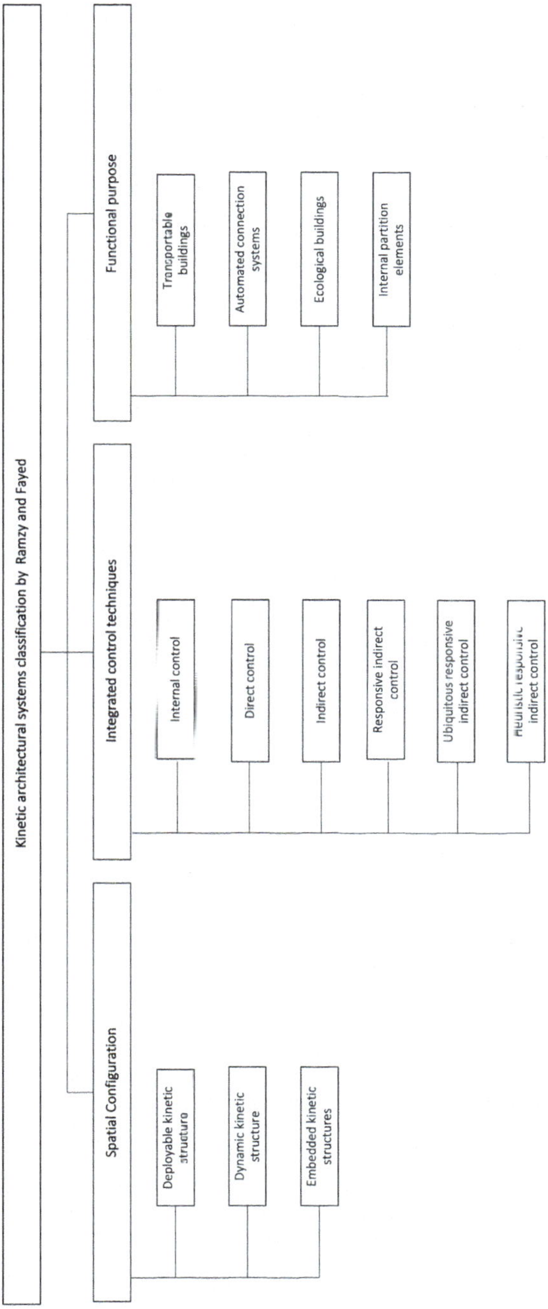

Fig. 26.2 Kinetic architectural systems classification by Ramzy and Fayed (2011)

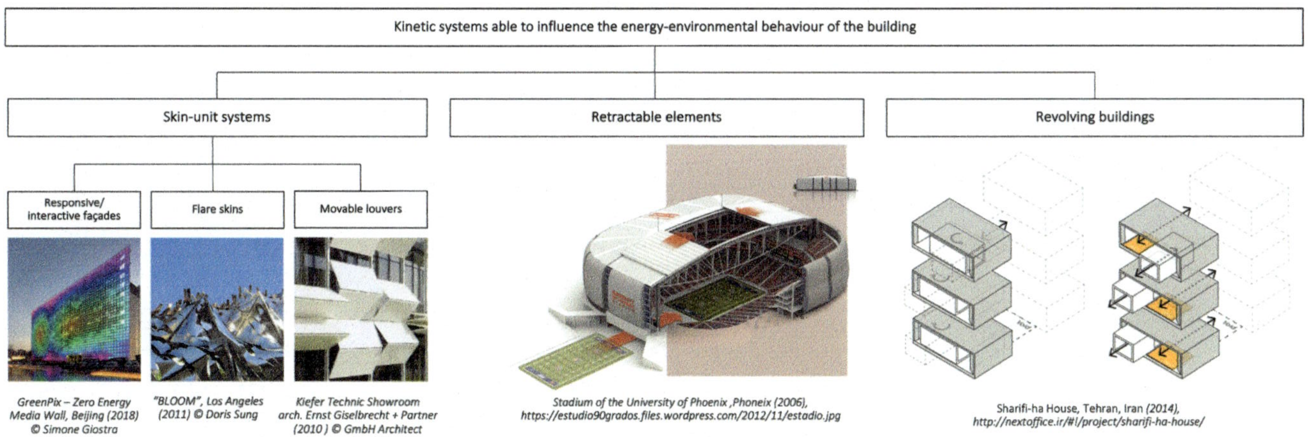

Fig. 26.3 Kinetic systems able to influence the energy-environmental behaviour of the building

(a) Skin-unit systems. These are envelope systems made with small modular units that have the ability to move, shift or rotate themselves without excessively influencing the shape and position of the building. This group includes: (i) "responsive/interactive façades" (e.g. the GreenPix—Zero Energy Media Wall, Beijing, 2018; the Arab World Institute, Paris, 1987; the Yale Sculpture Building, New Haven Connecticut, 2007); (ii) "flare skins", made of thermo-bimetals or other materials that can move when thermally and/or electrically stressed (e.g. "BLOOM", Los Angeles, 2011; SHAPESHIFT, Zurich, 2010); (iii) "movable louvers", often integrated on the outside of the envelope to protect it from incident solar radiation (e.g. the Campus Kolding, Kolding, 2014; the Kiefer Technic Showroom, Bad Gleichenberg, 2007; the ThyssenKrupp Quarter, Essen, 2010).

(b) Retractable elements. This category refers to retractable architectural elements (roofs, walls, floors, etc.) that can fold or expand as a whole element, according to users' needs, bringing about remarkable changes in the building's configuration with broader environmental performance ranges (e.g. the scenographic systems of Chuck Hoberman, from 1994 to the present; the Sliding House by the architect Alex De Rijke, 2009).

(c) Revolving buildings, in which the kinematic phenomenon, often imperceptible to the human eye and designed to optimise the energy performance of the building during the day and/or throughout the year, corresponds to a horizontal rotation or shift around or along an axis (e.g. the Dynamic Tower by David Fisher 2006, and the Sharifi-ha House by Next Office and Alireza Taghaboni, 2014).

Finally, it is important to remember how kinetic architectures are often inspired by the form or function (Pedersen Zari 2010) of other similar plant and/or animal systems (Hosseinia et al. 2019).

The first group (Fig. 26.4), includes technological solutions that unequivocally recall in morphological terms the biological system or living organism they wish to imitate, such as: the SMIT's Solar Ivy thin-film "photovoltaic leaves", designed to produce energy and shield the indoor space; the Meteorosensitive Architecture prototype developed by Reichert et al. (2014), clearly inspired by the morphology

Fig. 26.4 Bio-inspired kinetic façades: **a** The 'Solar Ivy' photovoltaic leaves by Brooklyn (https://inhabitat.com/solar-ivy-photovoltaic-leaves-climb-to-new-heights/); **b** The Meteorosensitive Architecture prototype developed by S. Reichert (https://claratresgallo.files.wordpress.com/2012/07/hygroscope_00.jpg); **c** The Living Glass façade system (https://www.architectmagazine.com/technology/products/six-responsive-products-that-have-a-mind-of-their-ownalmost_o)

and hygroscopic behaviour of the strobilus; the Living Glass façade system, which imitates the behaviour of fish gills, thanks to the presence in its transparent surface of parts made of memory material with a CO_2-sensitive shape, able to open and close fostering air exchange inside the building (Geiger 2010); the artistic installation Articulated Cloud by N. Kahn for the Children's Museum in Pittsburgh, where a curtain of polymeric elements integrated into the double skin façade is able to move, imitating the kinematics of a dragonfly's wings (Linn 2014).

The second group includes technological systems that, despite their appearance not necessarily being inspired by the natural world, can replicate the behaviour of plants or animals in kinetic and thermo-hygrometric terms. This is the case of the Flectofin shielding slats, patented by J. Knippers and integrated into the façade of the One Ocean pavilion created for EXPO 2012 in Yeosu, whose geometric shape and operation were inspired by the strelitzia reginae, commonly known as the crane flower (Knippers and Speck 2012).

26.4 Kinetic Façades. Definitions and Requirements

Definitions of kinetic systems are very broad (Fig. 26.5). Most commonly the words kinetic, kinematic retractable, convertible or simply adaptive are used to describe the non-stationary nature of the façade devices. In addition, terms like performative, responsive and dynamic are used to describe the functional interaction of envelope systems with their environment.

In order to clarify the use of some of these words Barozzi et al. (2016) suggest the following general definitions:

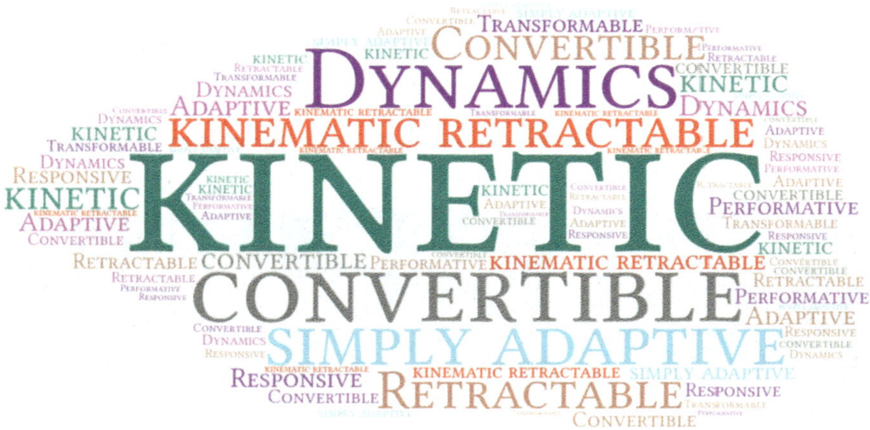

Fig. 26.5 Some of the most used words to define kinetic facades

- **Kinematics**: is the study of motion without considering the mass or external forces that may cause the movement. The study of kinematic systems starts with the description of the system geometry and the initial conditions related to the position values. In engineering disciplines, kinematics is used to describe the movement laws of systems made up of joined parts and is sometimes referred to as “the geometry of motion” (Meriam and Kraige 2012; Hibbeler 1974; Lienhard 2014).
- **Kinetics**: is the study of movement laws considering the forces and masses involved (Hibbeler 1974). If the cause of system motion is an elastic deformation the system falls under the category of Elastic Kinetics (Lienhard 2014).
- **Dynamics**: is the study of forces acting on an object and that results in a movement (Hibbeler 1974; Lienhard 2014). The word is often used to identify building systems or envelopes that can move.
- **Retractable**: the term is commonly used in architecture for textile membrane roofs in which the membrane is bunched or folded. As a simpler expression, such systems are also referred to as movable.
- **Convertible**: convertibility is considered, along with mobility, as one of the modes of variability in buildings for adapting to different functions (Otto 1971). It possible to identify two types of convertibility: external, i.e. the variability of the external envelope of the building, and inner convertibility, related to the interior spaces. Convertible systems are designed to change their form in relation to needs and in a short time (Otto 1971).
- **Transformable**: Similar to convertible, transformable can be associated with objects or structures that have the intrinsic property of controlled change. Transformable bodies can be foldable, retractable or shape shifting (Hoberman 2006). The response of transformable façades needs to be efficiently tuned to boundary conditions such as climatic conditions, different locations, varying functional requirements, or emergency situations. For this response, an actuation force is needed that generates movement. The transformation process goes from a compact to an expanded configuration or viceversa. The transformation phase must consist of controlled, stable movements and results in a rigid and secure structure, once it is locked in place (Chloë 2016).
- **Performative**: is the ability of building skin to mediate between the comfort required by users and the surrounding environments. Performative envelopes have the capacity to control external factors in relation to predefined architectural performances (Turrin et al. 2012; Wigginton and Harris 2002).
- **Adaptive**: “means the ability to adjust and adapt to changing circumstances by itself” (Loonen et al. 2013). Adaptive façades are technological systems that efficiently contribute to the energy balance of the building, limiting the need to use air conditioning devices, with the consequent reduction in energy consumption (Romano et al. 2018). In many cases, intrinsic dynamic façade systems are used which delegate the adaptive capacity to the smart materials. In other cases, the adaptivity is instead explicit in the façade system’s capacity to move all or some of its parts.

- **Responsive**: the term refers to a reactive system, i.e. a system that moves and it is opposed to the word manipulate, which means that the system is moved and controlled from outside and refers to passive environments (Velikov and Thün 2013). According to Negroponte (1976), "responsive means that the environment has an active role, initiating to a greater or lesser degree changes as a result and function of complex or simple computations". A responsive building skin includes functionalities and performance characteristics similar to those of an 'intelligent' building skin, including real-time sensing, kinetic climate-adaptive elements, smart materials, automation, and the ability for users to override it. However, it also includes interactive characteristics such as computational algorithms that allow the building system to self-adjust and learn over time, as well as the ability for inhabitants to physically manipulate elements of the building envelope to control environmental conditions (Schumacher et al. 2010; Kolodziej and Rak 2013).

In 1970, Zuk and Clark defined kinetic architecture as an architectural form that can be inherently displaceable, deformable, expandable, or capable of movement (De Marco Werner 2013). Therefore, a kinetic façade is a technological system in which there is a certain kind of motion (Loonen 2010) and that is able to guarantee variable locations or mobility and/or variable geometry to all or one of its parts (Fox and Yeh 1999). The term 'kinetic' in fact indicates an organism's response to a particular kind of stimulus in biology (Wang et al. 2012) and an ability to modulate energy in its primary forms: visible light and heat. A kinetic façade can respond to flows of energy, both natural and man-made, which primarily affect the building performance but also the comfort of the people in the buildings (Fortmeyer and Linn 2014). These types of envelope, in general, need to be efficiently tuned to boundary conditions such as climatic conditions, different locations, varying functional requirements, or emergency situations. In order to guarantee kinematic action, an actuation force is needed that generates the movement. Finally, the kinetic façade systems (Fox and Yeh 1999; Wang et al. 2012) are capable of changing by moving in space and taking on different structures and configurations over time (Fox and Miles 2009). Long-term changes are achieved through reversible and unique conversions in the context of a flexible structure, while short-term reversible adaptations can be brought about through mechanical solutions.

As suggested from Brzezicki (2017) we can systematize the functionalities of kinetic façades, dividing the latter into three macro-categories (Fig. 26.6), characterised by as many types of kinematic motion:

1. Façades with elements that move rigidly. This group includes façade systems made up of rigid modular elements that can move or rotate on an axis (e.g. the Campus Kolding façade system by Henning Larsen Architects, 2014; the solar shading system of ThyssenKrupp Quarter Corporate Headquarters by JSWD Architekte, 2010) or that can be combined to produce more complex movements, for instance: a directional rotation (the Generative Kinetic Sculpture 2009 made by Marilena Skavara) or a rolling; an opening or closing with respect to a scissor joint (pantograph system) which makes it scalable (the Institute Monde Arabe

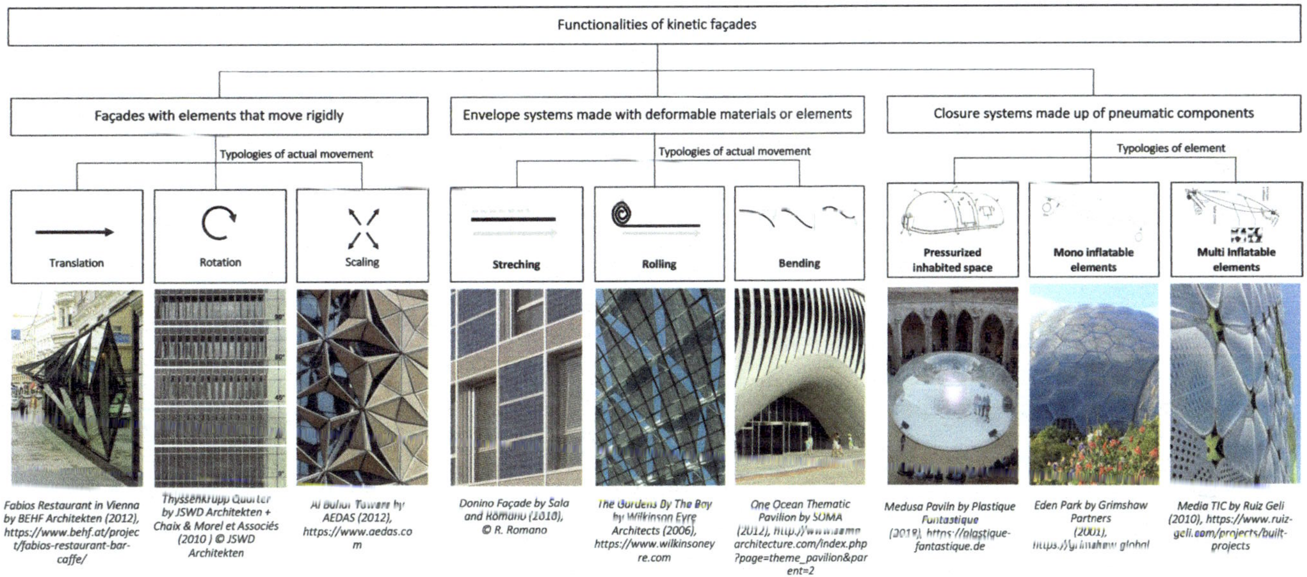

Fig. 26.6 Classification of kinetic façade to functionalities and typologies of movement

by Jan Nouvel Atelier, 1987; the Al Bahar Towers, by AEDAS, 2012); a vertical translation determined by the presence of a folding joint (e.g.: the Kiefer Technic Showroom by Ernst Giselbrecht + Partner).

2. Envelope systems made with deformable materials or elements (light and flexible). This category includes façade systems that move thanks to the integration of deformable materials that facilitate the transformation of the closing surface, reducing the use of mechanical actuators that characterise the façade systems of the first group, providing a wide range of complex yet efficient movements that are arguably more suitable for the design of freeform architectural envelopes of the future (e.g. the Domino Façade by Sala and Romano, 2010; the Gardens by the Bay by Wilkinson Eyre Architects, 2006; the One Ocean Thematic Pavilion by SOMA, 2012). The components that make up these façade systems move by means of: stretching; rolling; bending, fluttering; free form; gathering.
3. Closure systems made up of pneumatic components. Widespread in the construction sector from the Sixties thanks to experiments by Walter Bird, Frei Otto and Buckminster Fuller, these envelope systems were properly marketed thanks to the achievements of the US company ASATI (Air Structure America Technologies Incorporated), credited with developing the first inflatable roof in 1962 as roofing for a series of sports pitches. Pneumatic structures, which are light and consequently easy to transport, are made up of two main components: the membrane and the air or gas used to tension the membrane itself and reach the volume established in the project phase. Pneumatic systems can be classified into two main categories: (3.1) one in which the inhabited space itself is pressurized (e.g. the temporary structure of the Medusa or the Personal Protective Space for Doctors by Plastique Fantastique, 2020); (3.2) one in which the inhabited space—at atmospheric pressure—is enclosed within a structure made up of inflatable elements (e.g. the Silk Pavilion for the Expo Saragoza by Cloud 9, 2009, the Water Cube for the Chinese Olympic Games by PTW Architects, 2009). The most used materials to create membranes include: fibreglass, nylon, PVC, polyester, silicone rubber and polytetrafluoroethylene, better known as Teflon (Krauel 2013). In recent years, ETFE (a copolymer of ethylene-tetrafluoroethylene) combined with sophisticated electronic inflation systems, has become the most used material in the creation of kinetic pneumatic envelopes thanks to its aesthetic, thermal and mechanical properties. The possibility of inflating and deflating EFTE cushions with sophisticated electronic control systems means the energy performances of the building can be considerably improved, reducing overheating phenomena through the possibility of silk-screening the sheets of material making up the façade element with overlapping geometric patterns (e.g. the Media TIC in Barcellona by CLOUD 9, 2009; the Eden Project in Cornwall designed by Grimshaw Architects, 2001; etc.).

The potential of a kinetic façade can be attributed to a merely expressive and formal function (e.g. communicative), or to the capacity to respond to the environmental conditions and perform one or more of the following functions:

- Protect from bad weather, rain, wind and external temperatures;

- Regulate incident solar radiation;
- Manage the passage of thermal energy;
- Ensure adequate soundproofing;
- Ensure control of the building by increasing its level of security.

As described in the preceding paragraphs, the mobile elements of a kinetic façade must be designed so that they can respond efficiently to climatic and environmental factors that may affect the functioning and energy efficiency of the entire building (Sharaidin 2014).

As kinematics is strictly linked to the concept of movement, the requirements to be considered in the design of a kinetic façade system are: time, mass and/or weight; speed and acceleration. Moreover, given the propensity of kinetic architectures to interact (with both the environment and users) it is essential to develop a project that incorporates within the building a data detection and processing system, necessary to ensure the movement and adaptivity of the façade, as well as to ensure the security conditions in place (for users and for the device itself). In other words, to transform a static façade system into a kinetic system it is necessary to adopt solutions that can move one or all of its parts autonomously or as a reaction to human input. This is possible using elements, like mechanical joints, sensor and actuators, that are capable of transforming the external loads into internal forces or deformations.

Mechanical joints, in the form of kinematic chains, enable morphological configurations that vary substantially in relation to the degrees of freedom conferred on the constraint system. This is why the first examples of kinetic envelopes, equipped with purely mechanical devices, adopted simple constraints—the hinge and the trolley—respectively to define rotations and sliding, almost always linking the driving force to them directly.

The introduction of the variable element into the systems that control movement has made it possible to define new and more complex operating strategies, combining the interactiveness of visual communication, typical of mediating architectural envelopes, with a technological sensitivity oriented towards innovation. Consequently, it is important to recall that the automation and control systems (wind, electronic, mechanical, homeostatic, IT or pneumatic) used to create the kinetic façade belong to different types according to the production sector they come from. As a rule they are sensors and actuators: the former record the changes inside and outside the component, while the latter are mechanical elements that convert energy into movement and produce a visible reaction in the system, altering their geometry or the main characteristics, namely size or rigidity (Barozzi et al. 2016).

The most widespread actuators in the construction sector contain electrical microprocessors that allow their rotation and/or translation with respect to processes of transmission, electrical or hydraulic stress, and include: gears and transmission; toothed racks; cogged belts, winches; thrust, azimut and angle gears; chain actuators; swing door openers; roller drives; linear actuators; rack and pinion drives, sliding door openers.

Sensors can be categorised on the basis of the operating principles or the type of output signal, but more commonly they are defined on the basis of the type of physical dimension they measure (Mohan et al. 2003): light, sound, temperature, heat, electrical resistance, magnetism, pressure, gas and flow of liquids, movement, orientation, etc.

In many kinetic façade systems, actuators and sensors are managed by a processor, configured as a central control unit (IT system) capable of manipulating and processing incoming data to produce an adequate response of the façade system in line with the initial design requirements (set-point) (Sobek and Teuffel 2001; Sobek et al. 2000).

The Programmable Logic Controllers applicable to the kinetic mechanisms can be distinguished on the basis of the interactions that exist between sensors, actuators and the environment, and broken down into the following configurations (Musacchio 2009):

- Direct Control (Fig. 26.7) where the prevalent aspect is a univocal relationship that connects the actuator to the kinetic mechanisms;
- In-Direct Control (Fig. 26.8), where movement is achieved through a system of environmental sensors that individually send instructions to the control unit which, in turn, activates the movement through on/off inputs to the drive unit;
- Responsive In-Direct Control (Fig. 26.9), where the control unit can determine behaviour logics based on a comparative analysis of instructions from several sensors;
- Heuristic Responsive In-Direct Control (Fig. 26.10), in which the electronic system is capable of self-regulating according to self-learning logics.

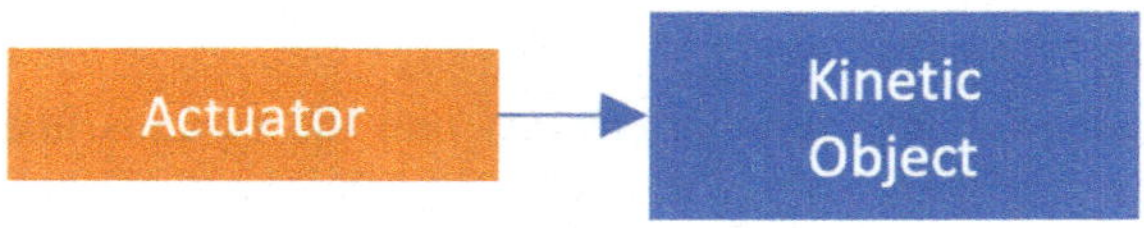

Fig. 26.7 Direct control (Musacchio 2009)

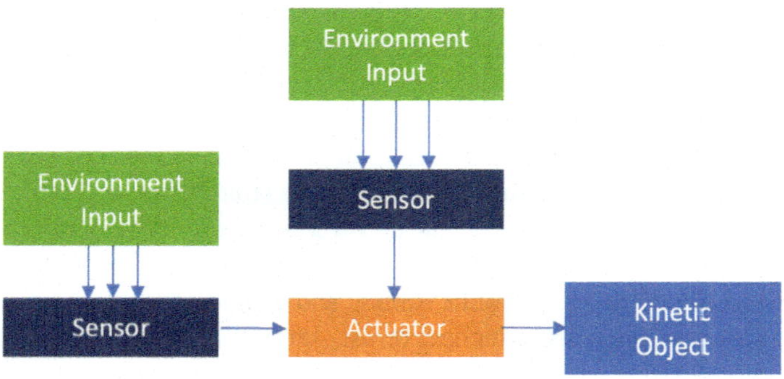

Fig. 26.8 In-Direct control (Musacchio 2009)

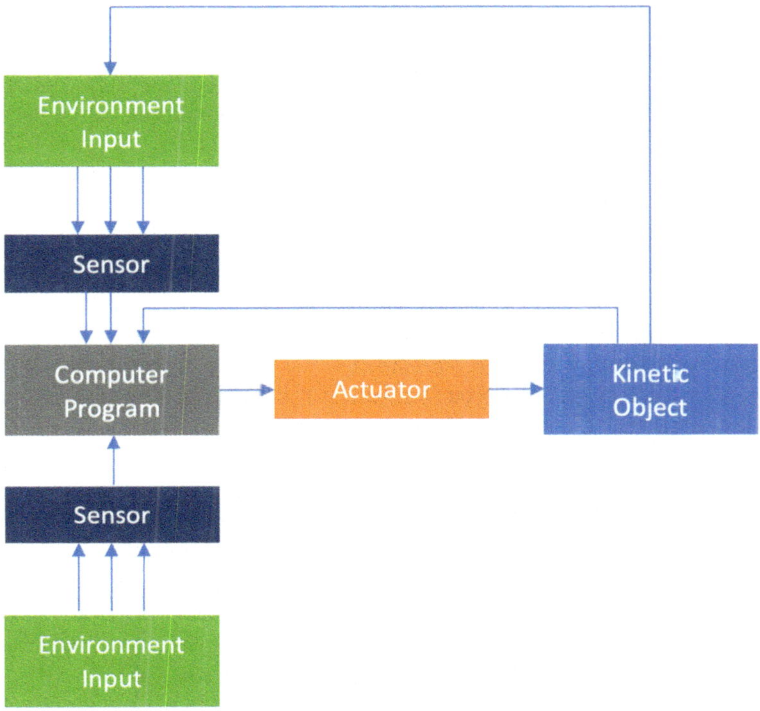

Fig. 26.9 Responsive In-Direct control (Musacchio 2009)

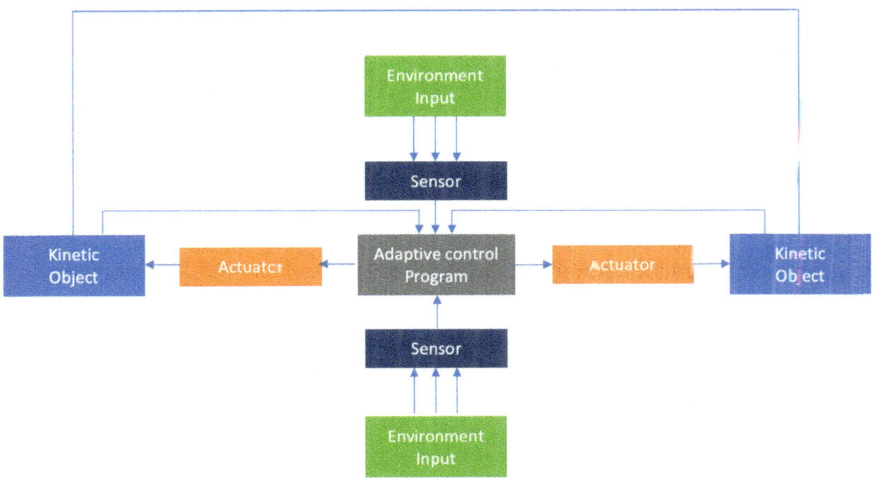

Fig. 26.10 Heuristic Responsive In-Direct control (Musacchio 2009)

In all cases, it is clear that the electronic element plays a fundamental role in the management of the architectural organism, and constitutes its behavioural model, potentially pushing the dynamic characteristics of a kinetic system beyond the logic of cause-and-effect connections and thereby making the building truly interactive.

26.5 Conclusions

The complexity of kinetic envelopes still requires a great deal of attention to be paid in the design and validation phase regarding problems linked to their maintenance and operation, which can sometimes result in obsolescence and consequent disuse. There are in fact numerous examples of kinetic façades built in the 1980s that are operationally inefficient due to the ageing of the movement systems and the consequent impossibility of automating their management.

Another consideration is the need for appropriate tools to analyse the impact of these façade systems, often used as solar screens, on comfort in terms of the lighting of the indoor space which sometimes (if the system is not controlled by sensors that determine its movement in relation to incident solar radiation) does not meet the standards set by international technical standards. Moreover, the psycho-perceptive reaction of users required to interact with the system must also be assessed in advance, since the movement of façade elements that cannot be managed manually could in the long run cause visual fatigue and, in workplaces, a drop in productivity. Finally, we must not forget that the kinetic façade systems only allow us to reach the nZEB standards required by European legislation if the automation and control systems that regulate the movement are powered by renewable energy sources. In this way, the technological system is truly efficient and does not have a negative impact on the overall energy balance of the building.

Consequently, in recent years, the spread of adaptive façade systems has been strongly influenced by the possibility of using new BIM and BEM modelling and verification tools as early as the design phase, which are capable of interacting with digital production systems (selective laser sintering, stereolithography, 3D printing, additive manufacturing systems), with a view to controlling not only the dimensional and performance characteristics of the building but also and above all its environmental impact in terms of LCA and LCC. The goal is to maximize product customization starting with the definition of the characteristics of the envelope component and the handling systems connected to it, ensuring the product is fully reversible.

In addition, the possibility of using nano-structured materials has made it possible to increase the energy and environmental performances of the kinetic façade systems, without affecting their weight or thickness, thereby favouring their use not only to create nZEB buildings but also during deep energy restructuring actions designed to drastically reduce the thermal consumption of existing buildings without compromising their structural characteristics.

Finally, current research at global scale indicates how the architectural sector is interested in developing and integrating increasingly sophisticated envelope systems capable of responding to the challenges set by the digital age. These are technological solutions which, just like living organisms, will be able to change their positions, automatically or dynamically correct any deformation in their structures, reduce their weights and physically reconfigure themselves to meet the needs of users over time, having a positive effect on global challenges linked to reducing climate-altering emissions to zero and safeguarding the environment for future generations.

Acknowledgements The author would like to gratefully acknowledge COST Action TU1403 "Adaptive Façades Network" for providing excellent research networking, who contributed to the creation of the database of adaptive façade systems, allowing the Author to broaden her knowledge on kinetic façade systems.

References

Ad van der Aa A, Heiselberg P, Perino M (2011) Designing with responsive buildings elements. IEA—ECBCS Annex 44, Aalborg University

Altomonte S (2004) L'involucro architettonico come interfaccia dinamica. Strumenti e criteri per un'architettura sostenibile. Editrice Alinea, Florence

Barozzi M, Lienhard J, Zanelli A, Monticelli C (2016) The sustainability of adaptive envelopes: developments of kinetic architecture. In: International symposium on "Novel structural skins: improving sustainability and efficiency through new structural textile materials and designs". Proc Eng 155:275–284. https://doi.org/10.1016/j.proeng.2016.08.029

Benham R (1969) The architecture of the well-tempered environment. Architectural Press, London

Brzezicki M (2017) Kinetic architecture and its relations with adaptivnes of the façade, lecture for EduPack by Grantee, COST TU 1403, WG 1, Training School Hamburg, HCU HafenCity University, 12–17 Sept 2016

Chloë M (2016) Structural adaptive façades. Master's dissertation, Ghent University

Davis M (1981) A wall for all seasons. RIBA J 88

De Marco Werner C (2013) Transformable and transportable architecture: analysis of buildings components and strategies for project design. Master's dissertation, Escuela Técnica Superior de Arquitectura de Barcelona

Fang Z, Li N, Li B, Luo G, Huang Y (2014) The effect of building envelope insulation on cooling energy consumption in summer. Energy Build 77:197–205. https://doi.org/10.1016/j.enbuild.2014.03.030

Feng G, Sha S, Xu X (2016) Analysis of the building envelope influence to building energy consumption in the cold regions. Proc Eng 146:244–250. https://doi.org/10.1016/j.proeng.2016.06.382

Ferguson S, Siddiqi A, Lewis K, De Weck O (2007) Flexible and reconfigurable systems: nomenclature and review. In: ASME 2007 international design engineering technical conferences and computers and information in engineering conference, Las Vegas, Nevada, USA, pp 249–263

Fortmeyer R, Linn CD (2014) Kinetic architecture: design for active envelope. The Images Publishing Group Pty Ltd., Australia

Fox MA, Yeh BP (1999) Intelligent kinetic systems in architecture. In: Nixon P et al (eds) Managing interactions in smart environments, pp 91–103

Fox M, Miles K (2009) Interactive architecture. Princeton Architect Press, New York

Friess W, Rakhshan K (2017) A review of passive envelope measures for improved building energy efficiency in the UAE. Renew Sustain Energy Rev 72:485–496. https://doi.org/10.1016/j.rser.2017.01.026

Geiger J (2010) The living: surface tensions. Architec Des 80(3):60–65. https://doi.org/10.1002/ad.1076

Hibbeler RC (1974) Engineering mechanics: statics and dynamics. McMillan Publishing Co., Inc., New York

Hoberman C (2006) Transformation in architecture and design. Trans Environ 3:70–73

Hosseinia SM, Mohammadia M, Rosemannb A, Schröderc T, Lichtenbergc J (2019) A morphological approach for kinetic façade design process to improve visual and thermal comfort: review. Build Environ 153:186–204. https://doi.org/10.1016/j.buildenv.2019.02.040

Knippers J, Speck T (2012) Design and construction principles in nature and architecture. Bioinspiration Biomim 7(1):1–10

Kolodziej P, Rak J (2013) Responsive building envelope as a material system of autonomous agent. In: Open systems: proceedings of the 18th international conference on computer-aided architectural design research in Asia, pp 945–954

Krauel J (2013) Architetture gonfiabili. Links Books, Barcellona

Lienhard J (2014) Bending active structures, form-finding strategies using elastic deformation in static and kinematic systems and structural potential therein. PhD's dissertation, University of Stuttgart

Linn C (2014) Kinetic architecture: design for active envelopes. Images Publishing, Australia

Loonen RCGM (2010) Climate adaptive building shells what can we simulate? Master's dissertation, Eindhoven University of Technology

Loonen RCGM, Trčka M, Cóstola D, Hensen JLM (2013) Climate adaptive building shells: state-of-the-art and future challenges. Renew Sustain Energy Rev 25:483–493. https://doi.org/10.1016/j.rser.2013.04.016

Loonen RCGM, Rico-Martinez JM, Favoino F et al (2015) Design for façade adaptability—towards a unified and systematic characterization. In: Proceedings of the 10th conference on advanced building skins, Bern, Switzerland, pp 1284–1294

Meagher M (2015) Designing for change: the poetic potential of responsive architecture. Front Architect Res 4:159–165. https://doi.org/10.1016/j.foar.2015.03.002

Meriam JL, Kraige LG (2012) Engineering mechanics: dynamics. Wiley, USA

Mohan N, Undel TM, Robbins WP (2003) Power electronics: converters. Applications and design. Wiley, New York

Musacchio A (2009) Architetture Cinetiche. Apparati meccanici ed elettronici nel progetto di architettura. Maggioli Editore

Negroponte N (1976) Soft architecture machines. MIT Press, Cambridge

Otto F (1971) Convertible roofs. Institute for Lightweight Structures, IL5, Druckerei Heinrich Fink KG, Stuttgart, pp 6–7

Pedersen Zari M (2010) Biomimetic design for climate change adaptation and mitigation. Archit Sci Rev 53:172–183. https://doi.org/10.3763/asre.2008.0065

Ramzy N, Fayed H (2011) Kinetic systems in architecture: new approach for environmental control systems and context-sensitive buildings. Sustain Cities Soc 1:170–177. https://doi.org/10.1016/j.scs.2011.07.004

Reichert S, Menges A, Correa D (2014) Meteorosensitive architecture: biomimetic building skins based on materially embedded and hygroscopically enabled responsiveness. Comput-Aided Des 60:50–69. https://doi.org/10.1016/j.cad.2014.02.010

Romano R, Aelenei L, Aelenei D, Mazzucchelli ES, Brzezicki M, Rico-Martinez MJ (2018) Case studies—adaptive facade network. TU Delft Open, Delft

Romano R (2020) Smart materials for nZEB BUILDING. Adaptive technological solutions for sustainable projects. AGATHÓN 7:59–72. https://doi.org/10.19229/2464-9309/572020

Schumacher M, Schaeffer O, Vogt M (2010) Move. Birkhauser, Berlin

Sharaidin K (2014) Kinetic facades: towards design for environmental performance. PhD dissertation, School of Architecture and Design RMIT University

Sobek W, Teuffel P (2001) Adaptive structures in architecture and structural engineering, smart structures and materials. In: Liu SC (ed) Proceedings of the SPIE 8th annual international symposium, Newport Beach, CA, USA, 4–8 Mar 2001. SPIE, Bellingham, pp 36–45
Sobek W, Haase W, Teuffel P (2000) Adaptive system. Stahlbau 69:544–555. https://doi.org/10.1002/stab.200001870
Stazi F (2019) Advanced building envelope components: comparative experiments. Butterworth-Heinemann, Elsevier
Tang C, Wu L (2011) Research on the application of building energy-saving measures in wall. In: Proceedings of the 2nd international conference on artificial intelligence, management science and electronic commerce (AIMSEC), Dengleng, China, 8–10 Aug 2011, pp 6361–6364
Turrin M, Von Buelow P, Kilian A, Stouffs R (2012) Performative skins for passive climatic comfort. A parametric design process. Autom Constr 22:36–50. https://doi.org/10.1016/j.autcon.2011.08.001
Velikov K, Thün G (2013) Responsive building envelopes: characteristics and evolving paradigms. High Performance Homes, Routeledge Taylor & Francis Group, London and New York, pp 75–92
Vermillion J. (2002) Phototropic architecture: intelligent responses to sunlight stimuli. A new school of architecture for Ogerlthorpe University. Master's dissertation, Ball State University
Wang J, Beltrán L, Kim J (2012) From static to kinetic: a review of acclimated kinetic building envelopes. Proc Solar Conf 5:4022–4029
Wigginton M, Harris J (2002) Intelligent skins. Architectural Press, Oxford

Printed by Printforce, the Netherlands